INTERNATIONAL UNION OF PURE AND APPLIED CHEMISTRY

**ENVIRONMENTAL ANALYTICAL AND PHYSICAL
CHEMISTRY SERIES**

Environmental Particles

Volume 1

T0155679

ENVIRONMENTAL ANALYTICAL AND
PHYSICAL CHEMISTRY SERIES

Environmental Particles

Volume 1

Volume Editors

Jacques Buffle
Department of Inorganic,
Analytical, and Applied Chemistry
University of Geneva
Geneva, Switzerland

Herman P. van Leeuwen
Laboratory for Physical
and Colloid Chemistry
Agricultural University
Wageningen, Netherlands

CRC Press
Taylor & Francis Group
Boca Raton London New York

CRC Press is an imprint of the
Taylor & Francis Group, an **informa** business

Foreword

This book reports critical reviews on sampling, characterization, and the role and behavior of particles in the environmental compartments: air, surface and ground waters, and sediments and soils. There is, presently, no clear definition of the word *particle* within the literature devoted to environmental science. It is often used with different meanings depending on the context and scientific discipline involved. For instance, biologists refer to bacteria or even phytoplankton as biogenic particles whereas, for chemists, the term particle is generally associated to nonliving entities. Most geologists and sedimentologists restrict the notion of particles to entities larger than 0.45 μm, whereas they may be much smaller for chemists and, even, subatomic for physicists. The situation is further confused since physical chemists and water treatment engineers often use the word colloid to refer to *particles* in the size range 1 to 100 μm, whereas biologists often use colloid for nonliving entities smaller than 0.45 μm. Finally, chemists often use the term macromolecule for certain small colloids or particles (<0.1 μm). All of these examples clearly show that there is an obvious basic language difficulty for such a book which intends to present a multidisciplinary approach of *particle* characteristics.

It is not the purpose of this book to try to establish the best definition of a particle, colloid, or macromolecule within the context of environmental sciences. The reader, however, should know that the word particle may be used mainly in this book with two different meanings:

Particle as a general term will refer to any entity with a size larger than ~ 1 nm. This is, for instance, the meaning used in the title of this book. Entities with sizes smaller than 1 nm have a molecular weight of less than ~ 1000, i.e., they include all small molecules and ions normally involved in classical chemical reactions. This book is not focused on these compounds. They are of interest here, however, because many of them may be toxic or vital and because their interactions with particles and the behavior of the latter largely control their circulation in the environmental compartments.

Particle in the restricted sense will be used in principle to refer to living or nonliving entities larger than 0.45 μm. In this case, this word will be used to discriminate between three different size-classes of entities (Figure 1): solutes (<1 nm), colloids or macromolecules (1 nm < size < 0.45 μm), and particles (>0.45 μm). There is, obviously, no well-defined size limit between the various environmental types of particles, and those mentioned here are arbitrary. The basis for the limit at 1 nm is cited here. The limit at 0.45 μm is traditionally used by limnologists and oceanographers. Although it is mostly based on experimental reasons, it has some rough meaning since (1) many organisms (except viruses) are larger, and (2) coagulation leading to sedimentation mostly occurs with larger particles (see Chapters 5 and 10).

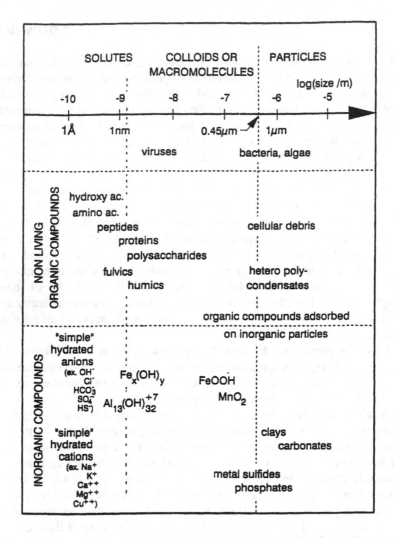

Figure 1. Nature and size domain of the most important particles of aquatic systems.

In specific cases the words particles and colloids might be used differently in the following chapters. But, in all cases, their meaning should be obvious from the context and by taking into account the previously stated remarks.

The term *environmental* particle (taken in its general sense) refers to particles of natural origin found in environmental compartments. In the air, the presence of particles is primarily due to erosion, marine aerosol formation, and combustion processes (either natural or anthropogenic). The description of the means allowing the determination of origin, nature, and fate of atmospheric particles is the subject of Chapter 1. In aquatic systems (oceans, lakes, rivers, ground water, and sediment pore water), particles (Figure 1) are mainly produced by weathering processes (e.g., clays and metal oxides) and direct or indirect formation (e.g., calcium carbonate, silica, Fe and Mn oxyhydroxydes, organic macromolecules, and cellular debris; see Chapters 6, 8, and 9) due to life processes. In these systems, therefore, living organisms play an important role in controlling the concentration and nature of particles. It is believed that they even indirectly control the behavior of nonbiogenic particles like clays. For instance, organic macromolecules may adsorb on their surface and strongly influence their reactivity towards adsorbable toxic or vital compounds, as well as their coagulation and sedimentation (see Chapter 6).

Little is known about the size distribution of environmental particles because of difficulties in sampling, sample handling (see Chapters 5 and 6), insufficient sensitivity and selectivity of analytical methods, and the lack of adequate theories which permit taking into rigorous account the physical and chemical heterogeneity of environmental samples. In all cases it must be realized that the size distributions of particles may be represented in different manners and give completely different pictures depending on whether particle mass, particle number, or particle surface area is considered as the dependent variable (Figure 2). In most aquatic systems the mass is dominated by particles larger than a few μm with smaller particles representing generally less than 10% (often <1%). If, however, the number of particles is considered for the same sample, particles smaller than 1 μm are the dominant class by several orders of magnitude. Consequently it is important to choose the most pertinent representation of the data, depending on what is the studied process. For instance, a size distribution based on particle mass is preferable for studying the fluxes of sedimenting particles (dominated by large particles), whereas coagulation processes (affecting mostly small particles) are better related to a size distribution based on particle number.

Similar considerations apply to the study of chemical reactivity of particles (Figure 3). In this case there is an additional difficulty since particles may react through the reactive groups located either on the external surface of the particle, for instance, with a crystallized iron oxide, or inside the particle itself, in the case of amorphous porous material, e.g., Fe oxyhydroxide, or of organic macromolecules, e.g., proteins and humic compounds. Depending on the degree of crystallinity of the particle, the surface area or the volume of the particles should be the dependent variable of interest for the size distribution, which may result in contradictory conclusions about the relative role of large and small particles. To make the choice a bit more difficult, large size particles more than small ones are often crystallized, which implies that the two size distributions should be considered simultaneously when studying the reactivity of particles. All these factors underline the urgent need of methods for specific chemical and physical characterization of environmental particles. Some of the analytical problems involved are discussed in Chapters 3, 4, 6, 7, 9, and 13. Development of such analytical methods

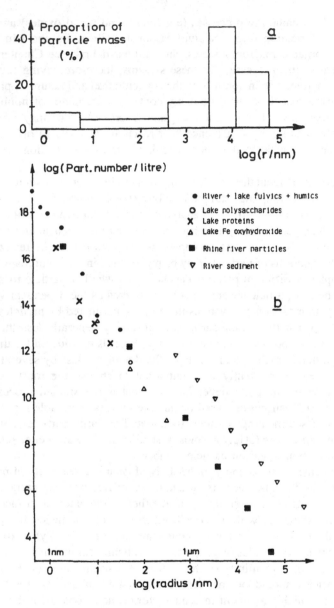

Figure 2. Example of size distributions of important aquatic particles. a–Size distribution, based on particle mass, of the Rhine river particles in Netherlands (same data as ■ in Figure b-) b–Size distributions of various types of particles, based on particle number. Results based on fractionation by filtration; particle numbers and proportions of particle mass are values obtained for each filtration fractions.

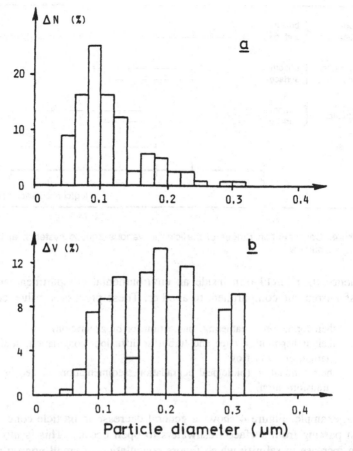

Figure 3. Size distribution of a well-characterized amorphous iron hydroxo phosphate formed *in situ*, by oxidation of Fe(II), at the oxic-anoxic boundary of a eutrophic lake. Determination made by transmission electron microscopy. Particles have spherical shape. a–size distribution based on particle numbers; ΔN = proportion of particle number in each class. b–size distribution of the same sample, based on particle volumes; ΔV = proportion of particle volume in each size class.

cannot be done without simultaneously studying the properties of particles. The latter are discussed in Chapters 2, 4, 9, 10, 11, and 12.

When compared to a homogeneous solution where classical chemical reactions occur, the peculiar role of environmental particles results from the fact that not only do they react chemically with toxic and vital compounds, but they also have important biophysico-chemical properties which strongly

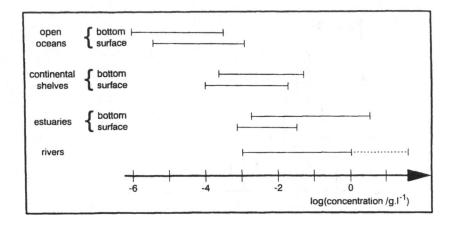

Figure 4. Concentration ranges of particles in various aquatic systems, at the earth surface.

influence their circulation inside an environmental compartment and their transfer from one compartment to another. These processes influence:

1. their formation (weathering, nucleation, or biogeneration)
2. their transportation (free and turbulent diffusion, coagulation, sedimentation, or water flow)
3. their elimination (biological degradation, sedimentation, or geochemical transformation)

As an example, Figure 4 shows a general decrease in particle concentration when passing from surface freshwaters to open oceans. This is mostly due to an increase in salinity which favors coagulation of small nonsettling particles to larger ones which are readily removed by sedimentation. The role of this particular process on the circulation of trace elements is discussed in Chapter 10.

Since vital and toxic compounds are, in most cases, associated with particles (from nm to hundreds of μm), the circulation and the role of these compounds in the environment cannot only be understood on the basis of their total concentrations or their own chemical properties. In addition to these data, all the processes mentioned in points 1 to 3 should be studied in detail and the corresponding, well-defined, physico-chemical parameters should be determined. Finally, these processes should be linked, using mathematical models, to the chemical properties of the toxic or vital compound of interest. Little is known in this field and a huge task remains to be done. This is particularly so if one considers that:

1. Environmental particles are chemically and physically very heterogeneous systems, and that their properties cannot be rigorously represented by physico-chemical parameters with a single, well-defined value. Most of

these chemical and physical properties must be represented by distribution functions (e.g., for size or complexation or acid-base properties, see Chapter 12). These functions must be combined together in the previously mentioned models which implies important conceptual and mathematical developments.

2. Environmental systems are generally not in equilibrium, but either in some steady-state or subject to change in time (particle formation, coagulation, sedimentation, etc.). The dynamic factors affecting the existence and properties of the particles must, therefore, be studied both to understand the behavior of environmental systems (see Chapter 10), and to allow correct interpretation of analytical signals (see Chapter 13). As mentioned in 1, for certain kinetic properties, the ultimate task will be to measure the distribution functions of the corresponding parameters, rather than a single value valid for a particular occasional situation.

Most of the work in this field remains to be done. This book only intends to contribute to it by reviewing part of our present and limited knowledge.

J. Buffle
H. P. van Leeuwen

Preface

Until recently the concern of most environmental scientists has been focused on air and water quality engineering and the development of analytical techniques limited to pollution control applications. In addition, the processes used for air and water quality improvement were often based on empirical grounds or on theoretical concepts, founded on oversimplified model systems, which are difficult to relate to real environmental ones.

In the early sixties a small number of scientists had already realized that environmental science has its own specificity, based on principles which must be studied and understood on rigorous grounds to allow the development of novel engineering and meaningful analytical measurements. But it is only during the last ten years, or so, that fundamental environmental science has been recognized as a basic field in its own right. It is now clear that the protection of our environment and the maintenance of the biosphere depends on the progress that will be made in this field just as, by analogy, medical advances depend on research in biochemistry, biophysics, and molecular biology.

Fundamental environmental science is multidisciplinary and must, therefore, incorporate both practice and theory drawn from all of the natural sciences: biology, chemistry, physics, geology, and mathematics. Consequently, subjects and aspects that are not of a purely chemical nature will show up in this series, even though it is prepared and published under the auspices of the International Union of Pure and Applied Chemistry (IUPAC), i.e., an organization devoted to chemistry. It should, however, be emphasized that chemistry has a key role in environmental science since it deals with the basic molecular reactions which underlie both biological and geological environmental processes. Progress in this field will depend on our capacity to relate reactions at the molecular level to processes occurring at the microscopic and, even, regional scale. This will only be possible through the development of new experimental methods and conceptual approaches which should:

1. avoid sample perturbation by performing the analysis *in situ*
2. develop remote-controlled continuous monitoring systems in order to study both spatial and time dependencies
3. measure not only concentrations of individual species but also thermodynamic and kinetic physico-chemical parameters (equilibrium and rate constants, diffusion coefficients, etc.)
4. interpret all of this data by considering the complexity and heterogeneous nature of the studied system

These considerations fully justify starting this new series on environmental analytical chemistry. The reader should, however, be aware that in order to attain the previously mentioned goals 1 to 4, environmental analytical chemistry must no longer be considered as a field limited to the mere determination of contaminant concentration. Analytical chemists are trained to use the concepts of physical chemistry to study real samples, by taking into account all of their complexities. A major goal of analytical chemistry, therefore, should be to exploit this potential to contribute to the development of environmental and biological sciences. Environmental analytical chemistry, in particular, should play an important role in the development of rigorous approaches and new concepts to investigate physico-chemical processes occurring in real environmental systems.

I would like to specially thank G. Nancollas and G. den Boef, former presidents of the Analytical Chemistry Division of IUPAC. They both had fully realized, at a time when it was not commonly accepted, the necessity of giving a wider dimension to analytical chemistry. Their openmindedness and spirit of enterprise has greatly contributed both to the founding of the IUPAC Commission on Environmental Analytical Chemistry and to the existence of this new series of multidisciplinary monographs. Finally, I would also like to thank J. Jordan whose encouragements have been very helpful.

J. Buffle
Chairman of the IUPAC Commission on
Environmental Analytical Chemistry

Jacques Buffle is an analytical and environmental chemist. He joined the staff of the Department of Inorganic, Analytical, and Applied Chemistry of the University of Geneva, Switzerland in 1969. He is also an Invited Professor at INRS-Eau, University of Quebec, Canada, since 1982 and has been Associate Professor of Electroanalytical Chemistry at the University of Lausanne Switzerland. He is presently the Chairman of the IUPAC Commission of Environmental Analytical Chemistry and a member of the Research Council of the Swiss National Foundation. He has authored 100 research papers, two books, and is a co-editor of two other monographs. He teaches under- and postgraduate courses in Analytical Chemistry and Environmental Chemistry and is actively involved in organizing postgraduate courses in limnology.

Dr. Buffle has received diplomas in biological chemistry, chemical engineering, and numerical calculation, and a Ph.D. degree in analytical chemistry. His research interests are the understanding of physico-chemical processes regulating the circulation of chemical components in environmental compartments, the relative influences of these processes and the development of biota, and the development of *in situ* sensors for the measurement of the corresponding key, physico-chemical parameters. He is particularly interested in contributing to the understanding of the control of macroscale effects, at the level of environmental compartments such as soils, lakes, etc., by microscale phenomena, at the molecular, colloidal, and microbial level, by considering multidisciplinary approaches.

Herman P. van Leeuwen is an electrochemist who got his education at the State University of Utrecht, The Netherlands. He obtained his degree in chemistry, with a main specialization in electrochemistry, in 1969. He then joined the electrochemistry group of Professor J. H. Sluyters where he prepared his thesis in the field of pulse methods in electrode kinetics. The Ph.D. degree was awarded cum laude in 1972. In the period from 1968 until 1973 he also was a parttime teacher of chemistry. In 1972 he joined the colloid chemistry and electrochemistry group of Professor J. Lyklema at the Wageningen Agricultural University, where he became a senior scientist, in 1978, and associate professor of electrochemistry, in 1986. He has been active in teaching analytical chemistry, physical chemistry, and electrochemistry.

His current research activities include electrodynamics of colloids (in relation to colloid stability), voltammetric speciation of heavy metals in environmental systems, and ion binding by synthetic and natural polyelectrolytes. He has published some 70 research papers, reviews, and book chapters in these fields. He has an intensive cooperation with colleagues from Czechoslovakia, Portugal, Spain, Switzerland, and the U.K. He is currently the secretary of the Electrochemistry Working Group of the Royal Dutch Chemical Society, associate member of the IUPAC Commission on Electroanalytical Chemistry, and titular member and secretary of the IUPAC Commission on Environmental Analytical Chemistry.

Environmental Analytical and Physical Chemistry Series
ENVIRONMENTAL PARTICLES — Volume 1
a project of the
IUPAC COMMISSION ON ENVIRONMENTAL ANALYTICAL CHEMISTRY
Managing Editor
P. D. GUJRAL
IUPAC Secretariat, Oxford, UK
INTERNATIONAL UNION OF PURE AND APPLIED CHEMISTRY
IUPAC Secretariat: Bank Court Chambers, 2-3 Pound Way,
Templars Square, Cowley, Oxford OX4 3YF, UK

CONTRIBUTORS TO THIS VOLUME

Dr. P. Artaxo
Department of Chemistry
University of Antwerp (U.I.A.)
B-2610 Antwerp-Wilrijk, Belgium

Dr. Jacques Buffle
Department of Inorganic,
  Analytical, and Applied
  Chemistry
Sciences II
30 Quai E.—Ansermet
1211 Geneva 4, Switzerland

R. Rob F. M. J. Cleven
Laboratory of Inorganic Chemistry
National Institute of Public Health
  and Environmental Protection
P.O. Box 1
3720 BA Bilthoven
The Netherlands

Dr. Lloyd A. Currie
Leader, Atmospheric Chemistry
  Group
National Institute of Standards and
  Technology
Gaithersburg, MD 20899

Professor William Davison
Institute of Environmenal and
  Biological Sciences
Lancaster University
Lancaster, LA1 4YQ,
  United Kingdom

Dr. Jan L. M. deBoer
Laboratory of Inorganic Chemistry
National Institute of Public Health
  and Environmental Protection
P.O. Box 1
3721 BA Bilthoven
The Netherlands

Dr. Richard De Vitre
Department of Inorganic,
  Analytical, and Applied
  Chemistry
Sciences II
30 Quai E.—Ansermet
1211 Geneva 4, Switzerland

Dr. Roy M. Harrison, Director
Institute of Aerosol Science
University of Essex
Colchester, CO4 35Q,
  United Kingdom

Dr. Bruce D. Honeyman
Environmental Sciences and
  Engineering Ecology
Colorado School of Mines
Golden, CO 80401

Dr. L. K. Koopal
Department of Physical and
  Colloid Chemistry
Wageningen Agricultural
  University
Dreijenplein 10,
6703 HB Wageningen
The Netherlands

Dr. Gary G. Leppard
Lakes Research Branch
National Water Research Institute
Environment Canada
Burlington, Ontario L7R 4A6,
  Canada

Dr. M. Newman
Department of Inorganic,
  Analytical, and Applied
  Chemistry
Sciences II
30 Quai E.—Ansermet
1211 Geneva 4, Switzerland

Dr. D. Perret
Department of Inorganic,
  Analytical, and Applied
  Chemistry
Sciences II
30 Quai E.—Ansermet
1211 Geneva 4, Switzerland

Dr. Peter H. Santschi
Department of Marine Sciences
Texas A&M University at
  Galveston
Galveston, TX 77553

Dr. Garrison Sposito
Department of Soil Science
University of California
Berkeley, CA 94720

Roger L. Tanner, Research
  Professor
Energy and Environmental
  Engineering Center
Desert Research Institute
Reno, NV 89506

Dr. André Tessier
INRS-Eau
University of Quebec
C.P. 7500
Saint-Foy, Quebec G1V 4C7,
  Canada

Dr. Anton van der Meulen
Laboratory of Air Research
National Institute of Public Health
  and Environmental Protection
P.O. Box 1
3721 BA Bilthoven
The Netherlands

Dr. R. Van Grieken
Department of Chemistry
University of Antwerp (U.I.A.)
B-2610 Antwerp-Wilrijk, Belgium

Dr. Herman P. van Leeuwen
Department of Physical and
  Colloid Chemistry
Wageningen Agricultural
  University
Dreijenplein 6
6703 HB Wageningen
The Netherlands

Dr. A. Van Put
Department of Chemistry
University of Antwerp (U.I.A.)
B-2610 Antwerp-Wilrijk, Belgium

Dr. W. H. Van Riemsdijk
Department of Soil Science and
 Plant Nutrition
Wageningen Agricultural
 University
Dreijenplein 10,
6703 HB Wageningen
The Netherlands

Dr. L. Wouters
Department of Chemistry
University of Antwerp (U.I.A.)
B-2610 Antwerp-Wilrijk, Belgium

Dr. Chris Xhoffer
Department of Chemistry
University of Antwerp (U.I.A.)
B-2610 Antwerp-Wilrijk, Belgium

Dr. Charles S. Yentsch
Bigelow Laboratory for Ocean
 Sciences
McKown Point
West Boothbay Harbor,
 ME 04575
  and
Department of Geography
Boston University Remote Sensing
 Center
725 Commonwealth Avenue
Boston, MA 02215

Dr. Clarice M. Yentsch
Bigelow Laboratory for Ocean
 Sciences
McKown Point
West Boothbay Harbor,
 ME 04575

Contents

PART I
SAMPLING AND CHARACTERIZATION OF ATMOSPHERIC PARTICLES

Chapter 1   Source Apportionment of Atmospheric Particles............ 3
               L. A. Currie

Chapter 2   Acid-Base Equilibria of Aerosols and Gases in
               the Atmosphere... 75
               R. L. Tanner and R. M. Harrison

Chapter 3   Characterization of Individual Environmental
               Particles by Beam Techniques............................107
               C. Xhoffer, L. Wouters, P. Artaxo, A. Van Put, and
               R. Van Grieken

Chapter 4   Characterization of the Cr(VI)/Cr(III) Ratio in
               Aerosols...145
               R. F. M. J. Cleven, J. L. M. de Boer, and
               A. van der Meulen

PART II
SAMPLING AND CHARACTERIZATION OF PARTICLES OF AQUATIC SYSTEMS

Chapter 5   The Use of Filtration and Ultrafiltration for
               Size Fractionation of Aquatic Particles,
               Colloids, and Macromolecules 171
               J. Buffle, D. Perret, and M. Newman

Chapter 6   Evaluation of Electron Microscopic Techniques
               for the Description of Aquatic Colloids 231
               G. G. Leppard

Chapter 7   Characterization of Particle Surface Charge 291
               G. Sposito

Chapter 8   Iron Particles in Freshwater 315
               W. Davison and R. De Vitre

Chapter 9   Characterization of Oceanic Biogenic Particles...........357
               C. M. Yentsch and C. S. Yentsch

PART III
REACTION AND TRANSPORT OF PARTICLES IN AQUATIC SYSTEMS

Chapter 10   The Role of Particles and Colloids in the
               Transport of Radionuclides and Trace Metals in
               the Oceans 379
               B. D. Honeyman and P. H. Santschi

Chapter 11   Sorption of Trace Elements on Natural Particles
              in Oxic Environments 425
              A. Tessier
Chapter 12   Ion Binding by Natural Heterogeneous Particles..........455
              W. H. van Riemsdijk and L. K. Koopal
Chapter 13   Dynamic Aspects of Metal Speciation in Aquatic
              Colloidal Systems 497
              H. P. van Leeuwen

Index 523

PART I
Sampling and Characterization of Atmospheric Particles

CHAPTER 1

SOURCE APPORTIONMENT OF ATMOSPHERIC PARTICLES*

Lloyd A. Currie

Atmospheric Chemistry Group, National Institute of Standards and Technology, Gaithersburg, Maryland

TABLE OF CONTENTS

1. Introduction 5
    1.1 Objective and Scope 5
    1.2 Definition: Complementarity of Source and
         Receptor Modeling 5
    1.3 Importance of Different Classes of Atmospheric
         Particles 6
         1.3.1 Extension 6
         1.3.2 Effects 7
    1.4 Anthropogenic vs. Natural Sources 8
         1.4.1 Global and Regional Budgets 8
         1.4.2 Historical Records 9

---

* Contribution of the National Institute of Standards and Technology. Not subject to copyright.

2.   Overview of Requisite Analytical Data 10
      2.1   Data Classes: Chemical, Meteorological,
             Biological, Visibility 10
             2.1.1   Response Variables 11
      2.2   Formation Processes and Compositional
             Information Content 11
             2.2.1   Compositional Information Content of
                     Combustion Aerosol (Inorganic) 13
             2.2.2   Information Content of Organic Species 15
             2.2.3   Information Content of Individual
                     Particles 20

3.   Source Apportionment Methodology 23
      3.1   Relation to Extensions of the Atmospheric
             Compartment 26
      3.2   Isotopic Tracers 27
             3.2.1   Carbon Isotopes 28
             3.2.2   Other Isotopes 33
      3.3   Chemical Mass Balance (CMB) 33
             3.3.1   Particle Class Balance 37
      3.4   Multivariate Techniques 37
             3.4.1   Tracer Multiple Linear Regression 38
                     3.4.1.1   Selection and Correction of
                               Source Tracers 39
                     3.4.1.2   Validation 40
             3.4.2   Cluster Analysis 43
             3.4.3   Principal Component and Factor Analysis 49
      3.5   Hybrid Techniques 52

4.   Validation of Source Apportionment Methodology 55
      4.1   Experimental Validation 55
      4.2   Mathematical (Model) Validation 56
             4.2.1   Long-Range Transport 56
             4.2.2   Linear (Receptor) Modeling 61

5.   Outlook and Conclusion 64

Acknowledgment 65

Notes 65

References 66

1. INTRODUCTION

1.1 Objective and Scope

The identification and quantification of sources of atmospheric particles directly relates to the Terms of Reference of the IUPAC Commission on Environmental Analytical Chemistry, viz., its responsibility to "assess analytical methodologies for the characterization of the environment," in this case its atmospheric compartment. The importance of this objective derives from the potential impact of atmospheric particles on health, visibility, and climate. In these respects we are particularly concerned with tropospheric concentrations and sources of toxic compounds (e.g., polycyclic aromatic hydrocarbons), sulfate, and graphitic carbon (soot). A primary objective of source apportionment of such species is to delineate the natural and anthropogenic components, since only the latter may be controlled. The scope of the study comprises local, regional, and global source apportionment.

The basic approach to particulate source apportionment involves chemical and physical characterization of ambient particles, together with their variations in space and time, plus relevant information on source (emission) composition, atmospheric transport, and chemical transformation. Univariate and multivariate statistical techniques form the link between the observed particle characteristics and the inferred source identification and quantification.

Specific topics to be considered include: (1) analytical (chemical) particle characterization methods, such as isotopic analysis, elemental and anionic analysis, patterns in homologous series of organic compounds, and individual particle assay; and (2) analytical (mathematical) source apportionment schemes, such as unique tracer regression, enrichment factors, cluster analysis, factor analysis, particle class and chemical mass balance, and wind trajectory analysis. Finally, essential means to control the quality of both the chemical and mathematical phases of analysis will be considered. These derive from source profile databases, representative field experiments, and standard (reference) materials and standard (simulation, multivariate) data.

1.2 Definition: Complementarity of Source and Receptor Modeling

The term "source apportionment" is frequently used as a synonym for atmospheric gas or particle "receptor modeling".[1] The latter expression, which unfortunately is used in an entirely different context in the biological sciences, was invented in opposition to "source modeling." Source modeling relates to the use of source emission inventories, together with models for atmospheric transport (dispersion modeling) and reaction, to *forecast* the impacts of sources as a function of time and location. By contrast, receptor modeling relates to the use of *observed* atmospheric concentrations of an array chemical species at a given time and location ("receptor site") to infer the impacts of the several sources at that site, usually by some form of multivariate analysis. Neither approach succeeds at representing the true state of Nature. Insufficient information always exists, so solutions are forced to rest on assumptions and approximations. During the last several years, the wisdom of bringing together the most reliable information from both methods has been realized; and increasingly one finds terms such as "composite receptor method"[2] and "hybrid receptor model"[3] employed in the source apportionment literature. The asymptotic meaning of *source apportionment* will surely take into account the complementary "source" and "receptor" viewpoints; employing observed time and space arrays of chemical species, prior knowledge concerning potential sources and their emission patterns, and data related to air mass trajectories and transformations. More explicit discussion of these issues, including the use of tracers for transport model validation, is given in Sections 3 and 4.

1.3 Importance of Different Classes of Atmospheric Particles

The impact of particles on the atmospheric environment depends qualitatively on their chemical and physical properties, and quantitatively on their lifetimes and abundances. These properties derive, in turn, from the particle sources and formation processes. Processes that result in the *direct* release of particles, such as dust storms and vegetative (particulate) emissions, tend to generate "large" or coarse particles — i.e., those having nominal diameters of 10 μm or more. The effects of such particles (mineral dust, pollen, etc.) tend to be localized, because of rapid sedimentation. Particles formed in combustion, or in secondary gas-to-particle reactions, on the other hand, are generally small, ca. 1 μm or less in nominal diameter. Furthermore, combustion (and photochemical oxidation) generally destroys the neutral balance of the starting material paving the way to "acid precipitation",[4] and generates substances (irritants, mutagens) with undesirable biological properties.

1.3.1 Extension

The extension of atmospheric particles in time and space, as related to size, has been treated by Jaenicke.[5] The essentials are given in Figure 1, which summarizes the relations among particle growth and removal processes, size, and lifetime. From the figure, and from an accompanying table in Jaenicke's article, we see that the decade of greatest consequence covers the size range of 0.1 to 1.0 μm. Such particles may be transported over distances of 8000 km and to altitudes of 20 km during their 2- to 3-week tropospheric residence times. Smaller particles are removed by coagulation, whereas larger particles undergo rapid sedimentation. (Fine particles injected into the stratosphere, e.g., from nuclear weapons tests or volcanic action, may linger for several years.) Thus, combustion and photochemical aerosol is clearly a matter of global (or at least hemispheric) concern. An illustration of such long range transport of combustion aerosol is the "Arctic Haze," which has been ascribed to industrial sources deep in the Soviet Union and Central Europe.[6]

Particle removal through cloud processing and chemical reaction can be extremely complex, and obviously highly dependent on particle composition. This is one reason that conservative tracers and properties, such as isotope

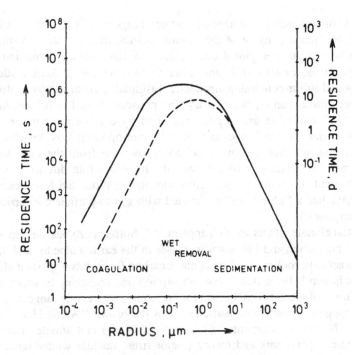

Figure 1. The residence time of aerosol particles. Two different cases have been calculated: — background aerosol. N = 300 cm^{-3}, τ_{wet} = 21 d; --- continental aerosol, N = 15,000 cm^{-3}, τ_{wet} = 8 d. In both cases, the height of the homogeneous aerosol layer was assumed to be 1.5 km. This only has effects on the sedimentation branch of the curve. (From Jaenicke, R. in *Aerosols: Anthropogenic and Natural Sources and Transport*, Vol. 338: 317–329. New York: New York Academy of Sciences, 1980. With permission.)

ratios, are so important for reliable source apportionment. Cloud cycling will not be discussed here; for background information and references, see (National Research Council;[7] and Rowland and Isaksen[8]). Transformation through chemical and photochemical reaction constitutes a major field of study, performed both in the laboratory and in polluted atmospheres. Accounting for such transformations, for example through trajectory analysis and hybrid (nonlinear) modeling permits one to (a) identify reactive precursors, and (b) build chemical transformations into the source apportionment (receptor) modeling.[9] Illustrations are given in Section 3 of this document.

1.3.2 Effects

The deleterious effects of carbonaceous particles, and more generally, combustion particles, serve as the driving force for research on the formation and atmospheric life-cycle of these particles. Primary health effects of these particles, and/or their gaseous precursors, relate to the formation of tropospheric ozone[10] and genotoxicity.[11] Ozone formation results from a complex series of photochemical reactions involving reactive hydrocarbons, nitrogen oxides, and the hydroxyl radical. Mutagenic and carcinogenic effects are associated with the biological activity of the organic combustion products. Visibility tends to be a local or regional consequence of fine particle formation and transport. In the eastern U.S. and other regions having at least moderate humidity and sulfur-containing air masses (originating from nearby or distant fossil fuel power plants), the primary fine particle aerosol is sulfate; in the western U.S. and other arid regions assaulted by combustion particles, the main component is carbonaceous soot.[12] In metropolitan Los Angeles, for example, one observes a brown, carbonaceous haze from the city. In the Great Smoky Mountains,[13] by contrast, the haze is white due to its major sulfate content. (A century ago, before industrialization, the Smokies were said to have had a "blue haze" associated with gas-to-particle conversion of natural terpenes.)

Potential climate effects are well appreciated. Sulfate particles are important for cloud nucleation and important increases in the earth's albedo, and therefore tropospheric cooling,[14] whereas black carbon (soot) has important effects on the radiation balance due to both absorption and scattering of solar radiation.[15] This is of special consequence for the Arctic, where long range transport of anthropogenic combustion particles forms the brown "Arctic Haze" each spring.[6] "Natural experiments" (volcanic emissions) and smaller scale anthropogenic experiments and theory (major fires, nuclear winter scenarios) provide semiquantitative data on temperature effects.[16] Sulfate aerosol is especially interesting and important on a regional and global scale because its dispersion can be very great, due to its gaseous precursor SO_2; and the sign of its potential climatic effects is opposite to that of both "soot" particles and the "greenhouse" gases (CO_2, CH_4).[17]

1.4 Anthropogenic vs. Natural Sources
1.4.1 Global and Regional Budgets

The influence of human activities on the production and life-cycle of atmospheric particles has become the focus of international research programs.[18] Apportionment of the anthropogenic component is very important; because of the limited lifetimes of atmospheric aerosols, we can influence the future by controlling their sources. An overview of global natural and anthropogenic aerosol sources is given in Table 1.[19] Here, we see that the global natural aerosol exceeds the anthropogenic component by about a factor of five, but each component is uncertain by at least a factor of two. Total source strengths can be a little misleading, however, in view of spatial and compositional heterogeneity. Spatially, anthropogenic sources are concentrated in inhabited regions, especially the cities or other densely populated areas; compositionally, the impacts are strongly dependent on chemical toxicity, particle albedo, cloud interactions, etc. With these factors in mind, we see from Table 1 that anthropogenic emissions are dominated by sulfates (secondary aerosol), while natural emissions are dominated by sulfates, sea salt, and dust. Carbonaceous aerosol derives primarily from hydrocarbon emissions and combustion, both of which have significant natural and anthropogenic sources. Hydrocarbon emissions have a major influence on visibility and regional ozone formation; combustion particles ("soot") have important albedo and toxicity effects. Carbon isotopes are most important for quantifying sources in each category. From the regional and local perspective, carbonaceous particles are quite important. Studies of the greater Los Angeles aerosol, for example, based on emissions inventories and receptor modeling, have shown that particulate carbon accounts for 40% of the average fine particle mass,[20] and that 40 to 70% of this arises from motor vehicles.[21] Direct measurement of the fossil fraction of particulate carbon, by means of radiocarbon, has shown typical values of 60 to 70% in a number of urban areas; particles from locations subject primarily to natural emissions or to woodburning show typically 80 to 90% biospheric carbon.[22]

1.4.2 Historical Records

Mankind's past contributions to atmospheric particles may be read from both human and natural archives. An interesting illustration of the former may be found in Brimblecombe.[23] Glaciers and sediments comprise the latter. Because the resident concentrations are generally quite low in these natural archives, single particle or very sensitive organic analytical methods are generally required. Illustrations of the records of combustion particles are found in the research of Hites[24] and that of Griffin and Goldberg.[25] In the former work, trace analysis of polycyclic aromatic hydrocarbons (PAH) was performed on sediment cores taken from Pettaquamscott River and the Grosser Pläner Sea. A dramatic rise in anthropogenic PAH was found in both of the locations, coinciding with the onset of the Industrial Revolution. The latter research utilized single carbonaceous particle morphology, as opposed to the former's bulk particle chemical (PAH) analyses. The same anthropological phenomenon (Industrial Revolution) was observed by Griffin and Goldberg, however, this time in a box core from Lake Michigan. Of special interest in

Table 1. Estimates of Global Particle Production (10^6 t/year)

	Manmade	Natural
Direct Production	30	780
	(Industrial processes — 12)	(Sea salt — 500)
	(Stationary power — 10)	(Windblown dust — 250)
	(Transportation — 2)	(Volcanoes — 25+)
		(Forest fires — 5+)
Secondary (from gases)	250	470
	(Sulfates — 200)	(Sulfates — 335)
	(Nitrates — 35)	(Nitrates — 60)
	(Hydrocarbons — 15)	(Hydrocarbons — 75)
Total	280	1250

Adapted from Prospero, J. M. in *Global Tropospheric Chemistry, A Plan for Action.* (Washington, D.C.: National Academy of Sciences Press, 1984) pp. 136–140. (Ref. 19) See the original publication for more detailed source information, and uncertainty ranges.

this case is the fact that the particle morphology preserved the structure or combustion fingerprint of the original fuel. Thus, one sees not only the record of accelerated industrialization, but also the distinctive transitions in fuel type.

2. OVERVIEW OF REQUISITE ANALYTICAL DATA
2.1 Data Classes: Chemical, Meteorological, Biological, Visibility

Successful source apportionment depends upon adequate observational data (ambient particle composition, meteorological data), model structure and parameterization (source profiles, kinetics data), and effects or response data (visibility, mutagenicity). No less important than the data, themselves, are reliable uncertainty estimates. (Further discussion of the very difficult matter of uncertainty estimation, and its impact on source strength estimates, follows in Sections 3.3 and 3.4.)

Aerosol compositional data (physical and chemical) comprise the fundamental material of source apportionment. A number of the multivariate techniques (Section 3.4) require no other input but ambient aerosol compositional patterns. Regression methods generally require more. For example, chemical mass balance (CMB, Section 3.3) is based on the (least squares) fitting of observed ambient aerosol chemical patterns to collections or "libraries" of emission patterns for potential sources. An important level of sophistication is sometimes found when such libraries provide source profiles as a function of particle size, or at least for the dichotomous "fine"/"coarse" classes.[26] Multiple linear regression (Section 3.4.1) is very effective when one has a complete set of "unique tracers" for the sources in question. These types of data suffice for conventional receptor modeling, if the tracers and source compositional patterns are conservative and additive — i.e., if one truly is dealing with a linear model. Such data are required in any case; in Section 2.2 we shall consider the compositional data more explicitly.

Meteorological data are mandatory for efforts involving dispersion modeling, as well as for long range transport modeling entailing back trajectories. Even on a more local scale, such data are valuable. For example, atmospheric stability data are most important for stratification of compositional data sets, and for designing field sampling plans. Wind speed and direction are at the heart of "wind trajectory analysis" (Section 3.5), and often contribute to the understanding of factor analysis results.

Chemical transformation data — such as unimolecular rate "constants" and the manifold reaction parameters related to photochemical pathways — must be available for certain types of nonlinear source apportionment. These data, and similarly important physical data (like volatility), are fundamentally different from the stochastic "field data" (chemical composition, meteorology). That is, we are dealing with basic physical and chemical properties, characterized by unique values. The inaccuracy of these fundamental data, however, can have a severe impact on our understanding and nonlinear modeling of the atmospheric system. For an insightful discussion of the nature of the associated error-propagation process, see Thompson.[27]

2.1.1 Response Variables

Response data constitute the final data class. Here, we address the question: Exactly *what* needs to be apportioned, and *why*? Classic receptor modeling methods treat the total particle mass as the response variable — i.e., the particle mass is apportioned quantitatively among the several contributing sources. When atmospheric aerosol *effects* are the major concern, then the respective effects become the response variables (observables) to be modeled or apportioned. The simplest is the dichotomous size variable. Consideration of health effects through inhalation has placed emphasis on the fine, or inhalable particles. Thus, for regulatory purposes, in the U.S., the fine particle mass (FPM) is the important response variable. (The regulation relates specifically to particles which have aerodynamic radii of 10 μm ["PM10"] or smaller.[1]) Other response (effect) variables of considerable interest at present are aerosol mutagenicity[28] and "visibility" or visible light scattering effects (extinction). Major visibility apportionment programs are currently underway in the U.S. National parks and other areas of the Southwest; and the Integrated Air Cancer Project of the U.S.E.P.A. has as its goal the apportionment of the genotoxicity of atmospheric aerosol. Closely related, because of the mutagenicity of certain organic species, are apportionment studies of the "extractable organic matter" of the atmospheric particles.[28]

In special cases it becomes important to apportion sources of individual compounds or classes of compounds. Examples are: CO, for wintertime urban pollution and indoor air pollution; volatile organic compounds (VOC), for their central role in the production of urban and rural ozone; and sulfate, for its relations to visibility, acid precipitation, and climate (tropospheric albedo).[10,29] Carbon monoxide has global apportionment implications as well, in that it is the primary consumer of the hydroxyl radical, which in turn controls the lifetime of atmospheric methane, an important greenhouse gas.[8] Global source apportionment (which would not be discussed under the more restricted topic of "receptor modeling") is crucial when one considers very long-range and long-term problems, such as stratospheric ozone depletion and global climate change.[8] Except for sulfate, the principal species of concern (our response variables) are nonreactive gases, including the halocarbons, methane, carbon dioxide, and nitrous oxide. Some brief comments about strategies for global source apportionment will be given in Section 3.1 of this report.

2.2 Formation Processes and Compositional Information Content

If each atmospheric particle carried a unique and absolute tracer, characteristic of its source, the task of source apportionment would become trivial but reliable conclusions would be assured. (See Note 1) In an effort to work toward such an ideal, it is useful to consider the compositional information content of atmospheric aerosols, particularly from the perspective of their formation. Empirical and theoretical studies of particle formation are ex-

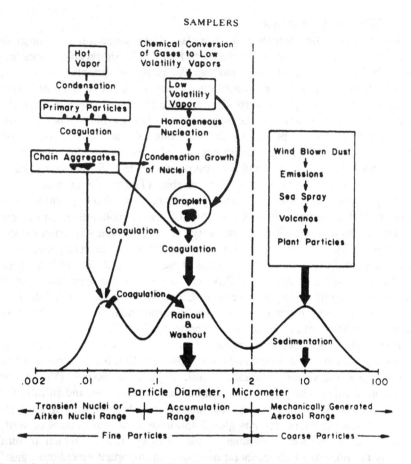

Figure 2. Idealization of the atmospheric aerosol surface area distribution showing the
principal modes, main sources of mass for each mode, and the principal
processes involved in injecting mass in each mode as well as primary removal
mechanisms. (From Hopke, P. K. *Receptor Modeling in Environmental Chemistry.* New York: John Wiley & Sons, 1985. With permission.)

tremely helpful in illuminating "informing features" for the identification
and quantification of sources. In short, they help us decide "What should we
measure?"[30]

As an initial focus it is useful to consider in Figure 2, an input/removal
diagram for aerosol mass, constructed by Whitby and Cantrell.[31] The smallest
particles result first, from high temperature processes, which means volcan-
ism, combustion (natural fires and anthropogenic processes), and certain industrial processing operations. The composition of the resultant primary aerosol is as diverse as that of the high temperature sources; and consideration
of those sources on an individual basis can provide considerable insight. Small
condensation nuclei also result from homogeneous nucleation from gases and
vapors. Photochemical conversion of reactive hydrocarbons (including natural

emissions), and oxidation of aerosol precursors (such as sulfur dioxide) fall into this category. Scientific understanding of the reaction processes and chemical and isotopic knowledge of the source processes are essential for intelligent source apportionment. Coagulation quickly transforms the condensation aerosol to the intermediate diameter ("accumulation") mode, where rainout and washout are the primary removal mechanisms. In this process, however, most of the primary elemental and less volatile organic emissions will have become entrenched in (or on) the fine (accumulation mode) aerosol.

From the other direction, we find the coarse (10 μm mode) particles produced by wind and wave erosion, including direct emission/suspension of mineral dust and vegetative material (plant lipids, pollen, etc.). Though the coarse particles tend to be more quickly removed by sedimentation, and are generally less reactive and less inhalable than the fine particles, they should not be totally discounted. As shown in Figure 2, the two classes do overlap, and both are found in the critical 0.1 to 10 μm size range. In certain respects, particles arising from erosion and direct emission processes are easier to identify, because they preserve the original structure and composition of the source material. Thus, microscopy and microanalytical techniques are especially pertinent for mineral and organic fragment identification; and homologous series organic analyis is useful for detecting the products of natural biosynthesis. Such compositional and structural information will be examined following discussion of some compositional consequences of combustion.

2.2.1 Compositional Information Content of Combustion Aerosol (Inorganic)

The elemental (non-organic) compositions of particles from the combustion of coal, fuel oil, gasoline (petrol), and wood have received considerable attention. (Gaseous fuels, such as C_1-C_4 alkanes, generate very little aerosol.) Coal, perhaps the most noxious in terms of inorganic aerosol and sulfate, has been extensively studied using the "enrichment factor" technique. To introduce this concept, it is useful to examine the coal fly ash formation process. Figure 3,[32] based on detailed investigations of the mechanism of coal combustion, shows multiple pathways taken by the different chemical and mineral components. Elements within the carbonaceous structure of the coal pass through volatilization and homogeneous nucleation, to form submicrometer "fume" particles that provide a large surface area for condensation of more volatile species. The residual ash particles are larger (0.5 to 10 μm or more), and often pass through a molten stage. (Outgassing of the molten ash generates the hollow "cenospheres" which may be characterized microscopically.) The several pathways from coal to aerosol necessarily produce compositional changes, or "enrichment", with respect to the initial fuel. The nature of the process ensures that the enrichment depends both on particle size and physicochemical properties. Actually there are two stages of enrichment: that which occurs in nature, during the coal formation process, and that which occurs during combustion.

The "enrichment factor" (EF) is generally defined with respect to crustal abundances,[33] using an appropriate element for normalization. An example is given in Figure 4, where

$$EF_{cr} = (X/Al)_a/(X/Al)_{cr} \qquad (1)$$

EF_{cr} denotes the enrichment factor for element X in the aerosol (subscript-a), with respect to its crustal abundance (subscript-cr). Aluminum is used here for normalization, because it is a readily measurable and relatively abundant, non-volatile (oxide) lithophile. Two sets of EF's are shown in Figure 4, one for coal combustion, the other for urban aerosol. With this pair of EF-patterns, we can speak to the *objective* of EF analysis: it is a qualitative, exploratory tool to suggest discriminating chemical species to employ in source apportionment. When EF_{cr} is significantly different from unity, it suggests useful tracers for discrimination from crustal aerosol. Arsenic and selenium (Figure 4) look especially attractive from this perspective — though volatility and measurability must also be considered. Comparison with the EFs for urban aerosol tells us more. Lead, in particular, was uniquely urban, compared to soil and coal, at the time of this study (1981). Less distinct in Figure 4, but nevertheless important, is vanadium. Just as lead has been a useful tracer for motor vehicles, vanadium has served as a useful tracer for oil combustion.[1] Were we to examine EFs for wood combustion, potassium would be outstanding. However, like the aforementioned EFs for coal, oil, and petrol, the *quantitative* relationship between potassium and wood combustion depends on many factors, such as the source and type of the wood-fuel. The quantitative aspects of aerosol composition and aerosol emissions

SURFACE ENRICHMENT OF FLY ASH

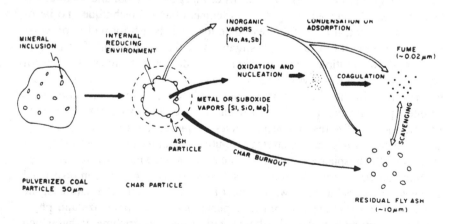

Figure 3. Schematic diagram of ash formation and behavior in coal combustion. (From Haynes, B. S. et al. *J. Colloid Interface Sci.* 87:226–278, 1982. With permission.)

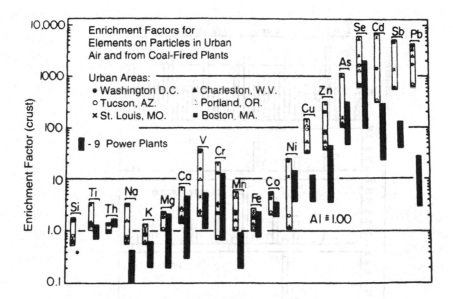

Figure 4. Enrichment factors with respect to crustal abundances for elements attached to urban particles from (●) Washington, D.C., (○) Tuscon, AZ, (x) St. Louis, MO, (▲) Charleston, WV, (△) Portland, OR, and (■) Boston, MA. (From Gordon, G. E. et al. in *Atmospheric Aerosol: Source/Air Quality Relationships.* Washington, D.C.: American Chemical Society Symp. Ser. 167, 1981, Ref. 123. With permission.)

lie at the heart of source apportionment, which forms the substance of Section 3 of this document.

In anticipation of quantitative source apportionment, and to illustrate the nature of elemental aerosol patterns from some common sources, we present Figure 5.[1] Here, the semiquantitative ranges of EFs for cities and coals have been replaced by aerosol mass normalized patterns ("source profiles") for the set of sources in a particular urban study. The development of such profiles is central to mass balance, multiple regression, and "absolute" factor analytic methods of analysis.

2.2.2 Information Content of Organic Species

In contrast to the elemental composition of atmospheric particles, the structure and composition of the carbonaceous fraction is extremely rich and diverse. Particulate organic carbon that has not undergone high temperature combustion or photochemical reaction is likely to contain structural clues ("molecular markers") of source-specific biosynthetic processes. Combustion carbon, on the other hand, generally has its biological memory erased, especially if it has passed through a gaseous/free radical state. Pyrosynthetic processes, however, generate their own imprints of the nature of the combustion process — especially through polyaromatization.[34] A glimpse of the pyrosynthetic process and its structural consequences is given in Figure 6.

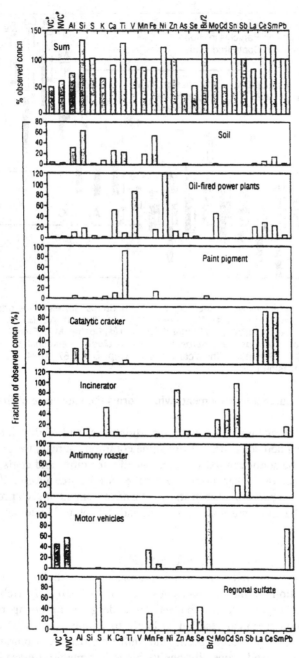

†The bromine values have been divided by two to keep them on scale. All Results are normalized to the observed concentration because of the wide range of concentrations of the various elements. If all elements were accounted for exactly, the sums of the contributions in the top panel would be 100% of the observed concentration.
‡VC and NVC refer to volatile and nonvolatile carbon. respectively.

Figure 5. Chemical mass balances for selected elements and species borne by fine particles in Philadelphia. (From Gordon, G. E. *Environ. Sci. Technol.* 22:1132–1142, 1988. With permission.)

High temperature combustion produces acetylenic free radicals that polymerize to form polycyclic aromatic hydrocarbons (PAH) characterized by larger and larger numbers of aromatic rings, terminating in graphitic microcrystals.[29,35] Such structures occur also as a result of diagenesis.[34,36] The distribution of ring sizes, and the extent and position of alkyl side chains, hold great promise for source apportionment and characterization (e.g., temperature) of the formation process.

Accompanying the diversity of organic structures is a diversity in reactivity. Bioemissions such as plant waxes, and the end product of high temperature but incomplete combustion, graphite, tend to persist; many other species are lost during the journey from source to receptor through reaction or volatilization. To achieve success with organic based source apportionment, therefore, one must either select conservative organic tracers, or employ reaction/volatilization modeling to account for the nonlinear (nonconservative) behavior. Such chemical modeling approaches have successfully taken organic source apportionment beyond the domain of conventional linear (receptor) modeling to what has become known as "hybrid" receptor modeling, a topic that we shall treat in Section 3.5. Note that the ultimate conservative source features, isotope ratios, have been very important for the apportionment of carbonaceous particles, as well as other critical parts of the atmospheric system. They will be treated separately in Section 3.2.

Carbonaceous particles are important. Carbon normally comprises 80% or more of the mass of particles coming from bioemissions or gaseous (carbon-

PYROSYNTHESIS OF PAH AND SOOT

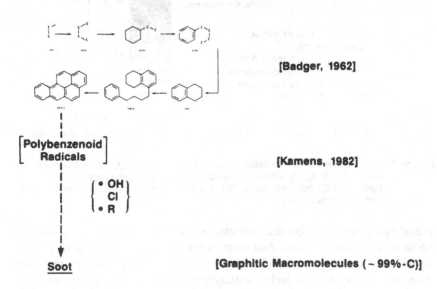

Figure 6. Pyrosynthetic formation of polycyclic aromatic hydrocarbons. (Adapted from References 35 and 124.)

CARBON IN LOS ANGELES

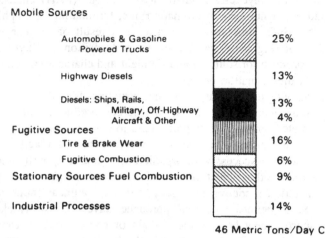

Total carbonaceous aerosol emissions. greater Los Angeles: Jan.-Feb. 1980.

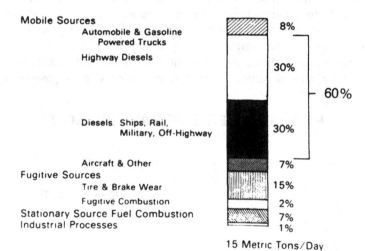

Sources of nonvolatile (elemental) carbon. greater Los Angeles: Jan.-Feb. 1980.

Figure 7. Los Angeles aerosol emissions, Jan., Feb. 1980. (Top) total carbonaceous emissions; (Bottom) elemental carbon emissions. (From Cass, G. R. et al. in *Particulate Carbon, Atmospheric Life Cycle.* New York: Plenum Press, 1982. With permission.)

aceous) precursors. Combustion particles, excluding coal fly ash, may contain up to 60% carbon by mass. Associated with the fine particle fraction, they travel great distances, reach the lungs, affect visibility and the radiation balance, and contain toxic and/or mutagenic compounds. Complementing the global aerosol source inventory of Table 1, we give an urban (Los Angeles) inventory of carbonaceous particles in Figure 7.[37] Extensive information on this most important component of atmospheric particles may be gained from the proceedings of a conference on the atmospheric life cycle of particulate carbon,[38] and the continuing series of quadrennial conferences on "Carbonaceous Particles in the Atmosphere".[39] An important review, with emphasis on the potential of polycyclic aromatic hydrocarbons for carbonaceous particle source apportionment, has been published by Daisey et al.[40]

Carbonaceous particle source identification, and to some extent quantification (apportionment) has been accomplished by the use of the biospecific "molecular markers" mentioned above. Noteworthy in this respect are: (a) the natural paraffins and lipids characteristic of vegetative emissions; (b) dehydrogenation products produced by gentle heating of softwoods; (c) characteristic woodburning compounds such as methoxylated phenols;[41] and (d) structural components of natural (and synthetic) polymers which can be detected with NMR chemical shifts,[42] microRaman spectroscopy,[43] and laser microprobe mass spectrometry (LAMMS).[44] In the first category, one finds paraffins (*n*-alkanes) with C_{23}-C_{35} predominant with a marked odd carbon number preference (Figure 8); the lipids (*n*-alkanoic acids) show C_{13}-C_{23} predominant with a marked even carbon number preference.[45] Two parameters, a "carbon preference index" and a "carbon number maximum" have been used by Mazurek et al.[46] to capture the biosynthetic structural information, and to apportion biogenic residues and (featureless) petroleum residues in Los Angeles aerosol. The dehydrogenation process, which carries some promise of woodburning source apportionment, is illustrated in Figure 9.[47] The end product, retene, has been widely considered as a unique identifier of softwood combustion aerosol;[48] and the entire set of dehydrogenation products (abietic acid to retene) has shown promise for the quantitative apportionment of woodburning aerosol.[49]

Saturated Hydrocarbon Fraction
(Gas Chromatogram)

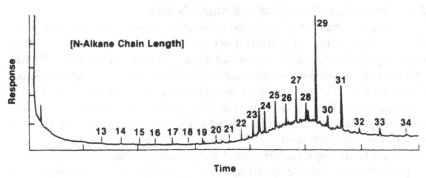

Figure 8. Gas chromatogram of saturated hydrocarbon fraction, Los Angeles aerosol; carbon-chain lengths for *n*-alkanes are shown. (From Currie, L. A. et al. *Radiocarbon* 25:603, 1983. With permission.)

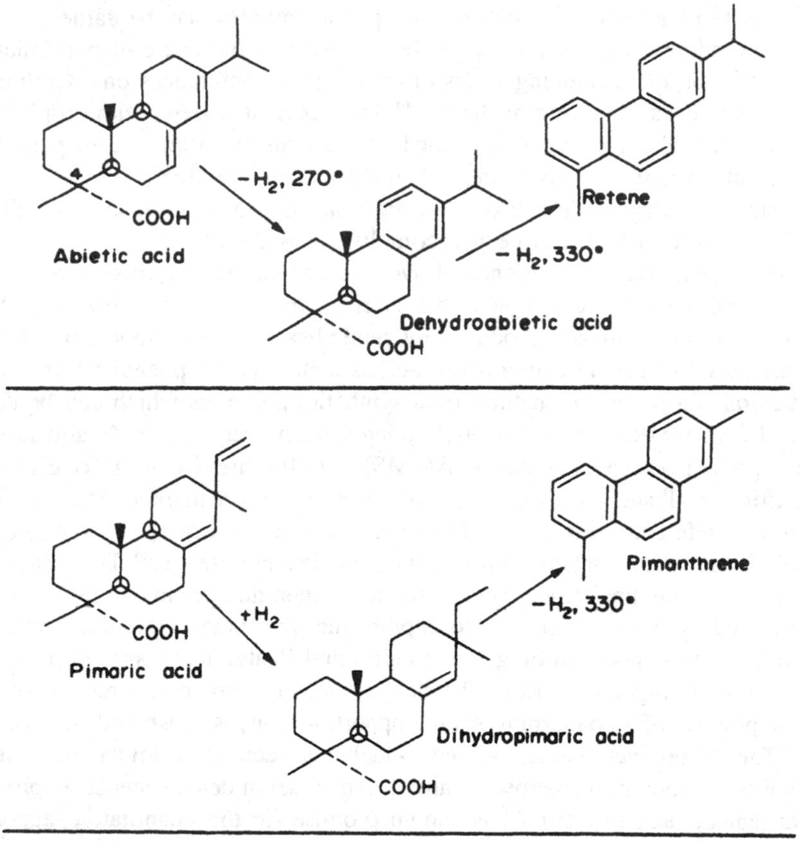

Figure 9. Dehydrogenation pathways for abietic and pimaric acids. (From La Flamme, R. E. and R. A. Hites. *Geochim. Cosmochim. Acta* 42:289, 1978. With permission.)

2.2.3 Information Content of Individual Particles

One might debate whether the forementioned organic patterns or the structure and composition of individual aerosol particles is richer. In fact, they come together in two of the techniques already referenced; for micro-Raman spectroscopy is suitable for investigating the carbonaceous (graphitic) structure of individual aerosol particles, as is laser microprobe mass spectrometry. An illustration of the latter is given in Figure 10, which shows a remarkable array of polycyclic aromatic hydrocarbons contained in the soot from an experimental oilshale retort.[36]

Optical microscopy and the several microanalytical beam techniques serve as powerful tools for characterizing the size, morphology, and composition of individual aerosol particles, sometimes yielding unique source identification. An illustration of the kind of information derived from one such technique is given in Figure 11.[26] This shows a scanning electron micrograph of a once-molten cenosphere from coal fly ash. Its identity is substantiated by the morphological data and the silica-alumina composition indicated by the electron beam induced X-rays. Another illustration of individual particle source identification comes from the work of Griffin and Goldberg.[25] Figure 12 presents data from a box core taken from Lake Michigan. Of special interest in this case is the fact we see not only the record of accelerated industrialization, we are able also to discern transitions in fuel type as a result of the unique morphologies of wood, oil, and coal combustion particles. Only a very brief view has been given of the information that can be derived from microanalytical techniques, because a comprehensive treatment of the topic has been prepared for this volume by Van Grieken.[50] For summaries of optical spectroscopy and scanning electron microscopy, as applied to atmospheric particle source apportionment, see Chapters 4 and 5 in Hopke's book.[26]

	Mass(amu)	Compound	
1	252*	benzo[a]pyrene[a,b]	
2		benzo[e]pyrene[a,b]	
3		perylene[a,b]	
4		benzo[k]fluoranthene[a,b]	
5	276*	anthanthrene[a,b]	
6		dibenzo[a,c]anthracene[a]	
7		benzo[ghi]perylene[a,b]	
8	300*	coronene[a,b]	
9	306*	a benzothiophene	
10	326*	dibenzo[a,ghi]perylene	
11	350*	benzo[a]coronene	
12	362*	a benzothiophene	
13	374*	dibenzo[a,bc]coronene	
14	380*	a benzothiophene	
15	398*	ovalene	
16	404*	a benzothiophene	
17	424*	benzonapthocoronene	
18	448*	benzo[a]ovalene	
19	472*	dibenzo[a,bc]ovalene	
20	496*	circumanthracene	
21	520	circumpyrene	
22	546	benzo[a]circumanthracene	
23	570	dibenzo[a,bc]circumanthracene	
24	594	circumtetracene	
25	620	tribenzo[a,bc,h]circumanthracene	

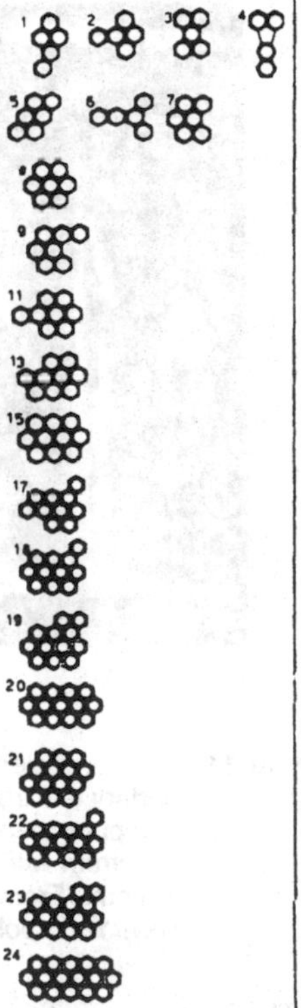

a - confirmed by HPLC analysis
b - confirmed by GC/MS analysis
* - confirmed by direct insertion EI and CI MS analysis

Figure 10. PAHs found by laser microprobe mass spectrometry in oil retort soot. (From Mauney, T. et al. *Sci. Total Environ.* 36:215–234, 1984. With permission.)

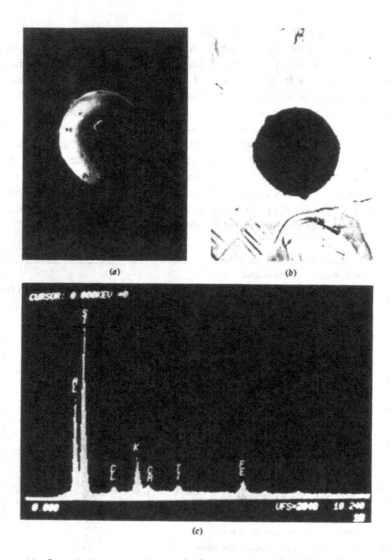

Figure 11. Scanning electron micrograph of a cenosphere (fly ash) particle using secondary electron image (a) and backscattered electron image (b) showing the computer generated image analysis lines for particle sizing. (c) The X-ray spectrum with the electron beam focused at the central spot is shown at the bottom. (From Hopke, P. K. *Receptor Modeling in Environmental Chemistry.* New York: John Wiley & Sons, 1985. With permission.)

The information that can be derived from individual particle characterization is so unique that it has given rise to its own method of source apportionment, "Particle Class Balance", the principles of which will be treated in Section 3.3. Anticipating the powerful multivariate approaches to data analysis (MVA, Section 3.4), it is worth noting that, in contrast to bulk analytical techniques, MVA can be applied to a sample from a single air filter when microanalytical techniques are employed.

3. SOURCE APPORTIONMENT METHODOLOGY

Source apportionment has developed from its parent, "receptor modeling." The latter began with attempts to employ a strictly linear model to quantify source impacts at a receptor (sampling) site, on the basis of measured multi-element concentrations. This technique was called by its inventor "chemical element balance".[51] The linear model of receptor modeling is in fact a direct analog of the model for multicomponent deconvolution of spectra in the laboratory. Both employ least squares estimation of component concentrations (**x**) dimension q, based on an observed vector of responses (**y**) dimension p, and a known calibration (or "sensitivity") matrix (**A**) — known as the "source profile" matrix in receptor modeling.

$$\mathbf{y} = \mathbf{Ax} + \mathbf{e} \qquad (2)$$

The last term in the matrix Equation 2 represents the measurement error vector.

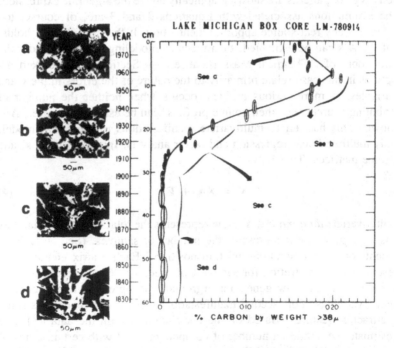

Figure 12. Charcoal particles in Core LM 780914 collected September 14, 1978 from southeastern Lake Michigan at 43°00′N and 86°22′W. The scanning electron micrographs show the particle morphologies common to four periods: (a) The 0–8 cm interval is representative of the particles in the sediments of the post 1960 period: (b) 12–14 cm, 1930–1960: (c) 27–28 cm, 1900–1930, and (d) 30–32, pre-1900 period. (From Griffin, J. J. and E. D. Goldberg. *Geochim. Cosmochim. Acta* 45:763–769, 1981. With permission.)

The well-known solution to Equation 2, for fixed measurement error (variance), is

$$\hat{x} = (A'A)^{-1} (A'y) \qquad \hat{V} = (A'A)^{-1} s^2 \qquad (3)$$

where $\hat{x}$ is the vector of estimated source component concentrations, $\hat{V}$ is the estimated variance-covariance matrix, and s^2 is the estimated measurement error variance. Over-determination (no. variables $p >$ no. sources q) is recommended for the estimation of variance and exploration of measurement and model errors. An extremely elementary illustration of a receptor model, for the purpose of elucidating the above quantities, is given in Figure 13. As shown in the figure, the first vector y, contains measured amounts of one isotopic quantity and four elemental concentrations; each column of the source profile matrix A represents the compositional profile or pattern for one of the pure sources; and the final vector x represents the unknown source contributions. Vector patterns are shown graphically above the algebraic expression.

The assumptions associated with Equations 2 and 3 are, of course, too stringent for any real source apportionment study. If the linear model holds, and if the A's can be estimated, then receptor modeling is treated under the general topic of "Chemical Mass Balance" — Section 3.3. Relaxed assumptions in this case relate primarily to the nature of the errors in the y and A variables. A more serious problem occurs when neither the number of contributing sources, nor their source profiles, can be estimated — i.e., A is unknown. This has led to multivariate manifestations of the linear model, notably multiple linear regression and factor analysis. In this case, y, x, and e become matrices. That is,

$$Y = XA + E \qquad (4)$$

The multivariate *data matrix*, Y, now represents the total observable data; its size is $n \times p$, where n represents the number of samples. (See Note 2 for comments on matrix notation and transposition.) Each matrix element, y_{ti}, represents the concentration for sample-t of species-i. The independent variable matrix, A, has now gone "underground"; it represents nonobservable "latent" variables. This second manifestation of the linear model has some very attractive features. Based on the correlations within the data (Y), one can estimate the unknown number of components; and with certain kinds of constraints or "scientific luck", one can even derive estimates for A. These topics will be covered in Section 3.4, together with selected comments on exploratory data analysis.

When one tries to better approach reality, limitations become apparent even in Equation 4. Unfortunately, in the absence of fairly rigorous constraints, this equation has no unique solution.[52] Further complications for receptor modeling are that distant sources may intrude on what is presumed to be a local multisource system (Long Range Transport), and chemical transfor-

POLLUTANT SOURCE RESOLUTION

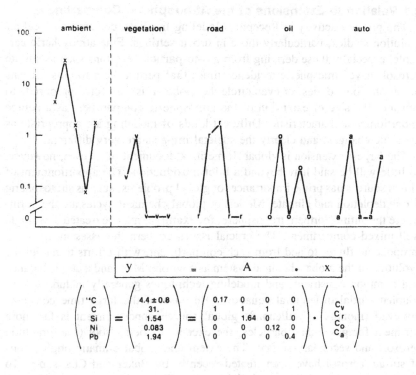

Figure 13. Pollutant source resolution: fine particle aerosol-individual source patterns and the receptor modeling (chemical mass balance) equation. Isotopic and chemical patterns are given for an ambient sample and four sources of carbonaceous particles. The source carbon contributions (C_j) are expressed as percent of total aerosol mass, and the subscripts refer to Vegetation (slash burning), Road dust, Oil (residual and distillate), and Auto exhaust. Units for the ordinate are also percent of total aerosol mass, except for ^{14}C which is expressed as dpm/g-aerosol. Abscissa points for each pattern are ordered as: ^{14}C, C, Si, Ni, and Pb. (Adapted from Currie, L. A. and G. A. Klouda. in *Nuclear and Chemical Dating Techniques Interpreting the Environmental Record.* Washington, D.C.: American Chemical Society, Symp. Ser. 167, 1982. With permission.)

mations and interactions may invalidate the assumption of model linearity. Fixes for these problems gave rise to the term, "Hybrid Receptor Modeling," but the variety of such model extensions makes it more appropriate now to consider Linear (Receptor) Modeling as a special case under the more comprehensive heading of Source Apportionment. Source Apportionment thus includes linear modeling or approximate linear modeling as appropriate, but it may include also "source modeling", back trajectory analysis, chemical transformation modeling, etc. This more general topic clearly applies to environmental systems having various space (extensions) and time constants, the topic of the next subsection.

3.1 Relation to Extensions of the Atmospheric Compartment

The primary activity in Receptor Modeling has been centered on local air pollution studies, particularly those in urban settings. Fine atmospheric aerosols, especially those deriving from gas-to-particle reactions such as sulfate aerosol, have atmospheric residence times that permit them to cross oceans or national boundaries, or even circle the globe. It is therefore appropriate to consider the size or extension of the atmospheric compartment as a source apportionment characteristic. Different kinds of modeling are appropriate as the scale changes, and clearly the societal impacts are very different.

The largest extension is global. It merits a document of its own, however, so little will be said here beyond a brief introduction. The apportionment of global sources has prime importance for global problems, such as stratospheric ozone depletion and climate. Models of global chemical cycles are the norm, where the entire Northern troposphere, for example, may be treated as a single well-mixed compartment. The crucial issues concern the assessment of anthropogenic fluxes to and from each compartment, with efforts to model the evolution of the global chemical system as a whole. Mass and isotopic balance is a common constraint, and modeling techniques generally include sets of coupled partial differential equations. Because of the large time constants involved (months to millennia), global source apportionment is far more pertinent for gases than particles, the exceptions being soot (fine graphitic aerosol) and secondary sulfate. The extent and global climatic implications of sulfate aerosol have been treated recently by Baker and Charlson.[53] To illustrate the general nature of global source apportionment modeling, we give in Figure 14 a "simple" six-compartment model for the carbon dioxide system.[54] The original publication by Bacastow and Keeling should be consulted for detailed information regarding the perturbation equations and transfer coefficients.

The intermediate extension of the atmospheric (now, tropospheric) compartment relates to *Long Range Transport* (LRT), which frequently carries aerosols across national boundaries. Although it has less far reaching consequences than the global category, *regional* source apportionment can have very significant sociopolitical implications. Since time constants are now in the realm of days to weeks, atmospheric particles can make significant contributions. An enormous range of potential sources, combined with nonlinearities and further modeling needs connected with transport and reaction, make LRT somewhat more complicated than the "simple" linear model of local particle pollution. Hence the forementioned "hybrid" modeling. This topic will be examined primarily in Section 3.5.

Local source apportionment is appropriately treated with linear modeling. The numbers of source types are smaller, generally their profiles may be more reliably estimated, and the complications of transport and reaction are significantly reduced. This topic is the substance of the next three subsections.

3.2 Isotopic Tracers

If each source of interest possessed a unique isotopic signature, we would be denied the opportunity to employ large databases and sophisticated multivariate modeling. In exchange, we would find it possible to obtain rather reliable, assumption-free estimates for each of the source contributions. The linear model, Equation 2, then has a diagonal source profile matrix, $\mathbf{A}$, so knowledge of the unique isotopic signature a_{jj} for source-j yields the result x_j by simple division, y_j/a_{jj}. This extremely simple, yet powerful, approach to receptor modeling is possible with a few, naturally occurring isotopes, notably ^{14}C. By selective use of the isotopic tracer *combined with* compositional data, one may (a) improve the overall resolution of sources, and (b) calibrate and/or validate elemental or molecular tracers for given sources. Finally, for a specific local (or even regional) study, it may be possible to purposely *add* a distinctive, non-natural isotopic tracer, to further improve apportionment of sources.

ATMOSPHERIC CARBON DIOXIDE AND RADIOCARBON: II

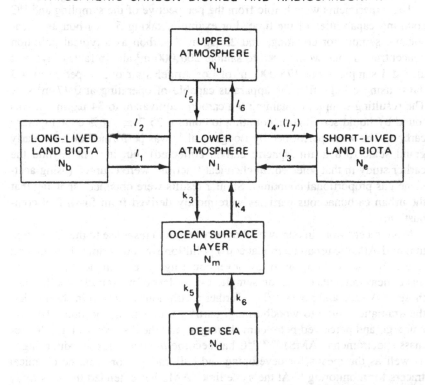

Figure 14. Diagrammatic representation of the six-reservoir model of CO_2 exchange. Arrows represent the direction of fluxes between reservoirs. The N_i denote carbon masses for each reservoir. In the five-reservoir model (not shown), the upper and lower atmosphere are combined to form a single atmospheric reservoir. Transfer coefficients k_j and l_j are defined in Appendix C of Reference 54.

3.2.1 Carbon Isotopes

[14]C has become the most reliable isotope for apportioning carbonaceous particles between fossil and biospheric ("living") sources. Very little atmospheric aerosol exists between the "dead or alive" dichotomy, given the eight millennium mean life of radiocarbon. The earliest measurements of ^{14}C in atmospheric particles actually took place nearly four decades ago, some 8 years after the discovery of radiocarbon dating. Two "heroic" experiments were performed in the late 1950s by two of the pioneers of radiocarbon dating, James Arnold and Hans Suess. It is noteworthy that Arnold's paper of 35 years ago begins with the statement that, "The Los Angeles smog is probably the most publicized of any air pollution problem;" and notes that, "There exists an unequivocal method for distinguishing between carbonaceous material arising from biological sources and that from fossil fuels ..." This historic paper suggests also the possibility of employing [7]Be to estimate the rate of subsidence of the Los Angeles atmosphere, and the application of isotopic tracers to label suspected (polluting) fuel supplies, topics to which we shall return later.

The experiments were heroic from the perspective of the sampling and ^{14}C counting capabilities of the time. For example, taking 5 g carbon as a reasonable sample for counting, and 25 μg.m^{-3} carbon as a typical pollution concentration, one would need to sample 200,000 m^3 air. In fact, Clayton et al.[55] did sample about 175,000 m^3 of Los Angeles air over a period of 4.3 days, using a huge filtering apparatus capable of operating at 0.47 m^3.s^{-1}. The resulting sample, containing 6 g carbon (equivalent to 34 μg.m^{-3}), was found by liquid scintillation counting to contain 25.7 $\pm$ 1.6% contemporary carbon. The work performed by Lodge et al.[56] was performed on a similarly grand scale (3.8 g atmospheric carbon collected), but it went beyond the earlier study in that selected subchemical fractions were counted (using acetylene gas proportional counting). Similar results were obtained, showing that the urban carbonaceous particles were mostly derived from fossil fuel combustion.

More recent work in our own laboratory began in response to the suggestion that mid-Atlantic urban (carbonaceous) pollution was due primarily to natural forest emissions.[57] Improved low-level counting systems made it possible to make measurements on carbon samples two orders of magnitude smaller than those of Arnold and Suess.[58] Two further revolutions followed in short order: the dramatic switch to woodburning, partly as a result of the mid-1970s oil embargo and perceived petroleum shortages; and the discovery of accelerator mass spectrometry (AMS).[59,60] [14]C has become *the* tracer for woodburning,[61] as well as the means for developing and calibrating more routine chemical tracers for monitoring.[62] At the same time, AMS has extended the sensitivity by an additional two to three orders of magnitude, so that we may now assay ^{14}C in individual compounds or classes of compounds in just 10 to 20 μg of atmospheric carbon.[63]

Table 2. Ambient Particle Samples (^{14}C)

Sample	Location	Contemporary C (%) (Median)
(Urban)	Denver	29 ("Brown cloud")
	Salt Lake City	28
	Houston	34
	Los Angeles	38
	Portland	77 (Vegetative burning)
	Elverum	69 (Wood burning)
(Rural)	Desert — Utah	88
	Forest — U.S.	92
	Forest — USSR	80
(Remote)	Point Barrow	27 ("Arctic haze")

A partial summary of ^{14}C atmospheric particle results from our laboratory is given in Table 2.[22] Four classes of environmental issues are represented: "normal" urban pollution, agricultural and/or woodburning pollution, rural concentrations (isotopic and chemical), and remote manifestations. The summary results, given as medians, tell only part of the story — viz., that (a) the carbon in (U.S.) urban particles is typically predominantly fossil in origin (Houston), (b) areas having a significant usage of woodstoves and fireplaces exhibit high contemporary carbon (Elverum, Norway — winter), (c) rural and forested regions have mostly contemporary carbon particles, and (d) special studies of remote areas can show large fossil carbon contributions (Barrow, Alaska). Not shown are the concentrations (μg.m^{-3}) of the carbon or the variations in percent contemporary carbon. In fact, the urban concentrations far exceed the rural and remote concentrations, and f_c (fraction of contemporary carbon) variability is also higher for urban regions, except for special cases of long range transport (Barrow). Figure 15 shows the rather different patterns of variability for the Houston and Shenandoah series. The ability of ^{14}C to reliably apportion fossil and biogenic carbon was extremely important in these two studies: it had been presumed that the carbon in Houston was essentially all fossil, due to refinery operations; the emissions inventory on which dispersion modeling was based showed no component for a strong, varying biogenic source; the ^{14}C "surprise" (of up to 60 + % biogenic carbon) led to a reexamination, and recognition of long range transport of soot from agricultural burning. The Shenandoah ^{14}C surprise was of the opposite sort; accompanying, large concentrations of coal burning sulfate led to the presumption that the C-particles would therefore be primarily fossil.

Figure 16 shows the ^{13}C-^{14}C plane for a number of isotopic standards plus the Barrow sample.[64] Isotopic reference materials are important not only for dating and isotope geology, but also for environmental studies, such as source apportionment. A vital point to be recognized in this latter application is the possibility of "isotopic heterogeneity" — i.e., that the isotopic composition of an environmental reference material may differ in different chemical fractions. This is illustrated by "S5", the urban dust standard reference material (SRM) #1648, where the elemental carbon fraction ("e") differs substantially in both ^{13}C *and* ^{14}C from the average isotopic composition. Interestingly, the isotopic composition of fraction "e" is remarkably similar to that of the Barrow sample that represents long range transport of industrial/urban soot. (The values on the ordinate of Figure 16 are expressed as percent modern carbon, where modern carbon is defined as having a ^{14}C/^{12}C ratio equal to 0.95 times that of standard *S3*. Values on the abscissa are expressed as parts per thousand deviation of the ^{13}C/^{12}C ratio from that of standard *S1*.)

New developments in environmental ^{14}C research take advantage of the multivariate nature of the source apportionment data. In much of the prior work with ^{14}C, for example, it has been employed to apportion the carbon *in* a given material or sample by *direct* measurement. The multivariate tech-

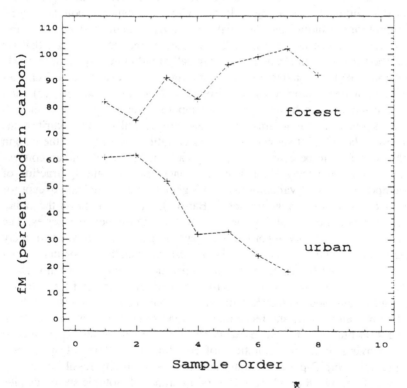

Figure 15. Aerosol radiocarbon: patterns of variability. SURPRISES: Forest area (Shenandoah) — high sulfate concentrations (fossil energy), yet mostly contemporary (biogenic) carbon; Urban area (Houston) — no biogenic carbon in the emissions inventory, yet up to 60% contemporary carbon in the aerosol. The abscissa indicates time-ordered sampling; the sampling campaign covered a period of about 2 weeks for the Shenandoah study, and about 1 week for the Houston study.

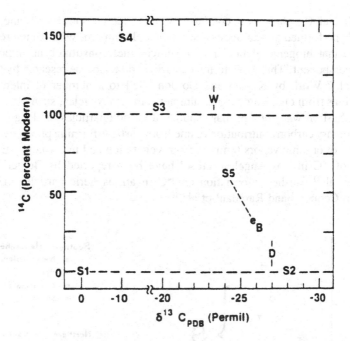

Figure 16. ^{13}C-^{14}C Plane — Standards. The NIST Urban particulate SRM #1648 illustrates isotopic heterogeneity: "S5" (total carbon), "e" (elemental or graphitic) carbon. The proximate sample "B" represents aerosol collected in Barrow, AK ("Arctic Haze"). Other standards shown: for ^{13}C, S1 (PBD), S2 (RM21 graphite); for ^{14}C, S4 (ANU sucrose), S3 (SRM #4990B, oxalic acid dating standard). Other samples: "W" (woodsmoke aerosol), "D" (diesel aerosol). (From Currie, L. A. in *Aerosols*. New York: Elsevier, 1984. With permission.)

niques differ in that isotopic data are used in *parallel* with other kinds of compositional data for calibration (e.g., using multiple linear regression) or for multivariate correlation studies (e.g., using factor analysis). The parallel approach is especially interesting, in that it (a) uses all of the compositional information, and (b) indirectly extends the influence of ^{14}C to the level of nanograms and even individual carbonaceous particles. An introduction to the direct (or serial) vs. indirect (or parallel) application of ^{14}C was presented at the Dubrovnik Radiocarbon Dating Conference.[62]

Measurements of ^{13}C form a very interesting-complement to those of ^{14}C, for they can indicate particles arising from marine sources or tropical vegetation, as opposed to petroleum or temperate zone wood. An illustration, showing the utility of ^{13}C for discriminating distant sources of carbonaceous aerosol in the marine atmosphere, is given in Figure 17.[65] Still more interesting, for environmental source discrimination, are dual isotope measurements. Important investigations of this type have been published by Court,[66] Kaplan,[67] and Tanner[68] and their respective co-authors. Polach, following the initial work with Court,[66] has employed the dual isotope technique for the discrimination of particles affecting the Sydney and Canberra atmospheres from C_3 and C_4 plants, oil combustion, and marine sources. Tanner and Miguel[68] performed an analogous study in Rio de Janiero, with the interesting outcome that biogenic alcohol motor vehicle fuel constituted an important source component. This work followed earlier, dual isotope research by Gaffney et al.[69] Work by Kaplan and Gordon[67] led to a number of interesting conclusions from dual isotopic measurements in Los Angeles, such as: natural gas combustion was *not* a primary source of carbon particles, and significant contemporary carbon contributions came from particle-forming photochemical reactions of organic vapors (emitted from vegetation and in cooking). Earlier studies of ^{14}C in Los Angeles aerosol have been reported by Berger[70] and Currie et al.[71] Further information on ^{14}C in atmospheric particles may be found in Cooper[72] and Ramdahl et al.[61]

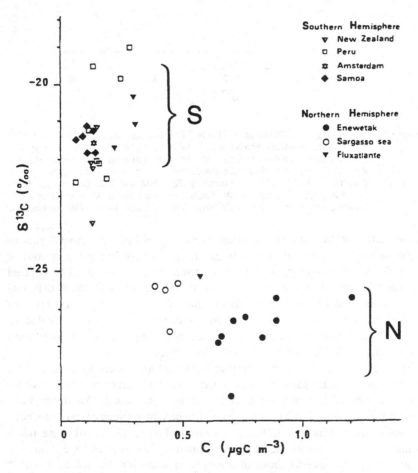

Figure 17. Bulk carbonaceous aerosol samples in the marine atmosphere. Particles originating in the northern (N) and southern (S) hemispheres show distinctive isotopic and concentration patterns. (Adapted from Cachier, H. *Aerosol Sci. Technol.* 10:379–385, 1989. With permission.)

3.2.2 Other Isotopes

Important apportionment information has been derived from isotopes of lead, neodymium, sulfur, oxygen, nitrogen, and beryllium, but we shall limit discussion to the isotopes Pb-206, 207. (Pb-210 also serves atmospheric chemistry well, through its direct link with a natural gaseous (radon) precursor.) Like $^{14}C/^{12}C$, the ratio $^{206}Pb/^{207}Pb$ has been used as a unique discriminator between pairs of atmospheric lead sources. Unlike ^{14}C which is cosmogenic in origin, the lead isotopes are radiogenic, being stable end products of natural uranium and thorium decay chains. The ratio differs in different ore bodies because of geologic age. A fascinating illustration of the application $^{206}Pb/^{207}Pb$ to the apportionment of lead aerosol between the U.S. and Canada has been given by Sturges and Barrie.[73] The aerosol lead sources, attributed to motor vehicles in both nations, showed significant differences in this isotopic signature, examples being: 1.213 ± 0.008 (SD) for the eastern U.S., vs. 1.148 ± 0.007 for Ontario sites. The differences were ascribed to different Pb ore bodies. In an extension of their earlier study, these same authors added a third source to their (pair-wise) scheme, allowing them to apportion lead between Canadian autos and Canadian smelters.[74] This work actually falls into the *regional* source apportionment category, because distances of the order of 1000 km or more were involved; also it could be labeled "hybrid" modeling in that an essential aspect was the construction of directional (wind) sectors from 48 h back trajectories.

A remarkably similar study took place in Turin, for the purpose of identifying aerosol lead sources in the local atmosphere. It differed from the above study, in that the stable isotope tracer was *purposely added* to the motor vehicle fuel (using again a different source for the lead ore), in order to test for consequent changes in aerosol lead and serum lead of inhabitants.[75] The change in the $^{206}Pb/^{207}Pb$ ratio in adult blood, following the tracer injection, showed that ca. $^1/_3$ of the serum lead was of local airborne origin.

3.3 Chemical Mass Balance (CMB)

Chemical mass balance (née "chemical element balance")[51] shares certain characteristics with isotopic tracer methods. It is designed to apportion the aerosol mass (or other property of interest) of each sample quantitatively among the contributing sources. This can be accomplished if there are at least as many variables (p) as sources (q), and each source has a unique chemical *pattern* (profile), as shown for example in Figure 13. The estimated source contributions and their uncertainties then follow (in the simplest case) from Equation 3. Although (isotopic) tracer source apportionment and CMB appear superficially the same (Equation 2), the latter differs in three, very important respects. (1) CMB generally relies on chemical patterns, rather than unique tracers, for source discrimination. (2) Patterns, as well as unique chemical tracers when they exist, usually vary among sources of the same class; therefore they must be measured or calibrated, for a given airshed (and time). Source-specific isotope ratios are generally fixed by Nature, so, once determined, they may be treated as "absolute". (3) Isotopic composition is conservative; chemical composition may not be. If a complete set of nearly constant, conservative chemical tracers can be found and calibrated, then CMB should be as effective as isotopic tracer apportionment. The optimal approach, of course, is to take advantage of the informing power of both. Illustrations of the resulting synergism will be given subsequently.

The breadth of source discrimination with chemical patterns (source profiles) certainly exceeds that of isotopic discrimination, but CMB faces some significant challenges, *all associated with the nature of the patterns* — i.e., the A matrix of Equation 2. These include the following:

- Profiles must be measured; that induces measurement error.
- Nontrivial measurement error becomes systematic error in profile databases.
- Profile variability within and between sources of the same type is not always negligible, nor is inter-variable covariance. This may be very difficult to (a) estimate, and (b) take into account. This distribution function for such variation (if it be random) is another matter.
- Profiles are seldom orthogonal (except for unique tracers); their *real* differences depend further on the forementioned measurement errors and profile variabilities. Lack of orthogonality leads to variance inflation; the information matrix (A'A) in Equation 3 becomes ill-conditioned, making source resolution increasingly difficult.
- The number of profiles (number of sources) to include in matrix A is commonly unknown. Careful scrutiny of the airshed — e.g., with emissions inventories — helps, as do the powerful "source-counting" multivariate techniques. A popular method of stepwise inclusion of sources, so long as they yield positive x estimates, should be approached with caution; its results are not robust to order of the trials, and it depends greatly on the skill (and luck) of the data analyst.
- Profiles measured at the source will differ from those at the receptor, except for strictly conservative tracers. Hybrid (physicochemical-statistical) modeling and multivariate techniques may be successful in bringing this problem under control.

Further discussion of some of these issues will be given in later sections, but for more complete development, including appropriate mathematics and references, the reader should refer to Hopke's book,[26] and selected reviews of CMB fundamentals, such as Stevens and Pace.[76] The reader should be aware also of databases for CMB, compilations of source profiles, as well as PC-versions of CMB code. Further information can be found, for example, in Hopke,[26] Gordon,[1] and Watson et al.[77]

A method devised to treat one of the problems cited above, random error in the source profiles, is generally represented in such code. That is "effective variance least squares".[78] This technique treats two complications ignored in the simplest formulation, Equation 2. First, the variance of the y_i is seldom constant; second, the measured a_{ij} have error. The more complete expression is therefore

$$y_j = \Sigma(a + e_a)_{ij} \cdot x_j + (e_y)_i \qquad (5)$$

The second error term is simply treated, by using weighted least squares. The errors in the profiles **A**, however, turn this into a nonlinear problem requiring an iterative solution. The maximum likelihood formulation for this problem may be found, for example, in Watson[78] and Hopke.[26] An approximate solution that has been widely used is

$$\mathbf{x} = (\mathbf{A'WA})^{-1} \cdot (\mathbf{A'Wy}) \qquad (6a)$$

where **W** is a diagonal matrix of weights having inverse variances as elements

$$1/w_{ii} = V(y_i) + \Sigma V(a_{ij}) \cdot x_j^2 \qquad (6b)$$

the summation extending over all q components. Iterative coupling of these two equations occurs because the weights in the latter involve the solution of the former, which uses the weights. (Equation 6b will be recognized as a simple propagation of error expression.) An interesting footnote on the solution of Equation 6 is that it was published by Parr and Lucas in 1964,[79] and independently derived by Currie in 1966[80] for a nuclear spectrum deconvolution program, "TVAR" (total variance least squares).

Despite the foregoing complications with CMB, which have been presented to stimulate careful thinking and research, the technique has been very successfully applied. An example of a success story is found in the work of Cooper and co-workers.[81,82] The Portland (Oregon) Aerosol Characterization Study was the first large urban CMB study to be designed from the outset to meet all of the essential data requirements of this means of source apportionment. The source profile matrix **A** was constructed from measurements of 27 chemical species in fine and coarse particle fractions for 37 source categories within the Portland air quality airshed, representing 95% of the emissions inventory. Some 1300 ambient samples, for the same size fractions, were collected over a 1-year period, and characterized for the same 27 species, representing about 90% of the ambient aerosol mass. Finally, CMB calculations were applied intensively to data from a subset of 32 days, selected by a stratification of meteorological regimes to assure a representative sample of the annual meteorological state. Apportionment results of the study are given in Figure 18, showing numerous quantified, specific source categories together with several nonspecific categories including anions and carbon. One interesting observation that springs from the two charts is that, in contrast to combustion and secondary aerosol, soil and road dust particles are largely contained in the coarse, noninhalable fraction.

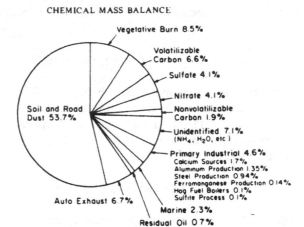

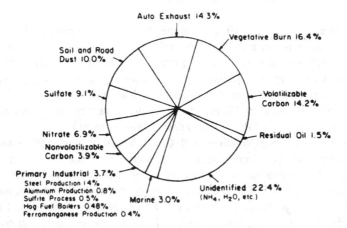

Figure 18. Source apportionment of the total suspended particulate matter (top) and fine (<2.5 μm) particulate matter (bottom) of the downtown Portland, OR, aerosol as given by Core et al.[82]

The Portland study was sufficiently carefully planned and executed that it had a very important, secondary benefit: it permitted the correction and calibration of the air quality (dispersion) model employed for making air quality management decisions.[82] Intercomparison of CMB results with those of the advection-diffusion predictions for similar sources, such as auto exhaust, residual oil, and road dust, showed some important discrepancies. This led to scrutiny of several of the assumptions involved in the predictive modeling. Erroneous assumptions were uncovered, and the final outcome was a significant increase in the quality of the predictions and hence, the quality of the decision-making.

3.3.1 Particle Class Balance

This method is a version of CMB suited to microanalytical (individual particle) data. It was developed by D. Johnson and co-workers to accommodate, in the linear model framework, the possible emission of a wide variety of particles from individual sources. Replacing the chemical element variables of bulk CMB, are particle class variables that form the profiles for individual particle mass balance. In special circumstances individual particles may serve as unique source tracers, such as cenospheres from coal combustion, or unique minerals that might be associated with just a single source type in a given airshed. In such favorable cases, statistical or mechanistic modeling is either unnecessary or enormously simplified; one needs only count the respective tracer particles to quantify their impact. (Estimation of the total source impact, of course, would also require knowledge of the relative emission rate of the tracer particles.)

One of the first applications of the Particle Class Balance Method was the apportionment of the ambient aerosol in Syracuse, NY. In this work, Johnson and McIntyre[83] used scanning electron microscopy with automated image analysis and X-ray energy spectroscopy to chemically and physically characterize more than 10,000 source and ambient particles. Particle class "fingerprints" are first obtained by sampling sources like those in the airshed under study. In the case of Syracuse, the authors included 12 source types, and 17 particle classes to apportion 23 ambient samples. Among the source categories represented were wood, oil, and coal fly ash, auto emissions, and steel mill emissions; among the particle classes were metal rich, Si only, wood ash, auto, feldspar like, and Al rich. Source and ambient particle class populations could then be used to derive both weight percent and number percent particle source apportionments. The technique offers greater resolution in comparison to bulk particle source apportionment, because the particle class fingerprints are generally much less collinear than the elemental fingerprints of chemical mass balance, partly because of such added microscale information as particle morphology and size. Another unique advantage of individual particle analysis is the opportunity to apply multivariate techniques to *single* ambient samples.[84] In contrast, conventional (bulk) analytical techniques require many samples to obtain the "multivariate advantage," which includes direct estimation of the number of active sources and their chemical fingerprints.

3.4 Multivariate Techniques

Multivariate approaches to the *linear* model begin with one simple, yet fundamental, extension of CMB. That is, one provides data in the form of a matrix rather than a vector. The input data matrix (DM), Y in Eq. (4), consists of column variables and row samples. A single row of Y is thus equivalent to the input vector y for CMB (See Equations 2 through 4.) The added sample dimension makes it possible to examine the variance in source contributions, and explore multivariate relationships among variables and among samples.

Using potent factor analytic tools one can estimate the number of sources *directly from the structure of the DM*, and in fortunate cases estimate source profiles and source impacts as well. We begin this section, however, with data vector(s) drawn from **Y**, for calibration/validation purposes, using the technique of "tracer multiple linear regression" MLR(T).

3.4.1 Tracer Multiple Linear Regression

The basic concept behind MLR(T) is that a dependent, response variable (**y**) of interest — such as fine particle mass, visibility, extractable organic matter — can be regressed against a *complete set* of unique tracer variables for the sources contributing to **y**. This can be viewed as the multivariable equivalent of the univariate "calibration curve" which is a staple of every laboratory. Algebraically, **y** and **a**, can be looked on as column vectors from the corresponding matrices in Equation 4. That is because, in contrast to CMB, the data vector **y** relates to a single variable observed in n samples, rather than (CMB case) p variables observed in a single sample. Consider, for example, the apportionment of extractable organic matter (EOM) between two sources having known, unique tracers. The general, and specific equations become

$$y_t = \Sigma y_{tj} = \Sigma x_{tj} a_j \qquad (7a)$$

where the sum is taken over all q contributing sources; x_{tj} represents the concentration of the unique tracer for source-j in sample-t; a_j is the regression coefficient (calibration constant) for source-j; and y_{tj} is what we're really interested in, the concentration of y in sample-t due to source-j. The simple least squares solution to Eq. (7a), in matrix form, is

$$a = (X'X)^{-1} \cdot X'y \qquad (7b)$$

which is the exact analog of Equation 3, except that the independent variables are given by the tracer matrix **X** rather than the source profile matrix **A**. The treatment of errors and weights follows just as in CMB.

Often, as with the familiar univariate calibration curve, an intercept y_o is included in the model to cover any background level of the property, or an average level from nonrepresented sources. A word of caution is necessary, however, regarding intercept models: if the intercept is used to compensate for nonrepresented sources, it can bias the regression coefficients a_j unless the contributions from those sources are random with respect to the modeled sources. The best use for an intercept is to check that it is negligible (statistically insignificant), implying that the model is complete (all sources represented). A second word of caution relates to design. Though one cannot control source mixes in a field study, major efforts should be made to collect samples that suitably span the range of likely occurrences. Otherwise disastrous interpolation or extrapolation may be invited, and model nonlinearities may go unnoticed — just as in the case of inadequately planned laboratory calibration exercises.

Returning now to the explicit, 2-source tracer regression study, Equation 7a takes the form

$$y_t = EOM_t = EOM_o + EOM_{t1} + EOM_{t2} \qquad (7c)$$

$$EOM_{t1} = K_t \cdot a_1, \qquad EOM_{t2} = Pb_t \cdot a_2 \qquad (7d)$$

These regression equations are taken from the investigation of Stevens et al.,[28] where EOM and mutagenicity were apportioned between motor vehicle and woodburning aerosol in Albuquerque, NM and Raleigh, NC. The studies were designed so that no other sources were likely, and potassium and lead were known to be unique tracers (in the study areas) for these two sources. EOM_o is thus the (presumed negligible) intercept for extractable organic matter, and the last two terms in Equation 7c represent the contributions of each of the sources for sample-t. (EOM_t, of course, is the observed total.) Equation 7d shows the relation of each of these terms to the regression coefficients and tracers for the respective sources — source-1 being woodburning, and source-2 being motor vehicles. The result of MLR(T) for the 44 Albuquerque samples, in units of $\mu g.m^{-3}$, was

$$EOM_t = (1.0 \pm 0.9) + (204 \pm 8) K_t + (10.6 \pm 3) Pb_t$$

Happily, the intercept differed insignificantly from zero.

Similar regressions were performed by Stevens et al. for observed mutagenicity, and the ratios of the two sets of regression coefficients led to the basic information sought: potency (revertants/μg-EOM) of woodburning aerosol as compared to that from motor vehicles. The results showed that the potency of motor vehicle aerosol was approximately three times that of woodburning aerosol. There are two additional, extremely important facets to MLR(T) studies of this sort: tracer selection, and tracer calibration/validation.

3.4.1.1 Selection and Correction of Source Tracers.
The ideal tracer is one that is unique to the source of interest, for the airshed in question. If one is interested in source-emitted substances other than the tracer per se, then it is also necessary to know the relative emission rates. In rare instances, in the case of absolute source tracers, this is known in advance. An example is the use of ^{14}C as a tracer for source carbon, where the absolute ratio, $^{14}C/^{12}C$, is fixed by the age of the source carbon. More commonly, tracers may be unique, but in need of calibration, either by direct measurement of source emissions, or by MLR(T). Successful application of the latter technique, however, requires a *complete set* of tracers for each of the contributing sources. That is the case for the preceding example, where the substance being apportioned

is EOM, and its emission comes from woodburning and motor vehicles. Our complete set of tracers for the two sources is: K for woodburning, and Pb for motor vehicles. Although an intercept was included in the MLR model, it must not be used to estimate EOM from other sources. Any covariation of nonrepresented sources will distort the meaning of the estimates for both the intercept and that of the source with which it is correlated. (Perfectly correlated, nonrepresented sources will automatically be included in the correlated source estimate. That makes possible the use of Pb to trace motor vehicles using *both* leaded and nonleaded fuel.) Other nonabsolute tracers that have proved useful for apportionment of urban and industrial sources include: Mn, Ti, Sc, Fe for soil; Pb for smelters; V, Ni for fuel oil fly ash; and F for aluminum plants.[85,86] Selected organic species also show some promise for tracing specific combustion sources. For an extensive review of organic tracers, especially polycyclic aromatic hydrocarbons, see the review by Daisey et al.[40]

Tracers may be apparently unique, and calibrated, but still lead to erroneous results. "Apparently" is the keyword, in that unsuspected or unaccounted for, secondary sources of the same tracer may be present. The problem of background and/or soil aerosol has been dealt with through element-ratio correction procedures, to yield "excess" tracer that then acts as a unique source indicator. Such correction techniques have been utilized effectively by Rahn[6] for the apportionment of regional aerosol, and by Lewis and Einfeld[87] for the apportionment of urban carbonaceous aerosol. The latter authors derived "excess potassium" tracer for woodsmoke, by using mineral-calibrated K/Fe ratios to subtract the soil-K component from the urban aerosol. Corrections of this sort are not foolproof, however. If long range transport brings tracer-laden aerosol from still another source, the uniqueness assumption is invalidated.[88] Another source of bias is found when the tracer is not conservative — i.e., when it does not maintain a fixed ratio to the substance being apportioned (EOM, mutagenicity, fine particle mass) during transit from source to receptor. This kind of problem was noted by Thrane[86] in connection with aerosol fluoride; it arises also whenever there is differential settling or chemical reaction. Modeling and correction for first order reactions has been successful in connection with certain organic compounds; this will be discussed under the heading of "Hybrid Techniques" in Section 3.5.

3.4.1.2 Validation. A classic approach to method validation is the application of (presumably) independent techniques. Comparative receptor modeling is a popular manifestation of this principle. It is exemplified by the application of a mass balance model together with a multivariate model (bulk or microanalytical) to the same airshed, or through the comparison of the results of source (dispersion) modeling to those of receptor modeling. Illustrations abound. See for example the book by Hopke,[26] or the forementioned articles by Thrane[86] and Pratsinis.[85] Agreement does not *prove* validity, however, and in the case of disagreement, it may not be clear which result, if either, to believe.

The alternate route to validation, and to the development of new tracer techniques, is by comparison to an absolute method or one of accepted validity. This method was employed by Stevens et al.[28] to validate the (excess) potassium, lead MLR(T) method for apportioning EOM and mutagenicity. [14]C was measured in parallel to the tracer elements and it was thus possible to compare the regression slopes with the direct, ^{14}C-derived values for wood carbon and fossil carbon. Excellent agreement was found between the MLR estimate for woodsmoke carbon, based on the potassium — lead tracers, and the direct, absolute estimate based on [14]C. The results are exemplified by the wood carbon regression coefficients ($\mu g - EOM/\mu g - K'$) quoted by Stevens et al.[28] — 204 ± 8 (Albuquerque), and 112 ± 4 (Raleigh) — as compared to ^{14}C results. The latter, given in Figure 19, yield consistent coefficients (206 ± 11 and 110 ± 8, respectively), and they also support the validity of the assumed linear relation.[30,89] An important corollary to this study is the obvious, airshed dependence of the regression coefficients; the absolute concentration of potassium in Albuquerque woodsmoke is approximately twice that of Raleigh woodsmoke. A second illustration of ^{14}C calibration/validation is given in Figure 20, for the case of organic tracers. Here, potential organic tracers are explored, by examining the correlation of selected polycyclic

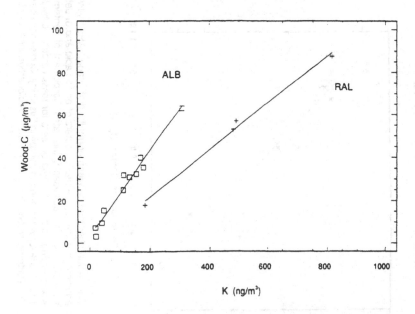

Figure 19. Linear regression, aerosol carbon from wood vs. mineral-corrected potassium; Albuquerque (□) and Raleigh (+). Intercepts are insignificant, but slopes differ by a factor of two. (From Currie, L. A. in *Four Decades of Radiocarbon Research: An Interdisciplinary Perspective. Radiocarbon* and Springer-Verlag, in Press, 1991. With permission.)

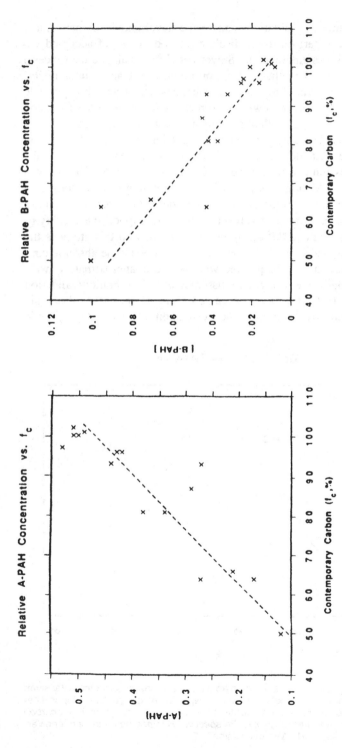

Figure 20. Correlation observed between normalized abietane PAHs (A-PAH) and normalized ^{14}C (f_c) (Top). Here, "A-PAH" includes retene and its principal precursors, the methyl esters of abietic acid and of dehydroabietic acid, normalized to the sum of all identified PAHs (ΣPAH) in each sample. The abscissa f_c indicates the percent of contemporary carbon in the ΣPAH fraction. f_c is proportional to the percent of woodburning carbon; it would be zero for fossil carbon, 100% for "living" carbon (in isotopic equilibrium with atmospheric CO_2), and ca 100—110% for wood burning carbon in this study. Correlation observed between normalized benzo(ghi)perylene ("B–PAH") and normalized ^{14}C (Bottom). (From Currie, L. A. et al. *Radiocarbon* 31:448–463, 1989. With permission.)

aromatic hydrocarbon ratios to the fraction of woodsmoke aerosol carbon and motor vehicle aerosol carbon determined by the absolute isotopic tracer.[62]

3.4.2 Cluster Analysis

The examination of multivariate atmospheric aerosol data for clusters, or groups of compositionally similar particles, can serve as an important initial step for exploratory data analysis. Partition of a large data set into more or less homogeneous subsets provides important information on the number of subsets or classes, and the nature of the subsets. Under favorable circumstances, the characteristics of individual classes can serve to identify a given source, or they may constitute the source class "fingerprints" needed for the Particle Class Balance technique of source apportionment (Section 3.3.1). Subdivision of heterogeneous data into homogeneous subsets is also important for quantitative multivariate methods involving linear models. Cluster analysis also is a very useful tool for the detection of outliers (singleton clusters) or "special" samples.

No universal technique exists for the definition of clusters. The basic steps in the process include: data transformation, computation of inter-sample (or inter-particle) "distances" based on a prescribed definition, assignment to individual clusters according to the algorithm selected, and cluster display. Perhaps the most common transformation applied to compositional data is *standardization*, where each variable is centered (mean subtracted) and "scaled" by dividing by the estimated standard deviation *s*. In some cases this may be preceded by a log transform. Among the distance measures, the most commonly used is the *Euclidean Distance* which is simply the root mean square distance in the multivariate space. Clustering algorithms are *hierarchical* or *nonhierarchical*. Hierarchical algorithms are divisive if one starts with a single cluster (the entire data set) and sequentially subdivides it into increasing numbers of clusters until all samples (or particles) are singletons. *Agglomerative* clustering is just the reverse. Nonhierarchical, or *partitioning* methods assign samples "optimally" into a prescribed number of groups. A number of rules exist for assigning samples to clusters, and deciding on the "correct" number of clusters. Exercise of judgment, and data exploration using alternative rules, are recommended. Hierarchical cluster display is generally rendered with dendrograms or "trees"; nonhierarchical clusters and, to a limited extent, hierarchical clusters may be displayed by two- or three-dimensional plots, examples of which will be given shortly.

Details concerning the several steps involved in cluster analysis are beyond the scope of this chapter; comprehensive treatments are found in Massart and Kaufman[90] and Hartigan.[91] For an introductory text, which emphasizes robust techniques, see Kaufman and Rousseeuw.[92] In view of the several clustering algorithms and distance measures, and, considering the fact that cluster analysis is primarily an exploratory technqiue, these last authors recommend the application of several techniques in the initial phases of a study.

Table 3. Multivariate Particulate Data: Elverum, Norway (Units: f_w, Percent Woodcarbon; C, $\mu g.m^{-3}$; All Others, $ng.m^{-3}$)

Sample	f_w	C	K	Pb	Fe	SO$_4$	Mn	V
6	63	15	121	263	20	730	4	1
7	78	50	424	607	33	1540	6	1
8	64	30	328	497	62	1620	10	2
9	64	9	197	213	27	2710	4.5	8
10	32	5.1	218	54	33	2800	6	6
11	62	6.9	140	160	72	2320	6.5	14
12	54	12	179	262	57	910	4.5	2
13	65	26	246	547	179	1300	8	2
14	83	22	231	325	56	1490	2.5	2
15	67	25	245	419	110	1410	4	2
16	26	11	339	94	150	5340	15.5	17
17	61	10	99	70	59	310	2	1

To illustrate the display and tentative interpretation of clustering results, we refer to an investigation of urban air pollution in Elverum, Norway.[88] Data for a series of air particulate samples, collected during the winter of 1982, are given in Table 3. The purpose of the study was to assess the impact of woodburning on the wintertime atmosphere in the town of Elverum, which was experiencing significant levels of carbonaceous pollution. Direct assessment, using ^{14}C, showed the average woodburning contribution to be about 65% for the period under study. Further source apportionment may be possible, however, using the multivariate data in the table. To explore this possibility, the following steps were taken: (1) data were normalized (to carbon) to create intensive variables, thus removing extensive effects related to atmospheric dilution volume or collective source strength; (2) hierarchical clustering was performed, following standardization, to reveal possible subgroups and singletons; (3) principal component analysis (PCA) was performed sequentially on the entire dataset and on the major subgroup, indicated by cluster analysis; (4) the final stage was to use the results of cluster and PCA data exploration to guide quantitative source apportionment. Although PCA is a topic of the next subsection of this report, it is introduced here because of its importance as a cluster display technique. In this regard, the principal components may be viewed as new axes in the multidimensional variable space that give the "best" view of the data — i.e., they project the maximum information (variance) in a small number of dimensions. This is accomplished by a mathematical process known as *singular value decomposition* (also known as eigenvector analysis) which creates the new coordinate axes from optimal linear combinations of real variable axes. The mathematics of the process is beyond the scope of this chapter; a very readable exposition may be found, however, in Massart et al.[93] The essential point is that the best two- or three-dimensional graphical display of clusters of data points and of variables is given by using the first two or three principal components as axes.

The two display techniques for the Elverum aerosol data are shown in Figures 21 (dendrogram) and 22 (PCA plot). Figure 21 displays the sequential linkage of samples based on the Euclidian distances in multivariable space. At level-6 of clustering, we find one large cluster plus 5 singletons. We give this the label "urban cluster," in part because all of the samples contained were collected within the town. Both background site samples — numbers 16 and 17 — lie outside of this cluster, as do three other samples having rather distinct compositions. The complementary, graphical display technique (Figure 22), cannot readily indicate linkage details but it has the great advantage of showing individual sample (and cluster) configurations with respect to the chemical variables. The grouping of points in the (encircled) "urban cluster" is now apparent. Also this cluster is seen to lie closest to the projection of the f_w (woodburning) variable with an internal dispersion lying between this variable and *Pb*, the motor vehicle tracer. Other samples are well removed, especially numbers 10 and 16 (at the upwind background site). The plot shows that these two samples are linked to relatively low amounts of wood carbon and relatively high amounts of the other species, excluding *Pb*. The complementarity of the two modes of display is also highlighted by samples 10 and 16: their apparent closeness in the plot (Figure 22) is belied by the dendrogram, which shows that they are rather weakly linked (large dissimilarity). Though cluster analysis, per se, operates on the full dimensional space, low dimensional projections can falsely create an illusion of proximity. Qualitative (exploratory) results of this graphical introduction to cluster analysis, as ap-

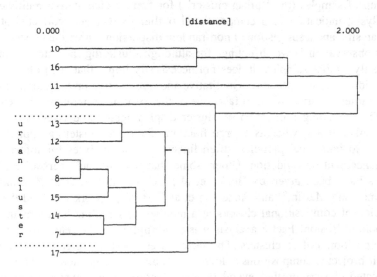

ELVERUM

Average Linkage Dendrogram (7 dimensions)

Figure 21. Dendrogram showing clustering of aerosol samples collected in Elverum, Norway.

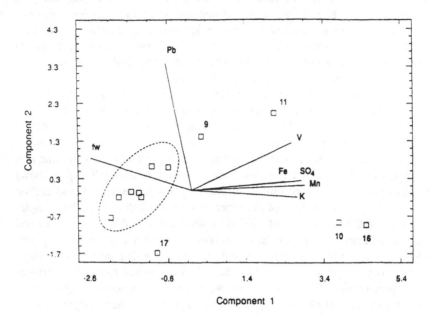

Figure 22. Principal component projection of 12 samples and 7 variables for Elverum, Norway. The dashed ellipse encloses samples comprising the "urban cluster".

plied to atmospheric particles, are: delineation of a relatively homogeneous group of samples (the "urban cluster") for further quantitative multivariate analysis; indication of a compositional outlier (#10), also collected at the urban site; and a suggestion of non-random dispersion within the cluster. This last observation is worth noting, for although clustering indicates compositionally similar samples, it does not necessarily imply that such clusters are devoid of internal structure. Qualitative and quantitative investigation of possible internal structure is, in fact, one of the exciting subsequent challenges.

The foregoing illustration of cluster display methods derived from bulk aerosol analysis, whereas a large fraction of current clustering applications relate to individual particles, often linked to bulk analyses for information enhancement or validation. Other, somewhat diverse, bulk aerosol applications have been given by Hopke et al.,[94] Gaarenstroom et al.,[95] Sanchez-Gomez and Martin,[96] and Anderson et al.[97] Hopke's work involved the formation of compositional clusters at a number of sampling locations in metropolitan Boston. Factor analysis was then applied to the centroids (average compositions) of 15 clusters. The factors were next interpreted according to their projected compositions ("loadings"), and the projections of each of the selected clusters on the several factors ("scores") was examined to assess source impacts. Major sources (e.g., incinerator, sea salt aerosol) were seen to be reasonably associated with various clusters, based on the geographic locations of the corresponding sampling sites. Gaarenstroom also sampled at different sites ("urban", "rural") in metropolitan Tucson, measuring 20 elements plus ammonium, nitrate, and sulfate ions. In this work, the variables, rather than the samples, were clustered. Inferences derived from this alternative clustering mode were, for example, that *Pb* clustered with nitrate at the urban site — due to coemission from motor vehicles, whereas it clustered with nonferrous metals (*Cd, Cu, Zn*) at the rural site — due to association with emissions of copper smelters found in the region. Gomez and Martin employed an interesting combination of tables of clusters and auxiliary variables to infer pollution sources in Valladolid, Spain. Measurements of nine metallic elements plus sulfate, in bulk aerosol samples collected over a 13-month period, were classified by a nonhierarchical method. The centroid compositions of six significant clusters were then compared with the tabulated external variables (primarily seasonal and meteorological) for the same six clusters. The comparison led to inferences concerning the origins of each of the clusters. Cluster-2, for example, was attributed to residual fuel burning, because it showed the greatest seasonal variation, and it was characterized high concentrations of sulfate, zinc, and cadmium. Anderson's work marks the transition to individual particle clustering, in that it is devoted to the classification of submicrometer particles, measured by X-ray spectrometry and secondary emission microscopy. Automated methods permit the chemical and morphological (size) characterization of hundreds to thousands of individual particles. Partition clustering was employed to generate particle classes, in order to construct data matrices representing particle class (population) variations with time, location and wind direction. The intent of the work was to combine microparticle clustering with factor analysis, to provide information in support of bulk particle analysis. Supporting bulk particle source apportionment with microparticle characterization is one of the most powerful means of achieving reliable conclusions. The combination can be especially powerful when the microanalysis indicates unique particles, characteristic of individual sources; bulk apportionment by CMB or factor analysis may then proceed with *reliable* knowledge of source components for inclusion in the quantitative data analysis model.[98,99]

Only a few additional comments on individual particle clustering will be given here, in light of the companion review by Xhoffer et al.[50] of individual particle characterization in support of source apportionment. One recent illustration of the technique, based on samples collected during the Amazon Boundary Layer Experiment, has been given by Artaxo et al.[100] The study is of particular interest because of the potential impact of emissions of vast tropical regions on the global biospheric aerosol budget. Through the use of electron probe X-ray microanalysis, Artaxo was able to obtain morphological and compositional data on approximately 500 individual particles on a series of coarse and fine particle filters. By subjecting the data to hierarchical cluster analysis, six to ten different particle groups could be discerned. Some groups were identified with specific sources, such as a group containing Si, Al, Ti, and Fe that was labeled soil dust aerosol. In a number of cases particles could be specifically identified as pollen grains, fungi, plant fragments, and algae. Among the important conclusions of the study was the fact that soil dust represented only 11% of the coarse mode aerosol, and 8% of the fine mode aerosol, whereas major fractions were inferred to arise from biogenic sources. Computer controlled scanning electron microscopy has been used effectively by Kim and co-workers[101,102] to derive chemical and morphological data for large numbers of particles, and subsequently to form source and ambient sample particle classes. Initially, clustering was used for classification, followed by principal component class modeling. More recently, Kim and Hopke[102] devised an expert system for classification, which appears to generate "homogeneous classes" — defined by unique, bit-coded combinations of variables — in a more reliable and efficient manner. This system of classification was applied in a Particle Class Balance (PCB) of ambient aerosol in El Paso, TX. The classification system, when applied to particles from 29 source samples, led to 283 homogeneous classes. The classes were based on a set of 25 variables, including 19 X-ray intensities and 6 morphological parameters. Source class profiles and ambient sample class patterns were constructed by analyzing about 700 particles per sample. Once the class profiles were constructed, PCB computations proceeded in the same manner (multiple linear regression) as CMB. (See Section 3.3.1.) Two significant differences from CMB noted by Kim and Hopke were that (1) source profile matrix collinearity problems were greatly reduced, due to the large number of variables (classes); and (2) unweighted least squares was more appropriate than the "effective variance" technique used with CMB, due to the nature of the mass classification uncertainties. Despite the great promise of the technique, some difficulties were observed, presumably due to aliasing by unidentified sources.

A clustering technique that, in a certain sense, bridges the gap between conventional classification and factor analytic apportionment is known as "fuzzy clustering" (FCV). In contrast to clusters that might be considered as indicative of specific sources, FCV recognizes the fact that observed sample characteristics may reflect a mixture of sources, and therefore be intermediate between end-member clusters. FCV takes this into account by characterizing each sample vector by its partial "membership" in each of a defined set of end member clusters. An example of FCV is the apportionment of polycyclic aromatic hydrocarbons (PAH) originating from an aluminum plant in Sunndalsøra, Norway.[86] Because of the absence of direct emission data for PAH, the program was used in its unsupervised pattern recognition mode, which meant that clusters had to be identified on the basis of external data. For the plant in question, external data from the unique tracer fluoride, together with wind direction, were used for cluster characterization. Measurements on 11 PAH variables for 37 samples over three seasons showed three levels of source impacts, depending on wind direction. In the most extreme case, where the wind direction was from the aluminum plant more than half the time, the plant contribution to collected PAH was 96%.

3.4.3 Principal Component and Factor Analysis

Principal Component Analysis (PCA) was introduced above as an efficient means to display multivariate data clusters in two- or three-dimensional projections. Its success for this application is based on the fact that linear combinations (rotations) of the original chemical variables can be constructed to project most of the information (variance) in a few dimensions. The optimal rotations, constructed by eigenvector analysis, are such that the maximum variance is projected on the first eigenvector (principal component); the direction of the second principal component, taken orthogonal to the first, corresponds to the maximum remaining variance, and so on. An equivalent way of viewing the process is to observe that the first eigenvector (or principal component) is the line of best fit to the multidimensional data array, where "best" is generally defined in the least squares sense with residuals being taken orthogonal to the fitted line. The second eigenvector is orthogonal to the first, and together, they define the plane of best fit to the data. The process continues, for example, with the third eigenvector, orthogonal to the first two, defining the volume (three dimensional "hyperplane") of best fit. Algebraically, the process results in a decomposition of the original data matrix **Y** into the product of a "score matrix" **U** and a "loading matrix" **V**.

$$\mathbf{Y} = \mathbf{UV} + \mathbf{E} \qquad (8)$$

The rows of **V** represent the rotation vectors (loadings) which convert the original chemical variable axes to the new, orthogonal coordinate system. The rows of **U** contain the projections (scores) of each sample on the new coordinates. The dimensionality of the matrix product **UV** (number of significant eigenvectors) ideally extends over that region of multivariable space corresponding to the number of sources q; **E**, then, represents the noise. The decomposition, or expansion of the data, in Equation 8 is fundamental both to principal component analysis and (abstract) factor analysis, the only difference being one of normalization: for factor analysis the rows of **V** are increased by the square root of the corresponding eigenvalues, and the columns of **U** are correspondingly decreased. A visual representation of PCA is given in Figure 23, where inherently two-dimensional data from an interlaboratory comparison is shown first in the original three-laboratory coordinate system, and then in the two-dimensional eigenvector projection which captures all of the nonrandom structure of the data in the plane of the first two principal components.

Besides its direct role in exploratory data analysis and multivariate data display, PCA is the first step in a collection of qualitative and quantitative data analytical techniques collectively labeled *Factor Analysis* (FA). As with cluster analysis, there is a large literature on the topic. An excellent intro-

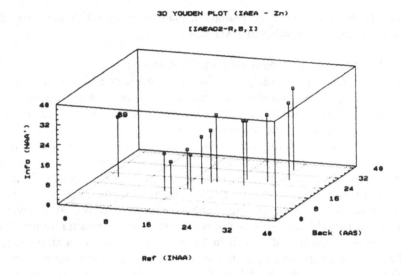

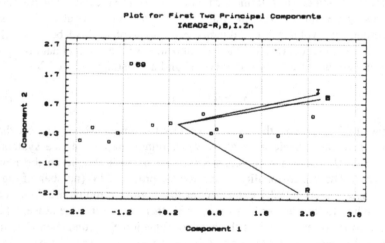

Figure 23. Visual representation of the relation between "real" variable projection and principal component projection of chemical data. Real (chemical) variable three dimensional plot of interlaboratory measurements of zinc in daily diet samples [Top]. Concentrations are given as mg/kg for each of the three laboratories. Principal component projection of three-dimensional interlaboratory zinc data (standardized) [Bottom]. I, B, and R are the three laboratory codes; the result for sample — #69 from laboratory — R is outlying. (From Currie, L. A. in *Biological Trace Element Research Multidisciplinary Perspectives.* Washington, D.C.: American Chemical Society, Symp. Ser. 445, 1991, [Ref. 125]. With permission.)

duction is given in Massart et al.;[93] the primary text for chemical applications is that by Malinowski;[103] and an introduction to aerosol apportionment applications is found in Hopke.[26]

Essentially three types of information are sought in the application of factor analysis to quantitative apportionment problems: the number of particle sources impacting the sampling site; the source identities, as reflected in the source profile matrix; and the source contributions to each sample. These are the three essential problems also for Chemical Mass Balance techniques, but there is one critical difference: sources and source profiles must be known in advance for the application of CMB, but FA makes it possible to estimate the quantities more or less directly from the data matrix. Like the assignment of particles to clusters, however, there are many potential pitfalls in deriving these estimates, and there is no universal agreement as to the best rules to employ. The number of significant factors r, for example, is popularly defined as that number for which the eigenvalues exceed unity. Although this is a good general guide, it tends to underestimate the number of factors (sources). Perhaps the best approach is that given by Malinowski,[104] where an F-test is used, subject to the constraint of constant error variance. An added problem occurs when model nonlinearities exist; then r may exceed the number of sources q, for r is an estimate of the number of *linearly independent* factors.

Deriving source profiles and source contributions (the $\mathbf{A}$ and $\mathbf{X}$ matrices of Equation 4) is another matter. The problem is lack of model uniqueness.[52] The point is that an infinite number of transformations (rotations) of Equation 8 are possible:

$$\mathbf{Y} = \mathbf{UV} = (\mathbf{UT})(\mathbf{T}^{-1}\mathbf{V}) = \mathbf{XA} \tag{9}$$

In Equation 9, $\mathbf{T}$ is a transformation matrix which produces rotated matrices $\mathbf{\check{X}}$ and $\mathbf{\check{A}}$ without altering the value of the product matrix. There is only one set of rotations that is correct: that for which $\mathbf{\check{X}} = \mathbf{X}$, and $\mathbf{\check{A}} = \mathbf{A}$; finding that set is the key. Space permits no more than a brief allusion to rotation techniques. The most common, a remnant from social science applications of factor analysis, is the VARIMAX procedure that seeks to create orthogonal factors with loadings that are primarily 1's or 0's ("simple structure"). The resulting factors are then more or less subjectively given names, which immediately creates the illusion of real source factors (profiles). Unless most of the sources have unique tracers, VARIMAX will lead to erroneous conclusions; for one thing real source profiles are rarely orthogonal. The Target Factor Analysis (TFA) technique of Malinowski[103] is an excellent method for testing possible sources, whose profiles are largely known, to see whether their profiles ("test vectors") lie within the factor space. Hopke's extension of TFA to "target transformation factor analysis" (TTFA) is based on an iterative fitting of initially unique vectors to the factor space, in order to develop source profile estimates.[26] Though TTFA is apparently successful, it is important to note Malinowski's caution that test vectors should be based on chemical insight and contain an initial number of input values at least equal to the rank of the UV matrix. TTFA yields more realistic solutions than orthogonal rotation techniques such as VARIMAX, however, for it does not impose orthogonality on the derived source profiles.

Another technique of profile or spectrum identification, based on variable clustering following PCA, has been developed by Windig and co-workers.[105] It has been successfully applied to the quantitative apportionment of multi-component chemical mixtures, especially in connection with pyrolysis mass spectrometry. The most satisfying approach for the estimation of unknown source profiles is that of Lawton and Sylvestre.[106] The method, based on an eigenvector directed search for unique variables in the data, has been successfully applied to a number of chemical problems,[107] and its potential for atmospheric studies has been outlined by Henry.[108] The basis for the technique is illustrated in Figure 24.[93] In this work, the score plot of eleven optical spectra of a two-component mixture coeluting from a chromatographic column is used, together with non-negativity constraints, to regenerate pure component spectra. Once pure component spectra — or equivalently atmospheric particulate source profiles — have been generated, multiple regression may be employed for source quantification. Regrettably, there have been few applications of this very important approach in the area of particulate source apportionment. Further review of factor analytic techniques and the question of assumption validity may be found in Currie.[109]

Among the most important results of factor analysis, as applied to source apportionment, is the ability to discover unsuspected sources and to refine presumed source profiles. These strengths are evident especially when different techniques are applied to the same data set. A classic illustration of an unsuspected component, revealed by factor analysis, is presented in the review by Gordon.[1] Referring to an early study by Hopke et al.[94] using neutron activation analysis, Gordon noted that factor analysis led to the identification of the following particulate sources in metropolitan Boston: crustal and soil dust, marine aerosol, oil combustion, motor vehicle exhaust, refuse incineration, and a mystery factor containing Mn and Se. The mystery was resolved when Thurston and Spengler[110] performed X-ray fluorescence analysis on aerosol samples collected in the region, and discovered a factor heavily loaded with sulfur — undetectable by neutron activation analysis. The unsuspected source, containing Mn, Se, and S, was then ascribed to long-range transport from coal-fired power plants in the Midwest.

3.5 Hybrid Techniques

The preceding methods of source apportionment were based essentially on statistical modeling of multivariate receptor data. Clearly, one should expect improved qualitative and quantitative understanding of the situation if additional knowledge is utilized. Such knowledge may include models and data for transport, sedimentation, and chemical reaction. The formal linkage of such mechanistic modeling with statistical receptor modeling constitutes one

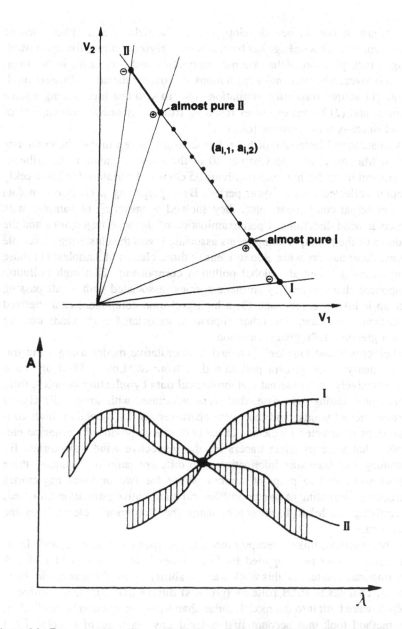

Figure 24. Representation of normalized spectra (same area) of two component mixtures (Top). (μ_{i1}, μ_{i2}) are the scores of spectrum (i) in the space defined by the two first eigenvectors of the variance-covariance matrix of the spectra. The solid lines represent the confidence regions of the two spectra. The almost pure spectra I and II, and the estimated spectra I and II (Bottom). The shaded area represents the confidence region for the true pure spectra. (From Massart, D. L. et al. *Chemometrics: A Textbook*. Amsterdam: Elsevier, 1988. With permission.)

of the most important new developments in the field of atmospheric source apportionment. This linkage has been labeled "Hybrid Source-Receptor Modeling". Incorporation of the external information and constraints in this manner may overcome two major limitations of strictly statistical (linear) modeling: (1) source resolution limitations due to collinearities among source profiles, and (2) biased estimates resulting from an erroneous assumption of model linearity (conservative tracers).

A qualitative illustration of hybrid modeling is given in the chemometrics text of Massart et al.[93] In Chapter 20 of their text the authors describe an experiment in the Netherlands involving 35 chemical variables for 150 weekly samples collected over a 3-year period. By superposing wind direction data on a principal components plot, they showed a clustering of samples with particular wind directions. Upon examination of the wind trajectories and the loadings of the principal components associated with the clustering of sample scores, these authors were able to identify three classes of samples: (1) those representing a "normal" global pollution component; (2) a high pollution component characterized by aromatic species, associated with winds passing over an industrial sector; and (3) a high pollution component characterized by decane, undecane, and other aliphatics, associated with winds coming from a greenhouse/highway direction.

Rheingrover and Gordon[111] created a quantitative model using wind trajectory analysis for aerosol pollution data from St. Louis. Their approach very effectively used external meteorological data by selecting samples, from a large pre-existing data base, that were associated with strong, directional (narrow sector) winds, and that were characterized by especially high concentrations of selected elements. These two criteria produced "criterion elements" that were in effect tracers for the respective wind trajectories. By combining such trajectory information from different sampling locations, these authors were able to pinpoint sources where the two or more trajectories intersected. Unfolding of source profiles and quantitative estimation followed, by applying the MLR(T) approach, using the "criterion" elements as the source tracers.

Hybrid methods linking receptor modeling, dispersion modeling, and chemical reaction have been applied by Friedlander[112] to emissions of PAH. A very important feature of this work was the ability to avoid linear model bias. That is, changes in PAH patterns (profiles) during transport from source to receptor were built into the model, rather than being assumed to be negligible. The method took into account first order decay reactions of selected PAH molecules, using rate constant and residence time data. An analogous study was reported by Lewis and Stevens,[3] this time incorporating in the CMB expression a transformation factor for the oxidation of sulfur dioxide during transport. An important series of investigations by Winchester and co-workers have utilized the hybrid approach to gain increased understanding of aerosol neutralization during transport.[9] Through a combination of absolute principal component analysis, time series analysis of resultant absolute factor scores, and chemical stoichiometry, these workers succeeded in evaluating the "atmospheric titration of particle alkalinity by SO_2" during long range transport. Multivariate data sets, developed from particles sampled at one site in Japan and two sites each in China and Hawaii, helped establish a model of alkaline particle "aging" by rapid and slow neutralization of carbonate and clay components, respectively. Neutralization-dissolution of alkaline aerosol components by atmospheric SO_2 and H_2SO_4, may be an important mechanism for the deposition of soluble mineral elements such as aluminum or iron.

The source-receptor hybrid concept continues to be refined and made more powerful, as detailed air mass trajectories are linked to multivariate, multi-receptor time series records plus chemical knowledge. A summary of recent developments has been given by Gordon.[1]

4. VALIDATION OF SOURCE APPORTIONMENT METHODOLOGY

Validation exercises fall into two main categories: those that test the accuracy, or at least intercomparability, of the measurements and conclusions in the actual atmospheric environment; and those designed to test modeling accuracy. Relevant to the former category are measurements of reference materials, and interlaboratory and intermethod measurements of common field samples. Blind and open quality control measurements fall into this class, but they should constitute an inherent part of *any* measurement program, so they will not be reviewed here. It should be mentioned, however, that the results of multivariate techniques are *very* sensitive to data quality, especially in view of the complicated natural environment. Partly because of this sensitivity, conclusion accuracy is strongly influenced by field study design. These "experimental" aspects of validation are briefly discussed in Section 4.1.

Erroneous models applied to excellent data yield only one kind of result! Therefore, again because of the great complexity of the natural system, model validation is equally important. Major efforts that have taken place nationally (in the U.S.) for validation of long range transport modeling, and for mass balance and multivariate receptor modeling, are summarized in Section 4.2.

4.1 Experimental Validation

Standard (or Certified) Reference Materials (CRMs) constitute the basic means for assuring measurement accuracy. A partial list of such materials, suitable for atmospheric particle measurement validation, is given in Table 4.1[13] Additional CRMs exist or are under discussion for isotopic, carbonaceous, and microanalytical measurements.[114] For example, a second urban particle standard, SRM 1649, was collected in Washington, D.C., and certified for a number of organic compounds, notably polycyclic aromatic hydrocarbons. This supplements the certification of SRM 1648, which was collected in St. Louis, for inorganic species. An additional material that is suitable for atmospheric particle quality control is SRM 1650, diesel soot. For measurement of ^{14}C, the basic standard is SRM 4990C; a certificate for this material is reproduced in Figure 25. Another material, RM21 graphite, is appropriate for air particulate stable isotope (^{13}C) quality control. Table 5[22] includes some (informational) results obtained for carbon isotope measurements on different chemical fractions of two of the above SRMs.

Designed field studies have been important in assuring the quality of a number of investigations of urban air particulate pollution. An excellent illustration is afforded by the Integrated Air Cancer Project of the U.S. Environmental Protection Agency.[115] A major objective of the study was to apportion sources of atmospheric particulate mutagens, especially in connection with wintertime urban activities. In a series of field experiments in Raleigh, NC, Albuquerque, NM, Boise, ID, and Roanoke, VA, a superb field design was constructed that balanced carbonaceous source mixes among motor vehicles, woodburning, and oil burning. The balance was achieved by the selection of urban region (airshed), sampling time (day, night) and sampling site (highway, residential); and the validity of conclusions was enhanced by selective use of both natural and artificial isotopic tracers. A recent review of study results may be found in Stevens et al.[28]

4.2 Mathematical (Model) Validation
4.2.1 Long-Range Transport

Dispersion and long-range transport modeling has long been one of the most difficult tasks faced by atmospheric scientists. Transport of air masses tends to adopt a chaotic element once several hundred kilometers have passed. The accuracy of numerical trajectories also has a very strong dependence on the type and density of meteorological data. Some very ambitious and very

Table 4. List of Currently Available Certified Reference Materials

Code	Producer	Matrix	Suitable for
SRM 1648	NIST	Urban particulate matter	INAA
SRM 1632b	NIST	Coal	INAA
SRM 1633a	NIST	Coal fly ash	INAA
CRM 038	BCR	Fly ash from pulverized coal	INAA
CRM 176	BCR	City waste incineration ash	INAA
EOP	IRANT	Coal fly ash	INAA
ECH	IRANT	Coal fly ash	INAA
ENO	IRANT	Coal fly ash	INAA
CTA-FFA-1	Poland	Fine fly ash	INAA
CRM 128	BCR	Fly ash on artificial filter	PIXE, XRF
SRM 1832	NIST	Film	PIXE, XRF
SRM 1833	NIST	Film	PIXE, XRF
SRM 1643b	NIST	Water	PIXE, XRF, INAA
SLRS	NRCC	River water	PIXE, XRF, INAA

Source: Parr[113]

NIST   National Institute of Standards and Technology (USA), formerly NBS.
BCR   Community Bureau of Reference (CEC).
IRANT Institute of Radioecology and Applications of Nuclear Techniques (Czechoslovakia).
NRCC   National Research Council of Canada.

National Bureau of Standards

Certificate

INTERNATIONAL STANDARD REFERENCE MATERIAL FOR CONTEMPORARY CARBON-14

OXALIC ACID SRM 4990C

This reference material consists of a one-half-pound portion of a 1000-lb lot of oxalic acid prepared by fermentation of French beet molasses, from the 1977 spring, summer, and autumn harvests, using <u>Aspergillus niger var</u>.

The mass spectrometric ratio of carbon-13 to carbon-12 in this material, and the corresponding ratio for the old contemporary carbon-14 standard SRM-4990, were measured by thirteen international carbon-dating laboratories. Measurements by twelve laboratories show that the ratio of carbon-13 to carbon-12 is slightly greater in the new standard. The <u>difference</u> in $\delta^{13}C$ is 1.49 ± 0.05 per mil, where the uncertainty is the estimated standard deviation.

The ratio of the radioactivity concentration of the new material (SRM 4990-C) to that of the old material (SRM 4990) was also measured by nine of the thirteen laboratories. The unweighted average value of the individual weighted-average results for this ratio, normalized to $\delta^{13}C$ values of -19.3 per mil for the old standard and -17.8 per mil for the new, is 1.2893 ± 0.0004. The weighted average of the individual weighted results, similarly normalized, is 1.2933 ± 0.0004. The uncertainties of each of these results is the estimated standard deviation of the average value. Using the method of Paule and Mandel [1] the weighted average of the weighted results is 1.2931 with one estimated standard deviation of the mean equal to 0.0005, and the estimated standard deviation between laboratories of 0.0008.

Systematic uncertainties can arise in the determination of the plateau threshold of a proportional counter, and in the measurements of pressure and temperature in the gas-handling systems. An estimate of the sum of these uncertainties in any measurement is of the order of 0.15 percent. Other uncertainties arise in liquid-scintillation counting systems. In this group of activity-concentration measurements from nine participating laboratories, these systematic uncertainties, from laboratory to laboratory, have a reasonable probability of being randomly distributed, some observations of, for example, temperature being high and some low. No account has therefore been taken of uncertainty other than that arising from counting statistics, in weighting the results submitted by the participating laboratories. The contribution of systematic uncertainty may be of the order of ± 0.05 percent.

This reference material was calibrated in an international comparison organized by L.M. Cavallo and W.B. Mann in the Center for Radiation Research, Nuclear Radiation Division, Radioactivity Group. A detailed report of this comparison has been published in RADIOCARBON. [2]

Washington, D.C. 20234
July, 1983

Stanley D. Rasberry, Chief
Office of Standard Reference Materials

*References on back

Figure 25. Certificate for SRM 4990C, C-14 Reference Material. [References on back of certificate: (1) Paule, R. C. and Mandel, J. "Consensus Values and Weighting Factors", *NBS J. Research*, 87, 377 (1982); and (2) Mann, W. B., "An International Reference Material for Radiocarbon Dating", *Radiocarbon*, 25, 519 (1983).]

Table 5. Urban Particle Standards

Reference Material	Percent Carbon	$\delta^{13}C_{PDB}$ (%)	^{14}C (% Modern)
SRM 1648 (St. Louis)	13.1 ± 0.3	−24.1	60 ± 3
PAH (BaP ~ 3.1 µg/g)	80 ± 2		22 ± 4
SRM 1649 (Washington, D.C.)	17.5 ± 0.4	−25.3	61 ± 4
PAH (BaP ~ 2.6 µg/g)	87 ± 2		17 ± 4

informative model validation studies have taken place, where an array of models were evaluated by comparison with the observed dispersion of conservative tracers. The most recent such study (ANATEX) involved transport to the U.S. East and Gulf Coasts of perfluorocarbon tracers released in Glasgow, MT and St. Cloud, MN.[116] A slightly smaller scale study (CAPTEX) was carried out in the Northeastern U.S. and Southeastern Canada in the early 1980s, with tracer releases from Dayton, OH and Sudbury, Ontario.[117] The outcomes of these experiments have significant implications for source apportionment on the regional scale — i.e., those approaches depending on accurate air mass trajectories.

The sampling network for the most recent, and most extensive, transport model validation exercise (ANATEX) is shown in Figure 26.[118] The experiment involved the release of three perflurocarbon tracers from Glasgow, MT (GGW) and St. Cloud, MN (STC) at 2.5 or 5.0 day intervals over a period of 3 months in 1987. Twenty-four hour sampling was then employed for

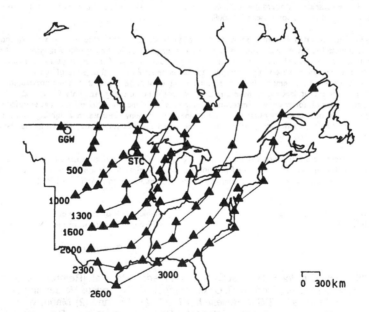

Figure 26. The 77-site ANATEX surface network for 24-h-average tracer concentration measurements and the location of the tracer release sites at Glasgow, MT (GGW) and St. Cloud, MN (STC). (From Clark, T. L. and R. D. Cohn. U.S. EPA Rep. 600/3–90/051, 1990.

Table 6. The 11 Long-Range Transport and Diffusion Models Evaluated in the Anatex Model Evaluation Study (AMES)

Model Acronym	Model Name	Model Developer
Single-layer Lagrangian (SLL) models		
SRL	Adjusted Geostrophic Model	Savannah River National Laboratory
TCAL	Trajectory Calculation Model	ENSCO, Inc.
VCAL[a]	Variable Layer Trajectory Calculation Model	ENSCO, Inc.
Multiple-layer Lagrangian (MLL) models		
ARL	Air Resources Laboratory Model	National Oceanic and Atmospheric Administration
BAT	Branching Atmospheric Trajectory Model	National Oceanic and Atmospheric Administration
GAMUT	Global Atmospheric Multi-layer Transport Model	ENSCO, Inc.
HY-SPLIT[b]	Hybrid Single-Particle Lagrangian Integrated Trajectories Model	National Oceanic and Atmospheric Administration
MLAM-FINE and MLAM-COARSE	Multilayer Air Mass Model	Pacific Northwest Laboratory
Multiple-layer Eulerian (MLE) models		
ADOM	Acid Deposition and Oxidant Model	Environment Canada-Atmospheric Environment Service
ADPIC	Atmospheric Diffusion Particle-In-Cell Model	Lawrence Livermore National Laboratory

[a] A model very similar to the TCAL model
[b] A model very similar to the ARL model

measurement of each of the tracers in the ambient (transported) air at the network's 77 surface sites, to allow a rigorous evaluation of long-range transport models and diffusion simulation of acid deposition models. Table 6 contains a brief description of the several models employed in the study. An illustration of some of the results, comparing measured with modeled concentrations of the transported tracers, is given in Figure 27. A complete analysis of the performance of the models in predicting plume widths, individual trajectories and ensemble averages is given in the Clark and Cohn report.[118] These authors found, for example, that the multiple layer Lagrangian models performed best in simulating the transport of the tracers, while the Eulerian models were best for simulating their ensemble concentration frequency distributions. At distances of 1000 to 2300 km downwind, ensemble mean concentrations were overpredicted, commonly by a factor of three, by the single-layer Lagrangian models; they were underpredicted by about 40% by the two Eulerian models. The study has been most important for quantifying model performance in a real, long-range transport setting, and in relating performance to model assumptions, meteorological scenarios, and complexity of air flow.

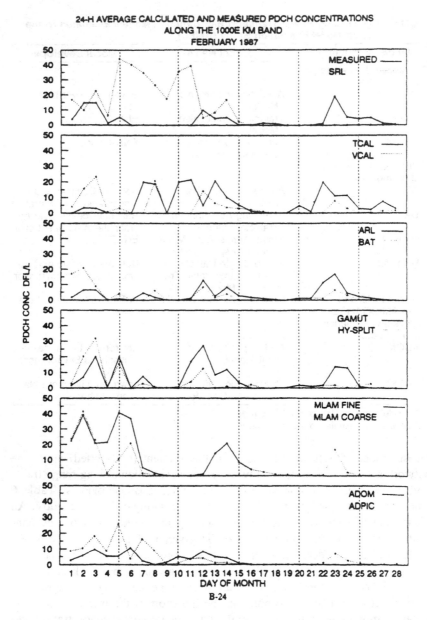

Figure 27. ANATEX Results: Measured vs. modeled perfluorocarbon concentrations along the 1000 km band (Feb. 1987). Definitions of model acronyms are given in Table 6. (From Clark, T. L. and R. D. Cohn. U.S. EPA Rep. 600/3–90/051, 1990).

4.2.2 Linear (Receptor) Modeling

Intercomparison of mass balance and multivariate receptor modeling on common field samples is an excellent means to check intercomparability, for the particular environment selected. Evaluation of model accuracy, on the other hand, requires knowledge of the "truth." One solution is to invent the "truth" in a realistic way. An exercise based on this premise took place in 1982, at the "Quail Roost II" Receptor Modeling Workshop. Simulated atmospheric aerosol sources, cities, and measurements (including errors) were created at the National Bureau of Standards, using actual meteorological data and the best available information on atmospheric dispersion and source compositions. The full results of the study, involving a number of U.S. experts, each armed with his favorite chemical mass balance or multivariate model, were published in Currie et al.[119] The following brief review of the exercise and its outcome is drawn primarily from Gerlach et al.[120]

Creation of the multivariate data set began with geographic placement of "sources" in one of two model "cities." Source/city geography for the first such placement is shown in Figure 28, where the locations of 9 different point and area sources are indicated by their respective symbols. Realistic emissions patterns and actual meteorological data were coupled with an appropriate dispersion model to transport emitted "particles" to the receptor site, denoted "R." The linear model of Equation 4 was then applied in the forward direction to construct a data matrix Y having 40 rows (diurnal sampling periods) and 20 columns (chemical element variables). Figure 28 contains an example of one of the source profiles a_{2i} and the corresponding source strength time series x_{t2} for the incinerator, source-2. The construction process included the addition of realistic measurement errors both to the source profile matrix A and to the computed data matrix Y. Chemical Mass Balance (CMB) and Factor Analysis (FA) methods were applied to three such simulated sets of data: set I having an unknown number of sources (nine); set II and set III having 13 known sources each. (Set III included the added complication of lognormal source profile variations.).

Results for the first data set are given in Table 7. The five sets of results were derived from: two slightly different CMB/FA combinations, involving different weighting schemes [CMB(EV)/FA and CMB(W)/FA]; tracer multiple linear regression combined with factor analysis [MLR(T)/FA]; and two applications of target transformation factor analysis [TTFA]. For the most part, the results lie within a factor of two or three of the correct values, though a major fraction show significant discrepancies, based on the reported standard errors. Results for the "sandblast" source were impressive, considering the fact that this source was not included in the source profile list provided to the participants. As expected, weak sources were missed by most of the participants. Comparisons of estimated (stated) standard errors with actual deviations from the "truth" were revealing. Figure 29 shows such a comparison for set II data analyzed by three presumably similar methods, all based on Chemical Mass Balance. It is interesting to note that only one of the three

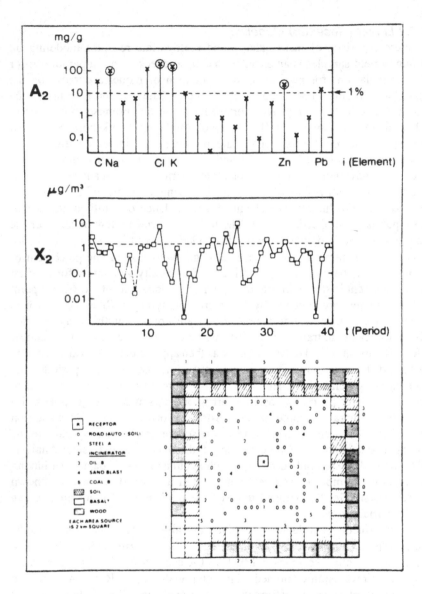

Figure 28. Source apportionment simulation (Data Set I). Upper portion shows one row of the source signature matrix **A**, and one column of the source intensity matrix **X** — both for source-2, INCINERATOR. A_2 has a discrete pattern (individual chemical elements), the most discriminating elements of which are marked by circles; the dashed line indicates which elements exceed 1% of the (Incinerator) particle mass. X_2 has a continuous underlying structure (time series) which is sampled at 40 equidistant points; the dashed line indicates samples for which the Incinerator source contributes more than 5% of the average aerosol mass. The lower portion of the figure displays the aerosol source emission map, with the source-type codes at the left. (From Currie, L. A. *NBS J. Res.* 90:409, 1986. With permission.)

Table 7. Estimated Source Impacts (µg·m⁻³) Compared With True Values for Set I[a]

Source	CMB(EV)/FA[b]	CMB(W)/FA[c]	MLR(T)/FA[d]	TTFA	TTFA[e]	Truth
Steel-A	1.0 ± 0.3	—	—	—	—	0.05
Steel-B						—
Oil-A	6.1 ± 1.0	8.7 ± 0.9	7.9 ± 0.8	2.3 ± 0.4	2.1 ± 1.0	—
Oil-B						2.0
Incinerator	1.5 ± 0.2	0.7 ± 0.3	2.4 ± 0.4	1.8 ± 0.1	1.9 ± 0.14	1.3
Glass	—	—	—	—	—	—
Coal-B	—	—	—	—	2.2 ± 0.73	—
Coal-A						—
Aggregate[g]	15.9 ± 1.1	15.9 ± 0.2	10.6 ± 0.9	12.7 ± 0.5	12.5 ± 0.75	—
Basalt						4.7
Soil[f]						8.0
Road[f]	5.3 ± 0.4	8.3 ± 0.2	5.0 ± 0.2	3.0 ± 0.1	4.0 ± 0.09	7.1
Wood	—	—	—	7.4 ± 0.7	4.3 ± 0.52	3.3
Sandblast[h]	—	1.9 ± 1.4	4.0 ± 0.3	5.2 ± 0.4	4.1 ± 0.20	4.2
Total	29.8	35.5	29.9	32.4	31.1	33.05

[a] Uncertainties represent one standard deviation.
[b] CMB result for the average of the observation periods.
[c] Average of CMB results for each observation period.
[d] Fitted intercept = 3.3 ± 1.5 µg·m⁻³.
[e] Results submitted after completion of Workshop.
[f] Data were generated using road as a source component, where road = 0.25 soil + 0.75 auto.
[g] A (composite) "crustal" component.
[h] Sandblast was an extraneous source, not included among the 13 profiles furnished to the participants.

had deviations that were consistent with the estimated standard errors. One must conclude that either standard errors or source strength estimates, or both, are subject to subtle, operator-dependent aspects of the computation. Perhaps the most likely such aspect is the somewhat subjective decision as to which source(s) to include in the regression equation.

The Quail Roost II data and the study live on. Since the models employed did not quite find the "truth", there has been continued interest on the part of receptor modelers to build better models and to evaluate the sensitivity of extant models to possible assumption weaknesses. A list of some of the new approaches that were developed and applied to the Quail Roost II data through 1985 is given in Table 8.[121] The interaction of these data with the search for better receptor models continues; the most recent requests for the database by receptor model developers came in September 1990.

5. OUTLOOK AND CONCLUSION

Accurate identification and apportionment of sources of particles in so complex a system as the urban, regional, or global atmosphere is as difficult as it is important. The impacts of aerosols on health, visibility, and climate

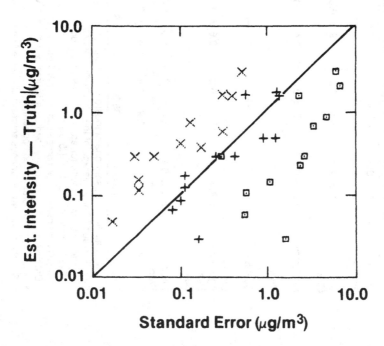

Figure 29. Absolute errors for each of the 13 source estimates for Data Set II, by three different laboratories applying Chemical Mass Balance, plotted as a function of the reported standard errors. (On the average about $^2/_3$ of the points for a correct method should fall below the diagonal.) (From Gerlach, R. W. et al. in *Proc. APCA Specialty Conf. Receptor Models Applied to Contemporary Air Pollution* SP-48:96, 1983. With permission.)

Table 8. Additional Studies of the Quail Roost II Data

Investigator	Topic
M-D. Cheng, P. Hopke[a]	Linear programming
L. Currie[a]	Detection, design, model error
I. Frank, B. Kowalski	Partial least squares
G. Gordon[a]	Student instruction, QA
R. Henry	Composite components (SVD)
P. Lioy	Student research
D. Lowenthal et al.	Special error propagation (covariance)
T. Pace	Sensitivity analysis

[a] Participants of the original intercomparison, performing advanced studies using the database. Others listed requested the data specifically for basic research on numerical facets of source apportionment.

Adapted from Currie, L. A. *NBS J. Res.* 90:409 (1986) (Ref. 121).

are so substantial that it is vital that we continue progress in apportionment modeling, measurement and model validation, and most of all, in fundamental *scientific understanding* of the basic processes leading to the formation, transport, and reaction of aerosols and their critical chemical components. Advances in measurement, ranging from morphological and chemical characterization of individual particles, to measurement of isotopic (carbon-14) and chemical (perfluorocarbon) concentrations at levels of 10^{-12} to 10^{-15} provide special opportunities to enhance our knowledge of source and transport processes, and at the same time refine dispersion and receptor models. Advances in computer simulation, database generation, and multivariate computations for qualitative (exploratory, graphical) and quantitative source apportionment have been critical to the present measure of success. Perhaps even more important for future progress, however, are the "failures" discussed in this chapter, indicative of inadequate accuracy of measurement or imperfect models or assumptions. More attention must be given to mechanistic, as opposed to purely statistical/empirical modeling; also, assumptions that treat natural processes as linear (conservative) or orthogonal deserve special attention. Research directed at such issues will surely improve our future understanding of the atmospheric system, and help resolve the effects on that system induced by human activities from those due to natural processes.

ACKNOWLEDGMENT

Material in Figures 1–7, 9–12, 14, 17, 18, and 24 has been reprinted with permission. Special credit goes to JanaRae Hintze for major assistance with a literature search and preparation of References and Figures. Grateful acknowledgment is due also to Donna Barraclough Klinedinst for assistance with the Figures.

NOTES

Note 1. This statement should be tempered slightly in two respects, because the tracer sampled *at the receptor site* gives a measure of the impact of the *specific source type* at that site. The two qualifications are (a) the tracer must be "conservative" (no losses or transformations) if it is also to give *a direct measure* of the emissions *at the source;* (b) multiple impacts from *different sources of the same source type* (along a trajectory, or from different directions) cannot be distinguished simply by measuring the accumulation of the tracer. Means exist for treating these two matters, however, through modeling, source labeling (added, source-specific tracer), and various time series approaches.

Note 2. Unfortunately, there exists neither consistent notation nor nomenclature for the linear model. One difficulty is that source apportionment activities span many fields of endeavor, each one of which has its "favorite" nomenclature. For the purposes of this document on source apportionment, we shall employ the notations of Equation 2 for chemical mass balance. A small problem arises with multivariate convention, in that the observation matrix (Y) uses rows for samples and columns for variables. To be totally consistent with that notation, all of the quantities in Equations 2 and 3 should be indicated as transposes (y, x, and e as row vectors).

Notation for indices is not affected by the transpose problem, though consistency across disciplines is lacking. We shall use: for samples, $t = 1 - n$; for variables, $i = 1 - p$; and for sources (components), $j = 1 - q$. "Unknowns" are x, for chemical mass balance; a, for tracer multiple linear regression; and X and A, for factor analysis. Scalars are italicized, vectors are boldface, and matrices are boldface and capitalized.

REFERENCES

1. Gordon, G. E., "Receptor Models," *Environ. Sci. Technol.* 22:1132–1142 (1988).
2. Dzubay, T. G., R. K. Stevens, G. E. Gordon, I. Olmez, A. E. Sheffield, and W. J. Courtney. "A Composite Receptor Method Applied to Philadelphia Aerosol," *Environ. Sci. Technol.* 22(1):46 (1988).
3. Lewis, C. W. and R. K. Stevens. "Hybrid Receptor Model for Secondary Sulfate From an SO_2 Point Source," *Atmos. Environ.* 19(6):917–924 (1985).
4. Tanner, R. and Harrison. "Acid-Base Equilibria of Aerosols and Gases in the Atmosphere," in *Characterization of Environmental Particles,* J. Buffle and H. P. van Leeuwen, Eds., IUPAC Environmental Analytical Chemistry Series, Vol. 1. (Chelsea, MI: Lewis Publishers, 1992), chap. 2.
5. Jaenicke, R. "Natural Aerosols," in *Aerosols: Anthropogenic and Natural, Sources and Transport,* Vol. 338, T. J. Kneip and P. J. Lioy, Ed., (New York: New York Academy of Science, 1980), pp. 317–329.
6. Rahn, K. A. and R. J. McCaffrey. "On the Origin and Transport of the Winter Arctic Aerosol," in *Aerosols — Anthropogenic and Natural, Sources and Transport,* Vol. 338, T. J. Kneip and P. J. Lioy, Eds. (New York: New York Academy of Science, 1980), pp. 486–503.
7. National Research Council, *Global Tropospheric Chemistry.* (Washington, D.C.: National Academy of Sciences Press, 1984).
8. Rowland, F. S. and I. S. A. Isaksen. *The Changing Atmosphere.* (Chichester: Dahlem Workshop, Wiley-Interscience, 1988).

9. Winchester, J. W. and M. Wang. "Acid-Base Balance in Aerosol Components of the Asia-Pacific Region," *Tellus* 41B:323–337 (1989).

10. Chameides, W. L., R. W. Lindsay, J. Richardson, and C. S. Kiang. "The Role of Biogenic Hydrocarbons in Urban Photochemical Smog: Atlanta as a Case Study," *Science* 241:1473–1475 (1988).

11. Tuominen, J., S. Salomaa, H. Pyysalo, E. Skyttä, L. Tikkanen, T. Nurmela, M. Sorsa, V. Pohjola, M. Saurl, and K. Himberg. "Polynuclear Aromatic Compounds and Genotoxicity in Particulate and Vapor Phases of Ambient Air: Effect of Traffic, Season, and Meteorological Conditions," *Environ. Sci. Technol.* 22:1228–1234 (1988).

12. White, W. H. and E. S. Macias. "Carbonaceous Particles and Regional Haze in the Western United States," *Aerosol Sci. Technol.* 10:111–117 (1989).

13. Shaw, R. W. "Air Pollution by Particles," *Sci. Am.* 257:96–103 (1987).

14. Hansen, J. E., A. A. Lacis, P. Lee, and W.-C. Wang. "Climatic Effects of Atmospheric Aerosols," in *Aerosols — Anthropogenic and Natural, Sources and Transport,* Vol. 338, T. J. Kneip and P. J. Lioy, Eds. (New York: New York Academy of Sciences, 1980), pp. 575–587.

15. Shaw, G. E. "Perturbation to the Atmospheric Radiation Field from Carbonaceous Aerosols," in *Particulate Carbon, Atmospheric Life Cycle,* G. T. Wolff and R. L. Klimisch, Eds., (New York: Plenum Press, 1982), pp. 53–73.

16. Fields, D. E., L. L. Cole, S. Summers, M. G. Yalcintas, and G. L. Vaughn. "Generation of Aerosols by an Urban Fire Storm," *Aerosol Sci. Technol.* 10:28–36 (1989).

17. Charlson, R. J., J. Langner, and H. Rodhe. "Sulphate Aerosol and Climate," *Nature.* 348:22 (1990).

18. Galbally, I. E., Ed. *The International Global Atmospheric Chemistry (IGAC) Programme,* Commission on Atmospheric Chemistry and Global Pollution of the International Association of Meteorology and Atmospheric Physics (1989).

19. Prospero, J. M. "Aerosol Particles," in *Global Tropospheric Chemistry, A Plan for Action.* (Washington, D.C.: National Academy of Sciences Press, 1984), pp. 136–140.

20. Larson, S. M., G. R. Cass, and H. A. Gray. "Atmospheric Carbon Particles and the Los Angeles Visibility Problem," *Aerosol Sci. Technol.* 10:118–130 (1989).

21. Pratsinis, S. E., T. Novakov, E. C. Ellis, and S. K. Friedlander. "The Carbon Containing Component of the Los Angeles Aerosol: Source Apportionment and Contributions to the Visibility Budget," *J. Air Pollut. Control Assoc.* 34:643–650 (1984).

22. Currie, L. A., G. A. Klouda, and K. J. Voorhees, "Atmospheric Carbon: The Importance of Accelerator Mass Spectrometry," *Nuclear Instrum. Methods* B5:371–379 (1984).

23. Brimblecombe, P. "Environmental Impact of Fuel Changes in Early London," in *Residential Solid Fuels,* J. Cooper and D. Malek, Eds. (Beaverton: Oregon Graduate Center, 1981), pp. 1–11.

24. Hites, R. A. "Sources and Fates of Atmospheric Polycyclic Aromatic Hydrocarbons" chap. 10 in *Atmospheric Aerosol: Source/Air Quality Relationships,* E. S. Macias and P. K. Hopke, Eds. (Washington, D.C.: American Chemical Society, Symp. Ser. 167, 1981), p. 187.

25. Griffin, J. J. and E. D. Goldberg, "Sphericity as a Characteristic of Solids from Fossil Fuel Burning in Lake Michigan Sediment," *Geochim. Cosmochim. Acta* 45:763–769 (1981).

26. Hopke, P. K. *Receptor Modeling in Environmental Chemistry*. (New York: John Wiley & Sons, 1985).

27. Thompson, A. M. and R. W. Stewart. "How Chemical Kinetics Uncertainties Affect Concentrations Computed in an Atmospheric Photochemical Model," *Chemometrics Intell. Lab. Syst.* 10:69–79 (1991).

28. Stevens, R. K., C. W. Lewis, T. G. Dzubay, R. E. Baumgardner, R. B. Zweidinger, V. R. Highsmith, L. T. Cupitt, J. Lewtas, L. D. Claxton, L. A. Currie, G. A. Klouda, and B. Zak. "Mutagenic Atmospheric Aerosol Sources Apportioned by Receptor Modeling," in *Monitoring Methods for Toxics in the Atmosphere*, W. L. Zielinski, Jr. and W. D. Dorko, Eds. (Philadelphia: Am. Soc. Testing Mtls., 1990), pp. 187–196.

29. Charlson, R. J. and J. A. Ogren. "The Atmospheric Cycle of Elemental Carbon," in *Particulate Carbon, Atmospheric Life Cycle*, G. T. Wolff and R. L. Klimisch, Eds. (New York: Plenum Press, 1982), pp. 3–18.

30. Currie, L. A., K. R. Beebe, and G. A. Klouda. "What Should We Measure? Aerosol Data: Past and Future," *Proc. 1988 EPA/APCA Int. Symp. Measurement of Toxic and Related Air Pollutants*, (Air Pollution Control Assoc., 1988), pp. 853–863.

31. Whitby, K. T. and B. Cantrell. *Int. Cong. Environ. Sensing and Assessment*, paper 29.1, (1976).

32. Haynes, B. S., M. Neville, R. J. Quann, and A. F. Sarofim. "Factors Governing the Surface Enrichment of Fly Ash in Volatile Trace Species," *J. Colloid Interface Sci.* 87:266–278 (1982).

33. Mason, B. and C. B. Moore, *Principles of Geochemistry*, 4th ed. (New York: John Wiley & Sons, 1982) 46–47.

34. Blumer, M. "Polycyclic Aromatic Compounds in Nature," *Sci. Am.* 234:34 (1976).

35. Badger, G. M. "Mode of Formation of Carcinogens in Human Environment," *J. Natl. Cancer Inst.* 9:1 (1962).

36. Mauney, T., F. Adams, and M. H. Sine. "Laser Microprobe Mass Spectrometry of Environmental Soot Particles," *Sci. Total Environ.* 36:215–234 (1984).

37. Cass, G. R., P. M. Boone, and E. S. Macias. "Emissions and Air Quality Relationships for Atmospheric Carbon Particles in Los Angeles," in *Particulate Carbon, Atmospheric Life Cycle*, G. T. Wolff and R. L. Klimisch, Eds., (New York: Plenum Press, 1982), pp. 207–244.

38. Wolff, G. T. and R. L. Klimisch. *Particulate Carbon, Atmospheric Life Cycle*. (New York: Plenum Press, 1982).

39. "Carbonaceous Particles in the Atmosphere," 1978, 1983, 1987, 1991. Conferences organized T. Novakov in 1978 [Proceedings, Univ. of California Press]; 1983 [H. Malissa, H. Puxbaum, T. Novakov, Eds., Sci. *Total Environ.* 36 (June 1984)]; and 1987 (T. Novakov, H. Malissa, A. D. A. Hansen, Eds., *Aerosol Sci. Technol.* 10(2), 1989]; the next in the series took place in April 1991 (Vienna).

40. Daisey, J. M., P. J. Lioy, and J. L. Cheney. "Profiles of Organic Particulate Emissions from Air Pollution Sources: Status and Needs for Receptor Source Apportionment Modeling," *J. Air Pollut. Control Assoc.* 36:17–33 (1986).

41. Hawthorne, S. B., D. J. Miller, R. M. Barkley, and M. S. Krieger. "Identification of Methoxylated Phenols as Candidate Tracers for Atmospheric Wood Smoke Polution," *Environ. Sci. Technol.* 22:1191–1196 (1988).

42. Cooper, J. A. and D. Malek, Eds. *Residential Solid Fuels.* (Beaverton: Oregon Graduate Center, 1981).

43. Schrader, B. "Micro Raman, Fluorescence, and Scattering Spectroscopy of Single Particles," in *Physical and Chemical Characterization of Individual Airborne Particles*, K. S. Spurney, Ed. (New York: John Wiley & Sons, 1986), pp. 358–379.

44. Denoyer, E., R. Van Grieken, F. Adams, and D. F. S. Natusch, "Laser Microprobe Mass Spectrometry 1: Basic Principles and Performance Characteristics," *Anal. Chem.* 54:27A (1982).

45. Hahn, J. "Organic Constituents of Natural Aerosols," in *Aerosols: Anthropogenic and Natural, Sources and Transport*, Vol. 338, T. J. Kneip and P. J. Lioy, Eds. (New York: New York Academy of Sciences, 1980), p. 361.

46. Mazurek, M. A., G. R. Cass, and B. R. T. Simoneit. "Interpretation of High-Resolution Gas Chromatography and High-Resolution Gas Chromatography/Mass Spectrometry Data Acquired from Atmospheric Organic Aerosol Samples," *Aerosol Sci. Technol.* 10:408–420 (1989).

47. La Flamme, R. E. and R. A. Hites. "The Global Distribution of Polycyclic Aromatic Hydrocarbons in Recent Sediments," *Geochim. Cosmochim. Acta* 42:289 (1978).

48. Ramdahl, T. "Retene — A Molecular Marker of Wood Combustion in Ambient Air," *Nature* 306:580 (1983).

49. Sheffield, A. E. "Application of Radiocarbon Analysis and Receptor Modeling to the Source Apportionment of PAH's in the Atmosphere," Ph.D. thesis, University of Maryland (1988).

50. Xhoffer, C., L. Wouters, and R. Van Grieken. "Individual Particle Characterization by Different Beam Techniques," in *Characterization of Environmental Particles*, J. Buffle and H. P. van Leeuwen, Eds., IUPAC Environmental Analytical Chemistry Series, Vol. 1, (Chelsea, MI: Lewis Publishers, 1992), chap. 3.

51. Friedlander, S. K. "Chemical Element Balances and Identification of Air Pollution Sources," *Environ. Sci. Technol.* 7(3):235–40 (1973).

52. Gleser, L. J. "Measurement Error Models," *Chemometrics Intell. Lab. Syst.* 10:45–57 (1991).

53. Baker, M. B. and R. J. Charlson. "Bistability of CCN Concentrations and Thermodynamics in the Cloud-Topped Boundary Layer," *Nature* 345:142 (1990).

54. Bacastow, R. and C. D. Keeling. *Carbon and the Biosphere.* (USAEC Symp. Ser. 30, 1973), p. 86.

55. Clayton, G. D., J. R. Arnold, and F. A. Patty. "Determination of Sources of Particulate Atmospheric Carbon," *Science* 122:751–753 (1955).

56. Lodge, J. P., G. S. Bien, and H. E. Suess. "The Carbon-14 Content of Urban Airborne Particulate Matter," *Int. J. Air Pollut.* 2:309 (1960).

57. Currie, L. A. and Murphy, R. B. "Application of Isotope Ratios to the Determination of the Origin and Residence Times of Atmospheric Pollutants," in *Proc. 8th Materials Research Symposium*, National Bureau of Standards, *NBS Spec. Publ.* 464:439 (1977).

58. Oeschger, H., B. Stauffer, P. Bucher, H. Frommer, M. Moll. C. C. Langway, B. L. Hansen, and H. Clausen. "¹⁴C and Other Isotope Studies on Natural Ice," in *Proc. 8th Int. Conf. Radiocarbon Dating*, T. A. Rafter and T. Grant-Taylor, Compilers, (Royal Society of New Zealand: Lower Hutt, 1972), pp. D70–D90.

59. Muller, R. A. "Radioisotope Dating With a Cyclotron," *Science* 196:489 (1977).

60. Purser, K. H., R. B. Liebert, A. E. Litherland, R. P. Buekens, H. E. Gove, C. L. Bennett, M. R. Clover, and W. E. Sondheim. "An Attempt to Detect Stable Atomic Nitrogen (−) Ions from a Sputter Ion Source and Some Implications of the Results for the Design of Tandems for Ultra-Sensitive Carbon Analysis," *Rev. Phys. Appl.* 12(10):1487–1492 (1977).

61. Ramdahl, T., J. Schjoldager, L. A. Currie, J. E. Hanssen, M. Møller, G. A. Klouda, and I. Alfheim. "Ambient Impact of Residential Wood Combustion in Elverum, Norway," *Sci. Total Environ.* 36:81–90 (1984).

62. Currie, L. A., T. W. Stafford, A. E. Sheffield, G. A. Klouda, S. A. Wise, R. A. Fletcher, D. J. Donahue, A. J. T. Jull, and T. W. Linick. "Microchemical and Molecular Dating," *Radiocarbon* 31:448–463 (1989).

63. Verkouteren, R. M., L. A. Currie, G. A. Klouda, D. J. Donahue, A. J. T. Jull, and T. W. Linick. "Preparation of Microgram Samples on Iron Wool for Radiocarbon Analysis via Accelerator Mass Spectrometry: A Closed-System Approach," *Nuclear Instrum. Methods Phys. Res.* B29:41 (1987).

64. Currie, L. A. "¹⁴C as a Tracer for Carbonaceous Aerosols: Measurement Techniques, Standards, and Applications", in *Aerosols*, B. Y. H. Liu, D. Y. H. Pui, and H. J. Fissan, Eds., (New York: Elsevier, 1984), pp. 375–378.

65. Cachier, H. "Isotopic Characterization of Carbonaceous Aerosols," *Aerosol Sci. Technol.* 10:379–385 (1989).

66. Court, J. D., R. J. Goldsack, L. M. Ferrari, and H. A. Polach. "The Use of Carbon Isotopes in Identifying Urban Air Particulate Sources," *Clean Air*, February 6-11 (1981). (This was the first of a series of Sydney and Canberra air pollution studies by Polach and co-workers.)

67. Kaplan, I. R. and R. J. Gordon. "Contemporary Carbon in Atmospheric Fine Particles Collected in Los Angeles During the 1987 SCAQS," Air and Waste Management Assoc. Conference, Anaheim (1989).

68. Tanner, R. L. and A. H. Miguel. "Carbonaceous Aerosol Sources in Rio de Janeiro," *Aerosol. Sci. Technol.* 10:213–223 (1989).

69. Gaffney, J. S., R. L. Tanner, and M. Phillips. "Separating Carbonaceous Aerosol Source Terms Using Thermal Evolution, Carbon Isotopic Measurements, and C/N/S Determinations:," *Sci. Total Environ.* 36:53–60 (1984).

70. Berger, R., D. McJunkin, and R. Johnson. "Radiocarbon Concentrations of California Aerosols," *Radiocarbon* 28:661–667 (1986).

71. Currie, L. A., G. A., Klouda, R. E. Continetti, I. R. Kaplan, W. W. Wong, T. G. Dzubay, and R. K. Stevens. "On the Origin of Carbonaceous Particles in American Cities: Results of Radiocarbon 'Dating' and Chemical Characterization," *Radiocarbon* 25:603 (1983).

72. Cooper, J. A., L. A. Currie, and G. A. Klouda. "Assessment of Contemporary Carbon Combustion Source Contributions to Urban Air Particulate Levels Using C-14 Measurement," *Environ. Sci. Technol.* 15:1405 (1981).

73. Sturges, W. T. and L. A. Barrie. "Lead 206/207 Ratios in the Atmosphere of North America as Tracers of US and Canadian Emissions," *Nature* 329:144–146 (1987).

74. Sturges, W. T. and L. A. Barrie. "The Use of Stable Lead 206/207 Isotope Ratios and Elemental Composition to Discriminate the Origins of Lead in Aerosols at a Rural Site in Eastern Canada," *Atmos. Environ.* 23(8):1645–1657 (1989).

75. Facchetti, S. and F. Geiss. *Isotopic Lead Experiment,* Rep. EUR 8352 EN, CEC Joint Research Centre, Ispra (Brussels: EEC, 1984).

76. Stevens, R. K. and T. G. Pace. "Overview of the Mathematical and Empirical Receptor Models Workshop (Quail Roost II)," *Atmos. Environ.* 18(8):1499–1506 (1984).

77. Watson, J. G., J. C. Chow, and C. V. Mathai. "Receptor Models in Air Resources Management: A summary of the APCA International Specialty Conference," *J. Air Poll. Control Assoc.* 39(4):419 (1989).

78. Watson, J. G., J. A. Cooper, and J. J. Huntzicker. "The Effective Variance Weighting for Least Squares Calculations Applied to the Mass Balance Receptor Model," *Atmos. Environ.* 18(7):1347–1355 (1984).

79. Parr, R. M. and H. F. Lucas, Jr. "A Rigorous Least Squares Analysis of Complex Gamma-Ray Spectra with Partial Compensation for Instrumental Instability," *Proc. 9th Scintill. Semicond. Symp.* (IEEE Trans Nucl Sci, NS-11 (No.3), 1964), p. 349.

80. Currie, L. A. "Nuclear Chemistry and Statistics in Nuclear and Analytical Chemistry, in Radiochemical Analysis: Nuclear Instrumentation, Radiation Techniques, Nuclear Chemistry, Radioisotope Techniques, July 1966 through June 1967," National Bureau of Standards Tech. Note 421:49–78 (1967), U.S. Government Printing Office, Washington, D.C. 20402.

81. Cooper, J. A. and J. G. Watson. "Portland Aerosol Characterization Study," Rep. to Oregon Dept. of Environmental Quality (1979).

82. Core, J. E., J. A. Cooper, P. L. Hanrahan, and W. M. Cox. "Particulate Dispersion Model Evaluation: A New Approach Using Receptor Models," *J. Air Pollut. Control Assoc.* 32(11):1142–1147 (1982).

83. Johnson, D. L. and B. L. McIntyre. "A Particle Class Balance Receptor Model for Aerosol Apportionment in Syracuse, NY," in *Receptor Models Applied to Contemporary Pollution Problems,* S. L. Dattner and P. K. Hopke, Eds. (Pittsburgh: Air Pollut. Control Assoc., 1983), pp. 238–248.

84. Johnson, D. L. and J. P. Twist. "Statistical Considerations in the Employment of SAX Results for Receptor Models," in *Receptor Models Applied to Contemporary Pollution Problems,* S. L. Dattner and P. K. Hopke, Eds. (Pittsburgh: Air Pollut. Control Assoc., 1983), pp. 224–237.

85. Pratsinis, S. E., "Receptor Models for Ambient Carbonaceous Aerosols," *Aerosol Sci. Technol.* 10:258–266 (1989).

86. Thrane, K. E. "Application of Air Pollution Models: A Comparison of Different Techniques for Estimating Ambient Air Pollution Levels and Source Contributions," *Atmos. Environ.* 22(3):587–594 (1988).

87. Lewis, C. W. and W. Einfeld. "Origins of Carbonaceous Aerosol in Denver and Albuquerque During Winter," *Environ. Int.* 11:243 (1985).

88. Currie, L. A., G. A. Klouda, J. Schjoldager, and T. Ramdahl. "The Power of ^{14}C Measurements Combined with Chemical Characterization for Tracing Urban Aerosol in Norway," *Radiocarbon* 28(2A):673 (1986).

89. Currie, L. A. "Mankind's Perturbations of Particulate Carbon," in *Four Decades of Radiocarbon Research: An Interdisciplinary Perspective*. (New York: Springer-Verlag, in press, 1992).

90. Massart, D. L. and L. Kaufman. *The Interpretation of Analytical Chemical Data by the Use of Cluster Analysis*, (New York: John Wiley & Sons, 1983).

91. Hartigan, J. A. *Clustering Algorithms*, (New York: John Wiley & Sons, 1975).

92. Kaufman, L. and P. J. Rousseeuw. *Finding Groups in Data*. (New York: Wiley-Interscience, 1990).

93. Massart, D. L., B. G. M. Van de Ginste, S. N. Deming, Y. Michotte, and L. Kaufman. *Chemometrics: A Textbook*. (Amsterdam: Elsevier, 1988).

94. Hopke, P. K., E. S. Gladney, G. E. Gordon, W. H. Zoller, and A. G. Jones. "The Use of Multivariate Analysis to Identify Sources of Selected Elements in the Boston Urban Aerosol," *Atmos. Environ.* 10:1015 (1976).

95. Gaarenstroom, P. D., S. P. Perone, and J. P. Moyers. "Application of Pattern Recognition and Factor Analysis for Characterization of Atmospheric Particulate Composition in Southwest Desert Atmosphere," *Environ. Sci. Technol.* 11(8):795–800 (1977).

96. Sanchez-Gomez, M. L. and M. C. Ramos-Martin. "Application of Cluster Analysis to Identify Sources of Airborne Particles," *Atmos. Environ.* 21(7):1521–1527 (1987).

97. Anderson, J. R., F. J. Aggett, P. R. Buseck, M. S. Germani, and T. W. Shattuck. "Chemistry of Individual Aerosol Particles from Chandler, Arizona, an Arid Urban Environment," *Environ. Sci. Technol.* 22(7):811 (1988).

98. Mamane, Y., J. L. Miller, and T. G. Dzubay. "Characterization of Individual Fly Ash Particles Emitted from Coal- and Oil-Fired Power Plants," *Atmos. Environ.* 20(11):2125–2135 (1986).

99. Dzubay, T. G. Personal communication (1989).

100. Artaxo, P., H. Storms, F. Bruynseels, R. Van Grieken, and W. Maenhaut. "Composition and Sources of Aerosols from the Amazon Basin," *J. Geophys. Res.* 93(D2):1605–1615 (1988).

101. Kim, D., P. K. Hopke, D. L. Massart, L. Kaufman, and G. S. Casuccio. "Multivariate Analysis of CCSEM Auto Emission Data," *Sci. Total Environ.* 59:141–155 (1987).

102. Kim, D. and P. K. Hopke. "Source Apportionment of the El Paso Aerosol by Particle Class Balance Analysis," *Aerosol Sci. Technol.* 9:221–235 (1988).

103. Malinowski, E. R. and D. G. Howery. *Factor Analysis in Chemistry*. (New York: John Wiley & Sons, 1980).

104. Malinowski, E. R. "Statistical F-Tests for Abstract Factor Analysis and Target Testing," *J. Chemometrics* 3:49–60 (1988).

105. Windig, W., W. H. McClennen, and H. L. C. Meuzelaar. "Determination of Fractional Concentrations and Exact Component Spectra by Factor Analysis of Pyrolysis Mass Spectra of Mixtures," *Chemometrics Intell. Lab. Syst.* 1:151–165 (1987).

106. Lawton, W. H. and E. A. Sylvestre. "Self Modeling Curve Resolution," *Technometrics* 13(3):617–633 (1971).

107. Hamilton, J. C. and P. J. Gemperline, "Mixture Analysis Using Factor Analysis. II. Self-Modeling Curve Resolution," *J. Chemometrics* 4:1–13 (1990).

108. Henry, R. C. "Self-Modeling Curve Resolution and Other Linear Constraints on Factor Analysis of Airborne Particle Composition Data," *Mathematics in Chemistry Conference*, College Station (1989). See also: Henry, R. C. and B. G. Kim, *Chemometrics*. Intell. Lab. Syst. 8:205–216 (1990).

109. Currie, L. A. "Metrological Measurement Accuracy: Discussion of 'Measurement Error Models' by Leon Jay Gleser," *Chemometrics Intell. Lab. Syst.* 10:59–67 (1991).

110. Thurston, G. D. and J. D. Spengler. "A Quantitative Assessment of Source Contributions to Inhalable Particulate Matter Pollution in Metropolitan Boston," *Atmos. Environ.* 19(1):9–25 (1985).

111. Rheingrover, S. G. and G. E. Gordon. "Wind-Trajectory Method for Determining Compositions of Particles from Major Air Pollution Sources," *Aerosol Sci. Technol.* 8:29 (1988).

112. Friedlander, S. K. "New Developments in Receptor Model Theory," chap. 1 in Atmospheric Aerosol: Source/Air Quality Relationships, *Am. Chem. Soc. Symp. Ser. 167*, E. S. Macias and P. K. Hopke, Eds. (Washington, D.C.: ACS, 1981).

113. Parr, R. M. Rep. of IAEA Advisory Group: Nuclear Techniques in Air Pollution Research and Monitoring. (International Atomic Energy Agency: Vienna, 1988).

114. National Institute of Standards and Technology (1991), Standard Reference Material Catalog.

115. Lewtas, J., L. T. Cupitt, V. R. Highsmith, R. Zweidinger, R. G. Merril, Jr., R. S. Steiber, R. C. McCrillis, L. A. Currie, C. W. Lewis, R. K. Stevens, R. Watts, and R. M. Burton "Integrated Air Cancer Project Study," in Proc. 1988 EPA/APCA International Symp. on Measurement of Toxic and Related Air Pollutants. (Air Pollution Control Assoc., 1988), pp. 799–895.

116. Draxler, R. and J. L. Hefter. "Across North America Tracer Experiment (ANATEX), Volume I: Description, Ground-level Sampling at Primary Sites, and Meteorology," NOAA Technical Memorandum ERL ARL-167, Air Resources Laboratory, Silver Spring, MD (1989).

117. Ferber, G. J., J. L. Heffter, R. R. Draxler, R. J. Lagomarsino, F. L. Thomas, R. N. Dietz, and C. M. Benkovitz. "Cross-Appalachian Tracer Experiment (CAPTEX '83) Final Rep.," NOAA Tech. Memorandum ERL ARL-142, Silver Spring, MD (1986).

118. Clark, T. L. and R. D. Cohn. "Across North America Tracer Experiment (ANATEX)," U.S. EPA Rep. 600/3–90/051 (1990).

119. Currie, L. A., R. W. Gerlach, C. W. Lewis, W. D. Balfour, J. A. Cooper, S. L. Dattner, R. T. DeCesar, G. E. Gordon, S. L. Heisler, R. K. Hopke, J. J. Shah, G. D. Thurston, and H. J. Williamson. "Interlaboratory Comparison of Source Apportionment Procedures: Results for Simulated Data Sets," *Atmos. Environ.* 18(8):1517 (1984).

120. Gerlach, R. W., L. A. Currie, and C. W. Lewis. "Review of the Quail Roost II Receptor Model Simulation Exercise," in *Proc. APCA Specialty Conf. Receptor Models Applied to Contemporary Air Pollution* SP-48:96 (1983).

121. Currie, L. A. "The Limitations of Models and Measurements as Revealed through Chemometric Intercomparison," *NBS J. Res.* 90:409 (1986).

122. Currie, L. A. and G. A. Klouda. "Counters, Accelerators, and Chemistry," in *Nuclear and Chemical Dating Techniques; Interpreting the Environmental Record*, L. A. Currie, Ed. (Washington D.C.: American Chemical Society, Symp. Ser. 176, 1982), pp. 159–185.

123. Gordon, G. E., W. H. Zoller, G. S. Kowalczyk, and S. W. Rheingrover. "Composition of Source Components Needed for Aerosol Receptor Models," chap. 3 in *Atmospheric Aerosol: Source/Air Quality Relationships,* E. S. Macias and P. K. Hopke, Eds. (Washington, D.C.: American Chemical Society, Symp. Ser. 167, 1981).

124. Kamens, R. M. "An Outdoor Exposure Chamber to Study Wood Combustion Emissions Under Natural Conditions," in *Proc. Residential Wood & Coal Combustion,* E. R. Frederick, Ed. (Mars, PA: The Air Pollution Control Association: Choice Book Manufacturer Company, 1982), pp. 207–225.

125. Currie, L. A. "The Importance of Chemometrics in Biomedical Measurements," chap. 6 in *Biological Trace Element Research Multidisciplinary Perspectives,* K. S. Subramanian, G. V. Iyengar, and K. Okamoto, Eds. (Washington D.C.: American Chemical Society, Symp. Ser. 445, 1991).

CHAPTER 2

ACID-BASE EQUILIBRIA OF AEROSOLS AND GASES IN THE ATMOSPHERE

Roger L. Tanner

Energy and Environmental Engineering Center, Desert Research Institute, Reno, Nevada

and

Roy M. Harrison

Institute of Aerosol Science, University of Essex, Colchester, England

TABLE OF CONTENTS

1. Introduction and Definition of Terms 76

2. Formation and Interaction of Atmospheric Acids and
    Bases 78
    2.1 Strong Acid Species 78
    2.2 Sources of Strongly Acidic Aerosols 79
         2.2.1 Primary Sulfuric Acid/Sulfate Emissions 79
         2.2.2 Secondary Acidic Sulfate Formation 80

2.3   Sources of Gaseous Strong Acids 81

3.   Measurement of Species Participating in Fast Equilibria 83
      3.1   Introduction 83
      3.2   Measurement Techniques................................. 84
             3.2.1   Filter Packs 84
             3.2.2   Diffusion Denuders 85
             3.2.3   Nonintrusive Techniques........................ 86
      3.3   Methods Intercomparisons................................ 86
      3.4   Conclusions 89

4.   Modeling and Laboratory Studies of Atmospheric Acid-
      Base Equilibria 89

5.   Ambient Studies of Atmospheric Acid-Base Phase
      Equilibria 93

6.   Kinetic Limitations to Atmospheric Phase Equilibrium 95
      6.1   Fundamental Considerations............................. 95
      6.2   Evaporation and Condensation of Solid
             Ammonium Salt Aerosol 96
      6.3   Evaporation and Condensation in Solution
             Droplets 97
      6.4   Comparison of Kinetic Expressions with
             Atmospheric Measurements 98

7.   Conclusions and Research Needs 99

References.. 99

1. INTRODUCTION AND DEFINITION OF TERMS

In the process of determining the chemical composition of atmospheric aerosol particles, it has become clear that some constituents are relatively inert and their composition depends little on the composition of the gaseous medium into which the aerosol particles are dispersed. For example, lead, sulfur, and alkaline earth metals can be determined in aerosol samples from the fluorescence induced by X-ray bombardment in a vacuum (X-ray fluorescence technique). The elemental composition found can then be related to the concentration of these elements in atmospheric particles prior to collection. However, other aerosol species are much more volatile, and difficulties have been encountered in successfully collecting them from the atmosphere for analysis. Sampling techniques are required which are consistent with observations that these species are distributed between gaseous and aerosol (aqueous or solid) phases in the atmosphere, and that this phase distribution depends on atmospheric conditions which are dynamically changing. This is particularly relevant to considerations of acid formation and neutralization in the boundary layer troposphere since ammonium salts of strong acids (and some weak acids) are among the principal "semi-volatiles" of concern.

The first cases in which multiphase equilibria of atmospheric acids and bases were documented derived from studies of the extent of atmospheric neutralization of sulfuric and nitric acids in aerosol droplets. It was suspected that aerosol strong acidity was due to the presence of sulfuric acid and its ammonia-neutralization products, since SO_2 oxidation should take place in the atmosphere in the presence of varying amounts of NH_3 to form sulfuric acid and ammonium sulfate salts which, due to their low volatility, would readily condense onto aerosol particles. Measured vapor pressures of gaseous ammonia over ammonium acid sulfates were very low (and much lower than observed atmospheric concentrations of NH_3) unless the ammonium/sulfate ratio was very close to 2.

Nitrate ion is also a widely reported aerosol constituent, although early reported data are strongly suspect due to measurement difficulties. Evidence summarized in this chapter strongly supports the view that nitrate is present in fine-mode aerosol particles due to absorption of nitric acid vapor onto sufficiently neutralized aerosol particles, that nitric acid having been formed principally by homogeneous gas-phase oxidation of NO_2. Other strong acids ($pK_a < 2$) which have been determined in ambient atmospheric aerosols or cloud droplets under some conditions (as acids or the corresponding dissociated anion) include hydrochloric acid, oxalic acid, hydroxymethanesulfonic acid (from the formaldehyde-sulfite reaction), and methanesulfonic acid (from the oxidation of dimethyl sulfide). Of these species, the ammonium salt of HCl is semi-volatile, and discussion of its distribution between gas and mixed nitrate-sulfate-chloride aerosol phases is included.

Weak acids present in the troposphere at significant levels include principally low molecular weight carboxylic acids and nitrous acid. Carboxylic acids are likely to be distributed between gaseous and aerosol phases under a range of atmospheric boundary layer conditions. There are a few reported laboratory and field studies of their gas-aerosol equilibria, but although the principles are the same, comprehensive discussion of phase equilibria of weak acids is not included in this chapter.

Strong and weak acids and their semi-volatile salts are part of a larger class of compounds which are semi-volatile under atmospheric conditions. For example, in the group of compounds known as the polynuclear aromatic hydrocarbons (PAHs), 2-, 3-, and some 4-ring compounds exist primarily in the gas phase, while some 4-ring species and all 5- and higher-ring PAHs are found predominantly in the aerosol phase.[95] This is due to their relative volatilities as modified by the sorptive capacities of other particulate constituents.[64,120] The difficulties in measuring semi-volatile compounds and in modeling their transport and deposition in the atmosphere have been well documented. However, we will restrict the discussion in this chapter to semi-volatiles whose atmospheric fate is predominantly dictated by acid-base equilibria existing between components of atmospheric particulate and gaseous phases.

Equilibrium has its normal thermodyanmic definition in this chapter, that is, it signifies that the activities of species represented in a chemical equation are not independent of each other, but are linked among themselves through a constant, K_{eq} (1), where a_x, a_y, and a_z are activities of species X, Y, and Z, respectively, γ_i are activity coefficients, [X], [Y], and [Z] are molar concentrations, and a, b, and c are stoichiometric coefficients.

$$aX + bY = cZ; \qquad K_{eq} = \frac{[a_Z]^c}{[a_X]^a[a_Y]^b} = \frac{\{\gamma_Z[Z]\}^c}{\{\gamma_X[X]\}^a\{\gamma_Y[Y]\}^b} \qquad (1)$$

Phase equilibrium in this chapter will represent a chemical equilibrium between species in more than one phase. In all cases, by definition, Le Chatelier's principle will be assumed to hold, such that removal of a given species at equilibrium will cause a shift in the activities of other species to approach and re-establish equilibrium as defined by the expression for K_{eq}. The usual complications in considering thermodynamic equilibria are in effect: activities are equal to concentrations only at infinite dilution, activities are those of free (unassociated) species only, and equilibrium constants are a function of many factors which dynamically vary in the ambient atmosphere, in particular, temperature, pressure, and water activity.

2. FORMATION AND INTERACTION OF ATMOSPHERIC ACIDS AND BASES
2.1 Strong Acid Species

The presence of strong acids in aerosol particles has been known for many years, and quantitated since the mid-1960s.[19,55,88] Extensive data sets on aerosol strong acid were reported in the mid-1970s, first by Brosset and co-workers,[11,12] followed by several groups in the U.S., e.g., References 102 and 112, and acidity was attributed to sulfuric acid and its partially NH_3-neutralized salts formed by oxidation (gas-to-particle conversion) of SO_2. Acquisition of quantitative data was dependent on the development and application of artifact-free, neutral filter media, and aided by the use of titration procedures which were able to distinguish between strong and weak acids in aerosol samples. Observation of free H_2SO_4 was reported in a few cases, but most strong acidity was due to the presence of hydrogen sulfate salts in aerosols.

Nitric acid and particulate nitrate are widely reported constituents of the ambient atmosphere, and the formation mechanisms are described below in Section 2.4. Phase equilibria considerations dictate that, contrary to sulfuric

Table 1. Strong Acidic Species in the Troposphere

Aerosol Species	Chemical Formula	Relative Abundance
Sulfuric acid	H_2SO_4	Minor
Ammonium hydrogensulfate	NH_4HSO_4	Major
Letovicite	$(NH_4)_3H(SO_4)_2$	Major
Oxalic acid	$(COOH)_2$	Trace
Methanesulfonic acid	CH_3SO_3H	Trace
Hydroxymethanesulfonic acid	$CH_2(OH)SO_3H$	Trace
Gaseous Species		
Nitric acid	HNO_3	Major
Hydrochloric acid	HCl	Minor

acid which exists with >99% abundance in the particle phase under all lower tropospheric conditions, nitric acid is distributed between gaseous HNO_3 and neutral nitrate salts in the particulate phase. The fraction present as nitric acid vapor depends on the temperature, the liquid water content, and the pH. Nitric acid concentrations in the 10–50 ppb(v) range have frequently been reported in the South Coast Air Basin of California and in other urban plumes under stagnant conditions. Concentrations are lower in other locations, but in many locations gaseous HNO_3 is very frequently the most abundant of strong acid species with respect to human exposure. Other strong acids ($pK_a < 2$) which could possibly be present in ambient atmospheric aerosols under some conditions include hydrochloric acid and oxalic acid. Their presence has been reported but data indicating widespread distribution at significant levels are lacking to date. A summary of strongly acidic species found in atmospheric aerosols, and their relative abundances, is given in Table 1.

2.2 Sources of Strongly Acidic Aerosols
2.2.1 Primary Sulfuric Acid/Sulfate Emissions

Primary acidic sulfates are emitted directly along with SO_2 from fossil fuel combustion sources and from a variety of industrial processes including petroleum refining, nonferrous smelting, pulp milling, and the manufacture of sulfuric acid. Goklany et al.[40] reported the acidic sulfate ratio to total emissions for various source types. Acidic sulfate is defined as $SO_3 + H_2SO_4$ for the purposes of this discussion. The total emissions of primary sulfate for U.S. sources have been estimated to be $0.6 \cdot 10^6$ t/y (ton/year, calculated as SO_2) for 1980, with a range for the fraction of acidic emissions of 0.57–0.83, depending on emission controls. Estimates of direct SO_2 emissions for 1980 are $2.45 \cdot 10^7$ t/y;[70] hence direct emissions of acidic sulfates constitute at most 2.0% of emitted sulfur from anthropogenic emissions.[111] Because it can be estimated that about 20% of SO_2 emitted from the U.S. is converted to sulfate during its atmospheric lifetime, it can be seen that secondary processes are the dominant source of acidic sulfates in the atmosphere, independent of the extent of their neutralization by ammonia (or other bases if present) after emission or formation in the atmosphere.

2.2.2 Secondary Acidic Sulfate Formation

The formation of sulfuric acid in the atmosphere, principally by oxidation of anthropogenic SO_2, is the major source of acidity in aerosols. Further, in the absence of complete neutralization of these acidic aerosols, it is a major source of strong acid in precipitation. Other processes may produce H_2SO_4 in the atmosphere, in particular, oxidation of natural sources of reduced sulfur compounds such as dimethyl sulfide from marine emission sources and H_2S from certain marshes. However, we will focus on the two major H_2SO_4-formation processes: homogeneous, gas-phase oxidation of SO_2, and aqueous phase oxidation of dissolved S(IV).

There is now abundant evidence that oxidation of SO_2 by the hydroxyl radical (Reaction 2) is the principal gas-phase route for H_2SO_4

$$SO_2 + {\cdot}OH \xrightarrow{(M)} HSO_3^-\qquad(2)$$

formation, greater than 98% of gas-phase oxidation proceeding thereby according to the estimate of Calvert et al.[16] where (M) indicates a molecule or atom which stabilizes a reactive intermediate by absorbing some of its excess internal energy. The reactions with other common radical and odd oxygen species (in particular, methylhydroperoxyl and $O(^3P)$) are much slower and not significant, given typical atmospheric residence times for SO_2. Conversion of HSO_3 to sulfuric acid is believed to proceed by Reactions 3 and 4; indeed, the kinetics of (3) have recently been

$$HSO_3^- + O_2 \rightarrow SO_3 + HO_2^-\qquad(3)$$

$$SO_3 + H_2O \rightarrow H_2SO_4\qquad(4)$$

reported.[39] The rate of oxidation of SO_2 is dependent on the concentration of $\cdot OH$, but with typical summer conditions, the rate is about 0.5%/h on a diurnally averaged basis. Winter oxidation rates are about an order of magnitude lower under mid-latitude tropospheric conditions.

The second type of process by which secondary acidic sulfate may be formed in the atmosphere is aqueous phase oxidation in cloud liquid water or precipitation. This process is important from two considerations — the acidic sulfate formed by aqueous phase oxidation may be directly deposited in precipitation or, as occurs in about 90% of cases,[80] the cloud evaporates, and acidic sulfate is restored to the aerosol phase in which it may undergo mass transfer between aerosol and gaseous phases, transport, and subsequent deposition at another location.

Several authors have reviewed the kinetics of various aqueous oxidation mechanisms that could be important in converting SO_2 to acidic sulfate under atmospheric conditions.[68,73,91] For example, the review of Calvert et al.[16] identified several potential oxidants on the basis of their calculated, instantaneous rates of reaction with dissolved S(IV) in equilibrium with 1 ppb(v) of $SO_2(g)$: O_3; H_2O_2 and its organic analogs, peroxyacetic acid and methyl hydroperoxide; O_2 catalyzed by Fe(III) or Mn(II); graphitic carbon; •OH and $HO_2^•$ radicals; and peroxyacyl nitrates. Rates were reported as a function of pH at two different temperatures. Rates for organic peroxides were calculated based on their reported aqueous solubility.[65]

Essentially, the results of this analysis are that H_2O_2-S(IV) reaction rates are high enough at all pHs to result in major conversions to acidic sulfate during the lifetime of most clouds. Oxidation by ozone is significant above about pH 5, a result of the strong positive pH dependence of the S(IV)-O_3(aq) reaction rate. No other reaction is fast enough under rural continental boundary layer conditions to convert significant amounts of S(IV), although some contributions are probably made by organic peroxides, by Fe(III)/Mn(II)-catalyzed autoxidation,[50] and/or by •OH and $HO_2^•$ (if mass accommodation coefficients (α) for these species are >ca. 10^{-3}; definition of α: see Equation 18). As has been frequently demonstrated,[66,91] the pH dependence of the rate of Reaction 5 in atmospheric liquid water is small because

$$H_2O_2 + H^+ + HSO_3^- \rightarrow H_2O + H_2SO_4 \qquad (5)$$

the general acid catalysis is offset by the acid-base dissociation of $SO_2^•H_2O$ in solution (Reference 68 and references therein).

The rates of oxidation of S(IV) by O_3 (30 ppbv) and H_2O_2 (1 ppbv) in the presence of liquid water clouds are shown by, e.g., Schwartz[91] to be about equal to each other and equal to several hundred %/h at about pH 5.6. Mass transfer limitations to attainment of these rates have been considered by Schwartz,[91] and found to be negligible under typical atmospheric conditions. Hence aqueous-phase conversion of SO_2 to sulfuric acid may proceed to exhaustion of the limiting reagent within the lifetime of liquid-water clouds. Lee et al.[62] have shown that the rate of Reaction 5 is similar (about 30% lower) in authentic rain samples compared to laboratory values in "purest possible" water. Thus, conclusions concerning extent of S(IV) oxidation by H_2O_2 based on laboratory studies are expected to be valid for atmospheric water samples.

A third source of secondary acidic sulfate must also be considered — sorption of SO_2 into aerosol-sized droplets and oxidation therein to acidic sulfate. Direct experimental evidence for this process has been very difficult to obtain. The work of McMurry and Wilson[67] on accumulation-mode aerosol growth certainly suggests that oxidation of SO_2 in aerosol particles occurs in the atmosphere (at least under polluted summertime conditions), but the relative importance of this acid-formation process is still not well established.

2.3 Sources of Gaseous Strong Acids

The division of atmospheric acid formation processes into gaseous and condensed-phase processes can now be seen to be somewhat arbitrary. Homogeneous oxidation of SO_2 to sulfuric acid is discussed under aerosols because acidic sulfate is found to be associated with aerosol particles (principally in the 0.1 to 1.0 μm diameter size range) in the atmosphere. Likewise we discuss nitric acid formation under gaseous strong acids not only because homogeneous gas-phase oxidation has been identified as the major route by which acidic nitrate is available for deposition in precipitation,[16] but also because at many sites inorganic nitrate is frequently found predominantly in the gas phase as HNO_3. Formation of atmospheric HCl(g) is also discussed briefly in this section, although the mechanism of its formation is not well established.

The formation of nitric acid through oxidation of NO_2 by •OH radical in Reaction 6 is about ten times more rapid than the corresponding

$$OH + NO_2 \xrightarrow{(M)} HNO_3 \qquad (6)$$

oxidation of SO_2 (Reaction 2), and is thus the major route of formation of nitric acid in the boundary layer troposphere during daylight hours (Reference 16 and references therein). The absence of seasonal dependence of nitrate deposition derived from nitric acid indicates that this process dominates in winter as in summer.[56] That is, the lower formation rate in winter is compensated for by lower dry deposition rates, or perhaps by enhanced below-cloud scavenging to precipitation in the form of snow.[81]

A second mechanism for nitric acid formation is now thought to be important at night. This involves oxidation of NO_2^-, the predominant form of NO_y in the absence of NO_2 photolysis, by ozone (Reaction 7). The nitrate

$$O_3 + NO_2 \rightarrow NO_3^- + O_2 \qquad (7)$$

radical, NO_3^-, is not stable during daylight hours due to photolysis and/or reaction in the presence of NO. These NO_3^- decomposition paths become slow at night, however, and other NO_3^- chemistry can become important, in particular, reaction with NO_2 and H_2O (Reactions 8 and 9) or with gaseous

$$NO_3^- + NO_2 \underset{}{\overset{(M)}{\rightleftarrows}} N_2O_5 \qquad (8)$$

$$N_2O_5 + H_2O \rightarrow 2HNO_3 \qquad (9)$$

aldehydes (Reaction 10) may form nitric acid during the night. Reaction

$$NO_3^- + RCHO \rightarrow HNO_3 + RCO^• \qquad (10)$$

9 is reported to be slow in the absence of water droplets;[116] hence it has been suggested that nitric acid may be formed by the $NO_3^- + NO_2$ reaction only in the presence of clouds or fog.

As noted above, the only other significant gas-phase strong acid in the atmosphere is likely to be HCl. Below we discuss the phase equilibrium between ammonium salts of HCl, HNO_3, and H_2SO_4 and the gas phase acids and ammonia. For the considerations of sources, it is sufficient to note that NH_4Cl is more volatile than NH_4NO_3 to the extent that with ambient levels of aerosol strong acid in fine particles, chloride originally in fine particles may be found predominantly in the gas phase as HCl.[61] Indeed, a major source of HCl(g) may be the disproportionation of Cl^- from fine particles as they are acidified by incorporation of secondary H_2SO_4 or HNO_3 (e.g., reaction 11).

$$Cl^- + HSO_4^- \rightarrow HCl(g) + SO_4^{2-} \tag{11}$$

Concerning the occurrence of other strong acids in the atmosphere, the presence of oxalic acid has been reported by Norton et al.;[71] methanesulfonic acid (CH_3SO_3H) has been reported in marine aerosols by Ayers et al.[6] and Saltzman et al.,[85] and both dimethyl sulfate and methyl hydrogen sulfate (CH_3OSO_3H) have been reported in near-source aerosol samples by Eatough et al.[27] Dissociation of oxalic acid into hydrogenoxalate ion is essentially complete at pH 4, likewise the dissociation of CH_3SO_3H into methane sulfonate; hence both qualify as strong acids. The occurrence of oxalic acid at significant levels in the troposphere has not been well documented, although it could exist, on the basis of its volatility and that of its ammonium salt, in both the particulate and gaseous phases under typical conditions. Methane-sulfonic acid is expected to be found only in the particulate phase. Dimethyl sulfate and CH_3OSO_3H are acute toxics which hydrolyze rapidly under most atmospheric conditions.

3. MEASUREMENT OF SPECIES PARTICIPATING IN FAST EQUILIBRIA
3.1 Introduction
As outlined in Section 2, ammonium nitrate and ammonium chloride coexist with their gaseous precursors in equilibria expressed in Reactions 12 and 13.

$$NH_4NO_3(s \text{ or } aq) \rightleftarrows HNO_3(g) + NH_3(g) \tag{12}$$

$$NH_4Cl(s \text{ or } aq) \rightleftarrows HCl(g) + NH_3(g) \tag{13}$$

Although, as discussed below in Section 6, there are likely kinetic limitations to the attainment of equilibrium, ambient measurements (Section 5) show that in many situations the concentration product, $[HA]_g[NH_3]_g$, in which HA represents the gaseous acid, approximates to that predicted from chemical thermodynamics. For solid phase NH_4A whose activity $= 1$, the value of the equilibrium constant, $K = [HA]_g[NH_3]_g$, is a function solely of temperature.

At higher humidities in which liquid droplets form by deliquescence, the value of the equilibrium constant is a function also of relative humidity, since the molality of the droplet solution responds to relative humidity to maintain equilibrium between water vapor pressures (activities) at the droplet surface and in the ambient atmosphere.

Equilibrium values of the product, $[HA]_g[NH_3]_g$, have been reported for solid NH_4NO_3 and for its aqueous solutions by Stelson and Seinfeld,[93] and for pure NH_4Cl by Pio and Harrison.[77,78] The values of K are influenced by other components of internally mixed aerosol particles (i.e., particles containing homogeneous mixtures of more than one salt). The influences of mixed salts and of variations in the pH of aerosol droplets have been considered by Stelson and Seinfeld.[100,101]

Because of the existence of a dynamic equilibrium, it is essential that measurement of any component substance of the equilibrium be made using a technique which disturbs the equilibrium as little as possible. Several sampling methods have been devised with this consideration in mind, and a number of methods intercomparisons with more traditional sampling methods have been carried out to test their performance. An important consideration is the time scale over which transfer between gas and aerosol phases occurs, in relation to the time scales inherent in the sampling procedures. This point is covered in some detail in Section 6.

Aerosol evaporation during sampling has been documented more often than condensation, and an indication of rates may be obtained from the recent work of Harrison et al.[45] When dry and aqueous NH_4NO_3 and NH_4Cl aerosols were passed through an annular denuder, measured evaporation rates expressed as rate of change of particle radius, r, were independent of particle size. Values of dr/dt were -0.105 nm·s^{-1} and -0.045 nm·s^{-1}, for dry NH_4Cl and NH_4NO_3 particles, respectively; values for solution droplets (expressed as equivalent dry particle size) of -0.452 nm·s^{-1} and -0.049 nm·s^{-1} for the two salts, respectively, were observed at 97% relative humidity and 20°C.

3.2 Measurement Techniques

Three main kinds of techniques have been used to measure the components of atmospheric species involved in kinetically constrained phase equilibria.

3.2.1 Filter Packs

These are assemblages of filters in series in which particulate matter is collected by impaction on the first filter (usually Teflon or quartz fiber), and gaseous components by diffusion and sorption onto subsequent filters (e.g., HNO_3 on nylon or NaCl-coated cellulose, SO_2 on carbonate-coated cellulose, and NH_3 on acid-impregnated cellulose filters).[23,48,96] The inherent problem with this technique is that changes in temperature or pressure can lead to evaporation from, or condensation onto, the first filter of gaseous ammonia and HCl(g) or HNO_3(g) from the volatile ammonium salts, and these effects (especially evaporation) are often encountered in practice.

A second problem with filter packs is that gas-particle and particle-particle reactions may occur on the first filter. Thus, ammonia may react with already collected acidic sulfate particles, leading to a negative ammonia and a positve ammonium artifact if atmospheric ammonia concentrations increase during the sampling period (Reaction 14). Another such reaction,

$$NH_3(g) + H_2SO_4 \rightarrow NH_4HSO_4(aerosol) \tag{14}$$

known to be of importance in the atmosphere and liable to occur on filter samples of marine aerosol, involves the reaction of $HNO_3(g)$ with NaCl(aerosol) (15). This reaction leads to an overestimation (positive artifact) of

$$NaCl(s \text{ or } aq) + HNO_3(g) \rightarrow NaNO_3(s \text{ or } aq) + HCl(g) \tag{15}$$

particulate nitrate and HCl and an underestimation of HNO_3 and particulate chloride. An example of a particle-particle interaction on filters is that of ammonium nitrate with sulfuric acid droplets (16).

$$NH_4NO_3 + H_2SO_4 \rightarrow NH_4HSO_4 + HNO_3 \tag{16}$$

3.2.2 Diffusion Denuders

The diffusion denuder depends upon passage of an air stream under conditions of laminar flow through a tube[26,32] or an annulus between two tubes,[79] which is coated with a substance providing a perfect reactive sink for the analyte gas. The denuder is designed to effect nearly quantitative removal of analyte gases, but allow passage of aerosols with negligible loss because of their much lower diffusion rate. The residence time in most denuder designs is of the order of 0.2–2 s, too small to allow significant evaporation of volatile salts in response to removal of gaseous components. After passage of air through one or more denuders in series, particles are collected on a filter, usually of Teflon. Volatile ammonium salts thereby collected are inherently unstable due to the removal of the gaseous species with which they were equilibrated, and must be stabilized or their evaporation products collected on subsequent filter stages.

A variant on the denuder procedure for nitric acid and nitrate sampling is the "denuder difference method",[92] in which two samplers are run in parallel. In one sampler both particulate and gaseous nitrate are collected on a nylon filter (or Teflon/nylon filter combination), while in the parallel sampler, $HNO_3(g)$ is removed by a denuder before collection of particulate nitrate on nylon or Teflon/nylon. The difference between nitrate collected on the two filters gives the nitric acid concentration. The advantage of this approach is that denuder tubes do not need to be recoated prior to each sampling period.[35] The technique may, of course, also be applied to other substances for which appropriate collection media are available.

Denuder samplers have some inherent problems. Although gas-particle interactions are overcome by prior removal of gases in the denuder, particle-particle reactions can still cause problems in measurements of constituents, for example, aerosol H^+ can be neutralized after collection by reaction with particles of, e.g., $CaCO_3$, NH_4NO_3, or NaCl. The denuders themselves can be subject to artifacts, in that particle deposition can arise if laminar flow is not fully developed, or from sedimentation if the denuder flow is not vertical. Since these problems affect mainly coarse particles (d $>\approx2.5$ μm), some workers have used an inlet cyclone or impactor[5,58] for their removal. The question of gas-particle interactions in the pre-separator (impactor stage or cyclone) does not appear to have been given sufficient thought, although losses of HNO_3 have been reported in such systems.[5] Other problems may also occur in denuders sampling air containing high levels of NO_x, since these gases may be collected as either HNO_3 and HNO_2 in a Na_2CO_3-coated denuder. This positive artifact can be corrected by the use of three denuders in a series.[53]

3.2.3 Nonintrusive Techniques

An ideal measurement procedure would involve a fast-response, non-intrusive (*in situ*) method for species involved in atmospheric acid-base equilibria. Spectroscopic techniques offer a partial solution for the gaseous components, and have been applied in a few cases to particulate samples.[14,53,54] Techniques applicable to determination of atmospheric reactive gases include tunable diode laser absorption spectroscopy (TDLAS),[89] Fourier transform infrared (FTIR) spectroscopy,[115] and where applicable, differential optical absorbance spectroscopy (DOAS).[9] Recent developments of the TDLAS demonstrate its applicability also to the analysis of ammonia and HCl[10] and for formaldehyde.[42] A current problem with the FTIR and DOAS spectroscopic methods is their limited sensitivity which usually restricts their limits of detection to rather high (several ppb) levels of analyte gases. The TDLAS technique has limits of detection below 1 ppb for several gases, but it does require the passage of sample air through an optical (white) cell. Hence, wall losses are a possibility, although they can be minimized using a high gas flow sampling rate.

3.3 Methods Intercomparisons

Since no method is currently accepted as a standard in the measurement of ammonia and acidic vapors, method performance has been evaluated primarily through intercomparison studies. When available, the results of nonintrusive spectroscopic methods are generally taken as the reference against which the accuracy of other methods is evaluated.

The earliest comprehensive intercomparison of measurement techniques for species of interest in this chapter was the nitric acid/particulate nitrate intercomparison held in Claremont, CA (South Coast Air Basin receptor site) in 1978, with the results reported by Spicer et al.[96] Under conditions of approximately equal contributions of nitric acid and nitrate to the total inorganic nitrate levels, ten methods were compared (two chemiluminescent, one in-

frared, two diffusion denuder, and five filter techniques). Five methods —
one chemiluminescent, one high-volume filter pack, one low-volume filter
pack, one denuder, and the long-path infrared spectroscopic method — ex-
hibited results in good agreement with the median concentrations. However,
two low-volume filter methods yielded best-fit regression slopes significantly
exceeding 1.0 vs. median concentrations, and three methods suffered operational problems which resulted in poor correlation with other techniques.

Anlauf et al.[3] compared measurements of HNO_3 by tungstic acid denuder,
Teflon-nylon-W41 filter pack, and tunable diode laser techniques. During
daytime sampling all three techniques were operating and the denuder and
filter pack techniques measured concentrations about 16% lower than the laser
method. The reason for this disparity was not clear. Nighttime tungstic acid
denuder results were about a factor of two larger than the filter pack method
(the diode laser was not operating). Nighttime concentrations were very low,
but the divergences were not explained by the authors. The denuder and filter
pack methods correlated well for particulate ammonium; for particulate nitrate
the correlation coefficient was 0.85 and the regression slope was 0.81, with
the filter pack giving the lower results.

Mulawa and Cadle[69] compared the denuder penetration method, a denuder
difference method, and a filter pack procedure for nitric acid and particulate
nitrate. At their site, nitric acid was the predominant form of inorganic nitrate,
and significant overestimation of $[HNO_3]$ and underestimation of particulate
nitrate by the filter pack were evident. In a more recent study, Dasch et al.[21]
from the same laboratory report quite different results comparing filter pack
and annular denuder samplers during a wintertime period. Agreement to within
$\pm 10\%$ was observed for several particulate species and gaseous HCl. Un-
expectedly, nitric acid concentrations were higher with the denuder sampler,
which was attributed to either losses of HNO_3 on the filter-pack particulate
filter, or to deposition of nitrate particles in the annular denuder. Similar
results were reported for HNO_3 by Tanner et al.,[113] using high volume filter
pack and denuder difference sampling methods.

Durham et al.[25] carried out a performance evaluation of a nylon diffusion
denuder. Relative to the predictions of the Gormley-Kennedy model, an excess
of HNO_3 was collected in the denuder and attributed to deposition of other
gaseous nitrogen compounds in the denuder and/or to release of nitric acid
from particles in the denuder. The latter possibility is, we believe, not likely
based on the kinetic data provided by Larson and Taylor[60] and the slow kinetics
of ammonium nitrate dissociation reported by Harrison et al.[45]

Ferm et al.[34] reported a field intercomparison study of measurement techniques for "total nitrate" ($HNO_3 + NO_3^-$), and "total ammonium" (NH_3
$+ NH_4^+$). Different experimental groups using the same technique obtained
results with relative standard deviations of ± 15 and $\pm 20\%$, respectively,
for total nitrate and ammonium. When different sampling technqiues were
compared (denuder, total filter, and filter pack), the same relative standard
deviations were found. It was concluded that all techniques gave comparable
results, i.e., no method-related systematic biases were observed.

An extensive data set comparing samplers for nitric acid and nitrate is available from the Nitrogen Species Intercomparison Study (NSIS) carried out in Claremont, CA (Los Angeles Basin site) in 1985, and reported by Hering et al.[48] Comparisons of a wide range of techniques did indicate some significant systematic differences. Filter pack measurements of nitric acid were higher than the denuder difference and transition flow reactor (denuder-based,)[57] methods by factors of 1.25–1.4 (depending on reporting group), except when quartz front filters were used. Results from denuder difference measurements were, in turn, higher than annular denuder and FTIR methods. Lower and varying annular denuder results for HNO_3 are likely due to losses in apparatus-specific inlet systems.

Solomon et al.[94] separately reported filter pack data on ammonia and nitric acid from the NSIS which showed an increase of 15 to 20% in both $[NH_3]$ and $[HNO_3]$ reportedly due to evaporation of particulate NH_4NO_3. Anlauf et al.[4] compared measurements made by a tunable diode laser method and by filter packs with those by other methods. During the daytime the filter pack $[HNO_3]$ exceeded the FTIR by 22%, whereas at night the filter pack results were 24% lower. Eatough et al.[28] compared six denuder methods with a filter pack for HNO_3. While some considerable variability was found between different denuder procedures, comparison of filter pack and denuder data showed excellent agreement when biased methods were eliminated. The authors considered that the use of acid-washed quartz filters, rather than Teflon, for particulate collection was a significant factor in obtaining good agreement, apparently due to a reduction in ammonium nitrate-acid sulfate reactions on the quartz filters. Recently, Wiebe et al.[119] reported an intercomparison of NH_3 and ammonium results from the NSIS, using FTIR results as the standard, which indicated that the scatter in ammonia results was most likely due to a combination of NH_3 release from ammonium nitrate evaporation, and sorption losses on the apparatus depending on inlet configuration and type of filter pack.

Other results for ammonia and nitric acid include the report of Appel et al.[5] in which oxalic acid denuder and filter pack methods for NH_3 were compared, with concentrations from the filter pack reported to be 50% higher than the denuder. Tanner et al.[113] compared four measurement techniques for nitric acid. Good correlation was found between results from a high volume filter pack and denuder difference methods, with a small absolute loss observed with the former. A dual-channel chemiluminescence technique measuring HNO_3 by difference using a nylon filter in one channel provided real-time data which when averaged, correlated well with denuder difference results in the daytime. It gave higher results during nighttime periods, possibly due to collection of other odd-nitrogen compounds by the nylon filters.

Harrison and Kitto[46] compared filter pack and denuder penetration methods for determination of particulate SO_4^{2-}, NO_3^-, Cl^-, NH_4^+, and gaseous HNO_3, HCl, and NH_3. The results for particulate sulfate were in excellent agreement (ratio of means = 0.99 ± 0.06, $r = 0.998$, $N = 30$), confirming the non-volatility of this species. Filter pack and denuder methods also correlated well for other species, the filter pack giving slightly higher results for the volatile species, HNO_3, HCl, and NH_3, compared to the denuder, with correspondingly lower results for particulate NO_3^-, Cl^-, and NH_4^+, attributable to volatilization of ammonium salts from the filter pack prefilter. Improving the inlet efficiency of the filter pack sampler also influenced the results.

3.4 Conclusions

An overall appraisal of the published comparisons of filter-pack and denuder methods leads to the view that while the denuder method has better overall performance, the filter pack can, under many circumstances, give acceptable results. The following conclusions are drawn from the work of Hering et al.,[48] Tanner et al.,[113] and Harrison and Kitto:[46]

- Reported results are site-specific. High particulate acidity, low relative humidity, and high ambient temperatures all produce enhanced HNO_3 volatilization from filter packs.
- Reported results are apparatus-specific. Not only does the filter-pack, front-filter material influence the data, it is likely that the construction of the inlet, the housing, and the siting of the equipment all have an influence. A filter pack which is heated above ambient temperatures is likely to suffer loss of volatile ammonium salts. Sampling flow rates are also important as they determine pressure drops within the filter pack, which in turn influences volatilization.
- There are differences between various denuder methods, but when only Teflon or other inert materials are in contact with the sampled reactive gases (and their small sorption capacities are quantified in calibration protocols), fine particle sampling comparisons of denuder methods with filter pack methods consistently show reduced sampling artifacts from evaporation of semi-volatile aerosol constituents for the denuder methods. Under these conditions the use of diffusion-denuder based methods is considered to provide the most accurate estimates of ambient nitric acid, particulate nitrate and ammonia concentrations.
- There is a general tendency for filter pack methods (especially with Teflon front filters) to overestimate gaseous HNO_3 and NH_3 concentrations by varying amounts depending on the sampling site and conditions.
- When ambient concentrations are high enough, FTIR and where applicable, DOAS serve as reference methods for comparison of aerosol and reactive gas measurement methods. TDLAS methods are also useful in this regard when the possibility of sorption losses in the white cell can be excluded.

4. MODELING AND LABORATORY STUDIES OF ATMOSPHERIC ACID-BASE EQUILIBRIA

It is difficult to separate modeling and laboratory or field studies of phase equilibrium in ambient inorganic aerosols (ammonium acid sulfates, nitrates, and chlorides). In fact, most modeling studies have incorporated some comparison of predicted and measured aerosol composition and concentrations, and all ambient field studies have used a "model" equilibrium expression to compare measured partial pressure products of $[NH_3][HNO_3]$, at minimum, with those predicted from water activity- and temperature-dependent equilibrium constants. In this section we will summarize efforts to provide models of aerosol formation and growth which take proper cognizance of thermodynamically driven gas-aerosol interactions and inter-phase equilibration dynamics. In a few cases the studies have also included the effects of phase re-equilibration on wet and dry deposition phenomena.

In all cases the model development has been greatly facilitated by the laboratory studies which have provided essential new or improved thermodynamic data on pure and mixed salt aerosols, in particular, vapor pressures over semi-volatile ammonium salts, deliquescence humidities of pure and mixed salts, and water activities and ionic activity coefficients in concentrated droplets. Studies of hysteresis phenomena,[109] of growth rates of suspended single particles,[107] and of water droplet accommodation coefficients for relevant gases[37] have all contributed to the development of more accurate models of aerosol formation and growth. Therefore the descriptions of model development will also include summary statements of the crucial laboratory study results where appropriate.

In a series of papers by Tang and co-workers,[105,106] the fundamental thermodynamics of ammonium salts of atmospheric relevance were considered, and measurements of water activities of several of these salts measured in concentrated solutions for which ionic activity data were previously sparse. This led to an improvement in the predictions of vapor pressures of gases (e.g., HNO_3,[98]) over aerosol droplets. Lee and Brosset[61] also considered the thermodynamics of mixed salt aerosols in attempting to understand the potential for gas-particle interactions and artifacts therefrom in atmospheric sampling of acidic aerosols.

Much of the development of models of aerosol formation which includes explicit treatment of gas-aerosol equilibria of semi-volatile inorganic aerosol constituents has been conducted by Seinfeld, by Cass, and by Seigneur, and their respective co-workers in California in response to the mesoscale photochemical smog problems of the South Coast Air Basin (SoCAB). The first comparisons of ammonia and nitric acid concentrations with predictions based on the presumed equilibrium between those species and ambient aerosol nitrate were reported by Stelson et al.[98] and Doyle et al.[24] The observations of Stelson et al.[98] led to a series of investigations of the thermodynamics of the ammonium nitrate–ammonia, nitric acid system, from which was determined how the ammonium nitrate-ammonia, nitric acid equilibrium was affected by dynamically changing atmospheric variables including the relative humidity (RH) and aerosol droplet pH,[99] how those variables affected the vapor pressures of ammonia and nitric acid at a fixed temperature,[100] and how these thermodynamic data could be used to predict the properties of mixed sulfate-nitrate droplets.[101] With this and other thermodynamic data for the mixed ammonium nitrate-sulfate system, it was then possible to construct a comprehensive model (EQUILIB) for the atmospheric production of sulfate/nitrate/ammonium/water aerosol[7] which could predict the aerosol chemical composition and physical state. An illustration of the use of this model is given in Figure 1, for the case of initial conditions, $[NH_3] = 5 \mu g/m^3$, RH = 50%, T = 25°C, and rates, R, of formation of nitric and sulfuric acids as follows: $R(HNO_3) = 20 \mu g.m^{-3}.h^{-1}$; $R(H_2SO_4) = 30 \mu g.m^{-3}.h^{-1}$. Further development of this comprehensive model of aerosol sulfate and nitrate distribution between phases with explicit treatment of the Kelvin effect[8] allowed this

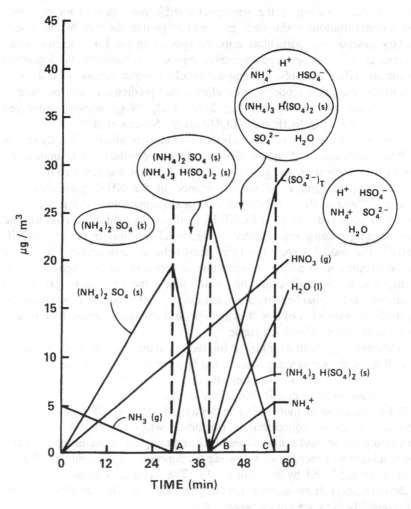

Figure 1. Evolution of gaseous and aerosol constituents. Up to point A the aerosol phase consists of solid $(NH_4)_2SO_4$. From point A to point B it consists of a mixture of solid $(NH_4)_2SO_4$ and solid $(NH_4)_3H(SO_4)_2$. From point B to point C the aerosol consists of an aqueous solution surrounding solid $(NH_4)_3H(SO_4)_2$. Finally, after point C the aerosol is an aqueous solution.

model (now KEQUILIB) to predict chemical composition and physical state as a function of particle size.

Russell et al.[84] used a vertically resolved Lagrangian model of transport (one in which the coordinate system is advected with the mean wind field within each layer), together with the thermodynamic formulations of Stelson and Seinfeld[99,100] and ammonia emissions inventory data to predict the location and maximum concentrations of ammonium nitrate aerosol in the Los Angeles basin. Sensitivity studies showed that temperature and NH_3 concentration were the most critical variables. Later, Russell et al.[83] incorporated a predictive model for ammonium nitrate aerosols into a gridded Eulerian model (based on numerical solution of the atmospheric diffusion equation for ensemble mean concentrations within each grid cell) to predict the distribution of secondary gaseous and particulate nitrogen species in the Los Angeles basin. Saxena et al.[86] also developed a similar approach to modeling the formation of nitrate/sulfate aerosols. A computationally simpler version of a thermodynamically explicit model was developed and predicted results on composition of concentration of secondary $SO_4^{2-}/NO_3^-/NH_4^+$ aerosols compared with those from EQUILIB and KEQUILIB by Saxena et al.[87]

Further development of comprehensive models to simulate the dynamics of both equilibrium and non-equilibrium aspects of sulfate/nitrate/ammonium aerosols by nucleation and gas-to-particle conversion was reported by Pilinis et al.[76] The recognition of the importance of the NH_4Cl/particulate Cl^- system[77,78] and of the variability of chemical composition with particle size led to an improved version of EQUILIB incorporating chloride equilibria and size-dependent composition called the sectional equilibrium model (SEQUILIB).[74] This model is considered the state-of-the-art in describing changes in aerosol composition during transport due to formation of new aerosol and to changes in atmospheric conditions which affect the position of gas-aerosol equilibria. It has also been incorporated into a more complex Eulerian model to study the evolution of size distributions and chemical composition of aerosols in the South Coast Air Basin.[75]

Although the potential for kinetic limitations to the attainment of gas-aerosol equilibria in the ambient atmosphere has been realized for some years,[38] only recently have explicit model treatments been attempted[93,118] to determine under what conditions these kinetic limitations are likely to be important. This topic will be discussed in more detail in Section 6. The significance of vertical profiles in gaseous concentrations (including water vapor) and temperature in altering gas-aerosol equilibria and requiring reequilibration during boundary layer mixing was recognized following the report of experimental profiles by Tanner et al.[114] and by Erisman et al.[29] The vertical resolution of current Eulerian models allows treatment of this phenomenon, if the kinetic constraints to re-equilibration are not too severe.

In a related area, the effects of dry deposition on the ammonium nitrate–nitric acid, ammonia equilibrium (and vice versa) may be profound, given the large differences in reported dry deposition velocities of gases and aerosols, and the uncertainty as to the direction of NH_3 fluxes in many areas. These effects have recently received significant attention in both experimental studies[47,51] and model formulations.[13]

5. AMBIENT STUDIES OF ATMOSPHERIC ACID-BASE PHASE EQUILIBRIA

As noted above in Section 4, most modeling studies have incorporated comparison of model-predicted and measured ambient aerosol composition, and all ambient field studies have used a "model" equilibrium expression to compare measured concentrations of ammonia and nitric acid with, at minimum, partial pressure products of $[NH_3]$ $[HNO_3]$ predicted from water activity and temperature-dependent equilibrium constants. In this section we will examine several examples of these comparisons with the goal of understanding under what atmospheric conditions the important acidic and basic gases are at equilibrium with ammonium salts in atmospheric aerosols.

Stelson et al.[98] made the first comparison of calculated equilibrium concentration products (based on two different reported values of the temperature-dependent equilibrium constant for Reaction 12), with hourly experimental data from the South Coast Air Basin. Data for the lowest relative humidity (RH) (<60%) are scattered, and the data for RH >90% are generally higher than predicted. Doyle et al.[24] also reported data from Riverside, CA for 54 cases in which short-time-resolution $[NH_3]$ and $[HNO_3]$ were detectable by the FTIR spectroscopic method used for the gas-phase measurements. Good agreement was obtained for some cases, with experimental concentration products generally less than calculated for higher temperatures.

Cadle et al.[15] obtained filter and denuder measurements of ammonia simultaneously with nitric acid data by two different filter techniques. Much of the data was for 24-h periods during which equilibrium conditions can undergo large changes. For the 4-h summer data in Louisiana (22–32°C), the observed concentration product, $[NH_3]$ $[HNO_3]$, was less than the equilibrium value, likely indicating that no ammonium nitrate phase was present. In contrast, for the winter data in Denver (− 10 to 5°C) the concentration product exceeded the equilibrium value, which the authors attributed to local ammonia sources or to kinetic limitations on the attainment of equilibrium.

Tanner[110] reported data from an east coast U.S. location for ammonia using a scrubber/fluorescence derivatization technique,[1] and nitric acid by a diffusion denuder difference technique,[35] both with 15-min time resolution to minimize changes in equilibrium conditions. The observed concentration products were in good agreement during daytime hours, but during the nighttime periods, NH_3 concentration was higher than predicted by up to an order of magnitude. There are surface sources of ammonia near the sampling site, and surface layer inversions may have restricted vertical mixing. Nevertheless, the observation of high $[NH_3]$ values coexisting with measured acid-to-sulfate molar ratios of the order of unity suggests that kinetic limitations to neutralization may also have influenced the results.

Harrison and Pio[44] used filter-pack sampling procedures to measure HNO_3 and NH_3 over 24-h periods, predominantly under cool, humid conditions. Comparison of $[NH_3]$ $[HNO_3]$ concentration products with theoretical prediction showed rather good agreement except at low ambient temperatures. Ferm[33] has reported measurements in Sweden in which concentration products were generally at or below the theoretical values. This suggests that for this site, ambient concentrations of NH_3 and HNO_3 are frequently too low to maintain a particulate ammonium nitrate phase — either it is not formed or evaporates thereafter.

Hildemann et al.[49] compared equilibrium calculations for the "pure" ammonium nitrate–nitric acid, ammonia system and for a system containing sulfate and other ions with 2- and 4-h filter pack measurements of $[NH_3]$ and $[HNO_3]$ at several sites in the Los Angeles basin. For inland sites, the calculated concentration products, assuming equilibrium with a pure ammonium nitrate phase, were an upper limit to those observed for the prevailing conditions, i.e., equilibrium generally prevailed when a pure ammonium nitrate phase was predicted to be present. For coastal sites, observations were often much lower than predicted, but good agreement was obtained if nitrate was assumed to be present in an internal, size-segregated mixture in non-volatile form on coarse particles, and as a semi-volatile mixture with ammonium sulfate in fine particles.

Lewin et al.[63] reported data for ammonia by a denuder collection method and nitric acid collected on nylon filters downstream from the particulate filter. Sampling was at an inland, northeastern U.S. site in wintertime. Good agreement was obtained for most sampling periods, except at temperatures $<0°C$. In an extensive study of gas-aerosol equilibria during humid wintertime conditions in the Central Valley of California, Jacob et al.[52] reported quantitative agreement of observations of $[NH_3]_g[HNO_3]_g$ with a dual filter pack method, as compared to calculated equilibrium values using a comprehensive thermodynamic model for the atmospheric H_2SO_4-HNO_3-NH_3-H_2O system. Application of this approach to gas phase-fog water equilibria gave satisfactory results if atmospheric conditions remained unchanging during the sampling period.

Erisman et al.[29] conducted a study to determine the profiles of aerosol species $NH_3(g)$ and $HNO_3(g)$ at a 237-m tower in the Netherlands, and to investigate whether the profiles were consistent with phase equilibria in the atmospheric H_2SO_4-HNO_3-NH_3-H_2O system. Over a wide range of conditions the data from 12-h day and night sampling showed good agreement with calculations based on the phase equilibrium assumption, except for $T < 0°C$ and RH $>80\%$, at which times the observed concentration products exceeded the equilibrium values.

A detailed study of both $[NH_3]$ $[HNO_3]$ and $[NH_3]$ $[HCl]$ concentration products was conducted by Allen et al.[2] using both filter pack methods at U.K. sites and denuder procedures at a site in the Netherlands. For both denuder and filter pack sampling procedures, the results showed a number of common features for both ammonia-nitric acid and ammonia-HCl concentration products. Twenty-four-hour averaged measurements generally gave better agreement with theory than 3- or 12-h data. Whereas the agreement between equilibrium predictions and the measurements was often good, some appreciable discrepancies were seen, especially at high relative humidities and low temperatures. As pointed out subsequently by Wexler and Seinfeld,[118] these are precisely the conditions under which kinetic factors severely reduce the rate of achievement of equilibrium (see Section 6). Ratios of [HCl]/[HNO$_3$] were also calculated and found to be in reasonable agreement with predictions of chemical thermodynamics for the NH$_4$NO$_3$ and NH$_4$Cl pure systems, ignoring other reactions, such as that between HNO$_3$ and NaCl to form HCl, which, if rapid, could profoundly perturb the system.

6. KINETIC LIMITATIONS TO ATMOSPHERIC PHASE EQUILIBRIUM

In Section 4 we have noted that observed deviations of the concentration products, [HA] [NH$_3$], measured in the atmosphere, from those predicted by chemical thermodynamics are frequently consistent with a kinetic limitation to the achievement of equilibrium. In the following section the theoretical basis for such a kinetic limitation will be described, and the rather sparse experimental evidence from laboratory studies reviewed.

6.1 Fundamental Considerations

Both ammonium nitrate and ammonium chloride are deliquescent salts; at 25°C they deliquesce at 62 and 77% relative humidity, respectively.[98,108] In internally mixed aerosol particles (i.e., particles containing uniform mixtures of more than one salt) they may deliquesce at lower humidities. Thus, because of hysteresis effects resulting from the slow kinetics of evaporation and especially crystallization, aqueous droplets may exist at humidities well below the deliquescence points of component salts.[36,109]

Let us then consider an aerosol particle composed of a single volatile solid or liquid phase, A, from which may be established a temperature-dependent equilibrium between A in particulate and gaseous phases (Reaction 17).

$$A \text{ (aerosol)} \rightleftarrows A \text{ (gas)} \qquad (17)$$

At equilibrium, the rate of transfer of A molecules into the particle from the gas phase equals the rate of loss of A from the particle surface, and the equilibrium vapor pressure above the particle surface, p_d, equals the gas phase partial pressure of A, p_A.

If the conditions depart from equilibrium, the particle may grow by condensation ($p_A > p_d$), or shrink by evaporation ($p_A < p_d$). The rate of evaporation or condensation is a function of particle size for two reasons. First, different kinetic laws govern the growth and/or condensation of large and very small aerosol particles (Kelvin effect). Second, for the two particle-size ranges the rates vary with particle diameter, d_p, via different functions. For particles much smaller than the mean free path of gas molecules (λ, about 0.066 μm at 20°C and 1 atm), the rates are determined by individual molecular bombardments. In this region controlled by gas kinetics in which $d_p \ll \lambda$, the rate of change of particle volume is given by (18):[36]

$$\frac{dV}{dt} = \frac{\alpha \pi V_m d_p^2 (p_A - p_d)}{(2\pi \mu kT)^{1/2}} \tag{18}$$

where V_m = molar volume
              μ = mass of molecule in the gas phase
              k   = Boltzmann's constant
              T   = absolute temperature

The parameter, α, is the accommodation or sticking coefficient, which expresses the probability of a colliding molecule to cross the phase boundary.

For particles much larger than the mean free path ($d_p \gg \lambda$), mass transfer is the rate-limiting process, and the analogous rate law is (19):[36]

$$\frac{dV}{dt} = \frac{2\pi D V_m d_p (p_A - p_d)}{kT} \tag{19}$$

where D = gaseous diffusion coefficient, $cm^2 \cdot s^{-1}$. For particles with $d_p \approx \lambda$, an intermediate, particle-diameter-dependent behavior pertains, for which descriptive equations have been proposed.[36]

6.2 Evaporation and Condensation of Solid Ammonium Salt Aerosol

The case of ammonium salt aerosols is considerably more complex than the simple rate behavior outlined above. At least three distinct chemical mechanisms may be postulated for growth of NH_4NO_3 or NH_4Cl (both termed NH_4A). Homogeneous gas phase reaction may occur to form NH_4A monomer which condenses onto the crystal surface (20). Alternately, individual

$$NH_3(g) + HA(g) \rightleftarrows NH_4A(g) \rightleftarrows NH_4A(s) \tag{20}$$

precursor molecules may be incorporated onto vacant crystal sites (21–24), where (a) denotes

$$HA(g) \rightleftarrows HA(a) \tag{21}$$

$$NH_3(g) + HA(a) \rightleftarrows NH_4A(s) \tag{22}$$

$$NH_3(g) \rightleftarrows NH_3(a) \tag{23}$$

$$NH_3(a) + HA(g) \rightleftarrows NH_4A(s) \qquad (24)$$

an adsorbed component. Finally, the reaction may be bimolecular on the particle surface (25), in which ammonia and HA are first adsorbed, then

$$NH_3(g) + HA(g) + -s-s- \rightleftarrows$$
$$NH_3 \; HA \rightleftarrows NH_4A \rightleftarrows -s-s-$$
$$\begin{matrix} | & | & | & | & | & | \\ -s- & -s- & -s--s- & -s-s- \end{matrix} \qquad (25)$$

react, and are finally incorporated into the NH_4A crystal structure. Different *chemical* kinetic expressions apply to each of these processes.[43]

The existence or not of monomeric NH_4A is critical to validating the first mechanism. Experimental results[20,30,31,41,59,93,117] as well as theoretical studies[17,18] have proven inconclusive on this point. The weight of evidence, however, is against the existence of appreciable concentrations of gaseous NH_4A monomer. The third mechanism, postulated by Schultz and Dekker,[90] gives reasonable agreement with experimental data[97] and gives rate dependencies consistent with experimental data for aerosols.

There have been a number of experimental studies of the growth and evaporation of ammonium salt aerosols. The theoretical expressions for physically limited processes (18 and 19) predict equal rates for evaporation and condensation for a given particle size and a given difference, p_A-p_d. Even if different chemical mechanisms were to apply to the evaporation and condensation of NH_4A, their rates must be nearly equal at or close to equilibrium.

The published studies of NH_4A aerosol particle growth[20,66,72] and evaporation[60,82,103] have been reviewed by Harrison and MacKenzie.[43] In addition, new data on the evaporation kinetics of NH_4Cl have been reported by Tang and Munkelwitz.[108] Allowing for differences in partial pressure driving force and recognizing the uncertainties in the measured partial pressures of solute gases required for estimates of rates, the evaporation rate studies appear to show faster rates than for condensation, despite the extreme difficulties in analyzing rates of condensation.

The evaporation studies gave rates of similar magnitude which were slower than the predictions based on purely physical processes (18, 19) and which had a different dependence on particle size. It was therefore concluded that a *chemical* kinetic rate limitation restricted the evaporation process. Wexler and Seinfeld,[118] in their consideration of this topic, also reach the conclusion that uptake into particles is substantially in the form of $NH_3(g)$ and $HA(g)$, rather than $NH_4A(g)$, and that both gas phase diffusion and surface reaction play a role in controlling the rates of ammonium salt condensation/evaporation processes.

6.3 Evaporation and Condensation in Solution Droplets

Evaporation/condensation in aqueous aerosol droplets is a more difficult problem than for solid aerosols. In the case of solid particles the time taken to reach equilibrium between gas phase partial pressure, p_A, and surface partial pressure, p_d, is simply that needed for sufficient inter-phase transfer to occur to alter p_A to equal p_d, the latter being invariant for a given temperature. In the case of liquid aerosol droplets, uptake or loss of solute affects the surface vapor pressure, p_d, tending to further equalize p_d and p_A. A further complication occurs if and when the liquid water content of the particle also changes in an attempt to maintain its molality and thereby its equilibrium state with atmospheric water vapor.

Wexler and Seinfeld[118] have recently calculated characteristic times for evaporation and condensation processes in solid particles and aerosol droplets. In the former case, the characteristic times depend upon transport and surface reaction parameters. Large particles ($d_p \approx 20$ μm) and low atmospheric concentrations (about 30 μg/m^3) correspond to long equilibration times, of the order of 1 day. This situation approximates to coastal conditions in southern California. For small particles (~ 0.1 μm) and high aerosol loadings (300 μg/m^3), such as might occur inland in the Los Angeles Basin, the characteristic time reduces to <1 min. In the case of liquid droplets, assuming no change in liquid water content occurs, the equilibrium constant between dissolved and gas-phase volatiles enters the calculation. Characteristic times depend most critically on particle size and ambient temperature. Low temperatures (~ 280 K) and large particles (~ 20 μm) again correspond to long equilibration times of about 1 day, whereas small particles (~ 0.2 μm) and high temperatures (~ 310 K) correspond to equilibrium times of <1 min.

6.4 Comparison of Kinetic Expressions with Atmospheric Measurements

After examining possible rate expressions of NH$_4$A evaporation/condensation processes, Harrison and MacKenzie[43] developed a simple numerical model of the atmospheric NH$_3$/HNO$_3$/NH$_4$NO$_3$/H$_2$O system containing parameterizations of the rates of gas/particle transfer, based upon laboratory studies of evaporation of NH$_4$A from both dry particles and solution droplets. The results were analyzed in terms of the deviation from thermodynamic equilibrium, which depended upon the atmospheric humidity, temperature, and aerosol concentrations, but were evaluated for conditions similar to those observed in the ambient atmosphere.

Wexler and Seinfeld's[109] calculations were restricted to characteristic equilibration times and estimation of the influence of equilibration processes upon atmospheric size distributions for ammonium salt aerosols. They conclude that, for atmospheric conditions and particle concentrations typical of inland Los Angeles Basin sites, equilibration times are short and that the NH$_3$/HA/ NH$_4$A systems are close to equilibrium. At the lower concentrations of larger-particle-size aerosols characteristically present at low temperatures and high humidities, equilibration times are large, which leads to significant departures from equilibrium. It is precisely these latter atmospheric conditions (lower concentrations, low temperature, high relative humidity) which were identified by Allen et al.[2] (and indeed, by Tanner[110] and Cadle et al.[15]) as being associated with the greatest departures from phase equilibrium.

7. CONCLUSIONS AND RESEARCH NEEDS

The influence of the equilibrium-driven distribution of certain key chemical species in the atmosphere between gaseous, aerosol particle, and liquid water droplets is of profound significance in several areas of atmospheric chemistry and physics. Inter-phase equilibration under dynamically changing atmospheric conditions affects the means by which one can measure atmospheric species, and techniques for minimizing measurement artifacts for phase equilibrated species were discussed in Section 3. The existence of multi-phase equilibria can also profoundly affect the atmospheric lifetimes of these species, because species in different phases have different transport, scavenging, and deposition properties. Thus, the range of impact of atmospheric emission sources of strong acid species and their ammonium salts (emitted directly or formed in secondary processes after their emission) can be influenced.

We have not discussed other atmospheric species, such as weak organic acids, carbonyl compounds, and PAHs, which also are reported to be distributed between aerosol and gaseous phases under atmospheric conditions, but clearly the analytical limitations and the transport considerations we mention apply to these species as well. Increased attention must be extended to all aspects of the sampling process to insure that reported concentrations of atmospheric species are specific to a phase and that the extent of re-equilibration during the sampling process has been quantified. This seems to the authors to require that additional knowledge of the kinetics of equilibration processes involving atmospheric strong acids and their ammonium salts be obtained. One additional recommendation is that further studies of the vertical distribution of strong acids and bases in the lowest layers of the atmosphere be obtained. Limited data suggest that feasible kinetic studies of the rates of strong acid neutralization in the ambient atmosphere can be conducted with the existing measurement science if appropriate attention is paid to the boundary layer mixing processes which may also be kinetically limiting under some circumstances.

REFERENCES

1. Abbas, R. and R. L. Tanner. "Continuous Determination of Gaseous Ammonia in the Ambient Atmosphere Using Fluorescence Derivatization," *Atmos. Environ.* 15:277–281 (1981).
2. Allen, A. G., R. M. Harrison, and J. W. Erisman. "Field Measurements of the Dissociation of Ammonium Nitrate and Ammonium Chloride Aerosols," *Atmos. Environ.* 23:1591–1599 (1989).
3. Anlauf, K. G., P. Fellin, H. A. Wiebe, H. I. Schiff, G. I. Mackay, R. S. Braman, and R. Gilbert. "A Comparison of Three Methods for Measurements of Atmospheric Nitric Acid and Aerosol Nitrate and Ammonium," *Atmos. Environ.* 19:325–333. (1985).

4. Anlauf, K. G., D. C. MacTavish, and H. A. Wiebe. "Measurements of Atmospheric Nitric Acid by Filter Method and Comparison With the Tunable Diode Laser and Other Methods," *Atmos. Environ.* 22:1579–1586 (1988).

5. Appel, B. R., Y. Tokiwa, E. L. Kothny, R. Wu, and V. Povard. "Evaluation of Procedures for Measuring Atmospheric Nitric Acid and Ammonia," *Atmos. Environ.* 22:1565–1573 (1988).

6. Ayers, G. P., J. P. Ivey, and H. S. Goodman. "Sulfate and Methanesulfonate in the Maritime Aerosol at Cape Grim, Tasmania," *J. Atmos. Chem.* 4:173–185 (1986).

7. Bassett, M. and J. H. Seinfeld. "Atmospheric Equilibrium Model of Sulfate and Nitrate Aerosols," *Atmos. Environ.* 17:2237–2252 (1983).

8. Bassett, M. and J. H. Seinfeld. "Atmospheric Equilibrium Model of Sulfate and Nitrate Aerosols. II. Particle Size Analysis," *Atmos. Environ.* 18:1163–1170 (1984).

9. Biermann, H. W., E. C. Tuazon, A. M. Winer, T. J. Wallington, and J. N. Pitts, Jr., "Simultaneous Absolute Measurements of Gaseous Nitrogen Species in Urban Ambient Air by Long-Pathlength Infrared and Ultraviolet-Visible Spectroscopy," *Atmos. Environ.* 22:1545–1554 (1988).

10. Brassington, D. J., "Measurements of Atmospheric HCl and NH_3 with a Mobile Tunable Diode Laser System," Rep. RD/L/3456/R89, Central Electricity Generating Board, Leatherhead, Surrey, U.K., 1990.

11. Brosset, C., K. Andreasson, and M. Ferm. "The Nature and Possible Origin of Acid Particles Observed at the Swedish West Coast," *Atmos. Environ.* 9:631–642 (1975).

12. Brosset, C. and M. Ferm. "Man-Made Airborne Acidity and Its Determination," *Atmos. Environ.* 12:909–916 (1978).

13. Brost, R. A., A. C. Delany, and B. J. Huebert. "Numerical Modeling of Concentrations and Fluxes of HNO_3, NH_3, and NH_4NO_3 near the Surface," *J. Geophys. Res.* 93:7137–7152 (1988).

14. Brown, S., M. C. Dangler, S. R. Burke, S. V. Hering, and D. T. Allen. "Direct Fourier Transform Infrared Analysis of Size-Segregated Aerosols: Results from the Carbonaceous Species Methods Intercomparison Study," *Aerosol Sci. Technol.* 12:172–181 (1990).

15. Cadle, S. H., R. J. Countess, and N. A. Kelly. "Nitric Acid and Ammonia in Urban and Rural Locations," *Atmos. Environ.* 16:2501–2506 (1982).

16. Calvert, J. G., A. Lazrus, G. L. Kok, B. B. Heikes, J. G. Walega, J. Lind, and C. A. Cantrell. "Chemical Mechanisms of Acid Generation in the Troposphere," *Nature* 317:27–35 (1985).

17. Clementi, E., "Study of the Electronic Structure of Molecules. II. Wavefunctions for the NH_3 + HCl → NH_4Cl Reaction," *J. Chem. Phys.* 46:3851–3875 (1967).

18. Clementi, E. and J. N. Gayles. "Study of the Electronic Structure of Molecules. VII. Inner and Outer Complex in the NH_4Cl Formation from NH_3 and HCl," *J. Chem. Phys.* 47:3837–3841 (1967).

19. Commins, B. T. "Determination of Particulate Acid in Town Air," *Analyst* 88:364–367 (1963).

20. Countess, R. J. and J. Heicklen. "Kinetics of Particle Growth. II. Kinetics of the Reaction of Ammonia with Hydrogen Chloride and the Growth of Particulate Ammonium Chloride," *J. Phys. Chem.* 77:444–447 (1973).

21. Dasch, J. M., S. E. Cadle, K. G. Kennedy, and P. A. Mulawa. "Comparison of Annular Denuders and Filter Packs for Atmospheric Sampling," *Atmos. Environ.* 23:2775–2782 (1989).

22. Daum, P. H., T. J. Kelly, S. E. Schwartz, and L. Newman. "Measurements of the Chemical Composition of Stratiform Clouds," *Atmos. Environ.* 18:2671–2684 (1984).

23. Daum, P. H. and D. F. Leahy. "The Brookhaven National Laboratory Filter Pack System for Collection and Determination of Air Pollutants," Rep. BNL 31381-R2, Brookhaven National Laboratory, Upton, NY, 1985.

24. Doyle, G. J., E. C. Tuazon, R. A. Graham, T. M. Mischke, A. M. Winer, and J. N. Pitts, Jr. "Simultaneous Concentrations of Ammonia and Nitric Acid in a Polluted Atmosphere and Their Equilibrium Relationship to Particulate Ammonium Nitrate," *Environ. Sci. Technol.* 13:1416–1419 (1979).

25. Durham, J. L., L. L. Spiller, and T. G. Ellestad. "Nitric Acid-Nitrate Aerosol Measurements by a Diffusion Denuder. A Performance Evaluation," *Atmos. Environ.* 21:589–599 (1987).

26. Durham, J. L., W. E. Wilson, and E. B. Bailey. "Application of an SO_2 Denuder for Continuous Measurement of Sulfur in Submicrometric Aerosols," *Atmos. Environ.* 12:883–886 (1978).

27. Eatough, D. J., V. F. White, L. D. Hansen, N. L. Eatough, and J. L. Cheney. "Identification of Gas-Phase Dimethyl Sulfate and Monomethyl Hydrogen Sulfate in the Los Angeles Atmosphere," *Environ. Sci. Technol.* 20:867–872 (1986).

28. Eatough, N. L., S. McGregor, E. Lewis, D. J. Eatough, A. A. Huang, and E. C. Ellis. "Comparison of Six Denuder Methods and a Filter Pack for the Collection of Ambient $HNO_3(g)$, $HNO_2(g)$ and SO_2 in 1985 NSMC study," *Atmos. Environ.* 22:1601–1618 (1988).

29. Erisman, J. W., A. W. M. Vermetten, W. A. H. Asman, A. Waijers-IJpelaan, and J. Slanina. "Vertical Distribution of Gases and Aerosols: Behavior of Ammonia and Related Components in the Lower Atmosphere," *Atmos. Environ.* 14:1153–1160 (1988).

30. Feick, G. "The Dissociation Pressure and Free Energy of Formation of Ammonium Nitrate," *J. Am. Chem. Soc.* 76:5858–5860 (1954).

31. Feick, G. and R. M. Hainer. "On the Thermal Decomposition of Ammonium Nitrate. Steady-State Reaction Temperature and Reaction Rate," *J. Am. Chem. Soc.* 76:5860–5863 (1954).

32. Ferm, M. "Method for Determination of Atmospheric Ammonia, *Atmos. Environ.* 13:1388–1393 (1979).

33. Ferm, M. "Concentration Measurements and Equilibrium Studies of Ammonium, Nitrate and Sulphate Species in Air and Precipitation," Rep. No. XXX, University of Goteborg, Dept. of Inorganic Chemistry, Goteborg, Sweden (1986).

34. Ferm, M., H. Areskoug, J. E. Hanssen, G. Hilbert, and H. Lattila. "Field Inter Comparison of Measurement Techniques for Total NH_4^+ and Total NO_3^- in Ambient Air," *Atmos. Environ.* 22:2275–2281 (1988).

35. Forrest, J., D. J. Spandau, R. L. Tanner, and L. Newman. "Determination of Atmospheric Nitrate and Nitric Acid Employing a Diffusion Denuder with a Filter Pack," *Atmos. Environ.* 16:1473–1485 (1982).

36. Friedlander, S. K. *Smoke, Dust and Haze.* (Chichester, U.K.: John Wiley & Sons, 1977).

37. Gardner, J. A., L. R. Watson, Y. G. Adewuyi, P. Davidovits, M. S. Zahniser, Worsnop, D. R., and C. E. Kolb. "Measurement of the Mass Accomodation Coefficient of $SO_2(g)$ on Water Droplets," *J. Geophys. Res.* 92:10887–10895 (1987).

38. Gill, P. S., T. E. Graedel, and C. J. Weschler. "Organic Films on Atmospheric Aerosol Particles, Fog Droplets, Cloud Droplets, Raindrops, and Snowflakes," *Rev. Geophys. Space Phys.* 21:903–920 (1983).

39. Gleason, J. F. and C. J. Howard. "Temperature Dependence of the Gas-Phase Reaction, $HOSO_2 + O_2 \rightarrow HO_2 + SO_3$," *J. Phys. Chem.* 92:3414–3417 (1988).

40. Goklany, I. M., F. Hoffnagle, and E. A. Brackbill. "Acidic and Total Primary Sulfates: Development of Emission Factors for Major Stationary Combustion Sources," *J. Air Pollut. Control Assoc.* 34:123–134 (1984).

41. Goldfinger, P. and G. Verhaegen. "Stability of the Gaseous Ammonium Chloride Molecule," *J. Chem. Phys.* 50:1467–1471 (1969).

42. Harris, G. W., G. I. Mackay, T. Iguchi, L. K. Mayne, and H. I. Schiff. "Measurements of Formaldehyde in the Troposphere by Tunable Diode Laser Spectroscopy," *J. Atmos. Chem.* 8:119–137 (1989).

43. Harrison, R. M. and A. R. MacKenzie. "A Numerical Simulation of Kinetic Constraints Upon Achievement of the Ammonium Nitrate Dissociation Equilibrium in the Troposphere," *Atmos. Environ.* 24A:91–102 (1990).

44. Harrison, R. M. and C. A. Pio. "An Investigation of the Atmospheric HNO_3-NH_3-NH_4NO_3 Equilibrium Relationship in a Cool Humid Climate," *Tellus* 35B:155–159 (1983).

45. Harrison, R. M., W. T. Sturges, A. M. N. Kitto, and Y. Li. "Kinetics of Evaporation of Ammonium Chloride and Ammonium Nitrate Aerosol," *Atmos. Environ.* 24A:1883–1888 (1990).

46. Harrison, R. M. and A. -M. N. Kitto. "Field Intercomparison of Filter Pack and Denuder Sampling Methods for Reactive Gaseous and Particulate Pollutants," *Atmos. Environ.* 24A:in press (1990).

47. Harrison, R. M., S. Rapsomanikis, and A. Turnbull. "Land-Surface Exchange in a Chemically-Reactive System; Surface Fluxes of HNO_3, HCl and NH_3," *Atmos. Environ.* 23:1795–1800 (1989).

48. Hering, S. V., D. R. Lawson, et al. "The Nitric Acid Shootout: Field Comparison of Measurement Methods," *Atmos. Environ.* 22:1519–1540 (1988).

49. Hildemann, L. M., A. G. Russell, and G. R. Cass. "Ammonia and Nitric Acid Concentrations in Equilibrium with Atmospheric Aerosols: Experiment vs. Theory," *Atmos. Environ.* 18:1737–1750 (1984).

50. Hoffman, M. R. and S. D. Boyce. "Catalytic Autooxidation of Aqueous Sulfur Dioxide in Relationship to Atmospheric Systems," *Trace Atmospheric Constituents. Properties, Transformations and Fates*, S. E. Schwartz, Ed. (New York: John Wiley & Sons, 1983) pp. 147–189.

51. Huebert, B. J., W. T. Luke, A. C. Delany, and R. A. Brost. "Measurements of Concentrations and Dry Surface Fluxes of Atmospheric Nitrates in the Presence of Ammonia," *J. Geophys. Res.* 93:7127–7136 (1988).

52. Jacob, D. J., J. M. Waldman, J. W. Munger, and M. R. Hoffman. "The H_2SO_4-HNO_3-NH_3 System at High Humidities and in Fogs. II. Comparison of Field Data with Thermodynamic Calculations," *J. Geophys. Res.* 91:1089–1096 (1986).

53. Johnson, S. A. and R. Kumar. "Current Acid Aerosol Measurement Techniques," in Acid Aerosol Measurement Workshop, R. J. Tropp, Ed., U. S. EPA Rep. EPA/600/9–89/056, Research Triangle Park, NC, 1989, pp. 36–38.

54. Johnson, S. A. and R. Kumar. "Composition and Spectral Characteristics of Ambient Aerosol at Mauna Loa Observatory," *J. Geophys. Res.* 96: 5379–5386 (1991).

55. Junge, C. and G. Scheich. "Determination of the Acid Content of Aerosol Particles," *Atmos. Environ.* 5:165–175 (1971).

56. Kelly, T. J. "Trace Gas and Aerosol Measurements at Whiteface Mountain, New York," ESEERCO Rep. EP83–13, Empire State Electric Energy Research Corp., New York, (1985).

57. Knapp, K. T., J. L. Durham, and T. G. Ellestad. "Pollutant Sampler for Measurements of Atmospheric Acidic Dry Deposition," *Environ. Sci. Technol.* 20:633–637 (1986).

58. Koutrakis, P., J. M. Wolfson, J. L. Slater, M. Brauer, J. S. Spengler, R. K. Stevens, and C. L. Stone. "Evaluation of an Annular Denuder/Filter Pack System to Collect Acidic Aerosols and Gases," *Environ. Sci. Technol.* 22:1463–1468 (1988).

59. de Kruif, C. G. "The Vapor Phase Dissocation of Ammonium Salts: Ammonium Bicarbonate," *J. Chem. Phys.* 77:6247–6250 (1982).

60. Larson, T. V. and G. S. Taylor. "On the Evaporation of Ammonium Nitrate Aerosol," *Atmos. Environ.* 17:2489–2495 (1983).

61. Lee, Y.-H. and C. Brosset. "Interaction of Gases with Sulfuric Acid Aerosol in the Atmosphere," Rep. B-504, Swedish Water and Air Pollution Research Institute, Gothenburg, Sweden, 1979.

62. Lee, Y.-N., J. Shen, P. J. Klotz, S. E. Schwartz, and L. Newman. "Kinetics of Hydrogen Peroxide — Sulfur (IV) Reaction in Rainwater Collected at a Northeastern U.S. Site," *J. Geophys. Res.* 91:13264–13274 (1986).

63. Lewin, E. E., R. G. de Pena, and J. P. Shimshock. "Atmospheric Gas and Particle Measurements at a Rural Northeastern U.S. Site," *Atmos. Environ.* 20:59–70 (1986).

64. Ligocki, M. and J. F. Pankow. "Measurements of the Gas/Particle Distributions of Organic Compounds," *Environ. Sci. Technol.* 23:75–83 (1989).

65. Lind, J. A. and G. L. Kok. "Henry's Law Determinations for Aqueous Solutions of Hydrogen Peroxide, Methylhydroperoxide, and Peroxyacetic Acid," *J. Geophy. Res.* 91:7889–7895 (1986).

66. Luria, M. and B. Cohen. "Kinetics of Gas to Particle Conversion in the NH_3-HCl system," *Atmos. Environ.* 14:665–670 (1980).

67. McMurry, P. H. and J. C. Wilson. "Growth Laws for the Formation of Secondary Ambient Aerosols: Implications for Chemical Conversion Mechanisms," *Atmos. Environ.* 16:121–134 (1982).

68. Martin, L. R. "Kinetic Studies of Sulfite Oxidation in Aqueous Solution," in *SO_2, NO and NO_2 Oxidation Mechanisms: Atmospheric Considerations*, J. G. Calvert, Ed. (Boston: Butterworths, 1984), pp. 63–100.

69. Mulawa, P. and S. H. Cadle. "A Comparison of Nitric Acid and Particulate Nitrate Measurements by the Penetration and Denuder Difference Methods," *Atmos. Environ.* 19:1317–1324 (1985).

70. National Acid Precipitation Assessment Program. C. N. Herrick and J. L. Kulp, Eds. Interim Assessment: The Causes and Effects of Acidic Deposition, Vol. II, U.S. Government Printing Office, Washington, D.C. (1987).

71. Norton, R. G., J. M. Roberts, and G. J. Huebert. "Tropospheric Oxalate," *Geophys. Res. Lett.* 10:517–520 (1983).
72. Olszyna, K. J., R. G. de Pena, M. Luria, and J. Heicklen. "Kinetics of Particle Growth IV-NH_4NO_3 from the NH_3-O_3 Reaction Revisited," *J. Aerosol Sci.* 5:421–434 (1974).
73. Penkett, S. A., B. M. R. Jones, K. A. Brice, and A. E. J. Eggleton, "The Importance of Atmospheric Ozone and Hydrogen Peroxide in Oxidising Sulphur Dioxide in Cloud and Rainwater," *Atmos. Environ.* 13:123–137 (1979).
74. Pilinis, C. and J. H. Seinfeld. "Continued Development of a General Equilibrium Model for Inorganic Multicomponent Atmospheric Aerosols," *Atmos. Environ.* 21:2453–2466 (1987).
75. Pilinis, C. and J. H. Seinfeld. "Development and Evaluation of an Eulerian Photochemical Gas-Aerosol Model," *Atmos. Environ.* 22:1985–2001 (1988).
76. Pilinis, C., J. H. Seinfeld, and C. Seigneur. "Mathematical Modeling of the Dynamics of Multicomponent Atmospheric Aerosols," *Atmos. Environ.* 21:943–955 (1987).
77. Pio, C. A. and R. M. Harrison. "The Equilibrium of Ammonium Chloride Aerosol with Gaseous Hydrochloric Acid and Ammonia Under Tropospheric Conditions," *Atmos. Environ.* 21:1243–1246 (1987).
78. Pio, C. A. and R. M. Harrison. "Vapour Pressure of Ammonium Chloride Aerosols: Effect of Temperature and Humidity," *Atmos. Environ.* 21:2711–2715 (1987).
79. Possanzini, M., A. Febo, and A. Liberti. "New Design of a High-Performance Denuder for the Sampling of Atmospheric Pollutants," *Atmos. Environ.* 17:2605–2610 (1983).
80. Pruppacher, H. R. and J. D. Klett. *Microphysics of Clouds and Precipitation.* (Hingham, MA: D. Reidel, 1978).
81. Raynor, G. S. and J. V. Hayes. "Variation in Chemical Wet Deposition with Meteorological Conditions," *Atmos. Environ.* 16:1647–1656 (1982).
82. Richardson, C. B. and R. L. Hightower. "Evaporation of Ammonium Nitrate Particles," *Atmos. Environ.* 21:971–975 (1987).
83. Russell, A. G., K. F. McCue, and G. R. Cass. "Mathematical Modeling of the Formation of Nitrogen-Containing Air Pollutants. I. Evaluation of an Eulerian Photochemical Model," *Environ. Sci. Technol.* 22:263–271 (1988).
84. Russell, A. G., G. J. McRae, and G. R. Cass. "Mathematical Modeling of the Formation and Transport of Ammonium Nitrate Aerosol," *Atmos. Environ.* 17:949–964 (1983).
85. Saltzman, E. S., D. L. Savoie, R. G., Zika, and J. M. Prospero. "Methane Sulfonic Acid in the Marine Atmosphere," *J. Geophys. Res.* 88:10897–10902 (1983).
86. Saxena, P., C. Seigneur, and T. W. Peterson. "Modeling of Multiphase Atmospheric Aerosols," *Atmos. Environ.* 17:1315–1329 (1983).
87. Saxena, P., A. B. Hudischewskyj, C. Signeur, and J. H. Seinfeld. "Comparison of the Performance of the MARS Model with the More Comprehensive EQUILIB and SEQUILIB Models," *Atmos. Environ.* 20:1471–1483 (1986).
88. Scarengelli, F. P. and K. A. Rehme. "Determination of Atmospheric Concentrations of Sulfuric Acid Aerosol by Spectrophotometry, Coulometry and Flame Photometry," *Anal. Chem.* 41:707–713 (1969).

89. Schiff, H. I., D. R. Hastie, G. I. Mackay, T. Iguchi, and B. A. Ridley. "Tunable Diode Laser Systems for Measuring Trace Gases in Tropospheric Air," *Environ. Sci. Technol.* 17:352A–364A (1983).

90. Schultz, R. D. and A. D. Dekker. "The Effect of Physical Adsorption on the Absolute Decomposition Rates of Crystalline Ammonium Chloride and Cupric Sulfate Trihydrate," *J. Phys. Chem.* 60:1095–1100 (1956).

91. Schwartz, S. E. "Gas-Aqueous Reactions of Sulfur and Nitrogen Oxides in Liquid-Water Clouds," in *SO_2, NO, and NO_2 Oxidation Mechanisms: Atmospheric Considerations*, J. G. Calvert, Ed., (Boston: Butterworths, 1984), pp. 173–208.

92. Shaw, R. W., Jr., R. K. Stevens, and J. W. Bowermaster. "Measurements of Atmospheric Nitrate and Nitric Acid: The Denuder Difference Experiment," *Atmos. Environ.* 16:845–853 (1982).

93. Shibata, S. "Structure of Gaseous Ammonium Chloride," *Acta Chem. Scand.* 24:705–706 (1970).

94. Solomon, P. A., S. M. Larson, T. Fall, and G. R. Cass. "Basinwide Nitric Acid and Related Species Concentrations Observed during the Claremont Nitrogen Species Comparison Study," *Atmos. Environ.* 22:1587–1594 (1988).

95. Sonnefeld, W. J., W. H. Zoller, and W. E. May. "Dynamic Coupled-Column Liquid Chromatographic Determination of Ambient Temperature Vapor Pressures of Polynuclear Aromatic Hydrocarbons," *Anal. Chem.* 55:275–280 (1983).

96. Spicer, C. W., J. E. Howes, T. A. Bishop, L. H. Arnold, and R. K. Stevens. "Nitric Acid Measurement Methods: An Intercomparison," *Atmos. Environ.* 16:1487–1500 (1982).

97. Spingler, K. *Z. Physik. Chem.* B52:90 (1942).

98. Stelson, A. W., S. K. Friedlander, and J. H. Seinfeld. "A Note on the Equilibrium Relationship Between Ammonia and Nitric Acid and Particulate Ammonium Nitrate," *Atmos. Environ.* 13:369–371 (1979).

99. Stelson, A. W. and J. H. Seinfeld. "Relative Humidity and Temperature Dependence of the Ammonium Nitrate Dissociation Constant," *Atmos. Environ.* 16:983–992 (1982).

100. Stelson, A. W. and J. H. Seinfeld. "Relative Humidity and pH Dependence of the Vapor Pressure of Ammonium Nitrate-Nitric Acid Solutions at 25°C," *Atmos. Environ.* 16:993–1000 (1982).

101. Stelson, A. W. and J. H. Seinfeld. "Thermodynamic Prediction of the Water Activity, NH_4NO_3 Dissociation Constant, Density and Refractive Index for the NH_4NO_3-$(NH_4)_2SO_4$-H_2O system at 25°C," *Atmos. Environ.* 16:2507–2514 (1982).

102. Stevens, R. K., T. G. Dzubay, G. Russwurm, and D. Rickel. "Sampling and Analysis of Atmospheric Sulfates and Related Species," *Atmos. Environ.* 12:55–68 (1978).

103. Sturges, W. T. and R. M. Harrison. "The Evaporation of Ammonium Chloride Aerosol." in *Aerosols: Their Generation, Behavior and Applications*, Proc. of the 2nd Conf. of the Aerosol Society, U.K. (1988).

104. Tang, I. N. "On the Equilibrium Partial Pressures of Nitric Acid and Ammonia in the Atmosphere," *Atmos. Environ.* 14:819–828 (1980).

105. Tang, I. N. and H. R. Munkelwitz. "Aerosol Growth Studies. III. Ammonium Bisulfate Aerosols in a Moist Atmosphere," *J. Aerosol Sci.* 8:321–330 (1977).

106. Tang, I. N., H. R. Munkelwitz, and J. G. Davis. "Aerosol Growth Studies. IV. Phase Transformation of Mixed Salt Aerosols in a Moist Atmosphere," *J. Aerosol Sci.* 9:505–511 (1978).

107. Tang, I. N., H. R. Munkelwitz, and N. Wang. "Water Activity Measurements in Single Suspended Droplets: the NaCl:H_2O and KCl:H_2O Systems," *J. Colloid Interface Sci.* 114:409–415 (1986).

108. Tang, I. N. and H. R. Munkelwitz. "Evaporation Kinetics of Ammonium Chloride Solution Droplets in Water Vapor," *J. Colloid Interface Sci.* 128:289–295 (1989).

109. Tang, I. N., W. T. Wong, and H. R. Munkelwitz. "The Relative Importance of Atmospheric Sulfates and Nitrates in Visibility Reduction," *Atmos. Environ.* 15:2463–2471 (1981).

110. Tanner, R. L. "An Ambient Experimental Study of Phase Equilibrium in the Atmospheric System: Aerosol H^+, NH_4^+, SO_4^{2-}, NO_3^-–NH_3(g), HNO_3(g)," *Atmos. Environ.* 16:2935–2942 (1982).

111. Tanner, R. L. "The Measurement of Strong Acids in Atmospheric Samples," in *Methods of Air Sampling and Analysis*, 3rd ed., J. P. Lodge, Jr., Ed. (Chelsea, MI: Lewis Publishers, 1989), pp. 703–714.

112. Tanner, R. L., R. Cederwall, R. Garber, D. Leahy, W. Marlow, R. Meyers, M. Phillips, and L. Newman. "Separation and Analysis of Aerosol Sulfate Species at Ambient Concentrations," *Atmos. Environ.* 11:955–966 (1977).

113. Tanner, R. L., T. J. Kelly, D. A. Dezaro, and J. Forrest. "A Comparison of Filter, Denuder, and Real-Time Chemiluminescence Techniques for Nitric Acid Determination in Ambient Air," *Atmos. Environ.* 23:2213–2222 (1989).

114. Tanner, R. L., R. Kumar, and S. A. Johnson. "Vertical Distribution of Aerosol Strong Acid and Sulfate in the Atmosphere," *J. Geophys. Res.* 89:7149–7158 (1984).

115. Tuazon, E. C., A. M. Winer, R. A. Graham, and J. N. Pitts, Jr. "Atmospheric Measurements of Trace Pollutants by Kilometer Pathlength FT-IR Spectroscopy," *Adv. Environ. Sci. Technol.* 10:259–300 (1980).

116. Tuazon, E. C., R. Atkinson, C. N. Plum, A. M. Winer, and J. N. Pitts, Jr. "The Reaction of Gas Phase N_2O_5 with Water Vapor," *Geophys. Res. Lett.* 10:953–956 (1983).

117. Twomey, S. "Nucleation of Ammonium Chloride Particles From Hydrogen Chloride and Ammonia in Air," *J. Chem. Phys.* 31:1684–1685 (1959).

118. Wexler, A. S. and J. H. Seinfeld. "The Distribution of Ammonium Salts Among a Size- and Composition-Dispersed Aerosol," *Atmos. Environ.* 24A:1231–1246 (1990).

119. Wiebe, H. A., K. G. Anlauf, E. C. Tuazon, A. M. Winer, H. W. Biermann, B. R. Appel, P. A. Solomon, G. R. Cass, T. G. Ellestad, K. T. Knapp, E. Peake, C. W. Spicer, and D. R. Lawson. "A Comparison of Measurements of Atmospheric Ammonia by Filter Packs, Transition-Flow Reactors, Simple and Annular Denuders, and Fourier Transform Infrared Spectroscopy," *Atmos. Environ.* 24A:1019–1028 (1990).

120. Zielinska, B. "Diesel-Derived Pollutants: Atmospheric Concentrations, Transport, and Transformation," Draft Rep. 8622.D1. Desert Research Institute, Reno, NV, 1990, 104 pp.

CHAPTER 3

CHARACTERIZATION OF INDIVIDUAL ENVIRONMENTAL PARTICLES BY BEAM TECHNIQUES

C. Xhoffer, L. Wouters, P. Artaxo, A. Van Put, and R. Van Grieken

Department of Chemistry, University of Antwerp, Antwerp, Belgium

TABLE OF CONTENTS

1. Introduction..108

2. Electron Microprobe Analysis 109

3. Particle-Induced X-Ray Emission.............................113

4. Laser Microprobe Mass Spectrometry.........................114

5. Secondary Ion Mass Spectrometry 116

6. Electron Energy Loss Spectroscopy...........................117

7. Fourier Transform Infrared Spectroscopy 120

8.   Micro Raman Spectroscopy.....................................120

9.   X-Ray Photoelectron Spectroscopy 121

10.   Applications 123
      10.1   Marine Aerosols.......................................123
      10.2   Remote "Continental" Aerosols 126
      10.3   Biogenic Aerosols.....................................127
      10.4   Urban Aerosols..128
      10.5   Volcanic and Stratospheric Aerosols 128
      10.6   Industrial and Workplace Aerosols 129
      10.7   Sediments, Suspension, and Estuarine/Riverine
             Particles 131

11.   Conclusions...133

Acknowledgments 134

List of Acronyms..134

References..134

1. INTRODUCTION

Individual particle characterization is nowadays one of the more challenging aspects of microbeam analysis. Several microanalysis beam techniques for individual particle characterization have been developed which can advantageously be applied to particulate environmental samples. Such microanalyses can reveal whether a specific element or compound is uniformly distributed over all the particles of a population or whether it is a component of only a specific group of particles. Sometimes even the element-distribution within a particle can be inferred. In this way it becomes more straightforward to assign particles to specific sources, while more refined information about source mechanisms and heterogeneous surface reactions can often be derived.

This chapter is concerned with techniques that are suited for the analysis of individual environmental particles by means of characteristic physical interactions between electrons, ions or photons, and the specimen. We shall give a brief overview of these techniques and their recent instrumental applications in environmental studies. Some aspects of micro- and surface analytical techniques for environmental studies as well as microanalysis applied to individual environmental particles have previously been reviewed by Grasserbauer.[1,2]

Table 1 gives an overview of the main analytical features of some micro-
and surface analytical beam techniques for the characterization and analysis
of environmental particles.

2. ELECTRON MICROPROBE ANALYSIS

The term "electron microscope" is generally used to describe the conventional transmission instrument that uses an electron beam for image formation.
Since the development in the early 1960s of newer types of electron microscope instruments, where no transmission signals are involved, this classification is somewhat superseded and the term "electron probes" seems more
convenient to use.

The most important types of electron probe instruments used for environmental studies can be classified as follows:

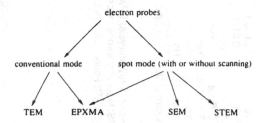

We shall limit the discussion to a brief description of EPXMA and SEM/
EDX since these two techniques are well developed for automated individual
particle characterization (see also Chapter 6 of this book for a discussion of
TEM/EDS).

All these instruments use an electron beam to excite various signals. By
means of coils, the exciting electron beam can be focused to a 10 nm diameter
probe. Both EPXMA and SEM/EDX instruments have one or more X-ray
spectrometer attachments for the measurements of X-ray photons emitted as
a result of the interaction of the electron beam with the specimen atoms. The
emitted X-ray photons can be detected by wavelength- or energy-dispersive
spectrometers (WDX and EDX, respectively). Figure 1 shows a schematic
drawing of an electron microprobe. Similar to PIXE analysis (Section 3), the
detected signals are transformed to electronic pulses and, after amplification,
stored in a multichannel device according to the corresponding wavelength
or energy. The result is a spectrum of intensity vs. wavelength or energy.
Characteristic X-rays are superimposed on a Bremsstrahlung continuum which
is rather intense as compared to, e.g., PIXE. This background is a result of
non-characteristic emissions from incident electrons interacting with the electrostatic field of the atomic nuclei and inner electron shells. In general, the
variety of signals that are available in the EPXMA can rapidly provide information about composition and surface topography in small areas of the
specimen. One of these signals is derived from backscattered electrons and
it gives origin to two types of images: a topographical image which shows
the roughness of the sample and a compositional image which is a visualization

Table 1. Comparison of Some Microprobe Techniques and Their Characteristics

	EPXMA SEM/EXD	micro-PIXE	LAMMS	SIMS	EELS	FTIR	micro-RAMAN	XPS
Source	Electron	Protons	Photons	Ions	Electrons	Photons	Photons	Photons
Detection	Photons	Photons	Ions	Ions	Electrons	Photons	Photons	Electrons
Elemental coverage (Z)	EDX: 11–92	EDX: 11–92	1–92	1–92	3–92	Func. groups	Func. groups	3–92
In-depth resolution (μm)	0.5–5	100	>1	10^{-3}	0.01–0.1	>10	10^{-3}	5.10^{-4}
Lateral resolution (μm)	0.1–5	5–10	1	<1	<0.01	5–10	<10	10
Detection limit (ppm)	1000	10	10	1	1000	10–30	>1000	1000
Quantization	Yes	Yes	Difficult	Difficult	Difficult	Yes	Yes	Difficult
Molecular information	Yes (WDX)	Yes (WDX)	Sometimes	Sometimes	Yes	Yes	Yes	Yes
Element mapping	Yes	Yes	No	Yes	Yes	Yes	Yes	Yes
Destructive	No	No	Yes	Yes	No	No	No	No

Note: Some of the values in this table can vary with the elements present in the sample, the sample itself, instrumental set-up and the goal of the analysis.

Abbreviations: SEM/EDX: scanning electron microscopy/energy dispersive X-ray analysis; EPXMA: electron probe microanalysis; micro-PIXE: micro-proton induced X-ray microanalysis; LAMMS: laser microprobe mass spectrometry; SIMS: secondary ion mass spectrometry; EELS: electron energy loss spectroscopy; FTIR: Fourier transform infrared spectroscopy; XPS: X-ray photo-electron spectroscopy.

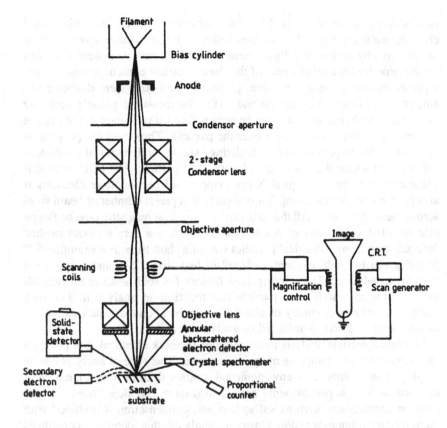

Figure 1. Schematic drawing of an electron microprobe X-ray microanalysis instrument.

of the atomic number variation with location in the sample. On the other hand, secondary electrons are mostly used for secondary electron imaging and electron micrographs. Both the backscattered and secondary electron signals can therefore be used for morphology studies.

The theory for EPXMA and SEM/EDX analysis is described in detail in textbooks by, e.g., References 3 through 5. In fact, there is no longer a sharp distinction between the two electron microprobe techniques EPXMA and SEM/EDX. SEM was originally used for high resolution imaging rather than for chemical analysis. EPXMA was primarily developed for achieving quantitative elemental information rather than for imaging purposes. This difference is more a matter of instrumental set-up and practical arrangement of the detectors. Both techniques are now converging to some extent for the purpose of chemical and morphological studies.

To study particulate samples by individual particle analysis requires the measurements of a large population set in order to obtain statistically meaningful data. At the University of Antwerp, a JEOL Superprobe JXA-733 electron probe X-ray microanalyzer is automated with a Tracor Northern TN-2000 system and controlled by an LSI 11/23 minicomputer. The following methodology is generally used for the automated particle recognition and characterization (PRC). An electron beam is raster-scanned over a preset sample area by means of a digital beam control. A particle is detected when the electron backscattered signal of the closed particle contour points exceeds a preset threshold value. The area, perimeter, and equivalent diameter (diameter of a circle which corresponds with the measured particle area) are calculated. An X-ray spectrum can be accumulated at the center of the particle or while performing cross scans over the particle. Thus the PRC program is set up in three sequential steps: localizing, sizing, and chemical characterization, after which the beam scans for the next particle. The PRC method is illustrated in Figure 2. Digital X-ray mapping of one or more elements is also possible by accumulating X-ray signals at a preset number of beam spots across the sample area. All the data can be stored on magnetic tape or floppy disc for off-line processing. Automated EPXMA is a very efficient method for analyzing many individual particles within a short time. For example, 500 particles can be multi-element analyzed in less than 2.5 h under optimized working conditions. The limiting time factors for such automated methods are the time to search for a particle and the time to analyze it. For such analysis, a relative accuracy of about 5% can be obtained. The relative detection limit of EPXMA using EDX analysis is about 0.1%.

Individual particle analysis partitioning of the bulk chemical data provides supplementary and complementary information. This type of analysis can be employed in a number of environmental studies dealing with source characterization and apportionment, particle transport and deposition, or with reaction mechanisms such as sulfatation and condensation. Combined with multivariate techniques and/or clustering analysis, this technique constitutes a powerful method for discriminating different particle types originating from the same source. Lee et al.[6] pointed out that, so far, only a limited number

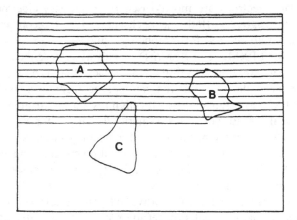

Figure 2. Localization of particles by the particle recognition and characterization method. Particle A has been localized and analyzed, particle C is being localized and particle B is localized and must be analyzed subsequently.

of studies have been published on the use of automated EPXMA in the field of environmental research.

3. PARTICLE-INDUCED X-RAY EMISSION

Ion beams collimated to micrometer size can become a very useful tool for the study of individual particle analysis. In the scanning proton microprobe (SPM) analysis, a well-focused beam of high energy protons is scanned over the sample. A variety of signals that can be collected and processed is used to obtain unique information on the composition of the sample. One of the major advantages of this technique is the excellent detection limit, of the order of 1 to 10 ppm, depending on the SPM design. Quantification is relatively easy to perform, with an accuracy of 10 to 20%.[7,8]

The proton beam with an energy from 1 to 3 MeV is finely focused in the sample by means of magnetic quadruples and/or electrostatic lenses. The high energy proton beam is obtained with cyclotrons or with nuclear electrostatic accelerators like Van de Graaffs. The recent availability of small commercial accelerators has greatly expanded the number of experimental facilities capable to do SPM analysis. The proton beam is focused to diameters of 0.5 to 10 μm, with currents of about 100 pA. A scanning system controls the beam positioning, and a computer system collects data from several detectors and beam information. The proton microprobe instrument is very similar to the electron microprobe. The only difference lies in the nature of the microbeam. The latter uses an electron beam whereas the former uses a proton beam.

Most SPM instruments provide routinely a spatial resolution of 0.5 to 10 μm. For aerosol particle analysis it is necessary to limit the proton beam current under 100 pA in order to avoid specimen damage. As with LAMMS and EPXMA, the sample is analyzed in a vacuum chamber, so possibly some of the organic components are lost.

There are several processes occurring during the interaction of the proton beam with the sample. When the generated X-rays are measured in particle induced X-ray emission (PIXE), elements heavier than sodium can be detected; PIXE has been reviewed in depth by Reference 9. Back-scattered particles provide information on light elements like carbon, nitrogen and oxygen through the Rutherford Back Scattering (RBS) analysis.[10] The proton beam also generates gamma rays from nuclear reactions, allowing to measure fluorine, sodium, and other elements by particle induced gamma emission or PIGE.[11] Frequently, PIXE and RBS analysis are done simultaneously, allowing the determination of carbon, nitrogen, and oxygen together with 10 to 15 trace elements heavier than sodium.

In PIXE analysis, a Si(Li) detector collects X-rays for each beam position while the beam scans over the sample, and an on-line sorting process makes it possible to obtain real time X-ray intensity imaging. These elemental maps of the sample being analyzed are constructed by the computer on a graphic terminal; they are similar to those obtained with EPXMA instruments. In several SPM setups, it is possible to observe simultaneously in real time 8 to 20 elemental maps. Also point analysis is possible, with the X-ray spectra stored for off-line quantitative analysis. As the proton beam generally goes through the sample and is collected in a Faraday cup, quantification is very easy and matrix effects are few. Accuracy of 10 to 20% is obtained for absolute analysis at detection limits down to 10 ppm.

In RBS the elastic back scattering of protons is used to obtain quantitative analysis of light elements like carbon, nitrogen, and oxygen. The sensitivity is not as good as in PIXE, but these elements occur generally in higher concentrations in aerosol particles; hence this is not a limitation for this application. A semiconductor surface barrier detector is often used to detect the back-scattered particles, and the energy spectra are analyzed by computer codes. It is also possible to make elemental maps of C, N, and O in real time, simultaneously with the trace element elemental maps. This has shown to be very useful in measuring the stoichiometry of compounds in atmospheric aerosol particles.[12] Sometimes the RBS analysis is made with a helium beam, instead of protons, because of the better kinematics of the helium beam scattering. It is possible to measure the depth profile of light elements in the elastic recoil detection or ERD technique. In ERD analysis the ion beam bombards the sample at grazing incident angles and the recoiling ions that escape from the surface of the sample are detected. Also hydrogen can be detected in a variant of the RBS technique called forward angle scattering technique or FAST. In FAST, the particle detector is placed in forward position relative to the beam, allowing the quantitative analysis of hydrogen.

In PIGE, a Ge(Li) detector is used to collect the prompt gamma rays produced in nuclear reactions due to the interaction between the proton beam and the sample. This technique is suitable to measure Li, B, F, Na, Mg, Al, and Si in particles. PIGE can be used simultaneously with PIXE and RBS, with a judicious choice of the proton beam energy.[13]

The SPM is a recent technique, and most of the applications are in biology, archaeology, geology, and material sciences. Artaxo et al.[12] showed the feasibility of a combined approach of EPXMA, LAMMS, and SPM analysis of individual aerosol particles. By exploring the advantages and disadvantages of each particular technique, it is possible to obtain a wealth of information for individual aerosol particles.

4. LASER MICROPROBE MASS SPECTROMETRY

Laser microprobe mass spectrometry (LAMMS) is based on the mass spectrometric analysis of ions, formed by the interaction of the sample with a high power density pulsed laser beam. Soon after their development in the early 1960s, lasers were applied in mass spectrometry. Originally, LAMMS was developed for analysis of biological samples, with high lateral resolution and extreme sensitivity. Nowadays, various instrumental set-ups have been developed for different analytical purposes. Figure 3 shows a schematic drawing of a typical LAMMS instrument. Van Vaeck and Gijbels[14] published an

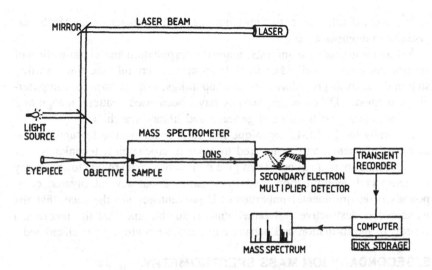

Figure 3. Schematic drawing of a LAMMS instrument.

overview of LAMMS techniques. Some of the instruments that have been constructed have also been commercialized, e.g., the LIMA-2A of Cambridge Mass Spectrometry, Cambridge, U.K. and the LAMMA-1000 and LAMMA-500 of Leybold-Heraus, Cologne, Germany. In the latter instrument, a Nd-YAG laser generates very short and intense light pulses for vaporization and ionization of a microvolume of the sample. The power density is 10^7 to 10^{11} W/cm^2 for a 1 μm laser focus and can be reduced to 2% of its initial value by a 25-step attenuating filter system. This is especially interesting for particle surface analysis. Depending on the spectrum polarity chosen, positive or negative ions which are formed by laser irradiation of the selected sample area are accelerated by a potential of 3000 V into a field-free drift region of the time-of-flight (TOF) mass spectrometer. The time-of-flight needed by an ion to traverse this region is related to its mass-over-charge ratio.[15] The signal is then fed into a 32 kbyte memory transient recorder and digitized. Spectra are stored in a personal computer for off-line data handling. Software packages are available for data processing and include a baseline correction algorithm, a peak integration routine, and spectrum averaging facilities. The commercial instruments differ in this geometry for ion collection from the specimen. Sometimes, other mass spectrometers are used in LAMMS, e.g., the single focusing magnetic sector instrument of the laser probe mass spectrograph (LPMS) and the double focusing Mattauch-Herzog type set-up of a scanning laser mass spectrometer (SLMS). A detailed description of these mass spectrometers can be found in the literature.[16]

Various review articles on several aspects of laser microprobe mass spectrometry are available in the literature. Early developments are discussed by Reference 17; laser desorption, Reference 18; applications in medicine, biology, and environmental research, Reference 19. Recently, Kaufmann et al.[20] evaluated achievements, shortcomings, and promises of LAMMS, especially in biomedical research.

Just as for microprobe analysis, manual interpretation and classification of several hundreds of individual particle spectra is certainly the time-limiting step in the analysis procedure. To speed up things, one searches for computer-aided methods. Different approaches have been used: pattern recognition techniques (e.g., artificial intelligence,[22] and library search[23]).

Generally the LAMMS technique has various interesting features: it can detect all elements, and, compared to other microchemical techniques, detection limits are quite good. It can give indications concerning stoichiometry or compound determination, and information about several organic compounds of environmental importance. Disadvantages are the facts that the technique is destructive and rather unreproducible and that the theoretical aspects of ion formation and behavior in the system are not yet elucidated.

5. SECONDARY ION MASS SPECTROMETRY

Secondary ion mass spectrometry (SIMS) is based on the bombardment of a sample surface by primary ions (Ar^+, F^-, O^-, etc.) or molecules generated in a duoplasmatron (energies in the keV range). Only a small fraction (typically 0.01 to 1%) of the sputtered atoms is charged. These secondary ions are attracted to a mass spectrometer where the separation according to their mass/charge ratio takes place. The separated positive or negative secondary ions are collected as mass spectra, as in-depth profiles or line-scans, or as distribution images of the sputtered surface. A mass spectrum can range from hydrogen to an m/z ratio of several hundreds. The mass spectrometer is based on electric/magnetic deflection fields or on the quadruple/time-of-flight principle; the latter being cheaper, but less satisfactory with respect to mass resolution and transmission. Figure 4 represents the principle of SIMS schematically. Lodding[24] distinguishes three classes of SIMS instrumentation: (a) non-imaging probes (static SIMS), used for depth profiling on laterally homogeneous specimens or for surface analysis, (b) imaging ion microprobes (dynamic SIMS), which use a narrow (<10 μm) beam of primary ions at energies of 5 to 20 keV, and allow imaging and microscopy by rastering the beam over the sample surface, and (c) direct imaging microscope microanalyzers, which use wide (5 to 300 μm) primary beams. Method (b) is best suited for determination of depth profiles, since the sputter rate is highest.

SIMS offers special capabilities for particle analysis. Heterogeneities in the composition in single particles are frequently observed from site to site. The distribution of constituents with depth in a particle is another microstructural feature of interest. SIMS is capable of obtaining signals from a depth of about 1 to 2 nm and this information originates from the surface of the sample. SIMS can also be used as a microanalysis technique with a minimum sampling volume of 0.01 μm.[3] The capabilities of SIMS for detection of all elements, compound detection, isotope ratio measurements, depth profiling, and ion imaging of specific constituents are described[25] with special reference to particle studies. Depth profiling can be successfully applied to individual particles; however, irregular topography of particles can degrade the depth resolution. Ion specific images of elemental or molecular constituents can be obtained in the ion microscope or ion microprobe.[26] The limiting lateral resolution is about 0.5 μm for the ion microscope and about 1 μm for the scanning ion microprobe.

The absolute detection limit for SIMS analysis is about 10^{-15} g for most elements and chemical compounds, and for anions down to 10^{-18} g.[27] For more detailed information, the reader is referred to the literature.[28] When molecular information of particles is needed, both LAMMS and SIMS can be used since they exhibit qualitatively the same positive ions. A comparative study of these two micro-techniques was performed on inorganic sodium sulfoxy salts.[29]

6. ELECTRON ENERGY LOSS SPECTROSCOPY

The use of electron energy loss spectroscopy (EELS) as a microanalytical technique was first discussed and demonstrated by Reference 30 and it is only recently that technological advances reactivated this technique for practical research. EELS studies changes in energy or momentum of electrons which have interacted with the specimen. As an electron beam penetrates through a microvolume of the sample, the initial electrons (having an energy E_o) suffer discrete energy losses due to ionization of inner orbitals or the excitation of oscillations of the electrons in the valence band of a solid (plasmon oscillation). The transmission signal consists of electrons in an energy range from E_o down to several thousand eV. After dispersion by an energy analyzer, electrons that have lost the same amount of energy are focused on the same point. By scanning this dispersion plane over a detector, the intensity I(E) of the transmitted signal can be plotted as a function of energy-loss. Figure 5 gives a schematic view of the beam path of a transmission microscope with

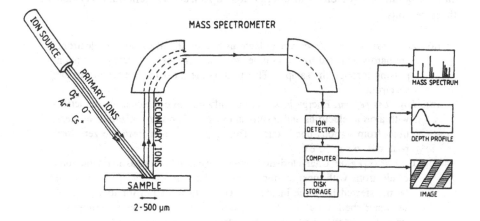

Figure 4. Schematic drawing of a SIMS instrument.

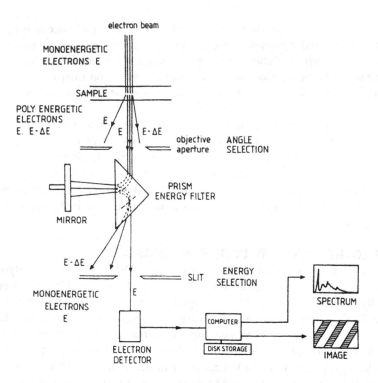

Figure 5. Schematic drawing of an EELS instrument with integrated prism spectrometer (type Castaing/Henry/Ottensmeyer).

an integrated imaging electron energy spectrometer (type Castaing/Henry/ Ottensmeyer) and upgraded by a computer image processing system. Figure 6 presents a typical EEL-spectrum from a thin amorphous carbon foil showing different spectrum regions, the interaction processes, and the information that can be obtained. An electron energy loss spectrum is generally divided in three regions:[31]

(a) *Zero-loss region*: Unscattered electrons and elastically scattered electrons are hardly affected by the sample atoms and no loss of energy occurs on passing through the sample. This peak is the most intense feature in the spectrum.

(b) *Low loss region*: Energy losses are mainly due to electrostatic interaction with atomic electrons and results in excitation and/or ionization of electrons from various bond states. This region extends from the zero-loss peak out to ca. 50 eV.

(c) *Core-loss region*: Characteristic features are caused by inelastic interactions of electrons with inner atomic shells. A discrete amount of energy must be transferred from the incident electrons to the atom for ionization of the inner shell to occur. These features show abrupt rises superimposed on a large background caused by valence shell excitations.

Egerton[32] and Maher[33] showed experimentally that the background decreases according to a logarithmic expression: $I(E) = A E^{-r}$. The constants A and r are dependent on the material, $I(E)$ is the intensity and E corresponds to the energy. The position of an edge-onset is thus characteristic for the presence of the elements in the interaction volume (qualitative analysis) and the amount of sample can be determined by calculating the net-peak intensities (quantitative information). One must know that EELS analysis is based on studying primary effects, i.e., the initial electrons give rise to the signal, whereas for X-ray analysis, secondary effects are observed.

EELS provides a way to identify and quantify elements present in a sample. Quantification can be performed without the use of reference materials but this is only applicable to very thin samples of uniform thickness and homogeneous element distributions.[34] The technique has some advantages compared to X-ray analysis in that low-atomic-number elements can be detected with high sensitivity. In addition, EELS also has an advantageous sensitivity in terms of detectable mass[35] and can provide information about the electronic state and chemical bonding of the sample. Contrarily, samples need to be very thin in order to transmit enough signal and not to introduce multiple scattering effects. However, the EELS technique has only recently been commercialized and is still in an experimental stadium for environmental studies and applications. Consequently, few applications of EELS analysis performed on individual particles have been reported in References 36 through 38.

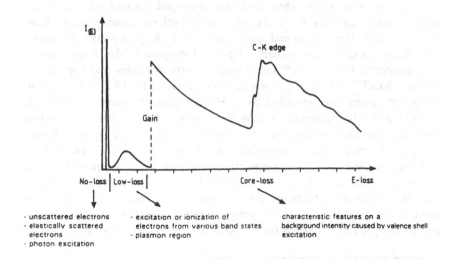

Figure 6. A typical electron energy-loss spectrum recorded at 80 keV from a thin carbon film.

7. FOURIER TRANSFORM INFRARED SPECTROSCOPY

This analytical technique is based on the fact that chemical compounds absorb infrared (IR) radiation provided there is a dipole momentum change during a normal molecular vibration or molecular rotation. Since the absorption of IR is unique for a chemical compound, its infrared spectrum can be used to identify and quantify a particular compound in a given unknown sample.

The usual set-up for this technique consists of an infrared radiation source, an interferometer, and a detector. Information on the chemical composition of a sample is extracted by analyzing the differences between two IR spectra: the one obtained without introducing an absorbing sample, which is characteristic for the IR radiation source, and the spectrum obtained when an absorbing sample is placed in the spectrometer's chamber. The spectrum analysis is performed using a mathematical algorithm, called Fourier transform, hence Fourier transform infrared spectroscopy (FT-IR).

The coupling of a reflecting microscope and an infrared spectrometer gave rise to a new analytical method: infrared microscopy, which was first introduced in the early 1940s. In fact, this type of set-up was used for recording infrared spectra on areas of a sample isolated using an optical microscope. In the early 1980s, infrared microspectroscopy started to be used again, mainly due to the advantages offered by FT-IR spectrophotometers, on a routine analysis basis. More details on infrared microspectroscopy can be found elsewhere.[39]

Recent applications of infrared spectroscopy and infrared microspectroscopy to surface analysis, polymers, and material science, among others, have been reviewed.[39] Environmental applications of FT-IR spectroscopy are somewhat scarce, and they are usually related to long-path FT-IR measurements of atmospheric trace gases,[40] monitoring of airborne gases and vapors,[41,42] whereas Small[43] and Gordon,[44] reported on the use of FT-IR spectroscopy to study atmospheric pollutants and characterize organic substances in aerosols.

As far as FT-IR microspectroscopy is concerned, the literature is much less abundant. Dangler[45] proposed a direct method for characterizing organic material in size-fractionated atmospheric aerosol deposits, using FT-IR. Kellner and Malissa[46] analyzed size segregated aerosol particles and aerosol clusters, but not on an individual particle base. They claim that this analytical technique has three major advantages for aerosol analysis: (i) little sample amount is needed, (ii) there is no need for sample dilution, and (iii) aerosol particles can be analyzed as deposited on the surface of the impactor stage.

8. MICRO RAMAN SPECTROSCOPY

Raman scattering has been used for bulk analysis since its discovery in 1928. The advent of lasers as sources for the excitation of Raman spectra and the developments in instrumentation optics now allow the analysis of discrete microsamples.

Raman spectrometry and microspectrometry are based on the Raman effect. When protons of frequency v_0 hit molecules, most of them are scattered elastically (Rayleigh scattering). The Raman scattering is caused by the inelastic collision of photons and molecules, resulting in photons with a series of frequencies, related to the original frequency v_0 by the expression $v_0 \pm v'$. The Raman frequency shift v' is independent of the incoming radiation and corresponds to certain rotational, vibrational, and electronic levels of the molecules under investigation. Depending on the symmetry of these molecules, vibrations are infrared active, Raman active, neither, or both, so sometimes infrared and Raman are complementary. Moreover, water is a suitable solvent for Raman spectroscopy but not for infrared spectroscopy.

Provided one can meet some specific technical requirements, Raman analysis of single microparticles becomes possible. The scope and limitations of single particle analysis by Raman microprobe spectroscopy is demonstrated.[48] Also Purcell and Etz,[49] and Etz and Roscasco[50] demonstrated a new spectrograph with a multichannel optical detector for the Raman characterization of microparticles and applications to the identification of airborne particulates.[51] In 1978, a Raman microprobe was developed at NBS specifically for the analysis of micrometer-sized particles. At that time, a commercial Raman microprobe/microscope MOLE[52] also became available. Rosasco[53] showed that Raman spectra from micrometer-size particles could also be collected with reasonable signal-to-noise ratio. Knoll[54] and Kiefer[55] reviewed micro-Raman spectroscopy of particles in the Mie-size range. It is possible to obtain Raman spectra of particles whose sizes are in the order of or larger than the wavelength of the exciting light. When these particles have a well-defined geometry (e.g., spheres, cylinders,...) the spectra can be greatly distorted by peaks which arise due to magnetic vibrations of the particle: morphology-dependent structural resonance modes occur. These additional peaks allow the user to determine the particle's size accurately, but, on the other hand, they complicate the assignment of peaks to molecular vibrational modes. Thurn and Kiefer[56] showed that this effect only occurs on particles with a well-defined geometry and not on microcrystals.

9. X-RAY PHOTOELECTRON SPECTROSCOPY

X-ray photoelectron spectroscopy (XPS), often called electron spectroscopy for chemical analysis (ESCA), has its origin in the investigation of the photoelectric effect in which X-rays are used as the exciting photon source. Its long history is intricately bound up with developments of the wave-particle duality principle and the early days of atomic physics. It was only after World War II that the first idea of using XPS as an analytical tool arose, particularly for studying surface chemical phenomena. Later, XPS contributed significantly to environmental particle analysis through its ability to determine surface elemental and chemical speciation information in a relatively nondestructive manner. There is a new trend towards XPS studies of smaller and smaller sample sizes down to the limit of single particle resolution.

In the XPS technique, inner shell electrons or electrons from the core levels are ejected from the target atoms by bombarding the sample with photons (usually Mg-Ka or Al-Ka X-rays). These ejected electrons (photoelectrons) are energy analyzed after collection. From the measurement of their kinetic energy (Ek) and knowing the initial energy of the bombarding photons (hv), the binding energy (Eb) of the photoelectron can be determined using the following equation:

$$Eb = hv - Ek - W$$

where W is the work-function of the spectrometer. Just as in electron energy loss spectroscopy, the intensity of the photoelectron signal as a function of the binding energy makes up an electronic spectrum that is unique for each element present in the interaction volume. Not only qualitative information can be obtained but speciation information from changes in the valence level due to oxidation, reduction, chemical bonding interactions, etc. have a small but definite influence on core-level electron binding energies which can be deduced from the spectra. Quantitative analysis by XPS is also possible but in practice, a number of phenomena have to be taken into account which makes it less straight-forward. Further details on the theory of XPS can be found elsewhere (References 57 and 58).

Spatially resolved XPS can in general be divided into two types: (1) small spot and (2) imaging. The former is the most developed and makes use of crystals to reflect and concentrate X-rays into a small area. In the latter, X-rays are generated within a small well-defined volume. By mounting the sample close to the X-ray source, the spatial resolution approaches 10 μm. A scanning XPS system with spatial resolution of about 10 μm was proposed[59] and is ideally suited for XPS analysis of selected particles. Also the photoelectron microscope developed (Reference 60) can detect 10 μm particles.

The surface sensitivity of XPS analysis is due to the very short depth (typically 0.5 to 2 nm) from which the secondary electrons (X-ray photoelectrons) can escape the sample interaction volume. XPS is therefore well suited for the evaluation of the surface enrichments of species which may thus have an environmental impact far greater than the bulk concentrations would indicate.[61] This can easily be understood considering the fact that airborne particles with diameters below 3 μm often have much higher surface area to volume ratio compared to larger particles.

Cox and Linton,[62] already discussed extensively the aspects of environmental applications of XPS involving airborne particle chemistry. Also a brief review of X-ray photoelectron spectroscopy for the characterization of surface regions of particulate pollutants from the atmosphere is given elsewhere.[63] However, the applications found in the literature are mostly restricted to bulk particle analysis.

For the near future, substantial improvements in the sensitivity and spatial resolution of XPS analysis can be expected. Further development is going on in the field of small area and imaging XPS that might ensure new applications in environmental and biological studies but also for individual environmental particle studies.

10. APPLICATIONS
10.1 Marine Aerosols

The investigation of the aerosol composition over different oceans and seas is important for several reasons: particles from remote regions can serve as a marine background aerosol, while transport, transfer, and wet and dry deposition of aerosol pollutants from continental sources to the marine environment can be assessed. In studies of the marine aerosol, Na and Cl are the most frequently used reference materials.[64] Their presence in the atmosphere can unambiguously be related to the ocean.

The marine aerosol composition has been studied by impactor sampling at coastal sites, well away from major land areas or by sampling at coastal sites under on-shore wind conditions, and in laboratories under simulated conditions.[65] For example, the North Sea is an important sink for dry and wet deposition of atmospheric particulate matter from the surrounding industrialized pollution sources. Laser microprobe mass spectrometry (LAMMS) and electron probe X-ray microanalysis (EPXMA) combined with an automated image analysis system have been used for the characterization of individual North Sea aerosols.[66] With the aid of EPXMA, about 2500 particles, sampled from a research vessel, were sized, chemically analyzed, and classified. Sea salt constituted the most abundant particle type when the collected air masses originated from over the Atlantic Ocean and traveled toward the continent. Contrarily, in air masses that spent longer residence times over the continent, high concentrations of aluminosilicate particles (mostly spherical fly-ash particles), carbonaceous particles, $CaSO_4$ and spherical iron oxides were observed.

LAMMS was applied to a number of representative particles from the different particle types in order to elaborate trace element contents and surface layer composition. Applying low energy laser shots to the spherical fly-ash particles revealed typical mass spectra that were interpreted as fingerprints for the desorption of polynuclear aromatic hydrocarbons (PAH). The most typical marine aerosol particles are of course sea salt particles, formed by the bubble bursting mechanism. Spectra of "pure" sea salt are dominated by Na, K, and typical Na/K/Cl cluster ions. Often, sea salt particles are found to have been transformed to some extent; in this case, nitrate and sulfate coatings are readily detectable.[67] The same authors found the amount of nitrate-coated sea salt particles to increase significantly from a beach site towards an industrialized area, 30 km downward from the ocean. Allegrini and Mattongo[68] also proved with XPS the existence of various nitrogen and sulfur species on particle surfaces. The detection of methanesulfonate, a biogenic airborne organic compound, above the Sargasso Sea, the Bahamas area,[69] and the coast of Antarctica[70] constitutes an excellent illustration of the occasional "organic successes'' of the LAMMS technique. Another advantage of LAMMS in aerosol research is its ability to detect ammonium compounds, which are very interesting from an environmental point of view. Otten[71] found the relative abundances of ammonium-rich particles in the North Sea aerosol to increase dramatically under the influence of polluted airmasses.

Analogous EPXMA characterizations were performed on a data set of more than 25,000 individual aerosols collected over the North Sea and the English Channel from a research vessel.[72] Differences between samples were studied on the basis of abundance variations using principal component analysis (PCA). Nine different particle types were classified and they all were source apportioned unambiguously. The release of sea salt into the atmosphere is dominated by the process of breaking waves and this is more effective as the relative windspeed increases. Transformed sea salt particles rich in Cl and S are formed by the conversion reaction of NaCl into Na_2SO_4 implying the release of HCl into the atmosphere. S-rich particles of various composition namely H_2SO_4, $(NH_4)_2SO_4$, $(NH_4)HSO_4$ and $(NH_4)_3H(SO_4)_2$ were assigned to originate from anthropogenic sources and they are probably formed by gas-to-particle conversion. Also, the $CaSO_4$ particles above the North Sea are predominantly emitted by anthropogenic sources such as combustion processes or eolian transport. $CaSO_4$ particles may be enriched in S and can therefore be partially identified as $CaSO_4(NH_4)_2SO_4$ resulting from coagulation of $CaSO_4$ with submicrometer sulfate aerosols.[73] These findings are also in agreement with LAMMS analysis.[66] On the other hand, Ca-rich particles can originate both from the marine environment and from continental sources. Various dissolved salts begin to crystallize sequentially as seawater evaporates. Calcite ($CaCO_3$) and dolomite $CaMg(CO_3)_2$ precipitate first followed by $CaSO_4$ and the Mg-salts. It is possible that $CaMg(CO_3)_2$ undergoes further reaction with gaseous S-rich components. EPXMA measurements support the existence of such transformed particle species. Aluminosilicate particles cannot be distinguished from fly-ash particles on the basis of their chemical composition. Only morphology can sometimes make the differentiation. Important differences of these typical, nearly perfect, spherical fly-ash particles were observed in samples taken over the North Sea as air-mass backtrajectories originated from above Eastern Europe. A minor fraction of the quartz can be emitted during combustion processes of coal in power plants.[74] The fact that quartz particles, just as Fe-rich particles, have very small diameters (<0.8 μm) supports the hypothesis that they are formed during various combustion processes. Titanium particles above the North Sea most probably find their origin on the continent and possible sources are paint spray, soil dispersion, asphalt production, and power plants.[75]

Continentally derived aerosols are often found in the marine environment. By use of EPXMA it was found that high number concentrations of aluminosilicates present as internal mixture with sea salt aerosols are more likely the result of coagulation of sea salt and silicate particles within clouds, including droplet coalescence[76] rather than resuspension of silicate particles from the sea surface as a result of bubble bursting processes. The fine aerosol mode contains higher concentrations of S, K, V, Ni, Cu, and Zn.

The atmospheric aerosol composition over remote ocean areas, such as the Atlantic Ocean, has been investigated fairly intensively, either from island-based sites or from research vessels. Aerosols collected over the Eastern Equatorial Pacific from a sailboat (from Panama to Tahiti) were analyzed both by bulk (PIXE)[77] and single particle (EPXMA)[74] techniques. Beside the major sea salt aerosols, a substantial crustal component and fine Cu and Zn, resulting from continental sources, were characterized. Long range transport from natural high-temperature sources such as volcanoes or emission from fuel combustion on the American continent, or enrichment at the sea surface are possible sources for the Cu and Zn concentrations found. Excess fine S sometimes reached levels similar to those reported for other remote regions. Excess fine K and partly fine S showed a tendency to exhibit higher values for samples collected near harbors, suggesting local island derived sources such as biomass burning and fossil fuel burning. EPXMA of more than 5000 individual particles, combined with cluster techniques, provided a classification of all particles on the basis of their chemical ~omposition. The most abundant particle type was rich in S (45% of all particles) and this in the absence of other detectable elements. Morphological inspection of these particles allowed differentiation between two groups, namely one group of S-rich particles in the sub-micrometer range that are unstable under the electron beam and are most probably $(NH_4)_2SO_4$, and one S-rich group in the micrometer range (mean particle diameter of 2 μm) showing more spherical contours. The latter group is much less affected by electron irradiation. The crustal component and also Cu and Zn are more abundant in those samples collected near the American continent in accordance with PIXE analysis. The mean diameter for the Cu- and Zn-rich particles is about 0.7 μm. An important fraction of particles only yielded characteristic Ca and P X-rays; their mean diameter varied between 0.4 and 0.8 μm. Their abundance shows a slight tendency to increase as the sample location approaches the continent, but this is insufficient to predict terrestrial sources to be responsible. It is known that the Pacific Ocean is slightly supersaturated in hydroxyapatite $(Ca_5(PO_4)_3OH)$,[78] but this is no sufficient reason to suggest that the ocean is responsible for the production Ca-P-rich aerosols found in the Pacific environment.

Aerosols collected from an aircraft in remote continental and marine regions at altitudes ranging from the boundary sea-air interface to the troposphere were analyzed with SEM/EDX and PIXE.[79] The continental aerosol population consisted of crustal particles with r $>$0.5 μm and sulfate aerosols with r $<$0.5 μm. No qualitative differences were noted as a function of altitude. Contrarily, Pacific marine measurements show large variations between the boundary layer and the troposphere. A decrease of the crustal component was observed from the North towards the South.

10.2 Remote "Continental" Aerosols

The areosol composition of extremely remote locations has always been a subject of interest. Indeed, due to the growing industrialization and the long-range transport of pollutants (e.g., linked to high-stack technology), it has become increasingly difficult to define the composition of baseline natural aerosols in the absence of any pollution. Several remote locations, most of the time located in the Southern hemisphere, have already been the subject of an extensive study.

The Antarctic continent is probably the most convenient place on earth to study the composition of background aerosols because it is the most distant area from the world's predominant pollution sources and local anthropogenic contributions are negligible. Single particle X-ray analysis of Antarctic aerosol samples from different locations has revealed the following particle types: sulfur-rich particles (which may be formed by gas to particle conversion), sea salt particles (formed by the bursting of gas bubbles that arise through wave action), aluminosilicates (earth crustal dust or particles originating from local sources like volcanoes, geysers, or other surface/ocean-floor disruptions), and particles whose X-ray spectra contain mostly Fe peaks (long-range transported anthropogenic or maybe meteoric dust particles).[80–84] Naturally, their relative concentrations vary with sampling site, and season and meteorological conditions. Especially in summer, sulfate particles tend to dominate the Antarctic aerosol by number and also by mass.[81,85] This sulfate should predominantly exist in the form of H_2SO_4, $(NH_4)HSO_4$, $(NH_4)_2SO_4$ or a mixture, possibly including more complex species as was also revealed by Raman microprobe measurements on identical particles.[48] Ito[82] and Bigg[86] detected H_2SO_4 and $(NH_4)_2SO_4$ particles. Hierarchical and nonhierarchical cluster analyses were performed on individual coastal Antarctic aerosols using EPXMA.[87] The results show a domination of marine components in both the fine and the coarse mode fractions. Only a minor crustal component was found and thus crustal weathering processes must be far less important than the strength of the marine source. LAMMS of similar aerosol particles[70] revealed the presence of salts derived from methane sulfonic acid (MSA). MSA is a reaction product of the photo-oxidation of dimethylsulfide (DMS), which originates from the metabolism of certain marine algae. This compound contributes significantly to the global sulfur cycle.

Sheridan and Musselman[88] performed LAMMS and EPXMA on particles sampled during flights over the Alaskan Arctic. Virtually all submicrometer particles yielded spectra that highly resembled those of an ammonium sulfate standard. Because the likelihood of finding appreciable amounts of ammonium vapor in the winter Arctic atmosphere is small, they concluded that those particles were collected as sulfuric acid and gradually transformed in the laboratory. K-rich particles in this aerosol were tentatively attributed to wood smoke.

Surkyn et al.[89] performed LAMMS on aerosol particles from the uplands of central Bolivia. The sampling station was located at 5230 m above sea level. The immediate surroundings are totally uninhabitated, and the ground over a wide area is stony, partly snow-covered year round, and without vegetation. Particles from this site were soil-derived aluminosilicates and/or Ca-rich particles and, in the smallest size fraction, ammonium sulfates. Occasionally, K- and C-rich particles were detected; most probably as a result of forest burning.

Bigg[86] analyzed individual particles from Cape Grim (Tasmania), Mauna Loa Observatory (Hawaii), and Point Barrow (Alaska) using an electron microscope and performed some chemical tests on them. This approach turned out to be very effective. The great majority of those particles were found to be composed of sulfuric acid or its reaction products with ammonia. The Barrow and Mauna Loa particles were predominantly sulfuric acid, while the South Pole and Cape Grim particles were predominantly ammonium sulfate. The latter aerosol was also analyzed with LAMMS,[89] and was found to contain sea salt-derived and exceptionally crust-derived particles.

10.3 Biogenic Aerosols

It has already been hypothesized that the forest is a major source of aerosols in both the fine and coarse fractions.[90] Plant leaves can be responsible for a large number of biogenic aerosol emission processes: large particles can be released by mechanical abrasion of the plant leaves as a result of wind action,[91] biological activity of leaf surface microorganisms can result in the production of airborne particles and windblown pollen, and during crop plant transpiration, migration of Ca^{2+}, SO_4^{2-}, Cl^-, K^+, Mg^{2+}, and Na^+ into the atmosphere can occur.[92] The forest is also a source of sulfates (especially during the rainy season), ammonium, and nitrates.[93] Despite the relatively vast area of tropical forests, relatively few authors reported studies involving aerosol trace element measurements in tropical regions.

As part of the Global Tropospheric Experiment (GTE) of NASA, individual aerosol particles sampled over the Amazon Basin were analyzed by automated EPXMA in order to study the processes of aerosol and gas emissions by the forest and to assess the chemical mechanisms occurring in the Amazon Basin atmosphere.[87] According to the methodology,[94] hierarchical and nonhierarchical cluster techniques were applied on the whole data set for the classification of all particles. About 27% of all particles showed no detectable elements with $Z > 10$. Most of the particle types could be related to two prevalent local sources, nl. soil dust and biologically derived material. The former type is typically composed of Al, Si, and Ti. The latter type is identified by the high Bremsstrahlung background (low Z elements) and the presence of elements as Na, S, K, Cl, P, Ca, Zn, and Cr or a combination of them. Particle types containing mainly S, K, and P can be related to aerosol emissions by vegetation. Also the Zn- and S-rich particle type can be attributed to forest emissions.[95] LAMMS spectra of Amazon Basin aerosols are very complex due to the presence of different organic compounds, fragmented to various extents: sulfate salts of amines, methane sulfonate, and fragmentation patterns of hydrocarbons, terpenes, and phospholipids have been identified up to now.[95,96] Part of this organic material was found to be associated with inorganic salt mixtures consisting of plant nutrients, which points to a plant-transpiration origin. Another interesting result is the association of some trace elements (e.g., Pb and Zn) with the organics. Micro-PIXE and RBS were used simultaneously on Amazon Basin aerosol particles. It was easy to measure elements like Rb, Zn, and Mn with concentrations below 10 ppm.

10.4 Urban Aerosols

The composition of an urban aerosol will of course largely depend on its geographical location, the activities performed locally, and the industries surrounding the sampling site. Still, some trends can be defined. Soil dust particles are usually abundant. Other particle types often found are sulfates ($CaSO_4$, fine and coarse sulfur-rich particles), auto exhausts (Pb-halides and sulfate derivatives), and different anthropogenic particles derived from various sources such as oil burning processes (S, V, Ni), abrasion processes (Fe-, Cr-oxides), and emissions of incinerators (Zn-, Pb-, Cr-, Zn-, and Sb-rich particles).[97] The principal source of particulate Pb in the urban atmosphere is the combustion of leaded petrol.

Automated EPXMA and LAMMS studies were performed on more than 15,000 individual Pb-containing particles sampled near the city of Antwerp, Belgium.[98] The results indicated that partial conversion of Pb-halide containing particles into Pb-sulfates often occurs by the reaction with ammonium sulfate present in the urban atmosphere. The ammonium sulfate can also be present as a coating on the Pb-containing particles. Keyser et al.[99] have reported elemental distributions as a function of the sputter depth by SIMS from large (>10 μm) automobile exhaust particles. Enrichments of Pb and Br and, less obvious, of S at the particle surfaces were found.

A comprehensive study of particulate material in the 0.1 to 30 μm size range in the urban aerosol of Phoenix, AZ, was conducted.[100] More than 8000 individual particles were analyzed by analytical scanning electron microscope. The coarse particle fraction (>1 μm) is mainly crustal material, i.e., clays, quartz, feldspar, calcite etc. A minority of biological material and S-compounds and Pb-salts from automobiles were observed as well. The submicron aerosol fraction comprises S-containing particles (60 to 80%) presumably present as $(NH_4)_2SO_4$. Some of these particles contain various amounts of elements such as Zn, Pb, Cu, Ca, Na, and K.

10.5 Volcanic and Stratospheric Aerosols

Some research efforts have been focused on natural aerosols released by several pathways. Analyses by SEM combined with XPS of volcanic ash particles have been reported by several authors.[101-103] Major elements detected by SEM/EDX were Al, Ca, K, and Si; minor ones Fe, Mg, and Ti. Volcanic eruptions are proven to be responsible for a fraction of terrestrial particles released directly into the stratosphere.[104] Particles present in the stratosphere can also be derived from sulfuric acid aerosols, sapphires, and meteorites.[105] They contain Al metal, Al_2O_3 spheres, and Al-prime particles. The submicron regime is dominated by sulfate aerosols of terrestrial origin.[106] Relatively high concentrations of sulfur are emitted during volcanic eruptions and the presence of thin sulfate gels on the surface of ash particles is probably the result of processes within stratospheric clouds.[107]

10.6 Industrial and Workplace Aerosols

Industrial aerosol research is often concerned with the quality of the working environment. As toxic effects are often correlated with specific particles, shapes, sizes, or element distributions, microanalysis is a very powerful tool to point out the dangerous parameters.

The superiority of microanalysis over bulk analysis techniques in this field is clearly demonstrated in the analysis of coal mine dust. The predominant role of quartz in the induction of "coal mine workers disease" (pneumoconiosis) is widely accepted, but there are still ambiguities about the existence of additional and/or modulating factors determining the degree of toxicity. Since coal mine dust is known to be a heterogeneous mixture, the use of single particle analysis seemed appropriate in this respect. Barths et al.[108] investigated airborne dust from 11 European coal mines. Correlating LAMMS results with toxicity data, they could confirm the role of quartz as a specific toxic agent for German, but not for French, coal samples. Cluster analysis of the element distribution patterns revealed factors which clearly modulate the quartz related toxicity and also factors with their own toxic potency. They seem to be mine dependent, or at least area dependent. The cytotoxicity of different silica dusts was found to be primarily determined by the incidence of silicon-dominated particles. The latter turned out to be a better cytotoxic parameter than the quartz content as determined by bulk analysis. Their results support the idea that some fraction of the quartz is toxicologically ineffective.

Different asbestos types, which are known to promote fibrosis and/or cancer, can be distinguished on the basis of their LAMMS spectra.[109] LAMMS and SIMS were also applied for the analysis of organic impurities at the surface of asbestos fibers.[110,111] The absorption behavior of different asbestos varieties for various organics was studied by LAMMS[112,113] and EELS.[37] Moreover, LAMMS revealed preferential leaching of elements in, e.g., biological liquids. These results are of importance in the sense that, next to fiber geometry, chemical properties and reactivities also determine the carcinogenic effect.

In an attempt to identify Ni-compounds emanating from pollution sources, different (standard) Ni-species were distinguished on the basis of their LAMMS-spectra.[114] A characterization algorithm using LAMMS and SEM was developed. Characterizing different compounds in airborne particles, however, turned out to be less straightforward, which is due to the complexity of the environmental aerosol.

Gondouin and Muller[115] and Poitevin et al.[116] used LAMMS to infer the oxidation state of chromium in dust particles formed during stainless steel machining and soldering operations. Their stoichiometric information, which is important from a toxicological point of view, was based on relative cluster ion intensities in the LAMMS spectra. Michaud,[117] who analyzed Cr-containing particles from pigmentation, soldering, and plating industries, showed that these ratios are extremely dependent on instrumental fluctuations, so standards should be analyzed on a regular basis. The anthropogenic Cr-particles appeared most of the time in the hexavalent (i.e., the most harmful) oxidation state. It should, however, be stressed that, in general, obtaining stoichiometric results with LAMMS is by no means straightforward. Cox[118] used XPS for the investigation of ferrochrome smelter dust particles. They showed that surface-detectable Cr(VI) can be removed by aqueous leaching and the remaining particle surface residue contains only insoluble Cr_2O_3.

Fly ash is a much investigated type of industrial aerosol. This name covers a variety of particles. They are emitted to the atmosphere by combustors mostly used for the generation of electric power. In 1968, the global fly-ash emission to the environment was estimated at 4×10^7 ton.[119] Knowledge of the bulk composition of fly ash is often insufficient because of the important internal composition heterogeneity within a particle population. Consequently, the impact and behavior differs largely from one particle to another and may be extremely complex. The characteristics of fly-ash particles furthermore depend on the mineral matter used, the thermal behavior of the coal in the furnace, melting and decomposition temperatures of the mineral matter, and possible chemical reactions and heterogeneous assemblages of the different emission products during their cooling in the atmosphere.[120] Several workers have used SEM and/or EPXMA to determine the morphological and chemical characteristics of fly-ash particles.[121-127] Energy dispersive X-ray analysis in the electron microscope provides semiquantitative element analysis of individual particles.[87,128-134] The main elements present in both the micron and submicron particles are Si, Al, K, Fe, Ti, Mg, and S. Ca, P, Na, Cl, and Ni are minor constituents.[134] Campbell et al.[135] used ion sputtering in combination with XPS to monitor the concentration and speciation of various elements in fly ash as a function of sputter depth. The concentrations of Ca, Na, C, O, and S were found to decrease with depth, whereas the concentrations of matrix elements such as Al, Fe, and Si increased with depth. Recorded SIMS depth profile studies of coal fly-ash particles (d > 4 μm) showed a significant surface enrichment of lead.[36] Similar studies[137,138] indicated strong surface enrichments of Pb and Tl; this implies that coal fly ash may have a more deleterious environmental impact than is apparent solely on the basis of conventional bulk analysis. Cox[139] investigated particles from a coal-fired power plant with a digital imaging system interfaced to an ion microscope. The setup they used permitted the simultaneous acquisition of spatially resolved mass spectral data for a number of single particles. These authors found substantial differences in the relative concentrations and/or depth profiles of the elements Ba, Pb, Si, Th, Tl, and U from particle to particle. Pb, Tl, and U were generally concentrated on the particle surfaces. The major part of fly-ash particles have a characteristic spherical geometry although irregularly shaped particles are observed as well.[74,133] One should differentiate between two fly-ash types according to the material (oil or coal) used for the operation of a power plant. Differences between oil fly ash and coal fly ash were reported by several authors.[74,140–143] Oil fly-ash particles vary in morphology from near-rounded spheres to lacy or spongy lumps, which indicate a long exposure history to heat and oxidants.[144] These spongy structures easily break down to smaller aggregates. Over 90% of the mass fraction occurs in the fine fraction. Oil combustion particles contain considerably more sulfur and substantial amounts of V and Ni. Coal fly ash predominantly consists of smooth mineral spheres and contains less cenospheres. Almost 90% of the mass fraction occurs in the coarse fraction and extensive measurements have been made on stark-suspended particulate material from oil- and coal-fired power plants using a Raman microprobe.[142] They found the existence of V_2O_5 as a principal component in the oil-ash particles but not in the coal-derived particles. LAMMS of fly-ash particles on some occasions leads to the detection of PAHs. Fly-ash particles with the SPM system of the Free University of Amsterdam have been analyzed.[141,145] A beam size of 7 by 10 μm was used with a current of 20 to 40 pA. It was possible to measure trace elements like Se, V, Cr, Ti, and Cu, and to obtain concentration profiles for large particles. The analysis was complemented by tube-excited X-ray fluorescence for bulk trace element measurements.

PAHs were also detected using LAMMS on soot particles from an experimental oil-shale retort.[146] SIMS analysis of single oil-soot particles (d $\geq$ 1.5 μm) showed that they are characterized by high levels of O, V, C, Na, Ca, and K.[147] Secondary ion micrographs of these elements provided information about their distributions throughout the particle.

Lang et al.[148] studied individual dust particles (ranging from 10 to 50 μm) from an office/lab environment using infrared and Raman spectroscopy. Raman spectra were obtained on dust specimens if the infrared spectra did not provide enough information to permit suitable characterization of the sample. Many of the dust particles identified could be linked to a paper product as the source.

10.7 Sediments, Suspension, and Estuarine/Riverine Particles

Suspended particulate matter from estuarine and marine environments is being investigated extensively in order to assess sedimentation processes, the interactions between sediments and the water column, and the physicochemical reactions that particles undergo. Most of the analytical techniques used for this study, such as instrumental neutron activation analysis, atomic absorption spectrometry, and X-ray diffraction, provide information on the bulk composition or mineralogy of the samples. Alternatively, only a few beam techniques for individual analysis of estuarine particles are invoked.

SEM/EDX were successfully applied.[149–152] These studies had in common that the particles were searched and analyzed manually for their chemical and morphological characteristics. The first results of automated EPXMA of marine suspended particulate matter was reported.[153] They applied this technique to individual particles from the nepheloid layer in the Atlantic Ocean, and classified the analyzed particles on the basis of their Si/Al ratio. A part of the data was also described.[154]

Later, various aquatic (estuarine and marine) environments were studied, all by using EPXMA. Suspended matter particles from the Scheldt river estuary (Belgium, the Netherlands) have also been studied with LAMMS.[94,155,156]

For all estuarine systems, EPXMA made it possible to evaluate the effect of mixing material from different origins and of separating the mixing process from other processes such as deposition and remobilization. The results on the Ems estuary (in Germany and the Netherlands) elucidated that the mixing with marine material occurs in the fresh water tidal area and that the suspended matter of marine origin is transported upstream the tidal area, across the salt wedge.[94] The same approach proved to be equally successful when applied to the sediment fraction of the Elbe estuary (Germany), for which the findings were in agreement with those of the Ems river.[157] For the Garonne and Rhône rivers (France), no evidence was found for a net flux of marine suspended particulate matter into the estuaries; this is a consequence of the different nature of these estuaries.[158,159]

For the Nile river (Egypt) no marine intrusion and thus no mixing process had to be taken into account; still significant relative abundance variations were observed for both the sediment and the suspended particulate matter. Information was gained on the impact of the Aswan Dam: the behavior of the particles in this area and their further course along the Sohag area downstream.[160]

For the Magela Creek river system (Australia) an advanced approach was achieved by analyzing both the suspension (>1 μm) and colloidal (1 to 0.1 μm) particles whereas most previous work was concentrated on the larger sized suspended particulate matter (generally >0.45 μm). The significance of these results lies in the fact that recent studies recognize an increasing importance of the role of this colloidal fraction for certain aquatic environments.[161] The inorganic mineral composition proved also to serve as an equally good tracer for the origin of suspended particulate matter of non-fluvial estuarine environments.

The abundance variations of the particle types correlated with hydrographical/hydrochemical and bulk data, provided information about geochemical and physical processes that influence the levels and distribution patterns of certain particle types throughout the Baltic Sea and the transient area to the North Sea.[156]

A better insight into the sources and lateral/depth dispersal of suspended matter in Makasar Strait and Flores Sea and around the Sumbawa Island

(Indonesia) was satisfactorily accomplished by EPXMA. It was found possible to differentiate between particles of terrestrial volcanic and biogenic origin of the sediment.[162,163]

Manual examinations by electron microprobe were found to be very valuable for recovery particle associations, e.g., $BaSO_4$ formation in recently dead siliceous plankton[164] and distinguishing different structures, e.g., for manganese[165] and species, e.g., for pyrite.[166]

The Hamburg SPM group has measured trace elements in particles from river sediments.[167] Using a 2 Mev proton beam of 2.3 by 3.0 mm and beam currents of 0.3 to 3 nA, detection limits were about 10 ppm. It was possible to detect Si, S, K, Ca, Ti, V, Cr, Mn, Fe, Ni, Cu, Zn, As, Br, Rb, Sr, Zr, and Pb. Using different absorbers, the SPM analysis can be optimized for a certain range of elements, further increasing the sensitivity for heavier elements.

11. CONCLUSIONS

Conventional studies of environmental systems are based on bulk sampling procedures in order to determine average concentrations. Such information is very important for environmental control purposes, but sometimes does not provide information about the identities and amounts of pollutants present in the microscopically small regions that constitute environmental interfaces.

Individual particle analysis has been proven extremely valuable as a complement of the bulk methods. Information concerning origin, formation, transport, reactivity, transformation reactions, and impact on the environment that would have been very difficult, if not impossible, to get with bulk analysis techniques, has been obtained.

Microanalysis techniques have their own specific problems that are often related to the extreme small sizes of the samples:

(1) Quantitative analysis is extremely troublesome because of the uncertainty in defining the precise analytical interaction volume and the difficulties encountered when preparing suitable standards on that level.

(2) In order to get statistically relevant information, huge numbers of particles need to be analyzed. This makes the analyses very time consuming. Therefore, much effort has been invested in automation and computerization. In this respect, EPXMA is certainly the most advanced.

Moreover, each technique has its own specific constraints, related to its principle (sample-beam interaction, experimental set-up, ...). As a consequence, they can complement each other with respect to lateral resolution, detection limits, detectable elements, etc.

Hence, the characterization procedure that uses several single particle analysis techniques (and preferably also a bulk method) will of course give the best results. Each of the above discussed instruments is however very expensive, so interlaboratory cooperation will most of the time be necessary.

ACKNOWLEDGMENTS

This work was partially supported by the Belgian Ministry of Science Policy (under contract Eurotrac EU 7/08), the Dutch Rijkswaterstaat, (under contract NOMIVE*2N°DGW-920) and the European Community via a subcontract of the EROS-program. One of us (P.A.) contributed to this work as a guest professor of the Belgian National Science Foundation.

LIST OF ACRONYMS

EELS:	Electron Energy Loss Spectroscopy
EPXMA:	Electron Probe X-Ray Micro Analysis
ERD:	Elastic Recoil Detection
ESCA:	Electron Spectroscopy for Chemical Analysis
FAST:	Forward Angle Scattering Technique
FTIR:	Fourier Transform Infrared Spectroscopy
LAMMS:	Laser Microprobe Mass Spectroscopy
LPMS:	Laser Probe Mass Spectrograph
micro-PIXE:	micro-Proton Induced X-Ray Emission
PIGE:	Particle Induced Gamma Emission
RBS:	Rutherford Backscattering
SEM/EDX:	Scanning Electron Microscopy/Energy Dispersive X-Ray Analysis
SIMS:	Secondary Ion Mass Spectrometry
SLMS:	Scanning Laser Mass Spectrometer
SPM:	Scanning Proton Microprobe
TEM:	Transmission Electron Microscopy
WDX:	Wavelength Dispersive X-Ray Analysis
XPS:	X-Ray Photo-Electron Spectroscopy

REFERENCES

1. Grasserbauer, M. "The Present State of Local Analysis: Analysis of Individual Small Particles," *Microchim. Acta (Wien)* 329 (1978).
2. Grasserbauer, M. "Micro and Surface Analysis for Environmental Studies, *Microchim. Acta (Wien)*, III:415 (1983).
3. Goldstein, J. I. and H. Yakowitz. *Practical Scanning Electron Microscopy* (New York: Plenum Press, 1987).
4. Holt, D., M. D. Muir, P. R. Grant and I. M. Boswarva. *Quantitative Scanning Electron Microscopy*, (London: Academic Press, 1974).
5. Reed, S. J. B. *Electron Microprobe Analysis*, (Cambridge: Cambridge University Press, 1975).
6. Lee, R. J., J. S. Walker, and J. J. McCarthy. "Evolution of Automated Electron Microscopy," *Microbeam Anal.* 485 (1986).
7. Watt, F. and G. W. Grime. *Principles and Applications of High-Energy Ion Microbeams.* (Bristol: Adam Hilger, 1987).
8. Vis, R. D. *The Proton Microprobe: Applications in the Biomedical Field.* (Boca Raton, FL: CRC Press, 1985).

9. Johansson, S. A. E. and J. L. Campbell. *PIXE — A Novel Technique for Elemental Analysis*. (New York: John Wiley & Sons, 1988).
10. Finstad, T. G. and W. K. Chu. "Ion Beam Techniques," in *Analytical Techniques for Thin Films*, K. N Tu and R. Rosenberg, Eds. (Boston: Academic Press, 1988).
11. Bird, J. R. in *Ion Beams for Material Analysis*, J. S. Williams, Ed. (Marrickville, Australia: Academic Press, 1989).
12. Artaxo, P., R. Van Grieken, F. Watt, and M. Jaksic. "The Microanalysis of Individual Aerosol Particles by Electron, Proton and Laser Microprobe," *Proc. of the 2nd World Congr. on Particle Technology*, Society of Powder Technology, Tokyo, Japan, 421 (1990).
13. Boni, C., E. Cereda, G. M. Braga-Marcazzan, and V. De Tomasi. "Prompt Gamma-Remission Excitation-Functions for PIGE Analysis of Li, B, F, Mg, Al, Si, and P in Thin Samples." *Nuclear Instrum. Methods Phys. Res. Sect. B*, B35, 1, 80 (1988).
14. Van Vaeck, L. and R. Gijbels. "Overview of Laser Microprobe Mass Spectrometry Techniques," *Microbeam Anal.* XVII–XXV, 35 (1989).
15. Cotter, R. "Time of Flight Mass Spectrometry: An Increasing Role in the Life Sciences," *Biomed. Environ. Mass Spectrom.* 18:513 (1989).
16. Davis, R. and M. Frearson. in *Mass Spectrometry*, F. Richard, Ed. (London: John Wiley & Sons, 1987).
17. Conzemius, R. and J. Capellan. "A Review of Applications to Solids of Laser Ion Source in Mass Spectrometry," *Int. J. Mass Spectrom. Ion Phys.* 34:197 (1980).
18. Cotter, R. "Laser and Mass Spectrometry," *Anal. Chem.* 56:485A (1984).
19. Verbueken, A. H., F. J. Bruynseels, and R. E. Van Grieken. "Laser Microprobe Mass Analysis: A Review of Applications in the Life Sciences," *Biomed. Mass Spectrom.* 12:438 (1985).
20. Kaufmann, R., P. Richmann, J. Tourmann, and H. Schnatz. "LAMMS in Biomedical Research: Achievements, Shortcomings, Promises," *Microbeam Anal.* 35 (1989).
21. Fletcher, R. and L. Currie. *Microbeam Anal.*, 303 (1989).
22. Harrington, P., T. Street, K. Voorhees, F. Radicati di Brozolo, and R. W. Odom. "Rule-Building Expert System for Classification of Mass Spectra," *Anal. Chem.* 61:715 (1989).
23. Wouters, L. "Laser Microprobe Mass Analysis of Environmental Particles," Ph.D. thesis, University of Antwerp, Antwerp, Belgium (1991).
24. Lodding, A. "Secondary Ion Mass Spectrometry," in *Inorganic Mass Spectrometry*, F. Adams, R. Gijbels, and R. Van Grieken, Eds. (New York: John Wiley & Sons, 1988).
25. Newbury, D. E. "Secondary Ion Mass Spectrometry for the Analysis of Single Particles," in *Characterization of Particles*, NBS Spec. Publ. 533, Proc. of the Particle Analysis Session of the 13th Annual Conference of the Microbeam Analysis Society, K. F. J. Heinrich, Ed., held at Ann Arbor, Michigan, June 22, 1978, 139 (1980).
26. Morrison, G. and G. Slodzian. "The Ion Microscope Opens New Vistas in Many Fields of Science by its Ability to Provide Spatially Resolved Mass Analysis of Solid Surfaces, *Anal. Chem.* 47:932A (1975).
27. Benninghoven, A. *Appl. Phys.* 1:3 (1982).

28. Klaus, N. in *Physical and Chemical Characterization of Individual Particles*, K. R. Spurney, Ed. (New York: John Wiley & Sons, 1986).
29. Marien, J. and E. De Pauw. "On the Identification of Sulphur Oxidation State in Inorganic Sodium Sulphoxy Salts by Laser Mass Analysis and Secondary Ion Mass Spectrometry," *Anal. Chem.* 57:361 (1985).
30. Hillier, J. and R. F. Baker. *J. Appl. Phys.* 15:663 (1944).
31. Joy, D. C. "The Basic Principles of EELS," in *Principles of Analytical Electron Spectroscopy*, D. C. Joy, A. D. Romig, Jr. and J. I. Goldstein, Eds. (New York: Plenum Press, 1986).
32. Egerton, R. F. "Inelastic Scattering of 80 keV Electrons in Amorphous Carbon," *Phil. Mag.* 31:199 (1975).
33. Maher, D. M., D. C. Joy, R. F. Egerton, and P. Mochel. "The Functional Form of Energy-Differential Cross Sections for Carbon using Transmission Electron Energy-Loss Spectroscopy," *J. Appl. Phys.* 50:5105 (1979).
34. Egerton, R. F. in *Electron Energy-Loss Spectroscopy in the Electron Microscope*, (New York: Plenum Press, 1986).
35. Jeanguillaume, C., M. Tencé, P. Trebbia, and C. Colliex. "Electron Energy Loss Chemical Mapping of Low Z Elements in Biological Sections," *Scanning Electron Microsc.* 11:745 (1983).
36. Wolf, B. "Elektronen-energie-verlust-spektroskopische Untersuchungen en Luftfiltern zu Bewertung der Luftqualität, *Z. Naturforsch.* 43c:155 (1988).
37. Xhoffer, C., P. Berghmans, I. Muir, W. Jacob, R. Van Grieken, and F. Adams. "A Method for the Characterization of Surface Modified Asbestos Fibres by Electron Energy Loss Spectroscopy," *J. Microsc.* 162, Pt 1:179 (1991).
38. Xhoffer, C., W. Van Borm, R. Van Grieken, W. Jacob, and J. Broekaert. "Study of Exhaust Aerosols in Slurry Nebulization Inductively Coupled Plasma Spectrometry for Ceramic Powders," 1991 Winter Conf. on Plasma Spectrochemistry, Dortmund (FRG) January 14–18, Or 19 (1991).
39. Messerschmidt, R. G. "Design and Performance Standards for Infrared Microscopes in Spectroscopy," *Microbeam Anal.* 169 (1987).
40. Nyquist, R. A., M. A. Leugers, M. L. McKelvy, R. R. Papenfuss, C. L. Putzig, and L. Yurga. "Infrared Spectrometry," *Anal. Chem.* 62:223R (1990).
41. Gosz, J., C. Dahm, and P. Risser. "Long-Path FTIR Measurement of Atmospheric Trace Gas Concentrations," *Ecology* 69:1326 (1988).
42. Levine, S., Y. Li-shi, C. Strang, and X. Hong-Kui. "Advantages and Disadvantages in the Use of FT-IR and FIR Spectrometers for Monitoring Airborne Gases and Vapors of Industrial Hygiene Concern," *Appl. Ind. Hyg.* 4:180 (1989).
43. Li-Shi, Y. and S. Levine. "Fourier Transform Infrared Least-Squares Methods for the Quantitative Analysis of Multicomponent Mixtures of Airborne Vapors of Industrial Hygiene Concern," *Anal. Chem.* 61:677 (1989).
44. Small, G., R. Kroutil, J. Ditillo, and W. Loerop. "Detection of Atmospheric Pollutants by Direct Analysis of Passive Fourier Transform Interferograms," *Anal. Chem.* 60:264 (1988).
45. Gordon, R., N. Trivedi, and B. Singh. "Characterization of Aerosol Organics by Diffuse Reflectance Fourier Transform Infrared Spectroscopy," *Environ. Sci. Technol.*, 22:672 (1988).
46. Dangler, M., S. Burke, S. Hering, and D. Allen. "A Direct Method of Identifying Functional Groups, in Size Segregated Atmospheric Aerosols," *Atmos. Environ.* 21:1001 (1987).

47. Kellner, R. and H. Malissa. "Fourier Transform Infrared Microscopy — A Tool for Speciation of Impactor-Sampled Single Particles or Particle Clusters," *Aerosol Sci. Technol.* 10:397 (1989).

48. Etz, E. and J. Blaha. in "Characterization of Particles," *Proc. of the Particle Analysis Session of the 13th Int. Conf. of the Microbeam Analysis Society,* K. Heinrich, Ed., Ann Arbor, MI, June 22, 153 (1980).

49. Purcell, F. and E. Etz. "A New Spectrograph with a Multichannel Optical Detector for the Raman Characterization of Microparticles," *Microbeam Anal.* 17:301 (1982).

50. Etz, E. and G. Rosasco. "Identification of Individual Microparticles with a New Micro-Raman Spectrometer," NBS Spec. Publ. (US), Vol. 464, National Bureau of Standards, Gaithersburg, MD, 343 (1977).

51. Etz, E. and G. Rosasco. "Application of a New Micro-Raman Spectrometer to the Identification of Airborne Particulates," Proc. Int. Conf. Raman Spectrosc., 5th, Hans Ferdinand Schulz Verlag Frieburg, Germany, AVAIL: E. Schmid, J. Brandmieller, and W. Kiefer, Eds. 776 (1976).

52. Dhamelincourt P., F. Wallart, M. Leciercq, A. N'Guyen, and D. Landon. "Laser Raman Molecular Microprobe," *Anal. Chem* 51:414A (1979).

53. Rosasco, G. J. in *Advances in Infrared and Raman Spectroscopy,* Vol. 7, R. J. H. Clark and R. E. Hester, Eds. (London: Heyden, 1980), p. 223.

54. Knoll, P., M. Marchi, and W. Kiefer. "Raman Spectroscopy of Microparticles in Laser Light Traps," *Ind. J. Pure Appl. Phys.* 26, February–March:268 (1988).

55. Kiefer, W. *Croatica Chem. Acta* 61(3):473 (1988).

56. Thurn, R. and W. Kiefer. "Raman-microsampling Technique Applying Optical Levitation by Radiation Pressure," *Appl. Spectrosc.* 38:78 (1984).

57. Ebel, M. F. in "Angewandte Oberflächenanalyse mit SIMS, AES und XPS," M. Grasserbauer, H.J. Dudrk, and M. F. Ebel, Eds. (Berlin: Springer-Verlag, 1986).

58. Briggs, D. and M. P. Seah, Eds. *Practical Surface Analysis by Auger and X-Ray Photoelectron Spectroscopy,* (Chichester: John Wiley & Sons, 1983).

59. Cazaux, J. "X-ray Photoelectron Microanalysis and Microscopy — Principles and Expected Performances," *Rev. Phys. Appl.* 10:263 (1975).

60. Dam, R., K. Kongslie, and D. Griffith. "Photoelectron Quantum Yields and Photoelectron Microscopy of Chlorophyl and Chlorophyllia," *Photochem. Photobiol.* 22:265 (1975).

61. Linton, R. W., D. T. Harvey, and G. E. Cabaniss. in *Analytical Aspects of Environmental Chemistry,* D. F. S. Natusch and P. K. Hopke, Eds. (New York: John Wiley & Sons, 1983), chap. 4.

62. Cox, X. B., III and R. W. Linton. "Particle Analysis by X-ray Photoelectron Spectroscopy," in *Physical and Chemical Characterization of Individual Airborne Particles,* K. R. Spurney, Ed. (New York: John Wiley & Sons, 1986), chap. 18.

63. Powell, C. in "Characterization of Particles," *Proc. of the Particle Analysis Session of the 13th Int. Conf. of Microbeam Analysis Society,* K. Heinrich, Ed., Ann Arbor, MI, June 22, 131 (1980).

64. Duce, R. A. and E. J. Hoffman. "Chemical Fractionation at the Air/Sea Interface," *Annu. Rev. Earth Planet. Sci.* 3:187 (1976).

65. Otten, P., F. Bruynseels, and R. Van Grieken. "Nitric Acid Interaction with Marine Aerosols Sampled by Impaction," *Bull. Soc. Chim. Belg.* 95, 447 (1986).

66. Bruynseels, F., H. Storms, R. Van Grieken, and L. Van Der Auwera. "Characterization of North Sea Aerosols by Individual Particle Analysis," *Atmos. Environ.* 22:2593 (1988).

67. Bruynseels, F. and R. Van Grieken. "Direct Detection of Sulphate and Nitrate Layers on Sampled Marine Aerosols by Laser Microprobe Mass Analysis," *Atmos. Environ.* 19:1969 (1985).

68. Allegrini, I. and G. Mattogno. "Analysis of Environmental Particulate Matter by Means of Photoelectron Spectroscopy, *Sci. Total Environ.* 9:227 (1978).

69. Kolaitis, L., F. Bruynseels, R. Van Grieken, and M. Andreae. "Determination of Methanesulphonic Acid and Non-Sea-Salt Sulfate in Single Marine Aerosol Particles," *Environ. Sci. Technol.* 23:236 (1989).

70. Wouters, L., P. Artaxo, and R. Van Grieken. "LAMMA of Individual Antarctic Aerosol Particles," *Int. J. Environ. Anal. Chem.* 38:427 (1990).

71. Otten, P., F. Bruynseels, and R. Van Grieken. "Study of Inorganic Ammonium Compounds in Individual Marine Aerosol Particles by Laser Microprobe Mass Spectrometry," *Anal. Chim. Acta* 195:117 (1987).

72. Xhoffer, C., P. Bernard, R. Van Grieken, and L. Van Der Auwera. "Chemical Characterization and Source Apportionment of Individual Aerosol Particles over the North Sea and the English Channel Using Multivariate Techniques, *Environ. Sci. Technol.*, in press (1991).

73. Harrison, R. H. and W. T. Sturges. "Fysico-Chemical Speciation and Transformation Reactions of Particulate Nitrogen and Sulphur Compounds," *Atmos. Environ.* 18:1829 (1984).

74. Xhoffer, C. "Electronenprobe microanalyse van vliegas en partikels uit het marine milieu," Master of Science thesis, University of Antwerp, Antwerp, Belgium, 1987.

75. Hopke, P. K. *Receptor Modelling in Environmental Chemistry*, (New York: John Wiley & Sons, 1985).

76. Andreae, M. O., R. J. Charlson, F. Bruynseels, H. Storms, R. Van Grieken, and W. Maenhaut. "Internal Mixture of Sea Salt, Silicates, and Excess Sulphate in Marine Aerosols," *Science* 232:1620 (1986).

77. Maenhaut, W., H. Raemdonck, A. Selen, R. Van Grieken, and J. W. Winchester. "Characterization of Atmospheric Aerosol over the Eastern Equatorial Pacific," *J. Geophys. Res.* 88:5353 (1983).

78. Riley, J. P. and R. Chester. *Introduction to Marine Chemistry*, (London: Academic Press, 1971).

79. Patterson, E. M., C. S. Kiang, A. C. Delany, A. F. Artburg, A. C. D. Leslie, and B. J. Huebert. "Global Measurements of Aerosols in Remote Continental and Marine Regions: Concentrations, Size Distributions and Optical Properties," *J. Geophys. Res.* 8:7361 (1980).

80. Parungo, F., B. Bodhaine, and J. Bortnak. "Seasonal Variation in Antarctic Aerosol," *J. Aerosol Sci.* 12:491 (1981).

81. Shaw, G. E. "X-ray Spectrometry of Polar Aerosols," *Atmos. Environ.* 17:329 (1983).

82. Ito, T. "Study of Background Aerosols in the Antarctic Troposphere," *J. Atmos. Chem.* 3:69 (1985).

83. Bodhaine, B. A. and M. E. Murphy. "Calibration of an Automatic Condensation Nuclei Counter at the South Pole," *J. Aerosol. Sci.* 11:305 (1980).

84. Bodhaine, B. A., J. J. Deluisi, and J. M. Harris. "Aerosol Measurements at the South Pole," *Tellus,* 38B:223 (1986).

85. Cunningham, W. and W. Zoller. "The Chemical Composition of Remote Aerosols," *J. Aerosol Sci.* 12:367 (1981).

86. Bigg, E. K. "Comparison of Aerosol at 4 Baseline Atmospheric Monitoring Stations," *J. Appl. Met.* 19:521 (1980).

87. Storms, H. "Quantification of Automated Electron Probe X-ray Analysis and Application in Aerosol Research," Ph.D. thesis, University of Antwerp (UIA), 233, Antwerp, Belgium (1988).

88. Sheridan, P. J. and I. H. Musselman. "Characterization of Aircraft-Collected Particles Present in the Arctic Aerosol, Alaskan Arctic Spring 1983," *Atmos. Environ.* 19:2159 (1985).

89. Surkyn, P., J. De Waele, and F. Adams. "Laser Microprobe Mass Analysis for Source Identification of Air Particulate Matter," *Int. J. Environ. Anal. Chem.* 13:257 (1983).

90. Bigg, E. K. and D. E. Turvey. "Sources of Atmospheric Particles over Australia," *Atmos. Environ.* 12:1643 (1978).

91. Beauford, W., J. Barber, and A. Barringer. "Release of Particles Containing Metals from Vegetation into the Atmosphere," *Science* 195:571 (1977).

92. Nemeruyk, G. E. "Migration of Salts into the Atmosphere during Transpiration," *Sov. Plant Physiol.* (English transl.) 17:560 (1970).

93. Delmas, R., B. Clairac, B. Cros, H. Cachier, and J. Servant. "Chemical Composition of Atmospheric Aerosols in an Equatorial Forest Area," 2nd Int. Symp. on Biosphere-Atmosphere Exchanges, Mainz FRG, March 16 (1986).

94. Bernard, P., R. Van Grieken, and D. Eisma. "Classification of Estuarine Particles Using Automated Electron Microprobe Analysis and Multivariate Techniques," *Environ. Sci. Technol.* 20:467 (1986).

95. Bruynseels, F., P. Artaxo, H. Storms, and R. Van Grieken. "LAMMA Study of Aerosol Samples Collected in the Amazon Basin," *Microbeam Anal.* 356 (1987).

96. Artaxo, P., H. Storms, F. Bruynseels, R. Van Grieken, and W. Maenhaut. "Composition and Sources of Aerosols from the Amazon Basin," *J. Geophys. Res.* 93:1605 (1988).

97. Van Borm, W., F. Adams, and W. Maenhaut. "Characterization of Individual Particles in the Antwerp Aerosol," *Atmos. Environ.* 23:1139 (1989).

98. Van Borm, W., L. Wouters, R. Van Grieken, and F. Adams. "Lead Particles in an Urban Atmosphere: An Individual Particle Approach," *Sci. Total Environ.* 90:55 (1990).

99. Keyser, T. R., D. F. S. Natusch, C. A. Evans, Jr., and R. W. Linton. "Characterizing the Surfaces of Environmental Particles," *Environ. Sci. Technol.,* 12:768 (1978).

100. Post, J. E. and P. R. Buseck. "Characterization of Individual Particles in the Phoenix Urban Aerosol Using Electron-Beam Instruments," *Environ. Sci. Technol.* 18:35 (1985).

101. Fruchter, J. S., D. E. Robertson, J. C. Evans, K. B. Olsen, E. A. Lepel, J. C. Laul, K. H. Abel, R. W. Sanders, P. O. Jackson, M. S. Wagman, R. W. Perkins, H. H. Van Tuyl, R. H. Beauchamp, J. W. Shade, J. L. Daniel, R. L. Erikson, G. A. Sehmel, R. N. Lee, A. V. Robinson, O. R. Mass, J. K. Briant, and W. C. Cannon. "Mount St. Helen Ash from the 18 May 1980 Eruption: Chemical, Physical, Mineralogical and Biological Properties, *Science*, 209:1116 (1980).

102. Wightman, J. P. "XPS Analysis of Mount St. Helens Ash," *Colloids Surf.* 4:401 (1982).

103. Gooding, J. L., U. S. Clanton, E. M. Gabel, and J. L. Warren. "El Chichón Volcanic Ash in the Stratosphere: Particle Abundances and Size Distributions after the 1982 Eruption," *Geophys. Res. Lett.* 10:1033 (1983).

104. Rampino, M. R. and S. Self. "The Atmospheric Effects of El Chichón," *Sci. Am.* 250:48 (1984).

105. Brownlee, D. E. "Terrestrial and Extraterrestrial Pollution in the Stratosphere," *Microbeam Anal.* 199 (1980).

106. Cadle, R. D. and G. W. Grams. "Stratospheric Aerosols and their Optical Properties," *Rev. Geophys. Space Physics* 13:475 (1975).

107. Mackinnon, I. D. R., J. L. Gooding, D. S. McKay, and U. S. Clanton. "The El Chichón Stratospheric Cloud: Solid Particulated and Settling Rates," *J. Volcanol. Geotherm. Res.* 23:125 (1984).

108. Barths, G., G. Schmidtz, R. Kaufmann, J. Bruch, and J. Tourmann. *Colloq. INSERM* (1987).

109. De Waele, J., P. Van Espen, E. Vansant, and F. Adams. "Study of Asbestos by Laser Microprobe Mass Analysis," *Microbeam Anal.* 371 (1982).

110. De Waele, J., E. Vansant, P. Van Espen, and F. Adams. "Laser Microprobe Mass Analysis of Asbestos Fiber Surfaces for Organic Compounds," *Anal. Chem.* 55:671 (1983).

111. Van Espen, P., J. De Waele, E. Vansant, and F. Adams. "Study of Asbestos Using SIMS and LAMMA," *Int. J. Mass Spectrom. Ion Phys.* 46:515 (1983).

112. De Waele, J., E. Vansant, and F. Adams. "Laser Microprobe Mass Analysis of N,N-Dimethylaniline and Catalytic Oxidation Products Adsorbed on Asbestos Fiber Surfaces," *Microchim. Acta (Wien)* 367: (1983).

113. De Waele, J., I. Verhaert, E. Vansant, and F. Adams. "Laser Microprobe Mass Analysis (LAMMA) and Adsorption Study of Alifatic Alkylammonium Ions and Alkylamines on Asbestos Fibre Surfaces," *Surf. Interface Anal.* 5, 186 (1983).

114. Musselman, I., R. Linton, and D. Simons. "The Use of Laser Microprobe Mass Analysis for Nickel Speciation in Individual Particles of Micrometer Size," *Microbeam Anal.* 337 (1985).

115. Gonduin, S. and J. Muller. "In Situ Identification of Chromium Oxidation States by LAMMA," Proc. of the 3rd Int. Laser Microprobe Mass Spectrometry Workshop, Antwerp, Belgium, 26–27 August (1986).

116. Poitevin, E., J. Muller, F. Klein, and D. Dechelette. "Application de la Détermination des Dégres d'Oxydation du Chrome à l'Hygiène Industrielle par Microsonde à Impact Laser," *Analysis* 17:47 (1989).

117. Michaud, D. Personal communcation (1991).

118. Cox, X. B., III, R. W. Linton, and F. E. Butler. "Determination of Chromium Speciation in Environmental Particles. Multitechnique Study of Ferrochrome Smelter Dust," *Environ. Sci. Technol.* 19, 345 (1985).

119. Seinfeld, J. H. *Air Pollution: Physical and Chemical Fundamentals.* (New York: McGraw Hill, 1975).

120. Ramsden, A. R. and M. Shibaoka. "Characterization and Analysis of Individual Fly Ash Particles from Coal-Fired Power Stations by Combination of Optical Microscopy, Electron Microscopy and Quantitative Electron Microprobe Analysis, *Atmos. Environ.* 16:2191 (1982).

121. Bonafede, G. and L. T. Kiss. "Study of Ash Deposits from Brown Coal Fired Boilers with the Aid of Scanning and Transmission Electron Microscopy," *Am. Soc. Mech. Eng.* 73-WA/CD-7 (1973).

122. Baker, J. E., C. A. Evans, A. Loh, and D. F. S. Natusch. "Microscope Investigations of Coal Fly Ash Particles, *Proc. Annu. Conf. Microbeam Analyt. Soc.* 10:33A (1975).

123. Jan de Zeeuw, H. and R. V. Abresch. "Cenospheres from Dry Fly Ash," Proc. 4th Int. Ash Utilization Symp., St. Louis, ERDA, Morgantown, WV, 386 (1976).

124. Chiu, A.S. "Characterization of Submicron Fly Ash in Cyclone Boilers," Thesis, Department of Chemical Engineering, University of New Hampshire, Durham (1978).

125. Gibbon, D. L. "Microcharacterization of Fly Ash and Analogs: The Role of SEM and TEM," *Scanning Electron Microscopy/* 1979/I, 501, SEM Inc., AMF O'Hare, Illinois (1979).

126. Carpenter, R. L., R. D. Clark, and Su Yin Fong. "Fly Ash from Electrostatic Precipitators and Characterization of Large Spheres," *J. Air Pollut. Control Assoc.* 30:679 (1980).

127. Lichtman, D. and S. Mroczkowski. "Scanning Electron Microscopy and Energy-Dispersive X-ray Spectroscopy Analysis of Submicrometer Coal Fly Ash Particles," *Environ. Sci. Technol.* 19:274 (1985).

128. Middleman, L. M. and J. D. Geller. "Trace Element Analysis Using X-ray Excitation with an Energy Dispersive Spectrometer on Scanning Electron Microscope," *Scanning Electron Microscopy* /1976/, 171, IITRI, Chicago, IL (1976).

129. Small, J. A. and W. H. Zoller. "Single-Particle Analysis of Ash from the Dickerson Coal-Fired Power Plant," *Natl. Bur. Stand. Monogr.* 464:651 (1977).

130. Fisher, G. L., B. A. Prentice, D. Silberman, and J. M. Ondov. "Physical and Morphological Studies of Size-Classified Coal Fly Ash," *Environ. Sci. Technol.* 12:447 (1978).

131. Parungo, F., E. Ackerman, H. Proulx, and R. Pueschel. "Nucleation Properties of Fly Ash in Coal-Fired-Power-Plant Plume," *Atmos. Environ.* 12:929 (1978).

132. Hayes, T. L., J. B. Pawley, and G. L. Fisher. "Effect of Chemical Variability of Individual Fly Ash Particles on Cell Exposure," *Scanning Electron Microscopy* /1978/I, 239, SEM Inc., AMF O'Hare, IL (1978).

133. Capron, R., P. Haymann, and F. Pellerin. "Preliminary Quantitative Analysis of Combustion Products by Electron Microscopy in Environmental Monitoring," *C. R. Acad. Sci. C* 289:313 (1979).

134. Kaufherr, N. and D. Lichtman. "Comparison of Micron and Submicron Fly Ash Particles Using Scanning Electron Microscopy and X-ray Elemental Analysis," *Environ. Sci. Technol.* 18:544 (1984).

135. Campbell, J. A., R. D. Smith, and L. E. Davis. "Application of X-ray Photoelectron Spectroscopy to the Study of Fly Ash," *Appl. Spectrosc.* 32:316 (1978).

136. von Rosenstiel, A. P., A. J. Gay, and P. J. Van Duin. Mikromorphologische und microchemische Untersuchungen von Flugasche, Beitr. elektronenmikroskop. *Direktabb. Oberfl.* 14:153 (1981).

137. Linton, R. W., A. Loh, D. Natusch, and P. Williams. "Surface Predominance of Trace Elements in Airborne Particles," *Science* 191:853 (1975).

138. Linton, R. W., P. Williams, C. A. Evans, and D. F. S. Natusch. "Characterization of Surface Predominance of Toxic Elements in Airborne Particles by Ion Microprobe Mass Spectrometry and Auger Electron Spectrometry," *Anal. Chem.* 49:1514 (1977).

139. Cox, X. B., III, S. R., Bryan, and R. W. Linton. "Microchemical Characterization of Trace Elemental Distributions within Individual Coal Combustion Particles Using Secondary Ion Mass Spectrometry and Digital Imaging," *Anal. Chem.* 59:2018 (1987).

140. Wagman, J. "Chemical Composition of Atmospheric Aerosol Pollutants by High Resolution X-ray Fluorescence Spectrometry," *Colloid and Interface Science,* Vol. II, (New York: Academic Press, 1976).

141. Denoyer, E., T. Mauney, D. F. S. Natusch, and F. Adams. "Laser Microprobe Mass Analysis of Coal and Oil Fly Ash Particles," *Microbeam Anal.* 191 (1982).

142. Mamane, Y., J. L. Miller, and T. G. Dzubay. "Characterization of Individual Fly Ash Particles Emitted from Coal- and Oil-Fired Power Plants," *Atmos. Environ.* 20:2125 (1986).

143. Raeymaekers, B. "Characterization of Particles by Automated Electron Probe Microanalysis," Ph.D. Thesis, University of Antwerp, Antwerp, Belgium, 1987.

144. McCrone, W. C. and J. G. Delly, Eds. *The Particle Atlas* 2nd ed. (Ann Arbor, MI: Ann Arbor Science, 1973).

145. Vis, R. D., A. J. J. Bos, V. Valkovic, and H. Verheul. "The Analysis of Fly Ash Particles with a Proton Microbeam, *IEEE Trans. Nuclear Sci.* Vol. NS-30, 2:1236 (1983).

146. Mauney, T. and F. Adams. "Laser Microprobe Mass Spectrometry of Environmental Soot Particles," *Sci. Total Environ.* 36:215 (1984).

147. McHugh, J. and F. Stevens. "Elemental Analysis of Single Micrometer-Size Airborne Particulates by Ion Microprobe Mass Spectrometry," *Anal. Chem.* 44:2187 (1972).

148. Lang, P., J. Katon, and A. Bonanno. "Identification of Dust Particles by Molecular Microscopy," *Appl. Spectrosc.* 42:313 (1988).

149. Dehairs, F., R. Chesselet, and J. Jedwab. "Discrete Suspended Particles of Barite and Barium Cycle in the Open Ocean," *Earth Planet. Sci. Lett.* 49:528 (1980).

150. Jedwab, J. "Rare Anthropogenic and Natural Particles Suspended in Deep Ocean Waters," *Earth Planet. Sci. Lett.* 49:551 (1980).

151. Skei, J. M. and S. Melson. "Seasonal and Vertical Variations in the Chemical Composition of Suspended Particulate Matter in an Oxygen-Deficient Fjord," *East Coast Shelf Sci.* 14:61 (1982).

152. Sundby, B. N., N. Silverberg, and R. Chesselet. "Pathways of Manganese in an Open Estuarine System," *Geochim. Cosmochim. Acta,* 45:293 (1984).

153. Bishop, J. K. B. and P. E. Biskaye. "Chemical Characterization of Individual Particles from the Nepheloid Layer in the Atlantic Ocean," *Earth Planet. Sci. Lett.,* 58:265 (1982).

154. Lambert, C. E., J. K. B. Bishop, P. E. Biscaye, and R. Chesselet. "Particulate Aluminum, Iron and Manganese Chemistry in Deep Atlantic Boundary Layer," *Earth Planet. Sci. Lett.* 70:237 (1984).

155. Wouters, L., P. Bernard, and R. Van Grieken. "Characterization of Individual Estuarine and Marine Particles by LAMMA and EPXMA," *Int. J. Environ. Anal. Chem.* 34:17 (1988).

156. Bernard, P., R. Van Grieken, and L. Brügmann. "Geochemical Composition of Suspended Matter in the Baltic Sea. I. Results of Individual Particle Characterization by Automated Electron Microprobe," *Mar. Chem.* 26:155 (1989).

157. Wilken, J. *Sci. Total Environ.*, Submitted (1991).

158. Eisma, D., P. Bernard, J. Boon, R. Van Grieken, J. Kalf, and W. Mook. "Loss of Particulate Organic Matter in Estuaries as Exemplified by the Ems and the Gironde Estuaries," *Mitt. Geol. Palaentol. Inst. Univ. Hamburg,* 58 (1985).

159. Eisma, D., P. Bernard, G. Cadee, V. Ittekkot, J. Kalf, R. Laane, J. Martin, W. Mook, A. Van Put, and T. Schumacher. Suspended Matter Particle Size in Some West-European Estuaries, *Neth. J. Sea Res.* Submitted, (1991).

160. Araújo, F., Z. Komy, A. Van Put, and R. Van Grieken. "Elemental Characterization of Sediments, Water and Groundwater in the Nile River Basin (Central Egypt): A Preliminary Study," *Sci. Total Environ.*, Submitted (1992).

161. Hart, B., G. Douglas, R. Beckett, A. Van Put, and R. Van Grieken. "Characterization of SPM in the Magela Creek System — Northern Australia," *Environ. Sci. Technol.*, Submitted (1991).

162. Eisma, D., J. Kalf, M. Karmini, W. Mook, A. Van Put, P. Bernard, and R. Van Grieken. "Dispersal of Suspended Matter in Makasar Strait and the Flores Basin," *Neth. J. Sea Res.* 24(4):383 (1989).

163. Eisma, D., A. Van Put, and R. Van Grieken. "Distribution and Composition of Suspended Particulate Matter around Sumbawa Island, Indonesia," *Neth. J. Sea Res.* in press (1991).

164. Bishop, J. "The Barite-Opal-Organic Carbon Andiatron in Oceanic Particulate Matter," *Nature* 332:341 (1988).

165. Middelburg, J., G. De Lange, H. Van der Sloot, P. Van Emburg, and S. Sophiah. "Particulate Manganese and Iron Framboids in Vau Bay, Halmahera (Eastern Indonesia)," *Mar. Chem.*, 23:353 (1989).

166. Luther, G., A. Meyerson, J. Krajewski, and R. Heres. "Metal Sulfides in Estuarine Sediments," *J. Sedim. Petrol.* 50:1117 (1980).

167. Grossmann, D., M. Kersten, M. Niecke, and A. Puskeppel. "Determination of Trace Elements in Membrane Filter Samples of Suspended Matter by the Hamburg Proton Microprobe," *SCOPE/UNEP Sonderband Heft* 58, 619 (1985).

CHAPTER 4

CHARACTERIZATION OF THE Cr(VI)/Cr(III) RATIO IN AEROSOLS

Rob F.M.J. Cleven and Jan L.M. de Boer

Laboratory of Inorganic Chemistry, NIPHEP, Bilthoven, The Netherlands

and

Anton van der Meulen

Laboratory of Air Research NIPHEP, Bilthoven, The Netherlands

TABLE OF CONTENTS

1.   Introduction..146

2.   Chromium Chemistry 147
      2.1   General...147
      2.2   Thermodynamics and Kinetics of Cr(VI)/Cr(III)..........148

3.   Pathways of Airborne Chromium 151
      3.1   Atmospheric Emissions......................................151
      3.2   Chromium Levels in Ambient Air............................151

3.3   Behavior of Airborne Chromium 153
       3.3.1 Chemical Forms 153
       3.3.2 Physicochemical Characterization 154

4.   Speciation of Chromium in Aerosols 155
       4.1   Sampling 156
       4.2   Sample Pretreatment 157
       4.3   Speciation Methods 157

5.   Concluding Remarks 161

References 162

1. INTRODUCTION

Under environmental conditions chromium in compounds exists in the trivalent Cr(III) or the hexavalent Cr(VI) state. Cr(III) is an essential trace element for mammals, including man, whereas it is presumed that Cr(VI) compounds are genotoxic and potentially carcinogenic in humans.[1] Evidence exists for the carcinogenicity of calcium chromate, strontium chromate, and zinc chromate. In the diet, chromium is predominantly present as Cr(III). Cr(VI) present in the diet will be converted partially to Cr(III) due to intragastric reduction. There is sufficient evidence, based on animal studies, that exposure to Cr(VI) compounds by *inhalation* will result in an increased risk for carcinogenic effects.[2] The unit risk for Cr(VI) is 40.10^{-3}.[1] This means that in a hypothetical population of 1000, in which all individuals are exposed continuously from birth throughout their lifetimes to a concentration of $1 \ \mu g.m^{-3}$ of hexavalent chromium in the air they breathe, 40 additional cases of cancer are expected to occur. Assuming an acceptable risk of one extra case of cancer per million persons exposed lifetime, this risk corresponds with an airborne concentration of $0.025 \ ng \ Cr(VI).m^{-3}$. Many people will be exposed to unacceptable chromium levels, *if* all airborne chromium would be in the hexavalent state, as concentrations of total Cr in ambient air amount to some $ng.m^{-3}$. Thus, there exists a need for speciation techniques to measure airborne Cr(VI).

Speciation of chromium in air is a challenge to environmental analytical chemists, because of (i) the rather low levels of airborne chromium in a world where that metal is ubiquitary, (ii) the probably low ratio [Cr(VI)]/[Cr(III)] in air, (iii) the behavior of chromium as a redox chameleon, and (iv) the difference between the rate of oxidation of Cr(III) and the rate of reduction of Cr(VI). In general, analysis and speciation of low concentrations of chromium are difficult. Any data from before 1987 are probably too high.[3] Data on chromium speciation, particularly in ambient air, are rather scarce. In this report, the possible oxidation of Cr(III) in air will be considered, existing data on chromium speciation in air will be reviewed, and analytical methods for valence-specific determination of chromium will be evaluated. It is noted that these methods pertain to aqueous solutions of chromium compounds. The instrumental analysis of airborne chromium is usually also performed in aqueous solutions of the species involved.

2. CHROMIUM CHEMISTRY
2.1 General

Chromium belongs to the group of VIA elements that show some 'noble metal' behavior, on which their corrosion resistance is based. The electron configuration of the outer electrons of chromium is $4s3d^5$. The chemistry of chromium compounds is complicated. Inorganic chromium compounds may occur in oxidation states ranging from $-$II to $+$VI. In *natural systems* Cr(III) and Cr(VI) are the most stable forms of chromium. Cr(III), that appears to be the dominant form, occurs as cationic $Cr(OH)_n^{(3-n)+}$ species (with n $=$ 0 . . . 4), whereas Cr(VI) always occurs in the anionic species CrO_4^{2-} or $HCrO_4^-$.

Cr(III), having an outer electron configuration d^3, has a strong tendency to form hexacoordinated octahedral complexes with ligands such as water, ammonia, halides, sulfates, and organic acids. These kinetically stable complexes can prevent precipitation of Cr(III) at pH values at which it otherwise would precipitate.[4] The chemistry of Cr(III) shows similarities with that of Al(III): Cr_2O_3 is amphoteric, albeit more basic than acidic. Many Cr(III) compounds are slightly soluble in water. The hydrated sulfates, chlorides, and nitrates are water soluble.

Cr(VI) is strongly acidic. All Cr(VI) compounds, except for CrF_6, are oxo-compounds: hydrochromate, chromate, and dichromate ionic species, which are all powerful oxidants. In solution, the proportion of each species is dependent on pH. Calculated abundances of Cr(VI) species in aqueous solution in the pH range from 1 to 8, an ionic strength of $\sim$1 mol.dm^{-3} and a total Cr(VI) concentration of 10^{-6} mol.dm^{-3} are given in Table 1.[5] It is noted that at an ionic strength of $\sim$0, the dichromate concentrations are about half those given in Table 1 and the concentrations of the other species are correspondingly greater. Under environmental conditions, dichromates are not formed at total chromium concentrations less than 0.01 mol.dm^{-3}. The redox potential of the Cr(III)/Cr(VI) couple is pH dependent: it increases with decreasing pH values. The acidity of $H_2Cr_2O_7$ appears to be stronger than that of H_2CrO_4. Zinc chromate and lead chromate show low solubility, whereas most other chromates, such as sodium chromate and potassium chromate, are very soluble in water.

Cr(III) is readily oxidized to CrO_4^{2-} by hypochlorite, peroxide, and oxygen under pressure at high temperature. Heating of chromium compounds in air in the presence of alkalies may yield chromate. The oxidizability of chromium

Table 1. Calculated Abundances of Cr(VI) Species in Aqueous Solution at pH 1–8, a Total Cr(VI) Concentration of 10^{-6} mol.dm^{-3} and an Ionic Strength of $\sim$1 mol.dm^{-3}

pH	Abundance (%)			
	CrO_4^{2-}	$Cr_2O_7^{2-}$	$HCrO_4^-$	H_2CrO_4
2	0.0	0.0	95.0	5.0
3	0.0	0.0	99.4	0.6
4	0.3	0.1	99.6	0.0
5	3.1	0.0	96.9	0.0
6	24.2	0.0	75.8	0.0
7	76.2	0.0	23.8	0.0
8	97.0	0.0	3.0	0.0

From Tandon, R.K. et al. *Talanta* 31:227–228 (1984). With permission.

has been studied by Zatka.[6] The chemical behavior of Cr(III) in alkaline solution is expressed by the following reversible reactions:

$$Cr^{3+} + 3OH^- \rightleftarrows Cr(OH)_3 \tag{1}$$

$$Cr(OH)_3 + OH^- \rightleftarrows [Cr(OH)_4]^- \tag{2}$$

Chromic hydroxide, precipitated according to Equation 1, dissolves in excess hydroxide reagent (2% NaOH/3% Na$_2$CO$_3$) with the formation of complex hydroxochromate(III) ions, Equation 2. The hydroxocomplex is heat sensitive and decomposes on digestion near the boiling point:

$$[Cr(OH)_4]^- \xrightarrow{\text{heat}} CrO(OH) + H_2O + OH^- \tag{3}$$

The newly formed precipitate is no longer soluble in alkaline solutions. In the presence of air, part of the hydroxocomplex may be oxidized, according to:

$$4[Cr(OH)_4]^- + 3O_2 + 4OH^- \xrightarrow{\text{heat}} 4CrO_4^{2-} + 10H_2O \tag{4}$$

Due to different *kinetics*, most of the hydroxocomplex will be hydrolyzed on digestion during 1 to 4 h, before a major oxidation to Cr(VI) will take place.[6]

In acid solution, Cr(III) is hard to oxidize. Cr(VI) can be readily reduced in the presence of organic matter. In the Cr(III)/Cr(VI) transformations, Cr(IV) and Cr(V) may function as intermediates. The lifetime of the unstable Cr(IV) is very short,[7] whereas that of Cr(V) is minutes to days.[8]

2.2 Thermodynamics and Kinetics of Cr(VI)/Cr(III)

In Figure 1, the occurrence of the stable chemical forms of chromium in water is presented as a function of the oxidation potential E$_H$ and pH for 10^{-6}

mol.dm^{-3} Cr in water.[9] For 10^{-8} mol.dm^{-3} Cr as well as 10^{-6} mol.dm^{-3} Cr in the presence of Cl$^-$ the diagram remains virtually the same.[10,11] In Figure 1, the relationship for water in contact with air is also indicated. In Table 2, the standard enthalpy of formation, the standard Gibbs free energy and the entropy of chromium species are given. A review of thermodynamic properties and environmental chemistry of chromium, pertaining to soil and water, has been given by Schmidt.[10] From the thermodynamic data, it appears that hexavalent chromium species can be expected to occur in ambient air.

According to thermodynamic calculations, *only* hexavalent chromium species are expected to be present in *oxygenated natural waters*, at equilibrium. For instance, if the pH of sea water is taken as pH $\simeq 8.1$ and the *theoretical* redox condition for sea water $E_H = 0.74$ V is used, the ratio [Cr(VI)]/[Cr(III)] is calculated to be $\sim 10^{21}$. Even if the redox condition, $E_H = 0.50$ V, measured for sea water is employed, the ratio still is to 10^9.[12,13]

The occurrence of Cr(III) in sea water has been explained assuming that the redox equilibrium is only *slowly* attained once Cr(VI) has been reduced to Cr(III). It is confirmed that Cr(III) is *easily* oxidized to Cr(VI) in the presence of manganese oxides under the conditions prevalent in natural sea water. On the other hand, it is shown that some organic materials can reduce Cr(VI) at the normal pH of sea water.[13] In thermodynamic calculations the presence of organic matter is usually not taken into account.

Schroeder and Lee[14] studied the Cr(III)/Cr(VI) transformation in natural *waters*. They found that only 3% of Cr(III) was oxidized by O_2 in 30 days at ambient temperature. In *alkaline* natural waters, particularly in sea water, oxidation of Cr(III) has been reported. The abundance of Cr(VI), as reported

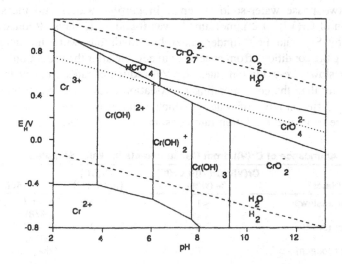

Figure 1. E_H - pH equilibrium diagram for 10^{-6} mol.dm^{-3} chromium in water.[9] ——— between two species indicates equal concentrations of each species. ------ represents the thermodynamic limits for the stability of water. ······· represents the relationship for water in contact with the atmosphere.

Table 2. Gibbs Free Energy ($\Delta G°$), Enthalpy of Formation ($\Delta H°$), and Entropy (S°) of Chromium Species, at 25°C and 1 Bar

Species	State	$\Delta G°$ (kJ/mole)	$\Delta H°$ (kJ/mole)	S° (J/degree-mole)
Cr	s	0	0	23.77
Cr_2O_3	s	-1065.41	-1147.00	81.17
CrO_3	s	-512.54	-591.49	66.53
$Cr(OH)_2$	s	-596.68	-679.57	81.17
$Cr(OH)_3$	s	-869.10	-997.84	95.40
Cr^{2+}	aq	-184.05	-143.51	29.16
Cr^{3+}	aq	-225.48	-254.39	-269.07
$Cr(OH)^{2+}$	aq	-440.28	-488.61	-101.00
$Cr(OH)_2^+$	aq	-641.62	-177.29	-41.97
$Cr(OH)_3$	aq	-834.33	-961.36	
$Cr_2(OH)_2^{4+}$	aq	-896.46	-1028.55	-321.21
$Cr_3(OH)_4^{5+}$	aq	-1578.67	-1801.88	-332.67
$Cr(OH)_4^-$	aq	-1017.88	-1176.16	
CrO_2^-	aq	-540.32		
CrO_4^{2-}	aq	-731.41	-882.53	57.66
$HCrO_4^-$	aq	-768.56	-878.77	194.89
H_2CrO_4	aq	-763.66		
$Cr_2O_7^{2-}$	aq	-1308.71	-1491.76	281.79

From Schmidt, R.L. Rep. PNL-4881 UC-1, U.S. Dept. of Energy, Contract DE-AC06-76RLO 1830 (1984).

for different natural and waste water samples, is presented in Table 3. Cr(VI) is the major chemical form in sea water. Saleh et al.[19] studied the kinetics of chromium transformation in the environment. Rates of the reduction of Cr(VI) and the oxidation of Cr(III) have been evaluated in single phase water systems and in two phase water-solid systems. In *aerobic* waters, no measurable reduction of Cr(VI), 0.2 mmol.dm^{-3}, was found in 41 days. Reductions of Cr(VI) by S^{2-} and Fe^{2+} under anaerobic conditions were instantaneous. Changing the conditions from anaerobic to aerobic after the reduction resulted in much slower re-oxidation rates. Cr(VI) reduction rates were at least ten times faster than the corresponding re-oxidation rates. The oxidation rate of Cr(III) in different types of water in which MnO_2 was present appeared to be slow. The estimated oxidation rate constants corresponded to 2.43 × 10^{-9}

Table 3. Abundance of Cr(VI), from Measurements in Water Samples

Type of Water	Cr(VI)/[Cr(VI) + Cr(III)] % (w/w)	[Cr(VI)] (μg.dm^{-3})	Ref.
Municipal wastewater	<1	50–750	Jan and Young (1978)[15]
Sea water (uncontam.)	~75	0.2	Jan and Young (1978)[15]
Sea water (ocean)	70	0.14	Nakayama et al. (1981)[16]
Sea water (coast)	20–40	2.2–4.2	Vos (1985)[17]
Factory discharge water	18	2	Kudoja (1986)[18]
River water	20–80	1–6	Vos (1985)[17]

s^{-1}.[19] In Table 4, a number of kinetic data on the oxidation of Cr(III) and on the reduction of Cr(VI) is presented.

Cr(III) is oxidized by manganese oxides in *soil*.[22] However, in most cases the percentage Cr(VI) formed will not exceed 15%. James and Bartlett[23] studied factors controlling the oxidation rate of Cr(III) in soils. Levels of water-soluble Cr(VI) increased for a month upon addition of Cr(III), despite the presence of reducing agents. The kinetics of oxidation of aqueous Cr(III) to Cr(VI) by reaction with manganese dioxide appeared to be not appreciably influenced by dissolved oxygen, indicating that Cr(III) reacts directly with the manganese dioxide, and not by catalyzed reactions with oxygen.[24] These findings support the idea that the concentration levels of oxidants other than oxygen present in the atmosphere, such as ozone, may influence the possible Cr(VI)/Cr(III) transformations.

3. PATHWAYS OF AIRBORNE CHROMIUM
3.1 Atmospheric Emissions

The use of chromium has increased dramatically in the 20th century. About 10^8 kg chromium were yearly emitted to the atmosphere in the 1970s from antropogenic sources.[25] Emissions from primary production to the air diminished substantially during the last decade, as a result of improved production techniques and pollution control measures. For the production of chromium metal and chromium compounds, chromite ($FeO \cdot Cr_2O_3$) is generally used. The main industrial compounds of chromium made directly from chrome ores are sodium chromate, sodium dichromate, potassium chromate and potassium dichromate, ammonium dichromate, and chromic acid, all Cr(VI) species.

Chromium compounds from industrial operations enter the ambient air from several sources. Kilns, smelting furnaces, boilers, leaching tanks, open boiling vessels, and plating tanks emit dusts and mists containing chromium. These sources are often small-scale industries. Fly ash emitted from coal-fired power plants may contain 10 to 600 mg $Cr.kg^{-1}$.[26] Also, domestic heating plants are suspected to contribute to the chromium concentration in urban air.[27] Catalytic emission control systems in automobiles using copper chromite catalysts represent another source of chromium emissions to the atmosphere.[28] Some chromate is also lost to the atmosphere as mist from chromate chemicals used as corrosion inhibitors in recirculating cooling waters and in spray painting of automobiles. Indirect sources also include waste incineration and cement plant emissions,[29] and tire dust. Besides, small amounts of chromium will enter the atmosphere as soil-derived aerosol.[30] Cr(VI) compounds in the environment are almost exclusively antropogenic. According to Bergbäck et al.,[31] the enormous amount of chromium in products (technosphere) is a potential source of release of chromium, predominantly as Cr(VI), to the environment in the future.

3.2 Chromium Levels in Ambient Air

Background concentrations in remote areas are about 0.3 ng.m^{-3}. The lowest concentrations determined at the Antarctic are about 0.01 ng.m^{-3},

Table 4. Kinetic Data for Oxidation and Reduction Reactions of Chromium at 25°C

Reaction	[Cr] μmol/dm^{-3}	[B]* mmol/dm^{-3}	pH	Medium mol.dm^{-3}	Rate or Half-life	Ref.
Cr(III) + O$_2$	10	bubbling	8.1	Seawater Borate:0.02	$t_{1/2} \geqslant 12$ days	Nakayama et al. (1981)[13]
MnO$_2$	2.4	0.3	8.6	KHCO$_3$:0.2	$t_{1/2} = 42$ min.	Schroeder and Lee (1975)[14]
		2.9			$t_{1/2} \sim 3$ min	Schroeder and Lee (1975)[14]
γ-MnOOH	10	0.3	8.1	Seawater Borate:0.02	5% in 2 days	Nakayama et al. (1981)[13]
Cr(VI) + Fe(II)	1.9	0.01	7.5	0.02 (buffer)	$t_{1/2} \sim 6$ min	Schroeder and Lee (1975)[14]
		0.02	7.1	0.02 (buffer)	$t_{1/2} \sim 2$ min	Schroeder and Lee (1975)[14]
S^{2-}	1.9	1.0	9.1	—	$t_{1/2} < 5$ min	Schroeder and Lee (1975)[14]
Ascorbic acid	10	0.001	8–12	Seawater Borate:0.02	$t_{1/2} < 72$ h	Nakayama et al. (1981)[13]
Ascorbic acid	10	0.001	2–8	Seawater Borate:0.02	$t_{1/2} \geqslant 72$ h	Nakayama et al. (1981)[13]
Humic acid	10	0.001	7–12	—	$t_{1/2} \geqslant 72$ h	Nakayama et al. (1981)[13]
Humic acid	10	0.001	2–7	—	$t_{1/2} < 72$ h	Nakayama et al. (1981)[13]
Fulvic acid	1.9	DOC: 5.8 (mg.dm^{-3})	5.9	Groundwater	$t_{1/2} \sim 5$ days	Stollenwerk and Grove (1985)[20]
H^{+} [b]	1.9	40	<2.0	Aqueous acid	1–10% loss per 29 days	Stollenwerk and Grove (1985)[20]
Hg	0.2	metallic	6.5	NH$_4$-acet.:0.1	2–6% loss in 15 min	Crosmun and Muller (1975)[21]

[a] [B] = concentration of second reactant.

[b] $4HCrO_4^- + 16H^+ = 4Cr^{3+} + 3O_2 + 10H_2O$; H$^+$ added as HCl, H$_2$SO$_4$, or HNO$_3$.

whereas at the Arctic the concentrations are six to ten times larger. Generally, chromium concentrations in rural areas do not exceed 10 ng.m^{-3}.[32] In cities, the concentrations may be two to four times larger than the local background. Concentration ranges (in ng.m^{-3}) in urban regions, reviewed by Schroeder et al.[33] are Canada 4–26, U.S. 2.2–124, Europe 3.7–277. Obviously, in industrial areas chromium levels will be elevated. Maximum annual average ambient (total) chromium levels within 20 km of industrial sources range from approximately 10 to 135 ng.m^{-3}.[29] In the heavily industrialized area of Liège, for example, a mean concentration of 170 ng.m^{-3} has been measured in 1985.[34] In the U.S., monitoring of the ambient air in many urban and non-urban areas has shown (total) chromium concentrations averaging in the range of approximately 5 to 157 ng.m^{-3}.[29]

Near cooling towers, ambient concentrations of chromium have been measured to be about 50 ng.m^{-3} at distances up to 200 m from the tower.[35] In the neighborhood ($\leq$50 m) of low sources, such as small-scale chrome plating industries, an additional concentration of about 5 ng.m^{-3} on top of the local background may be expected. Thus, near cooling towers and plating industries, substantial amounts of Cr(VI) may be present in the thoracic particulate fraction.

3.3 Behavior of Airborne Chromium
3.3.1 Chemical Forms

Chromium in air is always associated with particles. Little information exists regarding the chemical nature of chromium species present in the atmosphere, although it is assumed that part of the air-chromium exists in the hexavalent form, especially that from high temperature combustion.[36] Gaseous chromium species have never been observed in the atmosphere.[37]

The chemical forms of chromium in air depend on the source of emission. Chrome production, chrome plating, and cooling-tower drifts are primary examples of emissions of Cr(VI) to the atmosphere. Exhaust gas from chrome plating cells is considered to be a main source of Cr in the atmosphere. Prior to mist removal from the plating cells Cr in the exhaust gas exists mainly as Cr(VI), whereas after mist removal the proportion of Cr(VI) decreases.[38] Cox et al.[39] used various wet chemical extractions and different instrumental methods to characterize Cr in ferrochrome smelter dust. It was reported that roughly half of the total Cr in the primary smelter dust was extractable by routine acid/base leaching, of which 40% was Cr(VI). The Cr(VI) existed as $Cr_2O_7^{2-}$, or CrO_4^{2-}, and predominates in submicron particles probably formed during smelting. The remainder of the Cr is primarily insoluble Cr_2O_3 which is located in large particles similar to the original chromite ore. Lindberg et al.[40] report that in decorative chrome plating, Cr(VI) accounted for about half of the total emission of Cr. In polishing operations the percentage Cr(VI) appeared to be about 10 to 20. In welding fumes produced by manual metal arc/stainless steel, about half of the chromium content is in the Cr(VI) form.[41] In fumes generated from stainless steel operations, such as welding and cutting, the amount of hexavalent chromium content can evolve for some minutes, and appears to increase to a maximum level some time after the formation of the aerosol and then partly decay again.[42] According to Sullivan,[43] the most important compound in air may be chromium trioxide CrO_3.

Under normal conditions, Cr(III) and Cr(O) in the air will not undergo any reaction as these species are kinetically stable. However, as the concentration of oxidants like ozone rises, during sunny periods in urban areas, oxidation of airborne Cr(III) should not be excluded. It is noted that in many cities the ozone concentration diurnally varies over an order of magnitude.[44] Preliminary measurements were reported that indicated the possibility of the influence of the metereological conditions on the Cr(VI)/Cr(III) concentration ratio in the atmosphere.[45] Cr(VI) in air may react with dust particles to form Cr(III).[46,47] The exact nature of such transformations between Cr(III) and Cr(VI) in the atmosphere is not extensively studied. It has been reported[48] that preliminary measurements in California indicated the occurrence of 3 to 8% of chromium in the atmosphere as Cr(VI). However, it was annotated that, due to uncertainties in the method, a spread of 0.01 to 30% should be adopted. Speciation measurements in Singapore did not reveal any hexavalent chromium.[49] It is noted that the detection limit of the method used was reported to be 300 $ng.m^{-3}$. Results from measurements in Texas ranged from less than detectable (<5 ng $Cr(VI).m^{-3}$) to 94 ng $Cr(VI).m^{-3}$.[50]

3.3.2 Physicochemical Characterization

Aerosols are defined as particles in the solid or liquid phase with a diameter between 10^{-3} and 10^2 μm that are dispersed in the carrier gas. The aerodynamic mass mean diameter of chromium in airborne suspended particulate matter is in the range of 1.5 to 1.9 μm.[51,52] However, in later studies, mass mean diameters of 3 μm and of 1.1 μm have been reported.[53,54] The concentration of chromium in fly ash was found to increase markedly with decreasing particle size: from about 300 μg.g^{-1} in 11 μm-particles to about 1000 μg.g^{-1} at 1 μm-particles.[55,56] Thus, the chromium levels in particulates from controlled coal combustion may be relatively higher than that from uncontrolled coal combustion, as fine particles may escape the electrostatic precipitator. Moreover, the size of chromium containing particles may have decreased over the last decade as a result of pollution control measures. Chromium containing aerosol particles emitted by coal-fired plants are usually smaller than 1 μm, whereas the particles from open smelting furnaces are substantially larger, 5 to 50 μm.[57] The particles from chrome plating consist mostly of CrO_3 droplets of 100 μm diameter.[58] Cr(VI) in dusts from primary ferrochrome smelters predominates in the submicron particles, whereas Cr_2O_3, which accounts for most of the remaining forms of chromium, is located in large particles, similar to the original chromite ore.[59] The distribution of chromium as a function of the aerodynamic diameter in urban dust was reported to show a bimodal function with its minimum at 0.5 to 1 μm particle diameter.[60] Particles of aerodynamic diameter <10 μm may remain airborne for long (dry) periods, more than 2 weeks, and may then be transported over more than 1000 km. During air transport of particles, the size-distribution will shift to lower values. The aerodynamic mass mean diameters for urban distributions are indeed considerably larger than those of rural areas.[59] The (super)coarse mode will remain airborne for only short periods of time because of their large deposition rate. It is noted that only airborne particulates with its aerodynamic diameter in the range 0.2 to 10 μm are generally considered as respirable.

Between 44 and 96% of the total deposition of chromium occurs by wet precipitation. In general, urban areas have higher wet and total deposition than rural areas.[30] The elimination by wet deposition has been calculated to be about 3 to 4%/h.[61] Atmospheric fallout has become a major contributor to the chromium budget of many ecosystems.[59] For example, the outer leaves of cabbage at about 400 m from a smelter in Humberside contained 0.57 mg.kg^{-1} chromium (dry weight) as compared with <0.1 mg.kg^{-1} (dry weight) for control samples.[62] The total deposition from air ranged for different areas in Germany in 1988 from 0.1 to 2 mg.m^{-2} and appeared to be about 0.4 mg.m^{-2} yearly in The Netherlands during the period 1982 to 1988.[45,63]

In general, the solubility of chromium in particulate emissions from different sources is low in water and dilute acids. It was found to increase with the acidity of the extract.[59] Only 1 to 2% of airborne chromium is soluble in water. About 10 to 20% of chromium in particulates in urban air is exchangeable (using NaCl-solution), about 50% occurs as carbonate of hydrated oxide, 0 to 20% as metal or passivated oxide, and 20 to 30% appears to be organically bound.[64] The solubility of chromium is strongly dependent on chromium dispersion in the particles. The ratio of the chromium concentration in the surface (30 nm) microlayer to the bulk concentration is 7:2.[59]

4. SPECIATION OF CHROMIUM IN AEROSOLS

Methods for chemical analysis of *total* chromium have been reviewed by Griepink.[65] Decomposition methods for the analysis of atmospheric particulates have recently been compared by Yamashige et al.[66] Speciation of chromium started with occupational exposure studies, in which the concentrations of Cr(VI) are relatively high. Generally, speciation studies of chromium concern *valence-specificity*, although other types of speciation have been reported: the distribution of Cr(VI) over water-soluble and insoluble forms, as the soluble forms are considered to show a higher carcinogenic potential,[6] the occurrence of organically bound Cr(VI),[67] and the distribution over different forms of Cr(III).[68] It is noted from biological monitoring studies that the total chromium content in red blood cells,[69] and urinary total chromium,[70] are considered as indicative for exposure to forms of hexavalent chromium in the air. In this report speciation data are focused on valence specificity of instrumental analytical techniques.

4.1 Sampling

A commonly used sampler for particles in ambient air is the passing air collector. It consists of a filter holder and an air pump drawing an air stream through a filter of defined pore size. Particles are collected by impaction on the filter surface, which is weighed before and after exposure to determine the mass. Size selective provisions have been designed to enable the collection of different fractions. Additionally, denuders can be placed in front of the collector to scavenge gaseous oxidants such as ozone, nitrogen dioxide and nitric acid. However the scavenging capacity may be low. These aerosol collectors are used either as high-volume samplers passing 40 to 100 m^3 of air per hour and gathering 0.1 to 1 g of particulate matter for 24 h, or low-volume samplers allowing to collect 10 to 100 mg sample portions per 24 h. Another collecting device is the impinger, in which air is drawn through a solution scavenging the particles. The volume of air to be sampled is smaller than that using a high-volume sampler. The composition of the solution can be adapted for the determination of a particular species (see Chapter 2 of this book for more details on sampling devices for atmospheric particles).

An experimental approach for specific sampling of Cr(VI) in aerosols has been given by Neidhardt et al.[71] Cr(VI) species are selectively adsorbed in solution by human erythrocytes, immobilized on Ca-alginate beads. After a multiple-step clean-up procedure, the Cr(VI) collected by the erythrocytes is determined by graphite furnace atomic absorption spectrometry.

Sampling accounts for the *most* critical step in the determination of airborne Cr(VI).[72] The major sources of error in the sampling stage concern (i) the potential contamination of the sample, as chromium is ubiquitous in the technosphere, whereas the concentration of chromium in the atmosphere is low, (ii) the risk of reduction of Cr(VI) to Cr(III), and (iii) the risk of oxidation of Cr(III) to Cr(VI). Possible means to overcome difficulties due to redox reactions in this stage is drying the aerosol during collection.[73]

The chromium content of filter material widely varies. Gelman filters type GA (0.2 μm pores) contain 0.17 μg Cr per filter, thus 3 mg Cr.kg^{-1}, whereas glass-fiber filters appear to contain a high impurity level of about 30 mg Cr.kg^{-1}, and silver membrane filters about 24 mg Cr.kg^{-1}.[33,74] Cr(VI) brought on various types of filter materials (cellulose acetate, cellulose nitrate, glass-fiber, and polytetrafluoroethylene — PTFE) appeared to be reduced by the filter material with and without passage of air through the filter.[75,76] Coating of PTFE filters with NaOH did not prevent the reduction of added Cr(VI) during air passage through the filter. However, the application of an oxidant layer (PbO$_2$) diminished the reduction to a certain extent,[75] although the possible oxidation of Cr(III) to Cr(VI) could not be excluded. In case filter sampling of Cr(VI) is successful, the samples should be analyzed as soon as possible. On glass-fiber filter the loss of Cr(VI) was estimated to be 40% per month for the first month.[50] The risk of reduction or oxidation is also present when sampling is performed using impingers. Reduction of Cr(VI) might be minimized using an alkaline solution in the impingers.

4.2 Sample Pretreatment

An important part of the sample pretreatment for speciation studies is to ensure the stability of minute quantities of water-soluble Cr(VI) on the filter *and* during the dissolution step.[77] It has been recognized that Cr(VI) must be leached from sample specimen by *alkaline* solutions rather than by dilute sulfuric acid, to protect Cr(VI) from reduction to Cr(III) by reductants possibly present in the sample.[78] It has been recommended to use a solution containing carbonate buffer (pH 9) for the preservation of Cr(VI) in water samples, and a solution of EDTA for the fixation of Cr(III). In this way losses of Cr(VI) over 7 days were reported to be reduced from 60 to 80% to less than 20%.[79] Cr(III) appears not to be oxidized easily by air in alkaline solutions, despite the favorable redox potential of chromium and the excellent stability of the resulting chromate ion at pH >7.[6]

4.3 Speciation Methods

For the determination of *total* chromium a number of methods are available: atomic absorption and atomic emission spectrometry, neutron activation, X-ray fluorescence, gas and liquid chromatography, differential pulse polarography, and chemiluminescence techniques. Detection limits of the various techniques for the determination of *total* chromium associated with airborne particulate matter range from 0.05 to 5 $ng.m^{-3}$ air.[33] Few methods are available for *direct* valence-specific determination of chromium. The valence speciation of chromium in environmental samples presents intricacies due to the chameleon character of the oxidation state of chromium. In some cases sample pretreatment can be employed to eliminate reductants prior to final measurement.[80,81] It can be shown that voltammetric methods are most suitable to the determination of Cr(VI) compounds. Cr(VI) is electrochemically active over the entire pH range, so that medium pH can be selected for analysis, thus protecting samples effectively from undergoing redox reactions during the analytical procedure.[80] However, the detection limits of the known electrochemical methods are not sufficiently low. For the determination of Cr(VI) in aerosols very low detection limits are essential. Assuming a background level of total chromium of 5 $ng.m^{-3}$, and sampling 20 m^3 of air, a total amount of 100 ng of chromium is present in the sample. If 1% of this amount of chromium is hexavalent, a total amount of 1 ng of Cr(VI) is present in the sample, which should be measurable by the analytical method used. If 1 ng Cr(VI) is brought into 100 ml of solution a detection limit of 10 $ng.dm^{-3}$ is required. By sampling 80 m^3 of air and assuming that 25% of the amount of chromium is hexavalent, a detection limit of 1 $\mu g.dm^{-3}$ Cr(VI) in the sample solution would be sufficient. For the determination of Cr(VI), much research has been done in the area of welding fumes, and of chrome-plating mists. However the methods developed for those matrices show detection limits too high for the determination of atmospheric Cr(VI). Separation of Cr(III) and Cr(VI) species may be achieved by selective extraction or selective elimination of Cr(VI) species.

Table 5. Survey of Methods to Determine Cr(VI)[1]

Procedure	Detection Limit (μg.dm^{-3})	Ref.
EXTRACTION — AAS		
Complexation with Aliquat-336; extraction in toluene; direct analysis of organic phase with GF-AAS	0.01	de Jong and Brinkman (1978)[82]
Extraction of APDC-complex in MIBK; direct analysis of organic phase with flame-AAS	—	Ichinose et al. (1978) (cited)[83]
Extraction of DDTC-complex in MIBK; direct analysis of organic phase with flame-AAS	—	Ichinose et al. (1978) (cited)[83]
Complexation with Aliquat-336; extraction in MIBK or toluene; direct analysis of organic phase with GF-AAS	1.2	Chao and Pickett (1980)[84]
Extraction of Cr(VI)-thiosemicarbazide-complex in MIBK; direct analysis of organic phase with flame-AAS	30	Wang (1980)[85]
Precipitation of Cr(III) with hydrated iron(II) oxide; extraction of Cr(VI)-APDC-complex in MIBK; direct analysis of organic phase with flame-AAS	0.06	Mullins (1984)[86]
Complexation with DDTC and extraction of the complex in chloroform; back-extraction into a Hg(II)-solution for analysis with GF-AAS	0.01	Wai et al. (1987)[87]
Extraction of APDC-complex in MIBK from phthalate buffer; direct analysis of organic phase with GF-AAS	0.02	Subramanian (1988)[88]
ION-EXCHANGE — AAS/AES		
Collection of Cr(VI) on poly(dithiocarbamate) resin; digestion of the resin with HNO$_3$; detection: ICP-AES	36	Miyazaki and Barnes (1981)[89]
Extraction of filter with ethylenediamine in water; treatment with Dowex 1 × 8–100® anion-exchange resin after acidification; analysis with ICP-AES; calculation by difference with untreated portion	7	Ehman et al. (1988)[50]
Collection of Cr(VI) on Sephadex A-25; desorption by hydroxylammonium chloride solution; detection by GF-AAS	0.01	Hiraide and Mizuike (1989)[90]
COPRECIPITATION — AAS		
Coprecipitation of Cr(III) and Cr(VI) with hydrated iron(III) oxide {Cr(III)} or bismuth oxide {Cr(III + VI)}; dissolution of precipitate in hydrochloric acid; determination with GF-AAS. Cr(VI) calculated by difference (total Cr also determined after oxidation)	0.001[2]	Nakayama et al. (1981b)[22]
Coprecipitation of Cr(VI) with lead sulfate; dissolution in nitric acid; determination with GF-AAS	0.05	Oboils et al. (1987)[91]

In Table 5 a survey is given of different methods for the determination of Cr(VI). The detection limits given in Table 5 refer to *solutions* (extraction solutions of filters, or impinger solutions), and were derived from 3s levels in some cases. In fact, all methods except for one,[48] were developed for the determination of Cr(VI) in *water* samples. However, in principle, the methods

Table 5.(continued) Survey of Methods to Determine Cr(VI)[1]

Procedure	Detection Limit ($\mu g.dm^{-3}$)	Ref.
COPRECIPITATION — SPECTROPHOTO-METRIC METHODS		
Coprecipitation of Cr(VI) with barium sulfate; fusion of the precipitate potassium sodium carbonate; dissolution of the melt in hot water; determination spectrophotometrically with diphenylcarbazide	0.02	Yamazaki (1980)[92]
ELECTROCHEMICAL METHODS		
Addition of ammonium acetate buffer and ethylenediamine detection with differential pulse polarography	10	Crosmun and Mueller (1975)[21]
Buffering with ammonia buffer solution; detection with differential pulse anodic stripping voltammetry	1.6	Fuoco and Papoff (1975)[93]
Differential pulse polarography in 0.2 M NaF-solution	0.8	Cox et al. (1984)[94]
Separation of interfering cations with aluminum from phosphate-buffered solution; detection with DPP	30	Harzdorf and Janser (1984)[95]
Differential pulse voltammetry in 0.1 M dibasic ammonium citrate solution	2	Locatelli and Fagioli (1986)[96]
Polarographic determination in 0.1 M NH$_4$Cl + NH$_3$/KNO$_2$; catalytic current; pH 10.0	2	Khelfets et al. (1988)[97]
COPRECIPITATION — XRF		
Coprecipitation of Cr(III) with iron(III) hydroxide and of Cr(III) + Cr(VI) with iron(III) + iron(II) hydroxide; detection with XRF; calculation by difference	—	Kudoja (1986)[18]
Precipitation of Cr(VI) with DBDTC; coprecipitation of Cr(III) with hydrated iron(III) oxide; detection: XRF	—	Leyden et al. (1985)[98]
MISCELLANEOUS		
Electrodeposition of Cr(VI) on a graphite tube; detection by AAS	0.05	Batley et al. (1980)[99]
Removal of Cr(III) in AA graphite tube after addition of trifluoroacetylacton, tetramethylammonium hydroxide and sodium acetate to the sample prior to atomization step; detection of Cr(VI) by AAS	0.4[3]	Arpadjan and Krivan (1986)[100]
Preconcentration of Cr(VI) on a C-18 bonded silica column after addition of PIC A; detection: flame-AAS	1.2	Syty et al. (1986)[101]
Extraction of filter at pH of 4.5 in presence of APDC; aspiration of extraction solution through Sep-Pak C$_{18}$ cartridge; desorption of Cr(VI)-APDC complex with acetone; evaporation of acetone and dissolution in acid; analysis by GF-AAS	0.1	Air Resources Board (1986a)[102]
Preconcentration of Cr(VI) on C-18 column after complexation with sodium di(trifluoroethyl)dithiocarbamate; elution with toluene; detection by GC	0.05	Schaller and Neeb (1987)[103]

Table 5.(continued) Survey of Methods to Determine Cr(VI)[1]

Procedure	Detection Limit (μg.dm^{-3})	Ref.
Collection of Cr(III) on polystyrene-divinylbenzene column after complexation with quinolin-8-ol; elution followed by digestion; detection by GF-AAS; determination of Cr(III) + Cr(VI) after reduction of Cr(VI) with hydroxylamine; calculation by difference	0.004	Isshiki et al. (1989)[104]

[1] Abbreviations:
APDC	ammonium pyrrolidine dithiocarbamate
Aliquat-336	methyltricaprylylammonium
DDTC	diethyldithiocarbamate
MIBK	methyl isobutyl ketone
AAS	atomic absorption spectrometry
GF-AAS	graphite furnace atomic absorption spectrometry
GC	gas chromatography
ICP-AES	inductively coupled plasma-atomic emission spectrometry
XRF	X-ray fluorescence
DPASV	differential pulse anodic stripping voltammetry
DPP	differential pulse polarography

[2] Estimation, based on AAS-detection limit and concentration factor.
[3] Detection limit of GF-AAS for chromium.

may also be applicable for the determination of Cr(VI) in aerosols after a proper dissolution step. Table 5 is not exhaustive, it merely represents possibilities to determine Cr(VI) in samples in which also Cr(III) is present.

The efficiency of the extraction methods (including that using APDC or DDTC/MIBK) depends on the sample composition,[82] since interferences may be expected from other metal ions,[80] and should be checked beforehand. Application of GF-AAS enables a lower limit of detection in the extraction methods than flame-AAS.

The method developed for sea water by Nakayama et al.[22] shows a very low limit of detection for Cr(VI). Besides, Cr(III) and organically bound Cr are also determined. However, the amount Cr(VI) is calculated as the difference of Cr(III) + Cr(VI) and Cr(III), which is unfavorable for low Cr(VI)/Cr(III) ratios. The same holds for the methods described by Isshiki et al.[104]

A tailor-made procedure to determine airborne Cr(VI) has been given using complexation of Cr(VI) with APDC at pH 4.5.[48] However, the detection limit was reported to be 1–5 ng.m^{-3} Cr(VI) in air, and the procedure has not been validated in case of the presence of any metal other than hexavalent chromium.

An interesting combination of techniques has been proposed by Batley and Matousek:[99] a valence-specific electrodeposition was applied as a means of preconcentration of Cr(VI) on a graphite tube, further analyzed with an atomic absorption technique.

A different approach for chromium speciation in particulate matter collected on a suitable filter has been proposed by Klockow et al.[72] The method consists of the application of an extraction procedure on the filter, that is placed horizontally on top of a Weisz ring oven.[105] As the oven basically consists of a cylindrical heatable block, it allows the washing of the soluble components out of the material that has been deposited in the center of the filter, and it allows the concentration of the solutes in the heated ring zone, due to evaporation of the solvent used. Depending on the reagent used for leaching, the Cr(VI) and/or Cr(III) species either stay in the center of the filter or are concentrated in an 'outer' or 'inner' ring. After the extraction, the filter is concentrically cut into annular parts containing separated fractions.

5. CONCLUDING REMARKS

Knowledge of the occurrence of Cr(VI) in the atmosphere is still fragmentary, valence-specific analysis of chromium in ambient air appears to be difficult, and consequently, data on speciation of *airborne* chromium are scarce. There exists a lack of standardized collection and analytical methods with respect to the speciation of chromium. Although pollution control measures seem to have become effective the last few years,[45] the need for speciation still exists, because of the carcinogenic potential of chromates, and the still increasing use of chromium in our society of which future diffuse emissions are to be expected.

From the characterization of the present-day emissions of chromium from different types of sources into the atmosphere, it can be concluded that at least a substantial fraction is emitted as Cr(VI), and the Cr(VI)/Cr(III) ratio decreases away from the source. However, the redox history of chromium in air is probably dependent on the levels and types of oxidants and reductants locally present in the atmosphere. Considering thermodynamic conditions in the atmosphere, and considering the varying ratio Cr(VI)/Cr(III) determined in soils and waters under varying natural conditions, the occurrence of Cr(VI) in the atmosphere cannot be excluded, and periods in which oxidation of airborne Cr(III) will take place can be expected.

The required detection limit of the analytical method depends on the type of sampling applied and the ratio Cr(VI)/Cr(III) in aerosols. When high-volume sampling is applied, a number of existing methods are potentially able to meet the required detection limit to determine Cr(VI) in aerosols. However, up to now, filter sampling of Cr(VI) poses unsolved problems. When low-volume or impinger sampling is applied, Cr(VI) concentrations in the resulting sample solutions are much lower, and higher concentration factors are then required in the sample treatment. Therefore, it is recommended that any improvement of methods to determine Cr(VI) in aerosols should primarily be focused on sampling, sample treatment, and in particular on the necessary concentration step. A large concentration factor is needed in all methods. A further development of valence-specific electroanalytical techniques seems promising. In some of these techniques, the combination of a pre-concentration step of a chromium species on the electrode, and the subsequent reduction of that species, resulted in very low detection limits. With the application of adsorptive cathodic stripping voltammetry for the determination of chromium in sea water, for example, a detection limit of 0.04 μg Cr(VI).l^{-1} has been established.[106]

In general, the refining of methods to separate the small amount of Cr(VI) from the relatively large amount of Cr(III), before or during the actual measurement, is encouraged. In any case, methods in which the fraction Cr(VI) has to be calculated by subtraction of the amount Cr(III) from total Cr are not preferred, because of the low ratio Cr(VI)/Cr(III) expected in ambient air.

The high risk of reduction of Cr(VI) during sampling and preconcentration of the sample should be severely controlled: the pH during sample treatment should be alkaline, and appropriate nonreducing filter material should be used. The chemistry of chromium, in particular the difference in kinetics between the oxidation of Cr(III) and the reduction of Cr(VI), is to be further exploited to provide tools in the sample treatment.

It is generally recommended that the determination should be performed as soon as possible after the sampling. Due to the possibility of contamination of the sample, extreme care should be taken to ensure that during sampling and sample treatment no extra contributions are made to the chromium content.

The development of methods to determine very low levels of airborne Cr(VI) at unfavorable Cr(VI)/Cr(III) ratio, and the further elucidation of the kinetics of the Cr(VI)/Cr(III) transformations remain challenges to environmentalists.

REFERENCES

1. WHO (1987). "Air Quality Guidelines for Europe," World Health Organization, Regional Publications, Eur. Ser. No. 23, pp. 221–223.
2. Janus, J.A. and E.I. Krajnc. "Integrated Criteria Document Chromium: Effects," Appendix to RIVM Rep. 758701001, Bilthoven (The Netherlands, 1990).
3. Veillon, C. "Analytical Chemistry of Chromium," *Sci. Total Environ.* 86:65–68 (1989).
4. Cotton, F.A. and G. Wilkinson. *Advanced Inorganic Chemistry.* (New York: John Wiley & Sons), pp. 719–749 (1980).
5. Tandon, R.K., P.T. Crisp, J. Ellis, and R.S. Baker. "Effect of pH on Chromium(VI) Species in Solution," *Talanta* 31:227–228 (1984).
6. Zatka, V.J. "Speciation of Hexavalent Chromium in Welding Fumes Interferences by Air Oxidation of Chromium," *Am. Ind. Hyg. Assoc. J.* 46:327–331 (1985).
7. Nieboer, E. and A.A. Jusys. "Biological Chemistry of Chromium," in *Chromium in the Natural and Human Environments,* J.O. Nriagu and E. Nieboer, Eds. (New York: John Wiley & Sons, 1988), pp. 21–79.
8. Boyko, S.L. and D.M.L. Goodgame. "The Interaction of Soil Fulvic Acid, and Chromium(VI) Produces Relatively Long-Lived, Water Soluble Chromium(V) Species," *Inorg. Chim. Acta* 123:189–191 (1986).
9. Garrels, R.M. and C.L. Christ. *Solutions, Minerals, and Equilibria,* C. Corneis, Ed. (New York: Harper & Row, 1965), p. 381.
10. Schmidt, R.L. "Thermodynamic Properties and Environmental Chemistry of Chromium," Rep. PNL-4881 UC-1, U.S. Dept. of Energy, contract DE-AC06-76RLO 1830 (1984).

11. Angus, J.C., Bei Lu, and M.J. Zappia. "Potential-pH Diagrams for Complex Systems," *J. Appl. Electrochem.* 17:1–21 (1987).
12. Breck, W.C. *The Sea*, Vol. 5. (New York: John Wiley & Sons, 1974).
13. Nakayama, E., T. Kuwamoto, H. Tokoro, and T. Fujinaga. "Chemical Speciation of Chromium in Sea Water. II. Effects of Manganese Oxides, and Reducible Organic Materials on the Redox Processes of Chromium," *Anal. Chim. Acta* 130:401–404 (1981).
14. Schroeder, D.C. and G.F. Lee. "Potential Transformations of Chromium in Natural Waters," *Water Air Soil Pollut.* 4:355–365 (1975).
15. Jan, T.-K. and D.R. Young. "Chromium Speciation on Municipal Wastewater, and Sea Water," *Water Pollut. Control Fed. J.* 50:2327–2336 (1978).
16. Nakayama, E., T. Kuwamoto, H. Tokoro, and T. Fujinaga. "Chemical Speciation of Chromium in Sea Water. III. Determination of Chromium Species," *Anal. Chim. Acta* 131:247–254 (1981).
17. Vos, G. "Determination of Dissolved Hexavalent Chromium in River Water, Sea Water and Wastewater," *Fresenius Z. Anal. Chem.* 320:556–561 (1985).
18. Kudoja, W.M. "Determination of the Speciation of Chromium from a Bicycle Factory Discharge," *Int. J. Environ. Anal. Chem.* 26:77–81 (1986).
19. Saleh, F.Y., T.F. Parkerton, R.V. Lewis, J.H. Huang, and K.L. Dickson. "Kinetics of Chromium Transformations in the Environment," *Sci. Total Environ.* 86:25–41 (1989).
20. Stollenwerk, K.G. and D.B. Grove. "Reduction of Hexavalent Chromium in Waters Acidified for Preservation," *J. Environ. Qual.* 14:301 (1985).
21. Crosmun, S.T. and T.R. Mueller. "The Determination of Chromium(VI) in Natural Waters by Differential Pulse Polarography," *Anal. Chim. Acta* 75:199–205 (1975).
22. Rai, D., L.E. Eary, and J.M. Zachara. "Environmental Chemistry of Chromium," *Sci. Total Environ.* 86:15–23 (1989).
23. James, B.R. and R.J. Bartlett. "Behavior of Chromium in Soils. VI. Interactions between Oxidation-Reduction and Organic Complexation," *J. Environ. Qual.* 12:173–176 (1983).
24. Eary, L.E. and D. Rai. "Kinetics of Chromium(III) Oxidation to Chromium(VI) by Reaction with Manganese Dioxide," *Environ. Sci. Technol.* 21:1187–1193 (1987).
25. Lantzy, R.J. and F.T. Mackenzie. "Atmospheric Trace Metals: Global Cycles and Assessment of Man's Impact," *Geochim. Cosmochim. Acta* 43:511–525 (1979).
26. Block, C. and R. Dams. "Study of Fly Ash Emission during Combustion of Coal," *Environ. Sci. Technol.* 10:1011–1017 (1976).
27. Moriske, H.-J., I. Trauer, R. Kneiseler, and H. Rüden. "Schwermetallkonzentrationen im Stadtaerosol: Vergleich von Hausbrand- und Kfz.-Emissions — mit früheren Immissionsstaubproben in Berlin-West," *Forum-Städt-Hygiene* 38:58–65 (1987).
28. IARC. "Some Metal and Metallic Compounds," Monogr. Ser. of Int. Agency for Research on Cancer (IARC) on the Evaluation of the Carcinogenic Risk of Chemicals to Humans, Vol. 23, World Health Organization, (1980) pp. 205–323.
29. EPA. "Health Assessment Document for Chromium," U.S. EPA, Research Triangle Park, N.C., Rep. No. EPA-600/8-83-014F (1984).

30. Towill, L.E., C.R. Shrinre, J.S. Dury, A.S. Hammons, and J.W. Holleman. "Reviews of the Environmental Effects of Pollutants. III. Chromium," U.S. EPA, Cincinnati, OH. Rep. No. EPA-600/1-78-023 (1978).

31. Bergbäck, B., S. Anderberg, and U. Lohm. "A Reconstruction of Emission, Flow, and Accumulation of Chromium in Sweden 1920–1980," *Water Air Soil Pollut.* 48:391–407 (1989).

32. NAS. *Drinking Water and Health,* Vol. 3, (Washington, D.C.: National Academy of Sciences, 1980).

33. Schroeder, W.H., M. Dobson, D.M. Kane, and N.D. Johnson. "Toxic Elements Associated with Airborne Particulate Matter: A Review," *JAPCA* 37:1267–1285 (1987).

34. IHE. "Evaluatie van de gehalten aan zware metalen in de omgevingslucht in België," Institute of Hygiene and Epidemiology, Brussels, 6th Annu. Rep. ISSN 0773-0284 (1986).

35. Alkezweeny, A.J., D.W. Glover, R.N. Lee, J.W. Sloot, and M.A. Wolf. "Measured Chromium Distributions Resulting from Cooling-Tower Drift," in *Cooling Tower Environment,* S.R. Hanna, and J. Pell, Eds., U.S. Energy Research and Development Administration (1975), pp. 558–572.

36. WHO. "Chromium; Environmental Health Criteria 61," World Health Organization (1988).

37. Langård, S., Ed. *Biological and Environmental Aspects of Chromium.* (Amsterdam: Elsevier Biomedical Press, 1982).

38. PCKP. Pollution Center of Kanagawa Prefecture. "Present Situation on the Discharge of Chromium into the Atmosphere from Chrome Plating Plants," *Kanagawa-Ken Taiki Osen Chosa Kenkyu Hokoku* 19:131–137 (1977).

39. Cox, X.B., R.W. Linton, and F.E. Butler. "Determination of Chromium Speciation in Environmental Particles. Multi-Technique Study of Ferrochrome Smelter Dust," *Environ. Sci. Technol.* 19:345–351 (1985).

40. Lindberg, E., U. Ekholm, and U. Ulfvarson. "Extent and Conditions of Exposure of the Swedish Chrome Plating Industry," *Int. Arch. Environ. Health* 56:197–205 (1985).

41. Stern, R.M. "Indicators of Matrix Dependent Variation in Effective Dose for Chromium and Nickel Aerosol Exposures," *Sci. Total Environ.* 71:301–307 (1988).

42. Gray, C.N., A. Goldstone, P.R.M. Dare, and P.J. Hewitt. "The Evolution of Hexavalent Chromium in Metallic Aerosols," *Am. Ind. Hyg. Assoc. J.* 44:384–388 (1983).

43. Sullivan, R.J. "Air Pollution Aspects of Chromium and its Compounds," U.S. Dept. of Health, Education and Welfare, Raleigh, N.C. National Air Pollution Control Administration, Publ. No. APTD 69-34 (1969).

44. Steinberger, E.H. and E. Ganor. "High Ozone Concentrations at Night in Jerusalem and Tel-Aviv," *Atmos. Environ.* 14:221–225 (1980).

45. Slooff, W., R.F.M.J. Cleven, J.A. Janus, and P. van der Poel. "Integrated Criteria Document Chromium," RIVM Rep. No. 710401002, Bilthoven, The Netherlands (1990).

46. NAS. *Medical and Biological Effects of Environmental Pollutants: Chromium,* (Washington, D.C.: National Academy of Sciences, 1974).

47. Rühling, A., L. Rasmussen, K. Pilegaard, A. Mökinen, and E. Steinnes. "Survey of Atmospheric Heavy Metal Deposition in the Nordic Countries in 1985," The Nordic Counsil of Ministers, Kopenhagen (1987).

48. Air Resources Board. "Public Hearing to Consider the Adoption of a Regulatory Amendment Identifying Hexavalent Chromium as a Toxic Air Contaminant," State of California (1986b).
49. Goh, C.L., P.H. Wong, S.F. Kwok, and S.L. Gan. "Chromate Allergy: Total Chromium, and Hexavalent Chromate in the Air," *Dermatosen* 34:132–134 (1986).
50. Ehman, D.L., V.C. Anselmo, and J.M. Jenks. "Determination of Low Levels of Airborne Chromium(VI) by Anion Exchange Treatment, and Inductively Coupled Plasma Spectroscopy," *Spectroscopy* 3:32–35 (1988).
51. Cawse, P.A. "A Survey of Atmospheric Trace Elements in the United Kingdom (1972—1973)," AERE Harwell, Rep. R7669, H.M. Stationery Office, London (1974).
52. Lee, R.E., Jr. and D.J. von Lehmden. "Trace Metal Pollution in the Environment," *J. Air Pollut. Control Assoc.* 23:853–857 (1973).
53. Rahn, K.A. "The Chemical Composition of the Atmospheric Aerosol," Tech. Rep. Kingston (Rhode Island) Graduate School of Oceanography (1976).
54. Milford, J.B. and C.I. Davidson. "The Sizes of Particulate Trace Elements in the Atmosphere — A Review," *J. Air Pollut. Control Assoc.* 35:1249–1260 (1985).
55. Rinaldi, G.M., T.R. Blackwood, D.L. Harris, and K.M. Tackett. "An Evaluation of Emission Factors for Waste-to-Energy Systems," U.S. EPA, Cincinnati, OH, Rep. No. EPA-600/7-80-135 (1980).
56. Davison, R.L., D.F.S. Natusch, J.R. Wallace, and C.A. Evans, Jr. "Trace Elements in Fly Ash. Dependence of Concentration on Particle Size," *Environ. Sci. Technol.* 8:1107–1113 (1974).
57. Pacyna, J.M. and J.O. Nriagu. "Atmospheric Emissions of Chromium from Natural, and Anthropogenic Sources," *Chromium in the Natural and Human Environments*, J.O. Nriagu and E. Nieboer, Eds. (New York: John Wiley & Sons, 1988), pp. 105–124.
58. Stern, R.M. "Chromium Compounds: Production and Occupational Exposure," in *Biological and Environmental Aspects of Chromium*, S. Langård, Ed. (Amsterdam: Elsevier Biomedical Press 1982), chap. 2, pp. 5–48.
59. Nriagu, J.O., J.M. Pacyna, J.B. Milford, and C.I. Davidson. "Distribution, and Characteristic Features of Chromium in the Atmosphere," in *Chromium in the Natural and Human Environments*, J.O. Nriagu and E. Nieboer, Eds. (New York: John Wiley & Sons, 1988), pp. 125–172.
60. Zehringer, M., C. Hohl, A. Schneider, and M.R. Schüpbach. "Untersuchungen zur Verteilung von Metallen in Schwebestäuben," *Staub* 49:439–443 (1989).
61. van Jaarsveld, J.A. and D. Onderdelinden. "Luchtverontreiniging ten gevolge van kolengestookte installaties," Stichting Projectbeheerbureau Energieonderzoek (Nederland), Project No. 20.70-017.11; Rep. 4 (1986).
62. Hislop, J.S., E.M.R. Fisher, and C.J. Pickford. "Multi-Element Analysis of Cabbage Samples Grown Near a Heavy Metal Smelter," Harwell Rep. AERE-1643 (1980).
63. Führer, H.-W., H.-M. Brechtel, H. Ernstberger, and C. Erpenbeck. "Ergebnisse von neuen Depositionsmessungen in der Bundesrepublik Deutschland und im benachbaren Ausland," *DVWK-Mitteilungen* 14:76–83 (1988).
64. Obiols, J., R. Deseva, and A. Sol. "Speciation of Heavy Metals in Suspended Particulates in Urban Air," *Tox. Environ. Chem.*, 13:121–128 (1986).

65. Griepink, G. "Methods, and Sources of Error in Trace Element Analysis in Surface Waters Using the Determination of Chromium, and Mercury as Examples," *Pure Appl. Chem.* 56:1477–1498 (1984).

66. Yamashige, T., M. Yamamoto, and H. Sunahara. "Comparison of Decomposition Methods for the Analysis of Atmospheric Particulates by Atomic Absorption Spectrometry," *Analyst* 114:1071–1077 (1989).

67. Lum, K.R. "The Potential Availability of P, Al, Cd, Co, Cr, Cu, Fe, Mn, Ni, Pb and Zn in Urban Particulate Matter," *Environ. Technol. Lett.* 3:57–62 (1982).

68. Collins, K.E., P.S. Bonato, C. Archundia, M.E.L.R. de Queiroz, and C.H. Collins. "Column Chromatographic Speciation of Chromium(VI), and Several Species of Cr(III)," *Chromatographia* 26:160–162 (1988).

69. Wiegand, H.J., H. Ottenwälder, and H.M. Bolt. "Recent Advances in Biological Monitoring of Hexavalent Chromium Compounds," *Sci. Total Environ.* 71:309–315 (1988).

70. Mutti, A., C. Minoia, C. Pedroni, G. Arfini, G. Micoli, A. Cavalleri, and I. Franchini. "Urinary Chromium as an Estimator of Exposure to Different Types of Hexavalent Chromium-Containing Aerosols," in *Environmental Inorganic Chemistry*, H.J. Irgolic and A.E. Martell, Eds. (Deerfield Beach, FL: VCH Publishers, 1985), pp. 463–472.

71. Neidhardt, B., S. Herwald, Ch. Lipmann, and B. Straka-Emden. "Biosampling for Metal Speciation. I. Determination of Chromate in the Presence of Cr(III) by Separation with Cell Membranes and GFAAS," *Fresenius J. Anal. Chem.*, 337:853–859 (1990).

72. Klockow, D., R.D. Kaiser, J. Kossowski, K. Larjava, J. Reith, and V. Siemens. "Metal Speciation in Flue Gases, Work Places Atmospheres and Precipitation," in *Metal Speciation in the Environment*, J.A.C. Broekaert, S. Gücer, and F. Adams, Eds. (Berlin: Springer-Verlag, 1990), pp. 409–433.

73. Reith, J. "Beiträge zur Speziation von Kobalt und Chrom in Arbeitsplatz-Aerosolen," Doctoral thesis, University of Dortmund (1990).

74. Hammerle, R.H., R.H. Marsh, K. Rengan, R.D. Giauque, and J.M. Jaklevic. "Test of X-ray Fluorescence Spectrometry as a Method for Analysis of the Elemental Composition of Atmospheric Aerosols," *Anal. Chem.* 45:1939–1940 (1973).

75. Neidhardt, B., U. Backes, S. Herwald, Ch. Lippmann, and B. Straka-Emden. "Chromspeziesanalyse in Schwebstaubproben," *VDI Berichte* 838:485–498 (1990).

76. Blomquist, G., C.A. Nilsson, and O. Nygren. "Sampling, and Analysis of Hexavalent Chromium During Exposure to Chromic Acid Mist, and Welding Fumes," *Scand. J. Work Environ. Health* 9:489–495 (1983).

77. Sawatari, K. "Sampling Filters and Dissolution Methods for Differential Determination of Water-Soluble Cr(VI), and Cr(III) in Particulate Substances," *Ind. Health* 24:111–116 (1986).

78. Thomsen, E. and R.M. Stern. "A Simple Analytical Technique for the Determination of Hexavalent Chromium in Welding Fumes and Other Complex Matrices," *Scand. J. Work Environ. Health* 5:386–403 (1979).

79. Pavel, J., J. Kliment, S. Stoerk, and O. Suter. "Preservation of Traces of Chromium(VI) in Water and Wastewater Samples." *Fresenius Z. Anal. Chem.* 321:587–591 (1985).

80. Harzdorf, A.C. "Analytical Chemistry of Chromium Species," *Int. J. Environ. Anal. Chem.* 29:249–261 (1987).

81. Teasdale, P.R., M.J. Spencer, and G.G. Wallace. "Selective Determination of Cr(VI) Oxyanions Using a Poly-3-Methylthiophene-Modified Electrode," *Electroanalysis* 1:541–547 (1989).

82. de Jong, G.J. and U. A. Th. Brinkman. "Determination of Chromium(III), and Chromium(VI) in Sea Water by Atomic Absorption Spectrometry," *Anal. Chim. Acta* 98:243–250 (1978).

83. Ichinose, N., T. Inui, S. Rerada, and T. Mukoyama. "Determination of Chromium(VI) as Perchromic Acid by Solvent Extraction and Atomic Absorption Spectrometry," *Anal. Chim. Acta* 96:391–394 (1978).

84. Chao, S.S. and E.E. Picket. "Trace Chromium Determination by Furnace Atomic Absorption Spectrometry Following Enrichment by Extraction," *Anal. Chem.* 52:335–339 (1980).

85. Wang, W.J. "Determination of Traces of Chromium(VI) as Thiosemicarbazide Complex by Solvent Extraction and Atomic Absorption Spectrometry," *Anal. Chim. Acta* 119:157–160 (1980).

86. Mullins, T.L. "Selective Separation, and Determination of Dissolved Chromium Species in Natural Waters by Atomic Absorption Spectrometry," *Anal. Chim. Acta* 165:97–103 (1984).

87. Wai, C.M., L.M. Tsay, and J.C. Yu. "A Two-Step Extraction Method for Differentiating Chromium Species in Water," *Mikrochim. Acta* II:73–78 (1987).

88. Subramanian, K.S. "Determination of Chromium(III) and Chromium(VI) by Ammonium Pyrrolidinecarbodithioate-Methyl Isobutyl Ketone Furnace Absorption Spectrometry," *Anal. Chem.* 60:11–15 (1988).

89. Miyazaki, A. and R.M. Barnes. "Differential Determination of Chromium(VI)-Chromium(III) with Poly(Dithiocarbamate) Chelating Resin and Inductively Coupled Plasma-Atomic Emission Spectrometry," *Anal. Chem.* 53:364–366 (1981).

90. Hiraide, M. and A. Mizuike. "Separation Determination of Chromium(VI) Anions and Chromium(III) Associated with Negatively Charged Colloids in River Water by Sorption on DEAE-Sephadex A-25," *Fresenius Z. Anal. Chem.* 335:924–926 (1989).

91. Obiols, J., R. Deseva, J. García-Berro, and J. Serra. "Speciation of Chromium in Waters by Coprecipitation-AAS," *Int. J. Environ. Anal. Chem.* 30:197–207 (1987).

92. Yamazaki, H. "Preconcentration and Spectrophotometric Determination of Chromium(VI) in Natural Waters by Coprecipitation with Barium Sulfate," *Anal. Chim. Acta* 113:131–137 (1980).

93. Fuoco, R. and P. Papoff. "Individual Detection, and Determination of Chromium(VI), and Chromium(III) — Together with Cu, Pb, Cd, Ni, and Zn — in Tap Water at ppb Level," *Ann. Chim.* 65:155–163 (1975).

94. Cox, J.A., J.L. West, and P.J. Kulesza. "Determination of Chromium(VI) by Differential-Pulse Polarography with a Sodium Fluoride Supporting Electrolyte," *Analyst* 109:927–930 (1984).

95. Harzdorf, C. and G. Janser. "The Determination of Chromium(VI) in Waste Water, and Industrial Effluents by Differential Pulse Polarography," *Anal. Chim. Acta* 165:201–207 (1984).

96. Locatelli, C. and F. Fagioli. "Determination of Chromium(VI) in Dialysis Fluids by Alternating Current and Differential Pulse Voltammetry," *Mikrochim. Acta* III:269–276 (1986).

97. Kheifets, L.Y., A.E. Vasyukov, and L.F. Kabanenko. "Polarographic Determination of Chromium(III) and (VI) in Natural Waters from Catalytic Current," *Zh. Anal. Khim.* 43:458–464 (1988).

98. Leyden, D.E., K. Goldbach, and A.T. Ellis. "Preconcentration and X-ray Spectrometric Determination of Arsenic(III/V) and Chromium(III/VI) in Water," *Anal. Chim. Acta* 171:369–374 (1985).

99. Batley, G.E. and J.P. Matousek. "Determination of Chromium Speciation in Natural Waters by Electrodeposition on Graphite Tubes for Electrothermal Atomization," *Anal. Chem.* 52:1570–1574 (1980).

100. Arpadjan, S. and V. Krivan. "Preatomization Separation of Chromium(III) from Chromium(VI) in the Graphite Furnace," *Anal. Chem.* 58:2611–2614 (1986).

101. Syty, A., R.G. Christensen, and T.C. Rains. "Trace Determination of Cr(VI) by LC/AAS with On-Line Preconcentration," *At. Spectrosc.* 7:89–92 (1986).

102. Air Resources Board (1986). "Standard Operating Procedure for the Speciation and Analysis of Hexavalent Chromium at Ambient Air Levels. Method ADDL006," Aerometric Data Division Laboratory, State of California.

103. Schaller, H. and R. Neeb. "Gas-Chromatographic Elemental Analysis via Ditrifluoroethyldithiocarbamatochelates. X. Capillary Gas Chromatography at the pg-Level Determination of Cobalt and Chromium(VI) besides Chromium(III) in Riverwater," *Fresenius Z. Anal. Chem.* 327:170–174 (1987).

104. Isshiki, K., Y. Sohrin, H. Karatani, and E. Nakayama. "Preconcentration of Chromium(III) and Chromium(VI) in Sea Water by Complexation with Quinolin-8-ol and Adsorption on Macroporous Resin," *Anal. Chim. Acta* 224:55–64 (1989).

105. Weisz, H. *Microanalysis by the Ring Oven Technique,* 2nd ed. (Oxford: Pergamon Press, 1970).

106. van den Berg, C. M. G. "Adsorptive Cathodic Stripping Voltammetry of Trace Elements in Sea Water," *Analyst* 114:1527–1530 (1989).

PART II
Sampling and Characterization of Particles of Aquatic Systems

PART II
Sampling and Characterization of
Particles of Aquatic Systems

THE USE OF FILTRATION AND ULTRAFILTRATION FOR SIZE FRACTIONATION OF AQUATIC PARTICLES, COLLOIDS, AND MACROMOLECULES

J. Buffle, D. Perret, and M. Newman

Department of Inorganic, Analytical, and Applied Chemistry, Sciences II, Geneva, Switzerland

TABLE OF CONTENTS

1. Introduction 172

2. Application of Filtration and Ultrafiltration, in Natural
    Water Studies 176
    2.1 Discrimination between "Particulate" and
         "Dissolved" Material 176
    2.2 Sequential Size Fractionation 178
    2.3 Fractionation by Dialysis 180
    2.4 Application of Membrane Separation to Complex
         Stability Measurements 181

3. Experimental Factors to Consider when Applying
    Filtration and Ultrafiltration to Size Fractionation of
    Aquatic Samples 181
    3.1 Filter Characteristics Related to Size
         Fractionation Properties 182
    3.2 Sample Modification by Solute Adsorption on,
         or Contamination by the Filter 187
    3.3 Physico-Chemical Artifacts Produced by the
         Filtration Process 189

4. Physico-Chemical Factors Influencing Retention by
    Filters: Their Specific Nature and Quantitative Influence 192
    4.1 Flux and Retention Coefficients of Non-Reactive
         Compounds in Solution.................................... 193
    4.2 Concentration Polarization, Gel Formation, and
         Clogging: Compound-Compound Interactions at
         the Membrane Surface 194
    4.3 Solute-Membrane Interactions Inside Pores.............. 207

5. Conclusions.. 212
    5.1 Optimum Operating Conditions 212
    5.2 Checking the Validity of Results 216

Appendix.. 217

Glossary.. 219

References.. 220

1. INTRODUCTION

Although a large number of techniques can be used for particle size measurements,[1-3] only a few categories of techniques can be used to separate larger from smaller size components in natural waters with minimum perturbation. They are primarily

(a) Sieving, filtration, ultrafiltration and dialysis
(b) Sedimentation and centrifugation
(c) The various types of field flow fractionation (FFF)
(d) Hydrodynamic chromatography

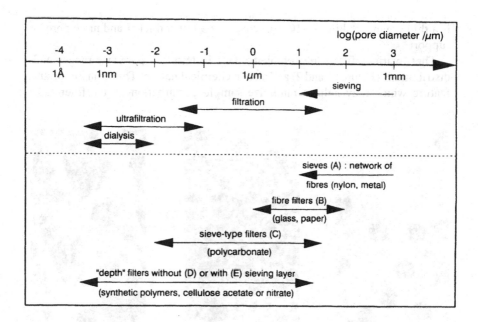

Figure 1. Size ranges for sieving, filtration, and dialysis techniques and for application of the most important filter types.

Until now, the last two categories of techniques have found limited application in natural waters. In the case of FFF, this is mostly because of experimental difficulties. Categories (b) and particularly (a), are by far the most widely used when studying natural waters, i.e., sea water, fresh water, sediment pore water, soil pore water, and ground water.

The various techniques in category (a) differ from each other by the size range of particles they can separate (see Figure 1), the nature of the filters used (Figure 1; see below) and the experimental conditions. They are discussed in detail in References 2 through 5. Sieves (A in Figure 1) are regular networks of nylon fibers or metal wires. Their use is discussed in References 2 and 3. They will not be described here. Filters may be separated into three groups (B through E in Figures 1 and 2):

(i)   Filters made of a random assembly of glass or paper fiber (B)
(ii) Polycarbonate filters (C) which are made of an impermeable organic material interspersed with discrete cylindrical holes, and which act as true sieves
(iii) So-called depth filters (D) which have a spongy structure.

Filters made of organic polymeric material are usually membranes. Polysulfone membranes (Figure 2E) are intermediate between types C and D filters: their filtering surface looks like that of a polycarbonate membrane (although the pore size distribution is broader), but the inside of the membrane body is similar to that of depth filters. For polysulfone membranes, the active filter is a thin organic "skin" (~10 μm) covering a much thicker and more porous support.

These various filters are complementary in terms of pore size ranges and distributions (Figures 1 and 2) and also in chemical nature. This is an important feature which may help to minimize sample perturbation by the filter (ad-

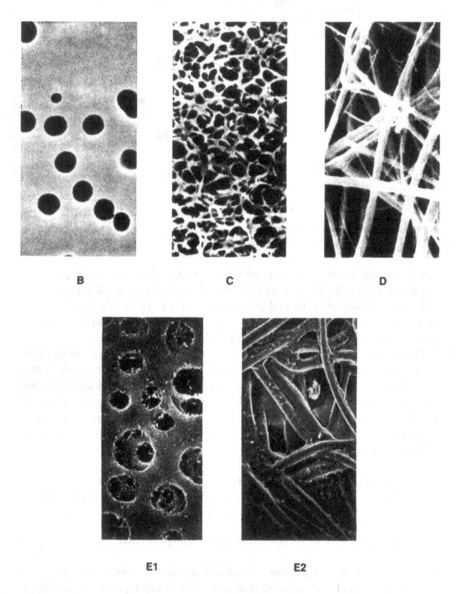

Figure 2. Electron microscopy images of the most important filter types. Letters refer to Figure 1. B: fiber filter; C: polycarbonate filter; D: depth filter; E: polysulfone filter; E1: top surface, E2: bottom surface. (From Reference 4 for B, C, and D; E from G.G. Leppard, Natl. Water Research Inst., Ontario).

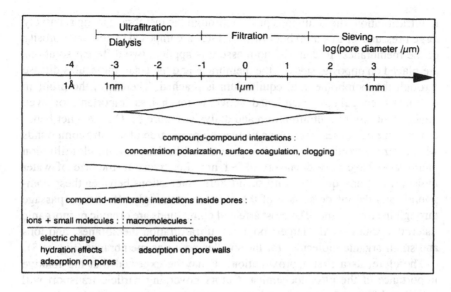

Figure 3. Schematic representation of size ranges where important filtration secondary effects are expected to play a significant role. The thickness of hatched zones reflects the relative importance of the corresponding factors. Note: the word "compound" designates any component different from water, either particulate, colloidal or dissolved (see glossary).

sorption on, contamination, or denaturation by the filters; Section 3). A large volume of literature exists on the preparation and characterization of synthetic membranes with a wide variety of properties. It is beyond the scope of this review to summarize this information, but the interested reader is referred to References 6 through 11, 90, 188, and 191 for preparation of the membranes and References 4, 12 through 23 for their characterization. Most of these synthetic membranes, however, are used for industrial purposes. Until now, almost all analytical applications to natural waters have made use of the membranes described in Figure 1.

It is also important to note the significant differences in experimental conditions between sieving, filtration, ultrafiltration, and dialysis (Figure 3). In the first two techniques, large solution flow rates can be achieved (and most often are used) thanks to the relatively large pore size of the filters, even if a rather small pressure (<1 atm) is applied on the above side of the filter. As will be discussed in details in Section 4, at such large flow rates, a so-called concentration polarization may develop at the filter surface (i.e., particle concentration is larger at the filter surface than in the bulk solution) which may induce coagulation at the surface, possibly resulting in the clogging of the filter. This effect is expected to be more important with filters than with sieves (Figure 3) because the porosity (i.e., the proportion of holes in the filter surface) decreases from loosely woven sieves to small pore size filters. Thus accumulation of compounds at the membrane surface and hence the formation of concentration polarization is favored by small pore size filters.

In ultrafiltration, the solution is pushed through the membrane using relatively large pressures (1 to 4 atm) but low flow-rates are imposed by the low porosity of the membranes. In dialysis, no pressure is applied; two different solutions are placed on opposite sides of the membrane and the small molecules diffuse through the membrane until equilibrium is reached. Therefore, the problem of surface coagulation mentioned above is much less important, or even nonexistent, both in ultrafiltration and dialysis (Figure 3). On the other hand, as pore size decreases, there is an increasing proportion of aquatic compounds whose size is similar to the filter pore size (the less porous ultrafiltration membranes have pore diameters of ~1 nm, i.e., close to the size of water molecules). Consequently, with such filters, interactions between these compounds and the inside surface of the pores may strongly affect their passage through the membrane. The most affected compounds are organic or inorganic macromolecules for the largest pore size ultrafiltration membranes, and ions and small organic molecules for the smaller pore size membranes (Figure 3).

Therefore, as a first approximation, it may be expected that the relative importance of the physicochemical factors governing particle transport will be rather different for sieving and filtration on the one hand, and ultrafiltration and dialysis on the other. Only filtration and ultrafiltration will be discussed hereafter, as they are the techniques most often used in natural water studies. This category of applications has been reviewed by Riley,[24] Grasshof,[34] de Mora et al.,[25] Hunt,[26] and Buffle.[27] Before discussing the physicochemical factors affecting filtration and ultrafiltration results (Section 4), their various modes of application to natural waters are presented in Section 2 and the various practical problems which have been reported in the literature are summarized in Section 3.

2. APPLICATION OF FILTRATION AND ULTRAFILTRATION, IN NATURAL WATER STUDIES
2.1 Discrimination between "Particulate" and "Dissolved" Material

Filtration is most widely used to separate the so-called "particulate" phase from the "dissolved" phase. Traditionally, filters with a pore size of 0.45 μm are used for this purpose, although this choice is arbitrary (see below), since size distributions of aquatic components extend in a continuous manner from tenths of nanometers to hundreds of micrometers (Figure 4). Hereafter, the term *particles* will be used for components with a size of >0.45 μm, whereas the terms "*macromolecules*" and "*colloids*" will be used for sizes between a few nanometers and 0.45 μm and "*solute*" will be used for ions and molecules in the size range of a few angstroms to a few nanometers (Figure 4). The term "*compounds in solution*" will be used to designate compounds of any size (particles, colloids, or solute). "*Permeate*" and "*retentate*" will be used hereafter to designate compounds of any size, either passing through or retained by the membrane, respectively. Figure 4 lists the most important types of aquatic particles, colloids, and macromolecules and their probable size ranges. The characteristics of particles larger than 1 μm are relatively well known both in terms of chemical nature and size distributions,[27,28] for lakes,[29] rivers,[30,31] and the oceans.[32] These characteristics are much less well known for particles smaller than 1 μm[27,33] because of experimental difficulties in their determination and theoretical difficulties in the interpretation of results.

The distinction between "dissolved" and "particulate" material was first operationally defined by Goldberg et al.,[35] who used filters with a nominal pore size of 0.5 μm. This limit is widely adopted for arbitrarily discriminating between only two size fractions, called particulate and dissolved. It is partly justified by the following three considerations:

(i)   Physically, the range of one to a few micrometers is a limit below which most natural particles (with densities between 1 to 3 g.cm^{-3}) do not settle

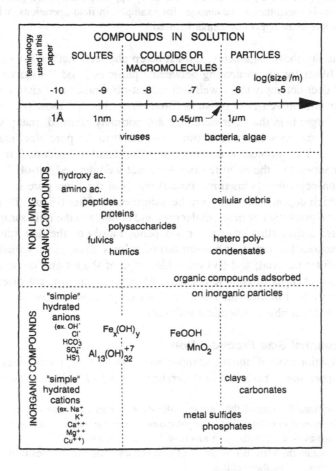

Figure 4. Schematic classification by size of important organic and inorganic aquatic components.

appreciably within a period of days.[28] It is also a limit below and above which the removal of particles from water bodies occurs by rather different processes,[36,37] irrespective of the aquatic system considered (water columns of lakes and oceans, ground waters, water treatment by filtration, water transport in pipes).

(ii) Biologically, one to a few tenths of micrometers is the minimum size range of most microorganisms (except viruses).

(iii) In practice then, the choice of a lower size limit for particles between 0.1 and 1.0 μm has the following advantages:

- The filtered sample is (at least partly) sterilized and is thus less prone to modification during any subsequent storage.
- Settling of individual particles does not occur in the filtered sample, and coagulation rate of the remaining colloids may even decrease since this process is accelerated by the presence of large particles.[36] Therefore alterations due to coagulation during storage is also minimized.
- Finally, 0.1 to 0.2 μm filters can be used by applying a rather low pressure. This is sometimes an advantage, for example, in field operations, where sophisticated apparatus cannot be used.

Despite the above arguments, it must be emphasized that the use of 0.2 to 0.5 μm filters for discriminating between "particles" and "dissolved compounds" does not imply that a well defined cut-off limit either exists in natural waters or is even achieved by such filtrations. Although most (but not all) particles larger than the filter pore size are normally retained, many smaller particles (sometimes 10 to 1000 times smaller than the pore size) may also be retained. This problem is discussed in more detail in Sections 3 to 4. Let us just mention that the main reasons for a not well-defined cut-off limit are (i) the non-negligible (sometimes broad) width of the filter pore size distribution which depends primarily on the nature of the filter (Figure 2); (ii) the coagulation properties of macromolecules and colloids in the bulk sample and at the filter surface (this latter process depends mostly on the flow rate); (iii) the interactions (in particular adsorption) of solutes, even of very small size, with the filter material; and (iv) the wide range of shapes and possible conformational changes of aquatic colloids. Effects (ii) and (iii) are often most important so that filtration and ultrafiltration cannot be viewed just as a sieving process, without physicochemical influence.

2.2 Sequential Size Fractionation

Size fractionation of aquatic components with filtering membranes of decreasing pore size is reviewed in References 25 and 27 and has been used to:

(i) Estimate the size of the various colloids and macromolecules (Figure 4)

(ii) Determine to what extent trace compounds or elements (particularly metals) are associated with these macromolecules or colloids

(iii) Separate the various categories of colloids and macromolecules from each other, for further studies

Application types (i) and (ii) have been reported, for instance, in References 30, 38 through 45, and 192 specifically in an attempt to better understand the association of trace metals with organic and inorganic colloids and macromolecules.[38,41-43] Reference 30 discusses a detailed size fractionation for particles >1 μm, whereas the others discuss fractionation of particles <0.45 μm. In most cases, sequential filtration has been used, each filtrate obtained with a particular membrane being further fractionated with a membrane of smaller pore size. Laxen et al.[43,44] however, proposed a scheme in which the initial water sample is filtered in parallel through 12 μm and 1 μm pore size filters and the filtrate of the 1 μm filter is itself filtered in parallel through filters with pore sizes in the range 1.6 nm to 0.4 μm.

Organic compounds and Fe and Mn oxyhydroxides (Figure 4) are among the most important and ubiquitous colloids in the size range 1 nm to 1 μm. Several workers have used fractionation by ultrafiltration to specifically estimate the size distribution of these compounds (e.g., References 46 through 51 for fulvic and humic compounds; References 52 and 53 for proteins; References 46 and 54 for polysaccharides; and References 55 through 58 for Fe(III) and Mn oxyhydroxides).

Type (iii) applications combine size fractionation by ultrafiltration with other methods, in order to determine more precisely the properties of the components of each size fraction, in particular the complexation capacity of organic compounds,[59,67,68] the stability of metal complexes with fulvic compounds,[60,61] the degree of lability of metal complexes,[43,62] the size distribution, morphology, and aggregation properties of fulvic molecules,[63] iron(III) oxyhydroxides,[64-66] and ground water silica,[192] and the electrophoretic mobility of iron(III) oxyhydroxides.[66]

Various fractionation procedures have been compared in Reference 51 for filtration of macromolecules with molecular weights smaller than 10^5 and in Reference 192 for the filtration of inorganic colloids of ground water. In particular one can use either washing (also called diafiltration) or concentration techniques and in each case either sequential (also called cascade) or parallel filtration (see also Reference 27). In the concentration technique, the solution to be filtered is pushed through the membrane by applying a gas (or piston) pressure. In the washing technique, a constant volume is maintained in the filtration cell by continuously compensating for the volume of filtrate by the addition of distilled water or any other "pure" solution (see Reference 27 for more details). It has been found that, although the washing technique is more time consuming, it yields more reproducible results because it avoids increasing the particulate concentration of the retentate in the cell, thus minimizing coagulation and aggregation problems. For the same reason, when the concentration technique is used, the volume reduction in the cell must be minimized as much as possible. In a sequential fractionation procedure the same solution is filtered successively through a series of membranes of decreasing pore size, the filtrate of one step being filtered on the following membrane. Part of the cell solution is withdrawn at each step for chemical analysis. In the parallel procedure, aliquots of the same initial sample are filtered through several membranes of different pore sizes. The content of each size fraction is calculated from the difference between the contents of each filtrate. Because reproducibility of membrane filtration is not better than 5 to 10% (Sections 3 and 4), the accumulated error becomes exceedingly large in both procedures when more than ~5 filtering steps are used. However, the sequential procedure is preferred because it minimizes the coagulation and aggregation processes which occur when colloid samples are stored for more than a few hours (Sections 3.3, 5). Indeed, the rate of aggregation decreases when the colloid concentration and chemical heterogeneity decreases[36] and therefore the ultrafiltration fractions are increasingly stable with respect to coagulation with decreasing pore size of the filters.

2.3 Fractionation by Dialysis

In this technique,[25,27] the test solution (compartment 1) is separated by the dialysis membrane, from the solvent or any other pure solution (compartment 2). In this way, only the solutes of the test solution, small enough to pass through the membrane, can diffuse into compartment 2 and they are left to do so until equilibrium is reached. When the solutes are metal ions, they can be accumulated in compartment 2 by means of an ion exchange resin.[69]

Since there is no pressure applied, as in ultrafiltration, dialysis is time consuming: usually 1 to 2 days are required to reach equilibrium (this was reduced to 5 h in Reference 69). This is probably the reason why dialysis has had only limited application in laboratory studies. Dialysis has been compared to other separation techniques by Beneš[70] and it has been used to determine the trace metal complexation capacity of fulvic and humic acids,[71-73] by using membranes impermeable to fulvic and humic compounds.

Dialysis has found its most fruitful application for *in situ* sampling of sediment pore water solutes. The technique was first introduced by Beneš et al.[79] for collecting selectively the "dissolved" components of river waters. Later on, it was improved[80,81] to determine the concentration profiles of "dissolved" components in the intersticial water of sediment (see also Chapter 11 of the present book). The device (a so-called pore water peeper) consists of a plexiglass plate into which rows of small compartments (~1 to 3 ml volume and 0.5 to 1 cm high, apart in a row) are machined. The compartments are separated from the external solution (sediment pore water) by a dialysis membrane and they are initially filled with deaerated demineralized water. The peeper is inserted into the sediment and left for 1 to 2 weeks for equilibration. After its retrieval, "dissolved" components are measured in each compartment. Several membranes have been tested and compared for their biodegradability by bacteria. Cellulose membranes have been found inadequate, whereas polysulfone membranes have been found resistant and are most often used. When the purpose is to collect, as far as possible, only the dissolved compounds, the pore size is not critical:[82] even with membranes of rather large pore size (0.2 μm), colloids larger than ~10 nm do not accumulate significantly in the compartments during a week equilibration period, because of their low diffusion coefficient. The choice of pore size is obviously much more important[72,74] when dialysis is used for complexation capacity measurements (see above) where free ions (~0.5 nm) must be discriminated from small size complexes like fulvic compounds (~1 nm).

2.4 Application of Membrane Separation to Complex Stability Measurements

In aquatic systems many complexing agents are polyelectrolytes, macromolecules, colloids, or particles[27] most of which can be retained by ultrafiltration membranes. When these agents form complexes with small molecules like metal ions, inorganic anions, or small organic compounds such as pesticides or herbicides, it is often possible to find an ultrafiltration membrane which retains the complexing agent and the complex but leaves the uncomplexed (or "free") small molecule to pass through. This provides a means to determine the free molecule concentration as a function of experimental conditions (complexant concentration, ionic strength, temperature, pH . . .) and, from these data, the corresponding equilibrium constant of the complexation reaction.[27]

Based on this principle, ultrafiltration has been applied to the determination of equilibrium constants for the distribution of organic compounds between water and synthetic micelles,[83] the binding of small compounds to macromolecules in biochemistry,[15] the complexation of trace metals by fulvic compounds,[84-86] and the complexation of atrazine, by fulvic compounds.[87-89] The role of the dissociation/association rates of the complexes has also been considered.[27] In this type of application, both the washing and concentration techniques can be applied equally well. Their advantages and limitations have been compared in References 27 and 88. Note that for this type of application it is essential to know precisely to what degree the complexing agent, the complexed species, and the free molecule can pass through the membrane[27] (since complete retention or passage are rarely achieved). This depends on factors described in Section 4 and which must be tested by a preliminary calibration of the experimental system.

3. EXPERIMENTAL FACTORS TO CONSIDER WHEN APPLYING FILTRATION AND ULTRAFILTRATION TO SIZE FRACTIONATION OF AQUATIC SAMPLES

An ideal filter, enabling us to perform particle and colloid fractionation based only on size, should satisfy a number of criteria which may be classified as follows:

(1) *Filter characteristics related to size fractionation properties* — The filter should have a well-defined average pore size and a narrow pore size distribution, and these characteristics should be reproducible from membrane to membrane. In addition, other properties which may affect the

passage of solute should also be well known and specified by the manufacturers. The properties include the pore density (number of pore/cm^2), the porosity (proportion of surface area occupied by pores), and secondary factors which may influence retention by adsorption, i.e., the chemical nature and degree of hydrophobicity, the electric charge and the physical structure of the filter (e.g., Figure 2B, C, D, or E).

(2) *Chemical composition of the sample* — The overall chemical composition of the sample should not be modified either by contamination with organic or inorganic impurities released from the filter, or by losses of dissolved trace elements or organic compounds due to their adsorption on the filter.

(3) *Physicochemical properties of the sample* — The size distribution of colloids and particles must not be modified in any way by the filtration process or the related sample handling. Important denaturation problems include changes in conformation, oxidation, or reduction of compounds followed by precipitation or degassing (e.g., FeS → Fe(OH)$_3$ + H$_2$S), colloid aggregation possibly linked to clogging of the membrane filter, and biological cell rupture.

(4) *Other filter characteristics, useful for practical applications* — Other filter properties useful to consider are the following:

- Its thickness, which should be large enough so that the filter can be easily handled, but small enough to minimize solute adsorption inside the pores
- Its mechanical strength
- Its possible use as a support for electron microscopy observations
- Its amenability to drying to constant weight so that the mass of retained material could be determined gravimetrically
- Its maximum permitted flow rate (Note that high flow rates are often desired for studies on water bodies containing low particle concentrations such as sea water, but that coagulation and clogging can be avoided only by using flow rates as low as possible; Section 4.2)

Criteria of types 1 to 3 are discussed below on the basis of experimental observations reported in the literature. However, the factors related to criterion type 3 are those which most drastically affect the results of fractionation by filtration and ultrafiltration. They, therefore, will be discussed in more detail in Sections 3.3 and 4.

3.1 Filter Characteristics Related to Size Fractionation Properties

Lists of commercially available membranes, their properties and analytical applications, are given in References 4, 7, and 27 and membrane technologies and properties are described in References 5, 6, 11, 15, 90, and 188 through 191. The charge density of membranes can be deduced from measurements of membrane potential[16,18,19,91] and will not be discussed here. Pore size characterization is often more problematic. Data reported by the manufacturer, for membranes with pore size larger than 20 nm, are most often deduced from methods based on gas pressure, solvent or gas permeability, gas adsorption/desorption studies, or scanning electron microscopy measurements.[5,6,13,23] The latter is not applicable to pore sizes smaller than 0.1 μm and, irrespective of the pore size, it is difficult to apply to fiber or depth filters (Figure 2B, D), for which pore geometry is ill defined. The other methods mentioned are indirect measurements, often difficult to relate to particle retention. The best procedures then consist in calibrating the membranes by filtration of a number of well-characterized particles or macromolecules.

3.1.1 Calibration Based on Retention Curves

For pore sizes larger than a few nanometers, calibration can be done by means of standard latex beads[12,93] which are commercially available[1,2,92] in the range 3 nm to 100 μm. These beads have a spherical shape and bear a large negative charge density which prevents them from coagulating. Plotting the percent retention as a function of the size of the standard particle gives a sigmoidal or sigmoidal-like curve (Figures 5 and 6). The size value at 50% retention corresponds to the average pore size, while the size range corresponding to the rising part of the sigmoid gives the width of the pore size distribution. Note that if the membrane is electrically charged, retention may be affected by particle charge and therefore standard particles with different charges should preferably be tested. Standard particles other than latex beads are cited in References 1, 2, and 92.

An elegant calibration method has been used by Sheldon et al.[94,95] which takes into account the possible influence of the solution composition. Indeed, adsorption of even small compounds on the filter may alter its charge and hydrophilicity and therefore its fractionation characteristics. Sheldon et al. determined the particle size distribution of a real water sample using a Coulter-Counter (in the range 0.5 to 15 μm) before and after filtration. The difference between the two curves enabled them to compute the retention curves mentioned previously. When such retention curves are determined with the same sample but different filters, they can be used to compare the efficiency of the various filters (Figure 5). In addition the curve of any particular filter can be used as a calibration curve for any water sample of the same type as that used for calibration. A similar approach has been used in References 55, 57, and 58 where electron microscopy has been used to check the size of particles retained on the filter. The efficiency of different membranes[55] or filtration modes[57,96] have thus been tested (see also Section 4.2). It has been emphasized[58] that verifying the filtration results by an independent size distribution detection method is absolutely required in determining the optimal conditions for well-controlled size fractionation by filtration.

Membranes with pore sizes smaller than about 10 to 20 nm are calibrated by determining the retentions of a number of organic compounds with different sizes and then plotting their retention as a function of their radius of gyration or, if this parameter is not available, their corresponding molecular weight[27,50,51,97,99] (Figure 6). Compounds with structures as similar as possible to those of the compounds to be tested should be used for calibration since charge or conformation may strongly affect permeation[27,90] (Figure 21). The

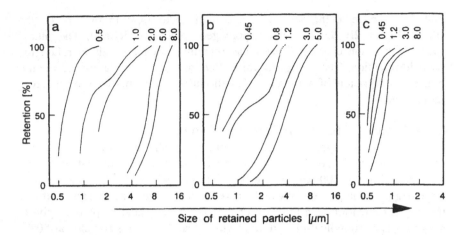

Figure 5. Retention curves for different types of membranes. Each curve is labeled with the nominal pore size (μm) indicated by the manufacturer a— Nuclepore (polycarbonate); b— Flotronic (metal membrane); c— Millipore (cellulose ester depth filters). (From Sheldon, R.W. *Limnol. Oceanogr.* 17:494–498 (1972). With permission.)

overall composition of the calibration solutions should also be as similar as possible to that of the test solutions, because, when the pore size of the membrane is similar to the size of the electrolyte ions, the nature of these ions may also affect the permeation of other solutes (Section 4.3.1).

3.1.2 Pore Size Characteristics of Different Membrane Types

In most cases, the values given by the manufacturers for the nominal pore size of membranes or their molar mass cut-off limit are the size or molecular weight values obtained from curves like those of Figures 5 or 6 (although often only a limited number of points are determined), corresponding to 90% rejection by the membrane. As these curves show, the pore size distributions of most filters are such that at least a few percent of the molecules or particles larger than the nominal pore size can pass through the filter. Comparisons between the nominal pore size, given by the manufacturers and the actual cut-off limit, have shown[55,93,94,98] that polycarbonate filters of type C give better and more reliable results than fiber (type B) or depth (D) filters (Figure 2). This is because in the latter two cases pore size is more ill defined and particles larger than pores are not only retained at the filter surface (as in type C filters), but may also be entrapped in the internal reticulum constituting the membrane. This structure is also less desirable for two additional reasons: (i) it provides a larger surface area for retention of small molecules by chemical adsorption (Section 3.2.2) and (ii) a significant amount of water may be retained in the pores; in the case of sea water this implies that a significant amount of salt is also retained by the filter,[93] which may pose problems for interpreting chemical analysis of the retained material. Because of these properties, it has often been suggested[24,25,43] that polycarbonate filters of type C are preferable to depth filters for size fractionation, even though their flow rate is lower and they are more quickly clogged (see below). Additional advantages of polycarbonate filters are that they show the smallest change in weight on washing, making them the best choice for gravimetric purposes, and that, thanks to their flat surface, they allow the clearest observation by electron microscopy.

3.1.3 Reproducibility of Membrane Production with Respect to Pore Size

There is little information on the reproducibility of membrane pore size distribution and porosity from one production set to another. This is partly

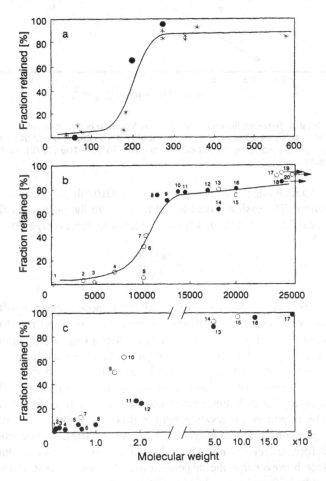

Figure 6. Calibration of ultrafiltration membranes with compounds of well-known molecular weights. Amicon membranes with nominal cut-off limits of 500 (a: UM05), 10,000 (b: PM10) and 300,000 (c: XM300). (From Reference 51 (Figure 6a) and 50 (Figures 6b,c); the compounds used for calibration are listed in the corresponding references).

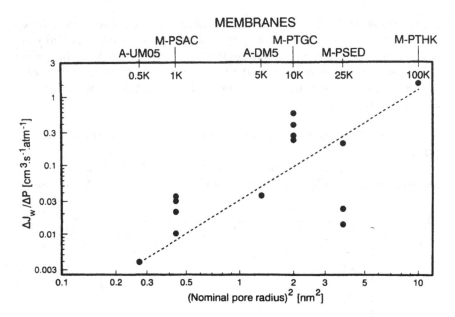

Figure 7. Relation between fluxes and pore sizes for various membranes. A = Amicon; M = Millipore. Other membrane symbols refer to the nature of the membrane. (From Macko, C. et al. *AIChE Symp. Ser.* 75:162–169 (1979). With permission.)

because a detailed check of these properties is difficult and time consuming for most filters. The easiest global test is to measure the pure water flow rate J_w (in $ml.cm^{-2}.s^{-1}$ or $m.s^{-1}$), which is related to the average pore radius, r_p by:[11,21,132]

$$J_w = \frac{\epsilon r_p^2 \Delta P}{8\eta\tau l} \qquad (1)$$

where ϵ = porosity of the membrane, ΔP = imposed pressure difference, η = viscosity of water, l = thickness of the membrane, and τ = tortuosity factor. Equation 1 with $\tau = 1$ rigorously applies to ideal membranes with cylindrical pores. $\tau > 1$ is used as an empirical factor to correct for the non-ideal geometry of depth filters. Figure 7 shows that Equation 1 is roughly followed by a number of Amicon® and Millipore® depth filters.[22] The dispersion of values for different membranes may be due to different values of τ and l. The important observation here is that the various points of Figure 7 for any particular membrane type correspond to different production sets. They therefore reflect the reproducibility in the manufacture of the membranes. Note however that the dispersion may be partly due to changes in τ (less probably in l) and not only to variations in pore size.

As shown in Equation 1, the water flow rate, J_w, is an interesting global parameter for monitoring any possible change in membrane characteristics. It has been observed for depth filters,[22] that the flow rate may change (generally decrease) with sequential uses of the same membrane, even when only water or pure solutions are used. Such flux decreases are generally attributed to plugging of the pores due to crushing of the membrane. Use of the minimum pressure compatible with an acceptable flow rate is therefore recommended. Normally the largest pressure used with less porous membranes is 3 atm. Use of the minimum flow rate is also required for other reasons (Section 4.2). When using flow rate for testing membrane characteristics it is important to realize that this parameter also depends on (Section 3.3 and 4) (i) membrane and particle charge and electrolyte nature;[27,134] (ii) dissolved organic compounds which may adsorb inside the pores;[133,135,136] and (iii) retained colloids and macromolecules which may form a gel layer at the membrane surface. Therefore, intrinsic membrane parameters, such as r_p, τ, and l can only be obtained from flow-rate measurements performed with pure water or well-controlled solutions.

3.2 Sample Modification by Solute Adsorption on, or Contamination by the Filter

3.2.1 Contamination

Because of the often very low concentrations of many metals and organic compounds in aquatic samples, care must be taken to minimize possible effects of contamination by the filter or filtering device. Metal contamination by filters can be expected since most filter material contains significant metal concentrations.[43,100-102] Systematic analysis of many filters are listed in References 102, 103, and 115, in the latter reference after various treatments. Polycarbonate filters seem to have the lowest metal content.[104] However, it has also been shown[114] that the metal content of Millipore® filters may vary by a factor of 10 from one set of filters to another. Therefore all membranes need to be carefully cleaned if trace metals are of interest.[50,105] Organic membranes may also release organic compounds, sometimes initially present as preserving agents,[50,106] which may cause errors in DOC, fluorescence, or ultraviolet absorbance measurements of the filtrate.[107-109] Low molecular weight hydrocarbons may also be leached from ultrafiltration membranes.[59] Glass fiber and paper filters may release inorganic and organic fibers, respectively. Finally, it has also been observed that contaminations from NH_4^+ and NO_3^- may occur with several filters.[109,110]

It has been proposed that metal decontamination of filters be achieved by rinsing with dilute nitric acid,[111] but Mart[112] found that to avoid contamination of sea water, which contains very low metal concentrations, membranes must be stored for at least a week in acid, followed by repeated washing with distilled water. It has even been suggested to avoid filtration of open sea water because the heavy metal content of particulate matter collected on filters is usually lower than the filter blank.[112] A simpler cleaning procedure proposed for trace metals, which also eliminates organic compounds, consists of filtering at least 50 ml·cm^{-2} of 0.01 M HCl through the desired membrane, followed by distilled water.[50,107,109] To find the best cleaning conditions, the decrease of contaminant concentration in the filtration should be traced, as a function of the filtrate volume. In dialysis experiments metal contamination may also be a problem but membranes can be cleaned by soaking in mineral acids.[72,79,113] The filtration apparatus must also be cleaned with acid as carefully as the storage bottles. Polycarbonate filter assemblies are easier to decontaminate than glass assemblies.[104]

3.2.2 Losses by Adsorption on the Filter and Filter Unit

Losses of solutes may occur during filtration due to adsorption onto the filter, the porous filter support, and the filtration vessel.[79,121-124] The latter problem is the same as losses by adsorption on any container (e.g., References 101 and 112) and will not be discussed here. An additional problem is the retention of small molecules, which should pass through the membrane, by adsorption on the material already deposited on the membrane. This is related to clogging problems and will be discussed in Section 3.3.

It is presently difficult to generalize the results reported for adsorption losses on filters and filter supports. Tests are often done with either synthetic solutions or real samples spiked or pretreated in order to characterize them as well as possible. However, adsorption reactions are highly dependent on the chemical nature and physical structure of the adsorbed species. Therefore, synthetic or spiked solutions may give unrealistic results. On the other hand it is difficult to generalize observations made with unspiked real samples because the results are specific to the nature of the adsorbed species which are often too difficult to characterize.[25,43] As a result, contradictory results have been reported. Florence,[104] for instance, states that serious adsorption losses have indeed been observed from synthetic solutions,[116,117] but not from natural waters,[112,116] and that all-glass filtration apparatus are therefore suitable for trace metal analysis. There are, however, reports of adsorption on glass apparatus from sea water[118,119] and tap water,[120] so that a polycarbonate filtration apparatus seems to be by far preferable.[26,105,127]

Despite the difficulty of making generalizations, the few following features are worth noting:

- Alkaline and alkaline-earth metal cations reduce adsorptive losses of trace metals on filters and filter units.[116,117] Based on this property, efficient preconditioning of filters and apparatus, to minimize adsorption of metals, can be accomplished with 0.1 M Ca(NO$_3$)$_2$.[43]
- In many cases, an important source of metal adsorption seems to be the filter support,[105,125] particularly glass frits. Adsorption of NH$_4^+$ on glass frits has also been reported.[126] Glass parts should be avoided for the filter apparatus, polycarbonate being by far preferable in particular for the frit.[26,105,127]
- The choice of membrane nature, with respect to adsorption problems, is more difficult to generalize. There seems to be an agreement that metal adsorption on polycarbonate membranes (Figure 2C) is low.[34,43,72,105,127] Cellulose-based depth filters apparently have significantly stronger ad-

sorption properties,[24,105,127] adsorptive effects being pH dependent and losses being very large at pH >9. Most of these observations are based on filtration with Millipore® filters. Adsorption of trace metals on Amicon® PM10 ultrafilters, which are made of similar material, is more controversial: negligible adsorption of trace metals is reported in References 128 through 130, whereas more than 60% adsorption was observed for Fe, Al, and Sc.[79] Hydrolysis properties of these metals might be an explanation of this behavior. Finally, trace metal adsorption was found to be strong on glass fiber filters.[72]

● Adsorption of organic compounds on filters has been much less studied. Fatty acids and lipids seem to adsorb strongly on Millipore® depth filters, but apparently not on paper filters.[131] Humic acids might adsorb on polycarbonate filters.[93]

3.3 Physico-Chemical Artifacts Produced by the Filtration Process

A number of artifacts may result directly or indirectly from the filtration process and must be considered in choosing the best filtration conditions.

3.3.1 Perturbations in the Bulk of the Sample: Field vs. Laboratory Filtration

In addition to the classical problems of contamination and loss by adsorption during water sampling and sample handling, a number of other perturbations may occur, where chemical compounds are not lost but undergo physico-chemical transformations (e.g., dissolution, formation of gas or solid, coagulation, or conformation changes of macromolecules[27]). Such problems are particularly important when studying anoxic waters where a number of compounds (Fe^{II}, Mn^{II}, $S^{(-II)}$) are easily oxidized by O_2 contamination resulting in concomitant precipitation reactions, pH changes, and speciation modifications. With natural samples containing colloids or particles, changes in the physical and structural characteristics occur *in most cases* even when the system is chemically stable, and stored in a fully closed container at its initial pressure.

For suspensions of hydrophobic colloids and particles (e.g., for most inorganic colloids) this is because they are inherently unstable systems which always tend to coagulate[137,138] (see also Chapter 10 in the present book). This is the more so for aquatic samples in which colloids and particles are highly polydisperse and chemically heterogeneous.[36] It has been shown theoretically that even for homodisperse synthetic systems the average particle radius of colloids increases due to aggregation, by a factor of 100 over 1 to 8 h (depending on the rate-limiting factor),[138] and it may be estimated that for the particle concentrations typically found in fresh waters (10^5 to 10^8 particles.cm^{-3}) coagulation of half of the particles occurs over a period of hours to days, depending on the chemical conditions. In the few cases where analytical techniques allowed the size distribution of particles in water samples to be followed as a function of time, it has indeed been observed that significant changes occur after a few hours in sea[75] and even in fresh waters[55,56,139] (see also Section 4.2). Another observation supports the importance of aggregation process, even with samples containing the rather small hydrophilic fulvic molecules which form most of the aquatic organic matter: in size fractionation of fulvics by ultrafiltration, the proportions of their lowest molecular weight fraction were found to be inversely related to the total organic matter concentration in the sample[51,63] and this could be related to the formation of aggregates by complementary measurements of electron microscopy, fluorescence, and surface tension.[63] These findings confirm that the tendency to aggregate is general, for most macromolecules, colloids, or particles, irrespective of their nature, even though the nature of the aggregation process may change from one type of colloid to another. Aggregation has been largely underestimated in the last decades by people using filtration, in spite of the fact that the main purpose of filtration was, and remains, to get realistic size fractionation.

On the basis of these considerations it is highly recommended, in order to obtain a size fractionation representative of the sample under study,[39,55-58] that filtration be performed in the field immediately after sampling with no or minimum preconcentration, and under well-controlled conditions. For anoxic waters or other samples which are easily denatured, it has even been shown that the sampling step itself must be avoided and that at-depth filtration[57] is required. An at-depth cascade filtration unit, controlled from the surface and allowing filtration of water through up to five filters, has been described in Reference 192.

3.3.2 Surface Coagulation and Clogging of the Membrane

A basic problem in filtration is the clogging of the membrane by the retained particles or colloids,[25,39,55,140,141] when the filter load is too large. During their accumulation at the membrane surface, particles gradually form a gel layer, less porous than the filter. The effective pore size then becomes smaller, and, consequently, at constant pressure, the water flow rate and the overall concentration of compounds passing through the filter (permeate) are reduced (Equation 1). Membrane clogging, therefore, can be detected by following the flow rate or the permeate concentration in the filtrate, as a function of filter load, at constant pressure[140,141] (Figure 8). This process is well known in industrial applications (Section 4.2.2) where high concentrations of colloids are filtered.

As mentioned above, gel formation results in a gradual decrease of the effective pore size, i.e., the basic characteristic of the filter is no longer maintained. In addition, most aquatic particulate matter has strong adsorption properties. Consequently, this gel layer retains, by both filtration and adsorption, ions or molecules which should pass through the filter. Gel formation can therefore be seen in two different ways:

- If total retention of particles, colloids, and even dissolved compounds is desired, gel formation should be favored.

● If size fractionation of particles representative of the real sample is desired, then gel formation must be avoided by fair means.

The factors affecting gel formation and clogging will be discussed in detail in Section 4.2. Surface coagulation is the key process leading to gel formation and may be minimized by using very low flow rates. This coincides with earlier empirical findings stating that correct size fractionations are obtained only (i) at low filter load,[140] (ii) at low particle concentration[94] (coagulation increases with particle concentration: Section 4.2.2), and (iii) at low pressure difference across the membrane[94] (this favors a low flow rate: Equation 1). Furthermore, filtration should be stopped well before noticing a decrease in flow rate (at constant pressure), or an increase in pressure (at constant flow rate).[55]

The evolution of the clogging process depends on the nature of the filter. Figure 8 shows typical examples for the filtration of iron(III) hydroxide particles. Nuclepore® membranes are often reported to clogg at lower loadings than other membranes. However, it can be seen in Figure 8 that filtration with this membrane is little affected by the process of gel formation before clogging. Millipore® and glass fiber filters have retention properties which change much more gradually and are therefore expected to give less reliable results. To check these properties, means for directly observing the behavior of the particles of interest must be used in addition to flowrate measurements.

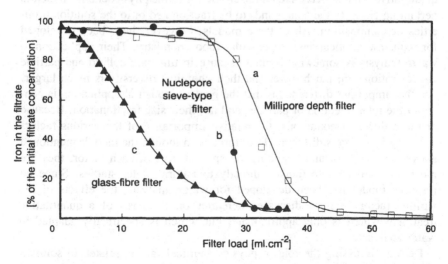

Figure 8. Effect of filter load on the retention of iron based colloids: a— Millipore HAWP depth filter (0.45 μm); b— Nuclepore polycarbonate filter (0.40 μm); c— Spectrograde glass fiber filter (0.7 μm). Because pore sizes are similar but not equal, the position of the curves on the horizontal axis is not relevant. Only the shapes of curves are comparable.

3.3.3 Biological Cell Rupture at the Membrane Surface

An additional filtration artefact is the possible rupture of biological cells when the pressure is too large.[24,25] Even if microorganisms are not the object of the measurements, ruptured cells will contribute to an increase in soluble organic matter, nutrients, and trace metals[142] in the test sample (concentration factors of 10^3 to 10^4 have been reported for many metals in plankton compared to sea water[143]), which may in turn significantly affect speciation studies. Furthermore, ruptured cells may liberate enzymes able to degrade organic compounds present in the sample. This has been shown to occur, for instance, during the determination of ATP in water.[144] Much less ATP was observed after filtration than without filtration, due to cell rupture and the subsequent release of enzymes.

To avoid cell rupture of phytoplankton, pressures less than 25 kPa must be used.[142] This is high enough for filtration with a pore size larger than 0.1 μm, on which all biological cells are retained, provided a high flow rate is not required. At any rate, maintaining a low flow rate is also preferable to minimize clogging.

4. PHYSICO-CHEMICAL FACTORS INFLUENCING RETENTION BY FILTERS: THEIR SPECIFIC NATURE AND QUANTITATIVE INFLUENCE

Many often sophisticated models have been developed in order to interpret quantitatively the effects on retention efficiency of the physicochemical factors pertaining either to the compounds to be fractionated or to the solution properties or composition. All of these models, however, have been developed for industrial applications, in particular food chemistry. Their application to water analysis is somewhat limited because in this case colloid and particle concentrations are much lower and their chemical diversity is much larger. Another important difference is that the goal in industrial application is generally the total retention of particles and not their size fractionation. Because of these different conditions, the relative importance of the various factors affecting filtration is different. Therefore, even though the models mentioned above are useful in understanding the specific role of each factor, they are most often difficult to transfer directly to natural water studies. Since no rigorous model has been developed for water analysis, the effects of the various factors will be discussed hereafter on the basis of a quantitative treatment which is only approximate, but which is specifically adapted to water analysis.

Before discussing the role of physicochemical factors related to solution composition (Sections 4.2 and 4.3), the relationship between the fluxes of water and the compound of interest must be established (Section 4.1) in the ideal case, i.e., without any influence of the solution properties or composition.

4.1 Flux and Retention Coefficients of Non-Reactive Compounds in Solution

Detailed thermodynamic and mechanistic theories explaining the separation of particles from solution with a polymeric membrane are given in References 5, 11, and 145. A recent comparison of the existing theories for the transport of water and compounds in solution through membranes is given in Reference 146. These theories differ in the assumptions made concerning the nature of the physicochemical process controlling transport through the membrane. For an uncharged compound and membrane, the two most important forces are often diffusion and convection due to the water flux, J_w (Equation 1). For such conditions the flux of solute X, J_x (in $mol.m^{-2}.s^{-1}$) through the membrane can be expressed by Equation 2:[146]

$$J_X = (1 - \sigma_X) J_w[X]_c + D_{m,X} \cdot \frac{d[X]_m}{d\chi} \qquad (2)$$

where the first term expresses the transport of solute X by the solvent (convection) and the second term its transport by diffusion in response to the concentration gradient existing within the membrane. (Note that Equation 2 is a simplified one. For more details see References 5, 11, and 146). $[X]_c$ and $[X]_m$ are the concentrations of X in the filtration cell and inside the membrane, respectively (expressed in $mol.m^{-3}$), $D_{m,X}$ ($m^2.s^{-1}$) is the diffusion coefficient of X through the membrane and $d[X]_m/d\chi$ is the concentration gradient of X in the membrane, which is assumed to be time-independent. $(1-\sigma_X)J_w$ represents the fraction of the total flux (J_w) that flows through the pores large enough for the passage of X: therefore $0 \le \sigma_X \le 1$: $\sigma_X = 0$ if X can pass through all pores indiscriminately and $\sigma_X = 1$ if X is completely retained. Although a rigorous calculation of the flux of X[11,148,19] is somewhat complicated because the diffusion process of X in the membrane must be solved, a good approximation may easily be obtained.[27] In most cases, however, convective flux is reported to be the predominant mechanism,[132] possibly with the exception of ultrafiltration membranes with the smallest pore size where diffusion might be non-negligible. J_X is then given by:

$$J_X = (1 - \sigma_X) \cdot J_w[X]_c \qquad (3)$$

which shows the dependency of J_X on pore and particle sizes: J_w depends on the effective pore size only, whereas σ_X is the so-called retention coefficient, which depends on the ratio of the effective sizes of pores to particles. Note that J_X may also be expressed as:

$$J_X = [X]_f \cdot J_w \qquad (4)$$

where $[X]_f$ = filtrate concentration of X. Equation 4 also follows from Equation 3 and the definition of σ_x:

$$\sigma_X = 1 - [X]_f/[X]_c \qquad (5)$$

The term *effective* size of pores (r_p) and particles (r) refers to the fact that size values which effectively control σ_X and J_w depend not only on the true values of r_p and r, but also on filtration conditions and more precisely on interactions between compounds in solution and between these compounds and the membrane. Essentially three types of such interactions may occur and are discussed below:

- Aggregation of particles and colloids at the membrane surface, possibly resulting in gel formation
- Adsorption of compounds of any size onto the membrane and pore walls
- Interactions due to electrostatic and hydration properties of small molecules and membrane pores

All these factors may influence the effective pore size and therefore the value of J_w. As mentioned in Section 3.1, monitoring changes in J_w is then a possible means of detecting the above interactions. However, for particle fractionation purposes, it is better to follow J_X (or preferably, both J_X and J_w) since the above effects may alter σ_X without greatly affecting J_w. A fractionation process can be considered as reproducible and well controlled only if J_X is constant during the filtration process (when performed by the concentration techniques) and independent of filtration conditions (in particular of colloid concentration and flow rate).

4.2 Concentration Polarization, Gel Formation, and Clogging: Compound-Compound Interactions at the Membrane Surface
4.2.1 Concentration Polarization Principle

Deposition of particles and gel formation at the membrane surface are well-known processes, particularly in concentrated solutions such as those used in industrial applications (solutions of several grams per liter organic macromolecules, or suspensions of a few percent by weight of colloidal particles). In such cases a gradual decrease of J_w with increasing macromolecule concentration is observed.[5,11] These observations are parallel to the decrease in J_w with increasing filter load which is sometimes observed in natural water studies. Such changes in water flux during the course of filtration have been modeled and it has been shown[141,150,151] by monitoring J_w as a function of filtration time and volume of filtrate, that two steps in the development of clogging and the formation of gel can be discriminated.

Such relationships, however, are not very useful for the application of filtration to well-controlled size fractionation, since such a fractionation is obtained only in the complete absence of clogging, i.e., when J_w is constant.

Understanding of the processes leading to clogging is therefore preferable in order to minimize it. Two main processes have been proposed and modeled:

- Specific interaction of particles with the membrane surface (adsorption)[155,156,159,160,168]
- Coagulation at the membrane surface due to the existence of a concentration polarization[21,152-154,157-159]

The first process should be rather specific to the nature of both the particle and membrane compositions, whereas this is not the case for aggregation processes.[138] The fact that clogging has been reported in many water analyses with almost all type of filters, despite of the broad diversity of particle and colloid nature in various water samples (Figure 4), then suggest that aggregation is the predominant process. This observation seems to be corroborated by all of the data reported specifically on these processes in the literature (e.g., References 154 through 159; see also discussion of Figure 15, Section 4.2.3). These references suggest that adsorption might play a role, but as a secondary factor affecting the strength of binding between the membrane and the gel formed by coagulation. As we are mostly interested in the factors affecting the initial steps of gel formation, adsorption of particles and colloids will not be considered here.

Concentration polarization is a fundamental effect of filtration: because of the partial ($0 < \sigma_x < 1$) or complete ($\sigma_x = 1$) rejection of the compound X by the membrane, its surface concentration, $[X]_{c,0}$, tends to increase relative to that in the bulk solution (Figure 9). This increase can be estimated by Equation 7.[76,77] At any distance, χ from the membrane, the flux of X carried along by the solvent is $J_w[X]_{c,x}$. Since $[X]_{c,0} > [X]_c$, a flux of back-diffusion is created: $D_x d[X]_{c,x}/dx$. After some transitory period a steady state is established at any χ such that:

$$J_w[X]_{c,x} = D_x \cdot \frac{d[X]_{c,x}}{dx} \tag{6}$$

This equation can be integrated with the following boundary conditions: (i) a gradient of [X] exists only inside the diffusion layer, of thickness δ, i.e., $[X]_{c,x} = [X]_c$ for $\chi > \delta$ ($[X]_c$ = concentration of X in the filtration cell); (ii) at $\chi = 0$, the flux of back-diffusion is equal (but of opposite sign) to the flux of the retained fraction of X carried along by the solvent, i.e.:

$$J_w \sigma_x [X]_{c,0} = D_x \left(\frac{d[X]_{c,x}}{dX} \right)_{X=0}$$

On integrating Equation 6 with these conditions, one gets:

$$\frac{[X]_c}{[X]_{c,0}} = (1 - \sigma_x) + \sigma_x \cdot \exp\left[- \frac{J_w}{D_x/\delta} \right] \tag{7}$$

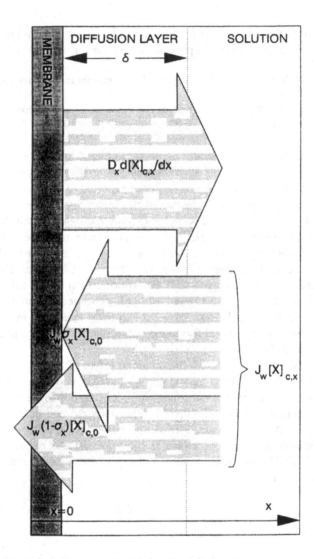

Figure 9. Schematic drawing of the fluxes of a compound, X, strongly rejected at the surface of a membrane causing concentration polarization. δ = diffusion layer thickness; $J_w(1 - \sigma_x)[X]_{c,0}$ = flux of X passing through the membrane; $J_w\sigma_x[X]_{c,0}$ = flux of X not allowed to pass the membrane; $D_x \cdot d[X]_{c,x}/d\chi$ = flux of X diffusing back in the solution. (From Buffle, J. *Complexation Reactions in Aquatic Systems: An Analytical Approach*. (Chichester: Horwood, 1988). With permission.)

where δ, the diffusion layer thickness, is constant if the filtration cell is stirred. Note that Equation 7 is approximate because its derivation implies a discontinuity at $\chi = \delta$. Nevertheless it is a good approximation for estimating $[X]_{c,0}$ values and it allows an easy discussion of the factors affecting $[X]_{c,0}$. Table 1 gives values of $J_w\delta/D_x$, for various particle sizes. The flow rate used to compute these values, 0.01 cm.s^{-1}, is typical of that obtained with the least porous membranes.[27] It is therefore a common flow rate for separating particles of 1 to a few tens of nanometers, but it is 100 times lower than usual for separation of particles larger than 0.1 μm by syringe filtration. Despite this very low flow rate, Table 1 shows that the term $J_w\delta/D_X$ is always very large. Therefore the exponential term of Equation 7 is most often negligible, except when $\sigma_x = 1$, i.e., for completely retained particles. For partly retained particles: $[X]_{c,0}$ is larger than $[X]_c$ but not dramatically: $[X]_{c,0}/[X]_c$ $\sim 1/(1 - \sigma_x)$. However, for completely retained particles, ($\sigma_x = 1$), one gets, irrespective of the filter pore size:

$$[X]_{c,0}/[X]_c = \exp\left[+ \frac{J_w\delta}{D_X} \right] \tag{7'}$$

Table 1 shows that, in such a case huge concentration factors may be reached for particles larger than a few tens of nanometers. The numbers computed in Table 1 for particles of 100 and 1000 nm are obviously not reached in practice for two reasons. First, for particles larger than a few tenths of microns, processes other than molecular diffusion, such as shear-enhanced diffusivity,[153,154] may reduce the concentration factor. Above all, when particle concentration becomes exceedingly large, coagulation occurs almost instantaneously.[137] The purpose of Table 1, however, is to suggest that for the fully retained particles extremely large concentration factors are reached during filtration, because of the concentration polarization effect, and that this is likely to affect the behavior of compounds at the membrane surface (Section 4.2.2).

4.2.2 Consequences of Concentration Polarization

The gel layer model — Considering Table 1, it is easily understandable that, for a given solution, when J_w is increased by increasing the applied pressure, $[X]_{c,0}$ may reach a concentration $[X]_g$ where the solution at the membrane surface is no longer fluid. A gel layer is thus formed, which has a hydraulic resistance opposing any further increase in J_w, and in which $[X]_g$

Table 1. Value of $[X]_{c,0}/[X]_c$ Computed From Equation 7, for Typical Values of Particle Size (r) and Flow Rates (J_w). $\sigma_x = 1$; $\delta = 10$ μm; $J_w = 0.01$ cm/s (= 2.4 ml.min^{-1} for a Membrane of 4 cm^2)

r (nm)	D_X (cm^2.s^{-1})	$J_w.\delta/D_X$	$[X]_{vc,0}/[X]_c$
1	$3.2\ 10^{-6}$	3.13	1.37
10	$3.2\ 10^{-7}$	31.3	23.0
100	$3.2\ 10^{-8}$	313.0	$3.7\ 10^{13}$
1000	$3.2\ 10^{-9}$	3130.0	$5.0\ 10^{135}$

is nearly constant.[158] Consequently, J_w approaches a limiting value J_{lim} obtained from Equation 7':

$$J_{lim} = \frac{D_x}{\delta} \cdot \ln \frac{[X]_g}{[X]_c} \qquad (8)$$

$[X]_g$ being constant, linear plots of J_{lim} vs. $\ln[X]_c$ are predicted by Equation 8, which is indeed observed in practice (e.g., References 164, 165, 167, and 169) for large values of $[X]_c$ (larger than fractions of $g \cdot dm^{-3}$).

The osmotic pressure model (case of ultrafiltration of small compounds) — Although the merit of the gel model is to give a simple qualitative representation of the hydraulic resistance, it has been shown not to be reliable or realistic in terms of gel properties.[158] For the retention of small compounds, by ultrafiltration membranes in particular, this resistance is better explained by the development of a high osmotic pressure, $\Delta\Pi$, inside the concentration polarization layer.[157,158,166] $\Delta\Pi$ constitutes a back-pressure opposing the applied pressure, ΔP, so that $J_w = $ constant.$(\Delta P - \Delta\Pi)$ (compare to Equation 1). When ΔP is increased, $[X]_{c,0}$ increases and so does $\Delta\Pi$. For small compounds $\Delta\Pi$ may become large enough to compensate ΔP completely; then J_w approaches 0. For large macromolecules $\Delta\Pi$ is small and the effect is negligible. It is therefore expected to occur mostly for separation of rather small macromolecules with smaller pore size ultrafiltration membranes. (Note, however, that this generalization must be taken with caution, because solute coupling may sometimes occur between small compounds and macromolecules.[166])

When the compound in solution is only partly rejected and penetrates the membrane (in particular in depth filters), the $\Delta\Pi$ thus produced inside the membrane may change the pore size. This might be the origin of the increased permeate concentration observed in the ultrafiltrate of natural waters when the retentate concentration in the cell becomes too large.[22,46,51] For this reason, it has been recommended not to ultrafiltrate more than 80% of the initial volume.[51]

Surface coagulation (case of filtration of colloids and particles) — Aggregation and coagulation of colloids and particles in well-mixed conditions are second order reactions,[36,137] whose rates increase strongly with particle concentration. Since very high concentrations are predicted for large particles at the membrane surface by Equation 7' (Table 1), coagulation is likely to occur at the membrane surface unless specific precautions are taken. Until now, no rigorous theory has been developed to quantitate the correct conditions for minimizing surface coagulation. Such a theory should consider:

- The concentration polarization effect (Section 4.2.1)
- The perikinetic coagulation process, i.e., coagulation resulting from collisions due to Brownian motion of particles. This process is controlled by temperature, T, viscosity of the solution, η, and the corresponding collision efficiency factor, α_p, which depends on the chemical composition of the particles and of the solution.

- The orthokinetic coagulation process, i.e., coagulation due to movement of the solution. This process depends on the velocity gradient G, the radius of particles, r, and the corresponding collision efficiency, α_o.
- The fact that particles in real samples have broad size distributions. This dramatically complicates computation, since (i) the concentration gradient in the diffusion layer depends on particle size (through the term J_w/D_x), (ii) all the concentration gradients are influenced by each other through the coagulation process, and (iii) coagulation of heterodispersed particles is much more complicated than that of a homodispersed system.[36]
- The fact that for particles larger than a fraction of micron, their size is comparable to that of the diffusion layer thickness itself. For such particles (which are the most important for filtration on 0.1 μm membranes), the concept of concentration gradient in the diffusion layer is no longer valid. In addition, their elimination from the diffusion layer is not only due to diffusion but also to the shear produced by stirring at the limit between the bulk solution and the diffusion layer.[154]

In the absence of a rigorous theory an order of magnitude estimate of the maximum flux, leading to the maximum concentration factor tolerable at the membrane surface in order to avoid coagulation, can be estimated by Equation 9 (Appendix 1):

$$J_w \leq \frac{D_x}{\delta} \cdot \ln\left(\frac{C_o}{f^2 \cdot [X]_c}\right) \tag{9}$$

with $C_o = 3D_x/4.\delta^2.\alpha.[4Gr^3 + kT/\eta]$ (with $\alpha = \alpha_p = \alpha_o$). The most important assumption of Equation 9 is that of homodispersed particles. Since it is known that coagulation is faster in heterodispersed systems than in homodispersed ones,[36,37] the values computed from Equation 9 must be considered as upper limits of J_w. Furthermore, as stated above, for large colloids or particles it is not correct to estimate the values of $[X]_{c,0}$ by introducing in Equation 7 the value of the diffusion coefficient valid in quiescent solution, since shear-enhanced diffusion must be considered. In other words, effective D values, larger than those valid for diffusion in quiescent solutions, must be used. The particle size corresponding to the limit between the application of true diffusion coefficients and shear-enhanced diffusion coefficients in Equation 7 seems to be between r ~0.1 to 1.0 μm. Figure 10, for instance, shows the change in water flux obtained by filtering suspensions of different particle sizes.[153] For sizes lower than 0.1 μm the water flux decreases as particle size increases due to the increased concentration polarization resulting from the decrease in the diffusion coefficient value introduced in Equation 7. For larger sizes, the water flux increases as particle size increases. This is attributed to a decreased concentration polarization due to the shear-enhanced diffusion of these larger particles; the shear enhanced diffusion coefficient indeed increases with the square of particle radius.[154]

With the above restrictions in mind, Equation 9 is useful to discuss the

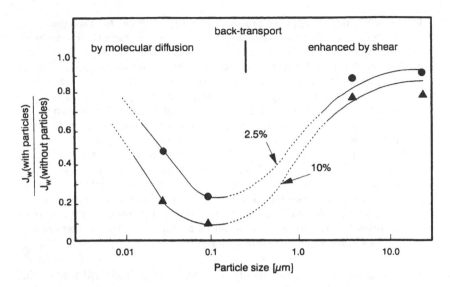

Figure 10. Change in water flux through a PM30 membrane, as a function of the size of retained particles from the filtered suspension. J_w (with particles) and J_w (without particles) = water fluxes in presence and absence of suspended particles, respectively. (From Fane, A.G. *J. Memb. Sci.* 20:249–259 (1984). With permission.)

maximum tolerable value of J_w, and the relative importance of the various filtration conditions. Values of Figure 11 are computed for various diffusion layer thicknesses (δ) and particle concentrations in the filtration cell, $[X]_c$, in the most restrictive condition, i.e., that corresponding to particles having the lowest diffusion coefficient, D_X. From Figure 10 it may be expected that this corresponds to particles having radii of a few tenths of microns, which corresponds to a shear-enhanced diffusion coefficient of $D_X \sim 3 \ 10^{-8} \ cm^2.s^{-1}$. The values of the velocity gradient, $G = 100 \ s^{-1}$, and of the coagulation collision efficiency factor,[137] $\alpha = 2.5 \ 10^{-3}$, have been used. The former is typical for well-stirred solutions and the second is a minimum value found for lakes.[161] Interestingly, Figure 11 shows that particle concentration does not have an important effect on the maximum usable value of J_w. This is due to the fact that $[X]_c$ is inside the logarithm of Equation 9. Similarly, α, G, and the thickness of the surface layer in which coagulation effectively occurs (see discussion in the appendix) will not have an important influence on the maximum tolerable value of J_w. On the other hand, D_X and δ are the two most important parameters. D_X primarily depends on the particle size, but δ may be controlled by experimental conditions. In order to allow the use of larger J_w values, δ must be minimized, for instance by means of efficient stirring. The exact role of stirring is, however, controversial since strong stirring will simultaneously increase the coagulation rate of particles in the bulk solution inside the filtration cell.[137]

At any rate, Figure 11 shows that for classical filtration cells, where δ is larger than a few microns, J_w must not exceed 10^{-4} to 10^{-3} cm.s^{-1}. This value (corresponding to 0.03 to 0.3 ml.min^{-1} for a 4 cm^2 membrane surface area) is orders of magnitude smaller than that classically used, at least for filters with pore size ≥ 0.1 μm. The fact that membrane clogging is often observed in these conditions is therefore not surprising. Furthermore, it is unlikely that δ might be decreased much below 1 μm, even with sophisticated filtration techniques (pulsed filtration, tangential flow filtration . . .). Therefore, even with these techniques, low J_w values must be used.

It may also be inferred from the above estimation and from Table 1 that surface coagulation will not be an important problem for fractionation of colloids less than a few tens of nanometers in size by ultrafiltration, *provided larger particles have previously been eliminated.* Thus it is always preferable to perform size fractionation by sequential cascade filtration, rather than by parallel sets of filtration.

4.2.3 Experimental Evidence of Surface Coagulation

Systematic tests have been performed to check the above considerations on the occurrence of surface coagulation induced by concentration polarization.[57,96,162,163] These tests have been performed by filtering under various conditions iron oxyhydroxophosphate particles formed at the oxic-anoxic boundary in a eutrophic lake[57,58,96] (see also Chapter 8 of the present book).

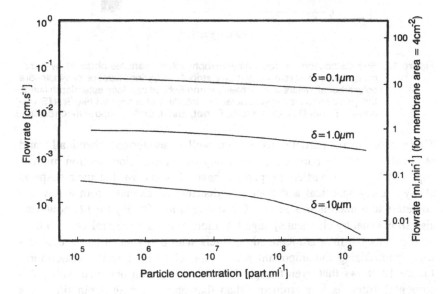

Figure 11. Theoretical maximum tolerable value of the flow rate, J_w, as a function of particle concentration in the filtration cell, in order to avoid surface coagulation at the membrane surface. Computed from Equation 9, with $\alpha = 2.5 \cdot 10^{-3}$ and $f = 10$ (i.e., 90% of particles are coagulated: see Appendix). $D_x = 3 \cdot 10^{-8}$ cm^2.s^{-1}. For other parameter values: see text.

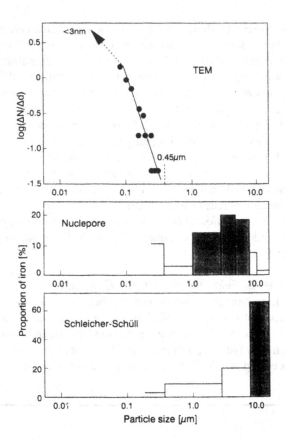

Figure 12. Size distributions of iron oxyhydroxophosphate particles obtained by transmission electron microscopy (true distribution), syringe filtration on Nuclepore
polycarbonate filters, and Schleicher and Schuell cellulose ester depth filters.
Iron particles formed at the oxic/anoxic interface of eutrophic Lake Bret (Switzerland). (From R.R. De Vitre, D. Perret, and J. Buffle. Unpublished results).

These natural iron particles have been well characterized chemically and
physically.[64,65,163] In particular, their study by transmission electron microscopy (TEM), on a particle by particle basis, has shown that their shape is
always nearly spherical and their size distribution extends from a few nanometers to a maximum size of ~0.3 μm (Figure 12). Figure 12, however,
also shows that by classical syringe filtration (flow rate ~20 ml.cm^{-2}.min^{-1}
= 0.3 cm.s^{-1}), they are retained on filters with pores much larger than 0.3
μm, emphasizing the important role of the clogging effect. Furthermore,
Figure 12 shows that even though the size distribution obtained with polycarbonate filters is less erroneous than that obtained with depth filters (as
pointed out in Reference 53), both size distributions depart from reality by
more than an order of magnitude.

In the following, F_m is defined as the proportion of iron (in percent of total
iron in the sample) "stuck" on the membrane surface, irrespective of the

nature of the "sticking process" (surface coagulation or chemical adsorption). Hereafter, F_m will be termed "adsorbed" iron to discriminate it from the proportion of the whole of retained particles which include "adsorbed" particles plus those remaining in solution in the filtration cell. Optimum conditions are those for which $F_m = 0$. F_m has been measured as a function of experimental parameters. Figure 13 shows F_m values for filtration on 0.2 µm membranes. As expected from TEM size distributions, F_m is low at low flow rates ($\sim 4 \cdot 10^{-4}$ cm.s^{-1}) but increases drastically at a larger flow rate ($7.4 \cdot 10^{-3}$ cm.s^{-1}) even though this latter is still 100 times lower than that normally used with this type of filter. Figure 13 also shows that this effect occurs with all the membranes tested. Although Figure 13 seems to suggest that F_m depends on the nature of the membrane this is misleading because the opposite relationship between F_m and the membrane nature was observed for the same sample with membranes of higher porosity (3 µm). In other words, after cascade filtration through 3 and 0.2 µm membranes, 81 ± 6% of total iron is adsorbed on the two membranes, irrespective of the nature of these membranes. This has been found for 11 different conditions (six types of membrane, with and without stirring, at $7.4 \cdot 10^{-3}$ cm.s^{-1}).

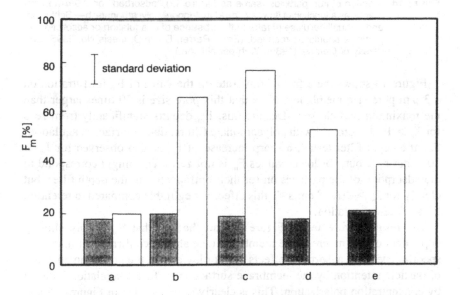

Figure 13. Fraction of iron particles (same as Figure 12) "adsorbed" on various 0.2 µm pore size filters. All solutions were initially filtered through 3.0 µm filters of the same nature. All filter surface areas were 4.5 cm². Fluxes: dashed bars = 3.7 10^{-4} cm.s^{-1} (= 0.1 ml.min^{-1}); white bars = 7.4 10^{-3} cm.s^{-1} (= 2.0 ml.min^{-1}). a— Gelman acrylic copolymer; b— Schleicher and Schuell mixed cellulose esters; c— Nuclepore polycarbonate; d— Schleicher and Schuell cellulose nitrate; e— Rhône-Poulenc polyvinyliden fluoride. (From References 96 and 162, see also Reference 57.)

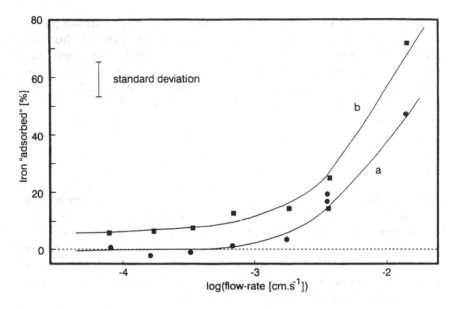

Figure 14. Fraction of iron particles (same as Figure 12) "adsorbed" on 3.0 μm membranes as a function of flow rate. a— Nucleopore polycarbonate; b— Schleicher and Schuell cellulose nitrate. In the absence of coagulation or adsorption no particles should be retained. (From Perret, D. Ph.D. thesis No. 2395, University of Geneva (1989). With permission.)

Figure 14 shows the effect of flow rate on the value of F_m for filtration on a 3 μm pore size membrane. Note that this pore size is 10 times larger than the maximum particle size. Despite this, F_m departs significantly from zero for $J_w > 1 \cdot 10^{-3}$ cm.s^{-1} with polycarbonate filters, due to surface coagulation. For the depth filter tested, a sharp increase of F_m is also observed for $J_w \geq 10^{-3}$ cm.s^{-1}, but, for lower values F_m is not zero. This might correspond to the adsorption of the particles on (or their entrapment in) the depth filter; but clearly for $J_w > 2 \cdot 10^{-3}$ cm.s^{-1}, this effect is negligible compared to retention by surface coagulation.

The result represented in Figure 14 and the fact that F_m is only slightly dependent on the nature of the membrane (see above) confirms that, although specific chemical adsorption effects might play a role, the predominant cause of particle retention by the membrane surface is surface coagulation induced by concentration polarization. This is clearly demonstrated in Figures 15 and 16. Figure 15 includes scanning electron microscopy (SEM) images, showing top views of the surface of polycarbonate membranes (0.2 μm) before and after filtration at different flow rates. Clearly particle retention decreases with flow rate, retention being very low at 50 μl/min (i.e., $1.9 \cdot 10^{-4}$ cm.s^{-1}). Interestingly, at large flow rates, a non-negligible proportion of particles have a size larger than 3 μm despite the fact that the sample was initially filtered on a 3.0 μm pore size polycarbonate filter. This results from the fact that only "apparent particles" which are in reality aggregates are seen in Figure

15 due to the low resolution of SEM. This can be seen in Figure 16 which shows a TEM image of a microtomic section of one of the above "apparent particles", cut perpendicularly to the membrane surface. Clearly it is a heterogeneous aggregate. Observations of large numbers of filters have always confirmed these observations. In most cases, aggregates are formed on the membrane surface and do not penetrate inside the pores, even with the so-called "depth filters" (only a few exceptions have been noticed). Chemical analysis by TEM-EDS of individual aggregates formed on the filter, has shown that in addition to Fe particles (which include P), the aggregates are composed mostly of Si, Al, and Ca, suggesting that they are composed of clays, silica, and some calcium carbonate. Organic carbon is probably an important additional constituent, but could not be determined. For a large number of aggregates, no correlation could be found between their Fe content and their Si, Al, or Ca content.[162] Once again this suggests that particle retention by

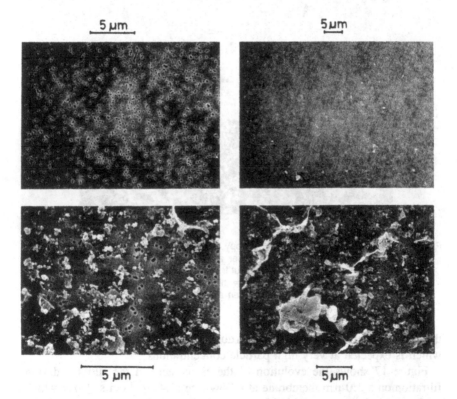

Figure 15. Scanning electron microscopy images of the surface of 0.1 μm Nuclepore polycarbonate membranes used to filter iron particles (same as in Figure 12) at different flow-rates. All bars represent 5 μm. a— membrane which was not used; b— $J_w = 2.1 \times 10^{-4}$ cm.s^{-1}; c— $J_w = 4.2 \times 10^{-3}$ cm.s^{-1}; d— $J_w = 4.2 \times 10^{-2}$ cm.s^{-1}. All solutions were previously filtered on 3.0 μm polycarbonate membranes with the same flow rate. (From Perret, D. Ph.D. thesis, University of Geneva (1989). With permission.)

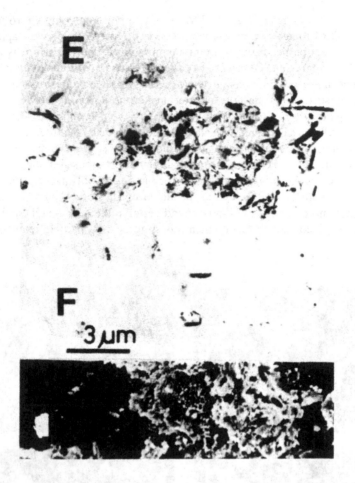

Figure 16. Transmission electron microscopy image of particles retained by filters as in Figure 15, but with a Schleicher and Schuell cellulose nitrate depth filter. Deposit and membrane were cut transversally. E = external solution on the top part of the membrane, F = filter material. Note that by SEM (bottom picture) this agglomerate is seen as only one single particle. (From References 96, 162 and 163.).

the membrane results from coagulation due to nonspecific particle interactions, which is expected at very high particle concentrations.

Figure 17 shows the evolution of the Fe content in the filtrate, during filtration on a 3.0 μm membrane at a flow rate ($3.4\cdot10^{-3}$ cm.s^{-1}) for which surface coagulation occurs. Without coagulation, the Fe concentration in the filtrate should be the same as the initial Fe concentration of the sample (7.5 μmol.dm^{-3}). Figure 17 clearly shows that for a polycarbonate membrane, surface aggregates gradually form an additional filter of increasing efficiency, which retains all Fe particles after a sufficient loading has been reached. For the cellulose nitrate depth filter clogging occurs even more easily, as is also seen in Figure 14.

The effect of solution stirring (i.e., decreasing δ; Equation 7) during filtration is shown on Figure 18 for various membranes. There is a systematic, though only slight, decrease in F_m. The weak influence of stirring has already been noted by Laxen et al.[55] The weakness of this effect may be partly due to the inefficiency of stirring, but it might also be due to the fact that a relatively large flow rate ($6.6 \cdot 10^{-3}$ cm.s^{-1}) was used in this experiment and that even a decrease of δ by a factor of 2 or 3 was not enough to decrease the surface concentration to a value low enough to avoid coagulation.

4.3 Solute-Membrane Interactions Inside Pores

When small pore size membranes are used for fractionation, rather small macromolecules are separated from each other. These small macromolecules are hereafter referred to as solutes (Figure 4). In such cases, fluxes of water and solutes are influenced by solute membrane interactions (Figure 3). The main types of interactions result from (i) electric and hydration properties of the solute and membrane; (ii) conformational changes of the solute in the membrane, and (iii) modifications of the membrane surface by adsorption of solute. The first two types of interactions are particularly important for solutes whose size is similar to the pore size.

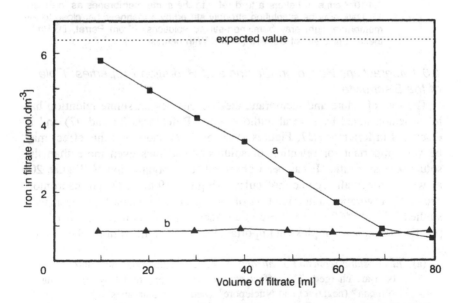

Figure 17. Fraction of iron particles, (same as in Figure 12), passing in the filtrates of a 3.0 μm Nuclepore polycarbonate filter (a) and Schleicher and Schuell cellulose nitrate filter (b). Note that iron particle size is much smaller than 3.0 μm. One therefore expects a constant Fe concentration as shown in the figure. Initial volume in the cell = 200 ml. Flow rate = J_w = 0.0037 cm.s^{-1}. (From Perret, D. Ph.D. thesis, University of Geneva (1989). With permission.)

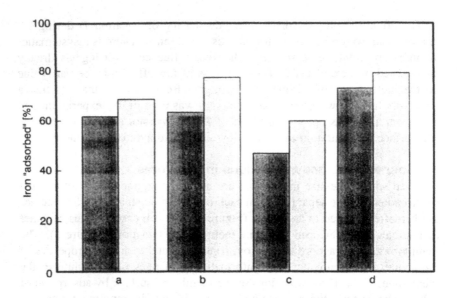

Figure 18. Effect of stirring on the fraction of iron particles (same as in Figure 12) "adsorbed" on various 3 μm pore size membranes during filtration at $J_w = 7.4 \cdot 10^{-3}$ cm.s^{-1}. Letters a to d refer to the same membranes as in Figure 13. Dark gray bars: solution strongly stirred by a magnetic bar close to the membrane; light gray bars: non-stirred solutions. (From Perret, D. Ph.D. thesis, University of Geneva (1989). With permission.)

4.3.1 Interactions Based on Electric and Hydration Properties: Role of the Electrolyte

The role of solute and membrane electric charges on solute retention has been demonstrated by several authors (e.g., References 22 and 97) and is discussed in Reference 27. Figures 19 through 21 show that this effect may be very important for retention of solutes, sometimes even more than the solute/pore size ratio. It has been observed for inorganic ions[97] (Figure 20) as well as for small organic molecules[97] (Figures 19 and 21) and macromolecules.[170] Theoretical interpretation of electrical phenomena have been well studied.[5,11,19,91,134,171,175] The observed effects have several causes which, in practice, result in a complex but important influence of the solution electrolyte:

(a) *Electrostatic repulsion or attraction* between membrane and solute when both are charged. This effect can be seen in Figure 19 by comparing Amicon® (negative) and Nuclepore® (neutral) membranes.

(b) *Hydration* of membrane pores (Figure 22) and solute:[90,172] in hydrophilic membranes, the pore surface is covered by a layer of highly ordered hydration water of low mobility, which diminishes the effective pore radius ($r_p^{eff} < r_p$). Solutes also have hydration layers which increase in thickness with charge density.[173] These two effects together explain (Figure 22) the following phenomena:

  • Compounds or inorganic ions of size $r < r_p$ may be retained. For membranes for which r_p is slightly larger than r, there is a good

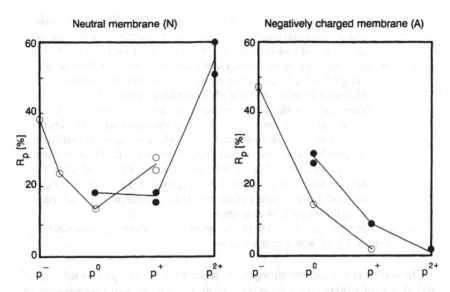

Figure 19. Role of membrane and solute electric charges on solute retention, R_p. Membrane A = Amicon UMO5 and N = Nuclepore MW500. Both are filters of type E (see Figure 2) with a nominal molecular weight cut-off limit given as 500 (but see also Figures 6a and 21). A = negatively charged membrane, N = neutral membrane (no net charge). ●: p = pyridoxamine; ○: p = pyridoxol. In both cases molecular size is the same and shape is similar. Charges on p are changed by varying pH. (From Staub, C. et al. *Anal. Chem.* 56:2843–2849 (1984). With permission.)

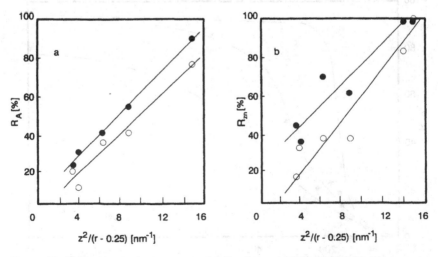

Figure 20. Role of ion hydration energy (abcissa) on retention coefficients of inorganic ions by the two membranes of Figure 19 (● = Amicon UMO5; ○ = Nuclepore MW500). (a) Retention R_A of the electrolyte anions A: Cl^-, NO_3^-, ClO_4^-, F^-, SeO_4^{2-}, SO_4^{-2} (in all cases cation = Na^+). (b) Retention of the minor ion Zn^{+2} R_{zn}, in presence of the above electrolytes, as a function of the hydration energy of the electrolyte anion z, r = charge and radius of A, respectively. Ionic strength = 3.10^{-3} mol.dm^{-3}. (From Staub, C. et al. *Anal. Chem.* 56:2843–2849 (1984). With permission.)

correlation (see Figure 20a) between the degree of ionic hydration and the retention coefficient[97]). Increasing retention as a function of the degree of hydration has also been observed for polymers.[174]

- Retention of the solute on neutral membranes increases with its electric charge (Figure 21), whatever its sign. For cations and anions, solute hydration increases with the absolute charge value.
- When r_p is only slightly larger than r, the diffusion coefficient value for a solute X in the membrane, $D_{m,x}$, decreases strongly with r_p, relative to its value in solution, D_x. Indeed being highly ordered, the pore hydration layer is also highly viscous (10 to 100 times more than in solution; Reference 90). So it exerts a significant frictional drag on the solute hydration layer (Figure 22). Low values of $D_{m,x}$ may influence J_x values (Equation 2) for membranes of small pore size, for which J_w is also small.

(c) *Electrokinetic effects.*[11,22] They occur when an ion is moving in the electric field of a pore with a charged surface.

These effects explain the stronger ion adsorption on charged membranes,[97] and above all explain the important role played by the *major electrolyte* on the retention of minor solutes by small pore ultrafiltration membranes:[27,86,97,176-178]

- It masks the surface charge inside the pores. The importance of electrokinetic processes therefore decreases with increasing electrolyte concentration.[22]

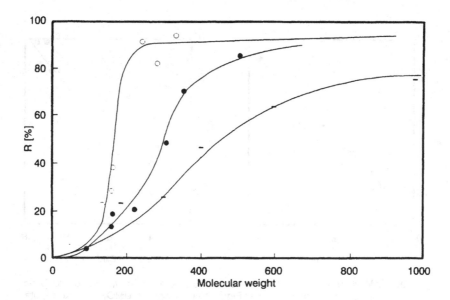

Figure 21. Retention of small organic molecules of different shapes on the neutral polycarbonate Nuclepore membrane MW500. ● = neutral molecules having rigid skeleton; ○ = negative compounds; — = neutral flexible linear polyethyleneglycols. Compounds are listed in Reference 97.

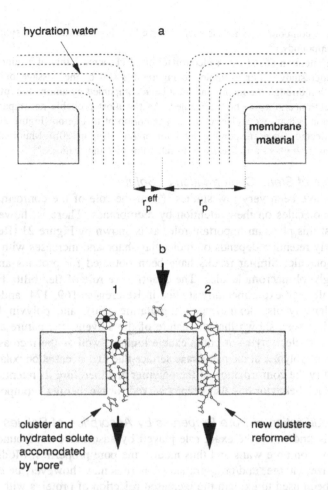

Figure 22. Schematic representation of (a) the water hydration layer in the membrane pores and (b) the effect of pore and solute hydration on solute retention. In pores No 2, pores are blocked by their hydration water and that of the solute. In order to permit the latter to pass, their hydration layers must be broken. (From Buffle, J. "Complexation Reactions in Aquatic Systems: An Analytical Approach." (Chichester: Horwood. 1989). With permission.)

- In the case of minor cation loss by adsorption (particularly on net negative membranes), the electrolyte cation can act simultaneously through masking and by competing for adsorption sites.[97] In general, therefore, adsorption decreases as the electrolyte concentration and the charge of the electrolyte cation increase. These results suggest that adsorption losses during filtration should be lower for hard waters or sea waters than for soft waters. This is indeed observed in practice and pretreatment with Ca^{2+} to minimize these losses (Section 3.2.2) is based on the same property.

- The electrolyte can also mask the charge of organic polyelectrolytic compounds in the solution and thereby facilitate their folding. An increase in electrolyte concentration leads, in this case, to a decreased retention of these compounds. This has been observed for biochemical[170] and synthetic compounds.[174]

- The retention of the major electrolyte ions can increase that of a minor ion under study.[97] For example, in Figure 20a and b, the passage of Zn^{2+} cation through the membrane must be accompanied by anions for solution electroneutrality to be maintained. As the most probable accompanying anion is that of the electrolyte, any retention of this anion (Figure 20a) is reflected by a corresponding retention of Zn^{2+} (Figure 20b). Major cations may also play an indirect role in retention of minor cations.[97]

4.3.2 Role of Steric Conformation of Solute

There have been very few studies[22,97] on the role of the conformation of organic molecules on their retention by membranes. There is, however, no doubt that this plays an important role, as is shown by Figure 21 (Reference 27): clearly retention depends on molecular shape and increases with rigidity of the molecule. Similar results have been obtained for proteins and poly-ethyleneglycol macromolecules. The quantitative role of flexibility has been theoretically and experimentally studied in References 169, 174, and 179 for polyethyleneglycols, dextran, poly(L-glutamic acid), and polyvinylpyrrolidone. It has been shown that the nature of the solvent, the nature and concentration of electrolytes or complexable ions, as well as the increased concentration of polymer at the membrane surface due to concentration polarization, may modify the conformation of the polymer and therefore its retention coefficient. Such behavior has also been observed for biochemical compounds.[170]

4.3.3 Modification of Pore Properties by Adsorption of Solutes

Little is known on the exact role played by adsorbable compounds when they adsorb on pore walls and thus modify the pore properties. Modification of pore size, charge, and/or hydration properties may, however, be expected and has been used to explain the increased rejection of proteins with time.[135] Another interesting effect results from the adsorption of (even small) amphiphile molecules onto the surface of (even large pore size) membranes.[133,136] The magnitude of the water flux may be changed by factors as large as 10 even with low concentrations ($\sim 10^{-4}\,M$) of adsorbable compounds. The water flux is increased when adsorption occurs on the low pressure side of the membrane and decreased in the opposite case.

5. CONCLUSIONS

Based on the previous results and discussion, it is possible to propose a number of optimum conditions for filtration as well as a few tests of the reliability of the results.

5.1 Optimum Operating Conditions
5.1.1 Choice of Membranes

For *size fractionation,* all the results reported in the literature suggest that polycarbonate sieve-type filters are preferable to depth filters, but this choice is by no means sufficient to get reliable results; the conditions listed below, particularly concerning low flow rate, should also be obeyed.

For *total retention* of particles and colloids, depth filters seem preferable as surface coagulation occurs at an earlier stage of filtration. But this choice must be combined with high concentration polarization conditions, in particular, high flow rate. For gravimetric purposes, however, polycarbonate filters are preferable because of their lower and more reproducible blank values.

New membranes should preferably be used for each new filtration experiment as (i) adsorption may modify membrane properties and no reliable membrane cleaning process has been reported up to now and (ii) change of porosity with time, due to slow plugging of the membrane by the applied pressure, may occur with some membranes.

Membrane dimension is a parameter to consider since pore size distribution is never infinitely narrow and may not be uniform over the same membrane. For this reason, membranes with small surface areas are preferable; but when low flow rate must be used (see Section 5.1.3), large surface areas may be necessary. For analytical purposes, areas ranging between 1 and 100 cm^2 are normally used.

5.1.2 Minimizing Adsorption and Contamination

This problem is most important for the analysis of trace compounds, either organic or inorganic.

- *Contamination* by the membrane may be minimized by washing, by filtering through it a sufficient volume (to be tested; at least 20 ml.cm^{-2}) of 10^{-2} mol.dm^{-3} HCl (for metal decontamination) followed by distilled water, or preferably by the water sample prefiltered on a less porous membrane. This last step may then also serve as a preconditioning step to minimize adsorption losses.
- *Adsorption losses* can be minimized by preconditioning the filter by passing through it 10^{-2} mol.dm^{-3} Ca^{2+} solution (for trace metal adsorption), or preferably the water sample prefiltered on a less porous membrane.

5.1.3 Minimizing Concentration Polarization and the Related Artifacts (Coagulation, Gel Layer Formation . . .)

The various factors which may be used to minimize concentration polarization are summarized below. Note that this is valid for size fractionation purposes. For total retention of particles, a concentration polarization as large as possible is desirable, and therefore the opposite conditions should preferably be applied.

Optimal condition
- Low concentration at the membrane surface

Possible approaches
a— Low J_w value obtained by means of:
→ low pressure for larger pore size membranes or independent control of J_w
→ naturally low J_w for smaller pore size membranes
b— Low δ value obtained by a high stirring rate

● Low concentration in the filtration cell   c— Low concentration factor during filtra-
                                                       tion
                                                   d— Low concentration in the initial sample

5.1.3.1 Flow Rate. From Figure 11 and the results reported in Section 4.2.3, an approximate maximum limit for J_w can be set at $J_w \sim 3 \cdot 10^{-4}$ cm.s^{-1} (corresponding to 0.02 ml.cm^{-2}.min^{-1}) (Figure 23). This rather low value has two practical consequences:

● Only relatively small sample volumes can be size fractionated in a reliable manner (at this flow rate, 1 h is required to filter 100 ml on a 100 cm^2 membrane). This implies that sensitive detection techniques must be used to study the fractionated material, when its concentration is very low in the initial sample, as is the case for sea water or ground water.
● When fractionation time becomes larger than ~ 1 h, aggregation and co-agulation in solution within the filtration cell becomes an important factor to consider (Section 3.3.1). This is shown in Figure 24 where Fe and Mn oxyhydroxide particles formed at the oxic/anoxic boundary layer of an anoxic lake have been filtered at different flow rates, all smaller than $3 \cdot 10^{-4}$ cm.s^{-1}. By comparing particle size with the pore size of the polycarbonate membrane used (0.2 μm), it is expected that only 10% of Fe and less than 10% of Mn particles should be retained. Clearly, retention increases with the duration of filtration plus storage, irrespective of the flow rate used for

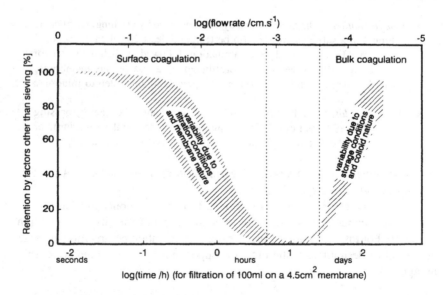

Figure 23. Semi-quantitative representation of the change in retention with flow rate due to concentration polarization (high flow-rate domain) and aggregation in the filtration cell (low flow-rate domain). Only an intermediate flow rate window is usable for size fractionation without artifact. (This figure is based on Figures 11, 14, 24 and additional data.)

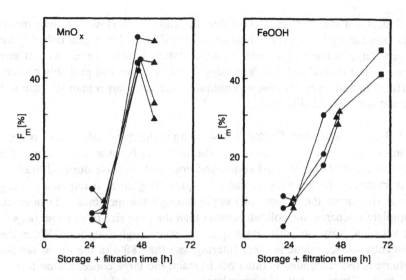

Figure 24. Effect of storage plus filtration time on the "adsorption" of iron particles (same as in Figure 12) and MnO_2 particles formed in the same lake (with size ≤0.2 μm) on 0.2 μm membranes. The various lines refer to five different membranes (Gelman acrylic copolymer, Nuclepore polycarbonate, Schleicher and Schuell cellulose nitrate and mixed esters, and DDS polysulfone filters). The various symbols refer to: ■ J_w = 3.3 × 10^{-4} cm.s^{-1}. ● 1.6 × 10^{-4} cm.s^{-1} and ▲ 0.3 × 10^{-4} cm.s^{-1}, respectively. Clearly at these low flow rates, membrane nature and flow rate values are unimportant. Considering the true particle size (<0.3 μm for iron particles; <0.2 μm for MnO_2) one expects not more than 10% of Fe and a few percent of Mn to be retained on the membrane in the absence of secondary factors. All solutions were previously filtered on 3 μm membranes and stored at 4°C in the dark.

filtration. This is due to aggregation in solution which, in this case, becomes significant after ~24 h. As a consequence

(i)   Too low values of J_w should be avoided (Figure 23), so that only a certain window of J_w is accessible for size fractionation. The lower limit depends on the volume to be filtered and the surface area of the membrane.

(ii) Filtration must be done as soon as possible after sampling (see below)

(iii) Laboratory dialysis should preferably be avoided for size fractionation because of the very long equilibration time required.

5.1.3.2 Minimization of δ (Stirring).

In principle, a high stirring rate is preferable to minimize δ and therefore concentration polarization. However, this may also have an adverse effect by favoring coagulation within the filtration cell, in particular when long filtration time (low flow rates) are used. Although stirring seems to slightly improve the results, more detailed studies are required in order to draw definite conclusions.

Ultrasonic stirring has been proposed, but it may damage the membrane. Filtration modes other than in stirred batch cells have also been proposed, such as cross-flow filtration,[154,185] pulsed filtration,[186,187] filtration on rotatory cylinders or discs,[181,184] and hollow fiber filtration.[182,183] However, these methods have mostly been applied to industrial purposes and they generally require a very large volume of sample, as recirculation is often necessary. At any rate, it must be realized that decreasing δ below 1 μm will probably be very difficult so that, even with these techniques, flow rates larger than 10^{-2} cm.s^{-1} should not be used (Figure 11).

5.1.3.3 Concentration Factor. The washing technique (diafiltration) is preferable to the concentration technique (Section 2.2) because in the former the concentration of the retained compound does not increase during filtration. Unfortunately the washing technique is quite long since it requires passing about five times the volume of sample through the membrane, in order to completely separate the colloids smaller than the pore size from larger ones.[51] At very low flow rate, this may require an exceedingly long time. When the concentration technique is used, filtering less than 80% of the initial sample volume is recommended in order not to reach too large concentration factors inside the filtration cell. For complete purification of the retained particles, the washing technique may be applied to the remaining solution.

5.1.3.4 Cascade Filtration. In order to minimize coagulation in the cell as well as at the membrane surface, sequential or cascade filtration is highly recommended, since aggregation rate is generally much larger in physically and chemically heterodisperse colloid suspensions (like those of natural waters) than in a homodisperse suspension. Filtration through more than four to five membranes is not useful because the reproducibility of filtration on a single membrane is not better than 5 to 10% and accumulation of experimental uncertainties may lead to unreliable results.

5.1.4 Minimization of Sample Perturbation

Several aspects of this problem are not specific to filtration.[24,27,57,112] The relevant point here is to ensure that size fractionation by filtration gives results which are realistic of the initial composition of the sample. For this purpose filtration must be done as soon as possible after sample collection (Section 5.1.3.1) without modification of temperature, pH, or electrolyte concentration in order to minimize artifacts such as aggregation processes. Therefore, filtration must be done in the field immediately after sampling and under thermostated conditions if filtration time is long. In very easily perturbed samples, cascade filtration must be done directly at depth.[57]

5.2 Checking the Validity of Results

A number of tests may be used to check the reliability of results. Several of them are easy to apply. They are listed below. A rule of thumb for discriminating between correct and incorrect retention when different conditions are applied to unknown samples, is based on the fact that many artifacts result in an increase of the retention of the test compound. Therefore, in many cases, conditions favoring passage through the membrane should be considered as giving more realistic results (this rule obviously must be used with caution by considering in particular all the effects mentioned in Section 4).

- *Flow-rate* may be varied at constant pressure by using a valve at the output of the cell. Retention must be independent of flow rate.
- *Stirring*: retention must be independent of stirring rate.
- *Filtrate concentration*: in the absence of secondary effects, (i.e., in the absence of a change of σ_x), filtrate concentration $[X]_f$ is given by:[27,22]

$$[X]_f = (1 - \sigma_x) \cdot [X]_c$$

Therefore $[X]_f$ should be constant during filtration and lower or equal to $[X]_c$ ($0 \leq \sigma_x \leq 1$). This has been observed for instance in Reference 22. In particular, a decrease of $[X]_f$ with filtration time is a much more sensitive indication of membrane clogging than the decrease in J_w, which is normally detectable only after a thick gel layer has already formed.

- *Scanning and trasmission electron microscopy* (SEM and TEM) are very powerful tools[96] for checking whether or not particles are retained on the membrane, what are their sizes and whether or not they are the result of aggregation process (Figures 15 and 16). Note, however, that SEM may be highly misleading. Because of SEM's low resolution, aggregates are often ''seen'' as single particles (Chapter 6 of the present book). As a consequence, size distribution should never be inferred from SEM observations of particles collected on filters, unless the absence of induced coagulation by concentration polarization has been proven. TEM is much preferable. Other techniques for characterizing particles on filters are described in Chapter 3 of the present book.
- *Electrolyte nature and concentration* may have important influences on diverse processes relevant in filtration and, consequently, on separation results. As a consequence generalization of this effect is difficult and its influence should be tested in each particular case.
  - Coagulation of colloids and particles
  - Change of conformation, size, and hydration of macromolecules
  - Adsorption of ions and molecules on filters
  - Retention of ions and small molecules on the least porous membranes, by flux coupling between these compounds and the ions of the electrolyte

APPENDIX

The maximum tolerable flux to avoid surface coagulation may be estimated as follows, by considering a homodispersed suspension of fully retained particles X ($\sigma_x = 1$). The limitation due to heterodispersion of real samples is discussed in the text. Due to the concentration polarization effect, a diffusion layer, in which the concentration of X is larger than in the bulk solution, is formed close to the membrane. X may then undergo two processes: back-diffusion into solution, and possibly, coagulation. Coagulation will be negligible, provided the corresponding elimination flux is much smaller than the elimination flux by back-diffusion:

$$\left(\frac{dN_X}{dt}\right)_{coag} \leq \frac{1}{f} \cdot \left(\frac{dN_X}{dt}\right)_{diff} \tag{A1}$$

where f is any arbitrary value larger than 1 (f = 10 is used in Figure 11).

The following treatment only deals with coagulation at the membrane surface. Since, however, the concentration of X is maximum at the membrane surface, and the coagulation rate of X depends on the square of its concentration (see below), this rate will also be maximum at the membrane surface. Conditions which minimize coagulation at the membrane surface will therefore also be applicable to the rest of the diffusion layer.

When concentration polarization occurs, the concentration of the fully retained particles X at the membrane surface, $[X]_{c,0}$, compared to that in the filtration cell, $[X]_c$, is given by Equation 7':

$$\frac{[X]_{c,0}}{[X]_c} = \exp(J_w \cdot \delta/D_X) \tag{7'}$$

The back-diffusion flux at the membrane surface is given by Fick's first law which may be simplified as:

$$\left(\frac{dN_X}{dt}\right)_{diff} = -A \cdot D_X \cdot \frac{[X]_{c,0} - [X]_c}{\delta} \tag{A2}$$

where the diffusion layer thickness, δ, is assumed to be constant, since the solution in the filtration cell is stirred. A is the membrane surface area.

Particles may also be eliminated from the diffusion layer by coagulation, either perikinetic (first term in Equation A3) or orthokinetic (second term in Equation A3). At the membrane surface, where concentration of X is $[X]_{c,0}$, the overall rate of concentration decrease is given by:[137]

$$\left(\frac{d[X]_{c,0}}{dt}\right)_{coag} = -\left(\alpha_p \cdot \frac{4kT}{3\eta} + \frac{16}{3} \cdot \alpha_o \cdot G \cdot r^3\right) \cdot [X]_{c,0}^2 \tag{A3}$$

The parameters are defined in the text (see Section 4.2.2). The corresponding flux of particles eliminated by coagulation is given by:

$$\left(\frac{dN_X}{dt}\right)_{coag} = \left(\frac{d[X]_{c,0}}{dt}\right)_{coag} \cdot A \cdot \mu \tag{A4}$$

where μ is the reaction layer thickness in which coagulation occurs. It is related to D_X and the mean life-time, τ_X, of X inside the reaction layer.[78] For a monomolecular reaction τ_X is itself equal to the reciprocal of the rate constant, k, of the reaction that inactivates X, i.e., the coagulation reaction:

$$\mu = \sqrt{D_X \cdot \tau_X} = \sqrt{D_X/k} \qquad (A5)$$

Coagulation is a bimolecular reaction (see Equation A3). However, after a short transition period of filtration, the concentration polarization effect is assumed to produce a stationary state in the diffusion layer (Section 4.2.1), and Equation 7' is only valid in this condition. $[X]_{c,0}$ being constant, Equation A3 may be written in the form of a pseudo first-order reaction:[137]

$$\left(\frac{d[X]_{c,0}}{dt}\right)_{coag} = -k \cdot [X]_{c,0} \qquad (A6)$$

with:

$$k = \frac{4\alpha}{3} \cdot (kT/\eta + 4Gr^3) \cdot [X]_{c,0} \qquad (A7)$$

where α_p and α_o are assumed to be equal ($\alpha = \alpha_p = \alpha_o$).

By combining Equations A1, A2, A4, A5, and A7 with the condition of a large concentration polarization (i.e., $[X]_{c,0} \gg [X]_c$), one gets the following expression for the maximum limiting value, $[X]_{c,0}^{max}$, of $[X]_{c,0}$, for which no coagulation occurs:

$$[X]_{c,0}^{max} \leq \frac{3 \cdot D_X/\delta^2}{4f^2 \cdot \alpha \cdot (4Gr^3 + kT/\eta)} = \frac{C_o}{f^2} \qquad (A8)$$

where C_o is the value of $[X]_{c,0}$ for which the fluxes of back-diffusion and coagulation are equal at the membrane surface.

Combining Equation A8 with Equation 7' gives the maximum tolerable flux for filtration:

$$J_w \leq \frac{D_X}{\delta} \cdot \ln\left(\frac{C_o}{f^2 \cdot [X]_c}\right) \qquad (A9)$$

GLOSSARY

In this chapter, the following definitions are used:

- *Particle*: any compound with size larger than 0.45 μm
- *Colloid*: any compound with size larger than a few nanometers and smaller than 0.45 μm

- *Macromolecule*: synonymous to colloid
- *Solute*: any compound or inorganic ion with size smaller than a few nanometers
- *Compound in solution*: any compound or inorganic ion, irrespective of its size (compounds in solution include particles, macromolecules, or colloids and solutes)
- *Permeate*: a compound or ion of any size passing through the membrane of interest
- *Retentate*: a compound or ion of any size retained by the test membrane, irrespective of the retention mechanism
- *Concentration polarization*: formation of a concentration gradient compounds, at the membrane surface, due to their retention by the membrane and their non-infinite rate of back-diffusion in solution

REFERENCES

1. Barth, H.G. *Chemical Analysis, Vol 73, Modern Methods of Particle Size Analysis*. (New York: John Wiley & Sons, 1984).
2. Barth, H.G., S.-T. Sun, and R.M. Nickel. "Particle Size Analysis," *Anal. Chem.* 59:142R–162R (1987).
3. Beyer, G.L. "Determination of Particle Size and Molecular Weight," in *Techniques of Chemistry*, Vol I. A. Weissberger and B.W. Rossiter, Eds. (New York: Wiley-Interscience, 1972).
4. Brock, T.D. *Membrane Filtration*. (Madison, WI: Science Tech. 1983).
5. Meares, P., Ed. *Membrane Separation Processes*. (Amsterdam: Elsevier, 1976).
6. Sourirajan, S., Ed. *Reverse Osmosis and Synthetic Membranes*. (Natl. Res. Council Canada, Ottawa 1977).
7. Pusch, W. and A. Walch. "Synthetic Membranes — Preparation, Structure and Application," *Angew. Chem. Int. Ed. Engl.* 21:660–85 (1982).
8. Piskin, E. and A.S. Hoffman, Eds. *Polymeric Biomaterials*. (Dordrecht: Martinus Nijhoff Publishers, 1986).
9. Sourirajan, S. and T. Matsuura, Eds. *Reverse Osmosis and Ultrafiltration*, ACS Symp. Ser. 281. (Washington, D.C.: American Chemical Society, 1985).
10. Michaels, A.S. "Polyelectrolyte Complexes," *Ind. Eng. Chem.* 57:32–40 (1967).
11. Tak, H.S. and K. Kammermeyer. *Techniques of Chemistry*, Vol. VII, A. Weissberger, Ed., (New York: John Wiley & Sons, 1975).
12. Simonetti, J.A., H.G. Schroeler, and T.H. Meltzer. "Membrane Testing. A Review of Latex Sphere Retention Work: Its Application to Membrane Pore-Size Rating," *Ultrapure Water* 3(4):46–51 (1986).
13. Bodzek, M. "Determination of Pore Size Characteristics of Polymer Ultrafiltration Membranes by Gas-Adsorption/Desorption Method," *Chem. Anal.* 30:563–575 (1985).
14. Nguyen, Q.T., P. Aptel, and J. Neel. "Characterization of Ultrafiltration Membranes. I. Water and Organic Solvent Permeabilities," *J. Membr. Sci.* 5:235–251 (1979).
15. Blatt, W.F. "Principles and Practice of Ultrafiltration," in *Membrane Separation Processes*, P. Meares, Ed., (Amsterdam: Elsevier, 1976), chap. 3.
16. Lakshminarayanaiah, N. "Measurement of Membrane Potential and Estimation of Effective Fixed-Charge Density in Membranes," *J. Membr. Biol.* 21:175–189 (1975).

17. Bitter, S.G.A. "Effect of Crystallinity and Swelling on the Permeability and Selectivity of Polymer Membranes," *Desalination* 51:19–35 (1984).

18. Hernandez, A., J.A. Ibanez, and A.F. Tejerina. "True and Adsorbed Charges in Passive Membranes. Surface Charge Density and Ionic Selectivity of Several Microporous Membranes," *Sep. Sci. Technol.* 20:297–314 (1985).

19. Ibanez, J.A. and A.F. Tejerina. "Surface Charge Density in Passive Membranes from Membrane Potential, Diffusion Potential and Ionic Permeating Species," *J. Non-Equil. Thermodyn.* 7:83–94 (1982).

20. Badenhop, C.T. and A.L. Bourguignon. "Membranfilter," *Chem. Anlagen Verfahren* 18:203–206 (1985).

21. Fane, A.G., C.J.D. Fell, and A.G. Waters. "The Relationship between Membrane Surface Pore Characteristics and Flux for Ultrafiltration Membranes," *J. Membr. Sci.* 9:245–262 (1981).

22. Macko, C., W.J. Maier, S.J. Eisenreich, and M.R. Hoffman. "Ultrafiltration Characterization of Aquatic Organics," *AIChE Symp. Ser.* 75:162–169 (1979).

23. Johnston, R.R. "Fluid Filter Media: Measuring the Average Pore Size and the Pore Size Distribution, and Correlation with Results of Filtration Tests," *J. Test. Eval.* 13:308–315 (1985).

24. Riley, J.P., D.E. Robertson, J.W.R. Dutton, N.T. Mitchell, and P.J. Le B. Williams. "Analytical Chemistry of Sea Water," in *Chemical Oceanography*, Vol. 3, 2nd ed. J.P. Riley and G. Skirrow, Eds. (London: Academic Press, 1975), chap. 19.

25. de Mora, S.J. and R.M. Harrison. "The Use of Physical Separation Techniques in Trace Metal Speciation Studies," *Water Res.* 17:723–733 (1983).

26. Hunt, D.T.E. "Filtration of Water Samples for Trace Metal Determinations," Water Research Centre, Technical Rep. TR-104 (1979).

27. Buffle, J. *Complexation Reactions in Aquatic Systems: An Analytical Approach.* (Chichester: Horwood, 1988).

28. Lerman, A. *Geochemical Processes.* (New York: Wiley-Interscience, 1979).

29. Jones, B.F. and C.J. Bowser, *Lakes: Chemistry, Geology and Physics*, A. Lerman, Ed. (Berlin: Springer-Verlag, 1978), chap. 7.

30. Tessier, A., P.G.C. Campbell, and M. Bisson. "Particulate Trace Metal Speciation in Stream Sediments and Relationships with Grain Size: Implications for Geochemical Exploration," *J. Geochem. Explor.* 16:77–104 (1982).

31. van de Meent, D., A. Los, J.W. Leeuw, and P.A. Schenck, "Size Fractionation and Analytical Pyrolysis of Suspended Particles from the River Rhine Delta," *Adv. Org. Geochem.* pp. 336–349 (1981).

32. Simpson, W.R. "Particulate Matter in the Oceans. Sampling Methods, Concentration, Size Distribution and Particle Dynamics," *Oceanogr. Mar. Biol. Annu. Rev.* 20:119–172 (1982).

33. Nomizu, T., T. Nozue, and A. Mizuike, "Electron Microscopy of Submicron Particles in Natural Waters — Morphology and Elemental Analysis of Particles in Fresh Waters," *Mikrochim. Acta* 11:99–106 (1987).

34. Grasshof, K. *Methods of Sea Water Analysis.* (Weinheim: Verlag Chemie, 1976).

35. Goldberg, E.D., M. Baker, and D.L. Fox. "Microfiltration in Oceanographic Research. I. Marine Sampling with the Molecular Filter," *J. Mar. Res.* 11:197–202 (1952).

36. O'Melia, C.R. "Aquasols: The Behaviour of Small Particles in Aquatic Systems," *Environ. Sci. Technol.* 14:1052–1060 (1980).

37. O'Melia, C.R. "Kinetic of Colloid Chemical Processes in Aquatic Systems," in *Aquatic Chemical Kinetics: Reaction Rates and Processes in Natural Waters.* W. Stumm, Ed. (New York: John Wiley & Sons, 1990).

38. Hoffmann, M.R., E.C. Yost, S.J. Eisenreich, and W.J. Maier. "Characterization of soluble and colloidal phase metal complexes in river water by ultrafiltration. A mass-balance approach," *Environ. Sci. Technol.* 15:655–661 (1981).

39. Salbu, B., H.E. Björnstadt, N.S. Lindström, and E. Lydersen. "Size Fractionation Techniques in the Determination of Elements Associated with Particulate or Colloidal Material in Natural Fresh Waters," *Talanta* 32:907–913 (1985).

40. de Haan, H., T. de Boer, J. Voerman, H.A. Kramer, and J.R. Moed. "Size Classes of "Dissolved" Nutrients in Shallow Alkaline Humic and Eutrophic Tjeukemeer, The Netherlands, as Fractionated by Ultrafiltration," *Verh. Int. Verein. Limnol.* 22:876–881 (1984).

41. Andren, A.W. and R.C. Harriss. "Observations on the Association Between Mercury and Organic Matter Dissolved in Natural Waters," *Geochim. Cosmochim. Acta* 39:1253–1257 (1975).

42. Schindler, J.E., J.J. Alberts, and K.R. Honick. "A Preliminary Investigation of Organic-Inorganic Associations in a Stagnating System," *Limnol. Oceanogr.* 17:952–957 (1972).

43. Laxen, D.H. and R.M. Harrison, "A Scheme for the Physico-Chemical Speciation of Trace Metals in Fresh Water Samples," *Sci. Total Environ.* 19:59–82 (1981).

44. Laxen, D.H. and R.M. Harrison. "The Physico-Chemical Speciation of Cd, Pb, Cu, Fe and Mn in the Final Effluent of a Sewage Treatment Works and its Impact on Metal Speciation in the Receiving River," *Water Res.* 15:1053–1065 (1981).

45. Eisenreich, S.J., M.R. Hoffmann, D. Rastetter, E. Yost, and W.J. Maier. in *Particulates in Water,* M.C. Kavanaugh and J.O. Leckie, Eds. (ACS, Washington, D.C., 1980), chap. 6.

46. Wheeler, J.R. "Fractionation by Molecular Weight of Organic Substances in Georgia Coastal Water," *Limnol. Oceanogr.* 21:846–852 (1976).

47. Wilander, A. "A Study on the Fractionation of Organic Matter in Natural Water by Ultrafiltration Techniques," *Schw. Z. Hydrol.* 34:190–200 (1972).

48. Gjessing, E.T. *Physical and Chemical Characteristics of Aquatic Humus.* (Ann Arbor, MI: Ann Arbor Science, 1976).

49. Ogura, N. "Molecular Weight Fractionation of Dissolved Organic Matter in Coastal Sea-Water by Ultrafiltration," *Mar. Biol.* 24:305–312 (1974).

50. Buffle, J., P. Deladoey, J. Zumstein, and W. Haerdi. "Analysis and Characterization of Natural Organic Matters in Fresh Waters. I. Study of Analytical Techniques," *Schw. Z. Hydrol.* 44:325–362 (1982).

51. Buffle, J., P. Deladoey, and W. Haerdi. "The Use of Ultrafiltration for the Separation and Fractionation of Organic Ligands in Fresh Waters," *Anal. Chim. Acta* 101:339–357 (1978).

52. Steinberg, C. "Schwer abbaubare stickstoffhaltige gelöst organische Substanzen im Schöhsee und in Algenkulturen," *Arch. Hydrobiol. Suppl.* 53:48–158 (1977).

53. Tuschall, J.R. and P.L. Brezonik. "Characterization of Organic Nitrogen in Natural Waters: Its Molecular Size, Protein Content and Interactions with Heavy Metals," *Limnol. Oceanogr.* 25:495–504 (1980).

54. Stabel, H.H. "Gebundene Kohlenhydrate als stabile Komponenten im Schöhsee und in Scenedesmus Kulturen," *Arch. Hydrobiol. Suppl.* 53:159–254 (1977).

55. Laxen, D.P.H. and I.M. Chandler. "Comparison of Filtration Techniques for Size Distribution in Fresh Waters," *Anal. Chem.*, 54:1350–1355 (1982).

56. Laxen, D.P.H. and I.M. Chandler. "Size Distribution of Iron and Manganese Species in Fresh Waters," *Geochim. Cosmochim. Acta* 47:731–741 (1983).

57. Buffle, J., R.R. De Vitre, D. Perret, and G.G. Leppard. "Combining Field Measurements for Speciation in Non-Perturbable Water Samples," in *Metal Speciation: Theory, Analysis and Application*, J.R. Kramer and H.E. Allen, Eds. (Chelsea, MI: Lewis Publishers, 1988) chap. 5.

58. De Vitre, R.R., J. Buffle, D. Perret, and R. Baudat. "A Study of Iron and Manganese Transformations at the $O_2/S(-II)$, Transition Layer in an Eutrophic Lake (Lake Bret, Switzerland): A Multimethod Approach," *Geochim. Cosmochim. Acta* 52:1601–1613 (1988).

59. Smith, R.G. "Evaluation of Combined Applications of Ultrafiltration and Complexation Capacity Techniques to Natural Waters," *Anal. Chem.*, 48:74–76 (1976).

60. Buffle, J., P. Deladoey, F.L. Greter, and W. Haerdi, "Study of the Complex Formation of Copper(II) by Humic and Fulvic Substances," *Anal. Chim. Acta* 116:255 (1980).

61. Buffle, J. and P. Deladoey. "Analysis and Characterization of Natural Organic Matters in Fresh Waters. II. Comparison of the Properties of Waters of Various Origins and their Annual Trends," *Schw. Z. Hydrol.* 44:363 (1982).

62. Hasle, J.R. and M.I. Abdullah. "Analytical Fractionation of Dissolved Copper, Lead and Cadmium in Coastal Sea Water," *Mar. Chem.* 10:487 (1981).

63. Leppard, G.G., J. Buffle, and R. Baudat. "A Description of the Aggregation Properties of Aquatic Pedogenic Fulvic Acids," *Water Res.* 20:185–196 (1986).

64. Leppard, G.G., J. Buffle, R. De Vitre, and D. Perret. "The Ultrastructure and Physical Characteristics of a Distinctive Colloidal Iron Particulate Isolated From a Small Eutrophic Lake," *Arch. Hydrobiol.* 113:405–424 (1988).

65. Buffle, J., R. De Vitre, D. Perret, and G.G. Leppard. "Physico-Chemical Characteristics of a Colloidal Iron Phosphate Species Formed at the Oxic/Anoxic Interface of a Eutrophic Lake," *Geochim. Cosmochim. Acta* 53:399–408 (1989).

66. Tipping, E., C. Woof, and D. Cooke, "Iron Oxide From a Seasonally Anoxic Lake," *Geochim. Cosmochim. Acta* 45:1411–1419 (1981).

67. Gillespie, P.A. and R.F. Vaccaro. "A Bacterial Bioassay for Measuring the Copper Chelation Capacity of Sea Water," *Limnol. Oceanogr.* 23:543–548 (1978).

68. Ramamoorthy, S. and D.J. Kushner. "Heavy Metal Binding Components of River Water," *J. Fish Res. Bd. Can.* 32:1755–1766 (1975).

69. Hart, B.T. and S.H.R. Davies. "A New Dialysis-Ion Exchange Technique for Determining the Forms of Trace Metals in Water," *Aust. J. Mar. Freshwater Res.* 28:105–112 (1977).

70. Beneš, P. and E. Steinnes. "Migration Forms of Trace Elements in Natural Fresh Waters and the Effect of Water Storage," *Water Res.* 9:741–749 (1975).

71. Weber, J.H. in *Aquatic and Terrestrial Humic Materials*. R.F. Christman and E.T. Gjessing, Eds. (Ann Arbor, MI: Ann Arbor Science, 1983).

72. Truitt, R.E. and J.H. Weber. "Determination of Complexing Capacity of Fulvic Acid for Copper(II) and Cadmium(II) by Dialysis Titration," *Anal. Chem.* 53:337–342 (1981).

73. Rainville, D.P. and J.H. Weber. "Complexing Capacity of Soil Fulvic Acid for Cu^{2+}, Cd^{2+}, Mn^{2+}, Ni^{2+} and Zn^{2+} Measured by Dialysis Titration: A Model Based on Soil Fulvic Acid Aggregation," *Can. J. Chem.* 60:1 (1982).

74. Truitt, R.E., J.H. Weber. "Cu (II) and Cd (II) Binding Abilities of some New Hampshire Freshwaters Determined by Dialysis Titration," *Env. Sci. Technol.* 15:1204 (1981).

75. Kranck, K. *Handbook of Environmental Chemistry. Vol. 2*, O. Hutzinger, Ed. (Berlin: Springer-Verlag, 1980), pp. 47–59.

76. Sherwood, T.K., P.L.T. Brian, R.E. Fisher, and L. Dresner. "Salt Concentration at Phase Boundaries in Desalination by Reverse Osmosis," *Ind. Eng. Chem. Fundam.* 4:113 (1965).

77. Harris, F.L., G.B. Humphreys, and K.S. Spiegler. "Reverse Osmosis (Hyperfiltration) in Water Desalination," in *Membrane Separation Processes*, P. Meares, Ed. (Amsterdam: Elsevier, 1976), chap. 4.

78. Heyrovsky, J. and J. Kuta, *Principles of Polarography*, (New York: Academic Press, 1966), pp. 345–346.

79. Beneš, P. and E. Steinnes. "In Situ Dialysis for the Determination of the State of Trace Elements in Natural Waters," *Water Res.* 8:947–953 (1974).

80. Carignan, R. "Interstitial Water Sampling by Dialysis," *Limnol. Oceanogr.* 29:667–670 (1984).

81. Hesslein, R.H. "An In-Situ Sampler for Close Interval Pore Water Studies," *Limnol. Oceanogr.* 21:912–914 (1976).

82. Carignan, R., F. Rapin, and A. Tessier. "Sediment Porewater Sampling for Metal Analysis: A Comparison of Techniques," *Geochim. Cosmochim. Acta* 49:2493–2497 (1985).

83. Dongherty, S.J. and J.C. Berg. "Distribution Equilibria in Micellar Solutions," *J. Colloid Interface Sci.* 48:110–120 (1974).

84. Buffle, J. and C. Staub. "Measurement of Complexation Properties of Metal Ions in Natural Conditions by Ultrafiltration: Measurement of Equilibrium Constants for Complexation of Zinc by Synthetic and Natural Ligands," *Anal. Chem.* 56:2837–2842 (1984).

85. Lee, J. "Complexation Analysis of Fresh Waters by Equilibrium Diafiltration," *Water Res.* 17:501–510 (1983).

86. Tuschall, J.R. and P.L. Brezonik. "Application of Continuous Flow Ultrafiltration and Competing Ligand/Differential Spectrophotometry for Measurement of Heavy Metal Complexation by Dissolved Organic Matter," *Anal. Chim. Acta* 149:47 (1983).

87. Gamble, D.S., M.I. Haniff, and R.H. Zienius. "The Solution Phase Complexing of Atrazine by Fulvic Acid. II. A Batch Ultrafiltration Technique, *Anal. Chem.* 54:727 (1986).

88. Gamble, D.S., M.I. Haniff, and R.H. Zienius. "The Solution Phase Complexing of Atrazine by Fulvic Acid. III. A Theoretical Comparison of Ultrafiltration Methods," *Anal. Chem.* 54:732 (1986).

89. Wang, Z.D., D.S. Gamble, and C.H. Langford. "Interaction of Atrazine with Laurentian Fulvic Acid: Binding and Hydrolysis," *Anal. Chim. Acta* 232:181–188 (1990).

90. Kesting, R.E. *Synthetic Polymeric Membranes*. (New York: McGraw-Hill, 1971).
91. Ibanez, J.A. and F. Tejerina. "Surface Charge Density, Membrane Potential and Salt Flux," *Phys. Lett.* 88A:262–264 (1982).
92a. du Pont de Nemours, E.I. and Co. (Inc.) "Ludox Colloidal Silica: Properties, Uses, Storage and Handling" Publ. E-94776 (1988).
92b. Interfacial Dynamic Corporation, Product Guide and Price List (February 1990).
93. Cranston, R.E. and D.E. Buckley. "The Application and Performance of Microfilters in Analyses of Suspended Particulate Matter," Report Series/BI-R-72-7/October 1972, Bedford Institute of Oceanography, Dartmouth, Nova Scotia, Canada.
94. Sheldon, R.W. "Size Separation of Marine Seston by Membrane and Glass-Fibre Filters," *Limnol. Oceanogr.* 17:494–498 (1972).
95. Sheldon, R.W. and W.H. Sutcliffe. "Retention of Marine Particles by Screens and Filters," *Limnol. Oceanogr.* 14:441–444 (1969).
96. Perret, D., R.R. De Vitre, G.G. Leppard, and J. Buffle. "Characterizing Autochtonous Iron Particles and Colloids. — The Need for Better Particle Analysis Methods," *Large Lakes; Ecological Structure and Function.* M.M. Tilzer and C. Serruga, Eds. (Brock/Springer Series in Contemporary Bioscience, Berlin: Springer-Verlag, 1990), chap. 12.
97. Staub, C., J. Buffle, and W. Haerdi. "Measurement of Complexation Properties of Metal Ions in Natural Conditions by Ultrafiltratioin: Influence of Various Factors on the Retention of Metals and Ligands by Neutral and Negatively Charged Membranes," *Anal. Chem.* 56:2843–2849 (1984).
98. Wagemann, R. and G.J. Brunskill, "The Effect of Filter Pore-Size on Analytical Concentrations of Some Trace Elements in Filtrates of Natural Waters," *Int. J. Environ. Anal. Chem.*, 4:75–84 (1974).
99. Aiken, G.R. "Evaluation of Ultrafiltration for Determining Molecular Weight of Fulvic Acid," *Environ. Sci. Technol.* 18:978–81 (1984).
100. Robertson, D.E. "Role of Contamination in Trace Element Analysis of Sea Water," *Anal. Chem.* 40:1067–1072 (1968).
101. Zief, M. and J.W. Mitchell. *Contamination Control in Trace-Element Analysis.* (New York: John Wiley & Sons, 1976).
102. Robertson, D.E. "Contamination Problems in Trace-Element Analysis and Ultrafiltration," in *Ultra-purity: Methods and Techniques.* M. Zief and R.M. Speights, Eds. (New York: Marcel Dekker, 1972), chap. 12.
103. Wagemann, R. and B. Graham. "Membrane and Glass Fibre Filter Contamination in Chemical Analysis of Fresh Water," *Water Res.* 8:407–412 (1974).
104. Florence, T.M. "The Speciation Of Trace Elements in Waters," *Talanta* 29:345–364 (1982).
105. Truitt, R.E. and J.H. Weber. "Trace Metal Ion Filtration at pH 5 and 7," *Anal. Chem.* 51:7–2059 (1979).
106. Olson, W.P., R.D. Briggs, C.M. Garanchon, M.J. Ouellet, E.A. Graf, and D.G. Luckurst. "Aqueous Filter Extractables: Detection and Elution From Process Filters," *J. Parent. Drug Assoc.* 34:254 (1980).
107. Cooney, D.O. "Interference of Contaminants from Membrane Filters in Ultraviolet Spectroscopy," *Anal. Chem.* 52:1068 (1980).
108. Chiou, W.L. and L.D. Smith. "Adsorption of Organic Compounds by Commercial Filter Papers and its Implications on Quantitative-Qualitative Chemical Analysis," *J. Pharm. Sci.* 59:843 (1970).

109. Hwang, C.P., T.H. Lackie, and R.R. Munch. "Correction for Total Organic Carbon, Nitrate and Chemical Oxygen Demand when Using the MF-Millipore Filter," *Environ. Sci. Technol.* 13:871–872 (1979).

110. Marvin, K.T., R.R. Proctor, and R.A. Neal. "Some Effect of Filtration on the Determination of Nutrients in Fresh Water and Salt Water," *Limnol. Oceanogr.* 17:777–784 (1972).

111. Batley, G.E. and D. Gardner. "Sampling and Storage of Natural Waters for Trace Metal Analyses," *Water Res.* 11:745–756 (1977).

112. Mart, L. "Prevention of Contamination and Other Accuracy Risks in Voltammetric Trace Metal Analysis of Natural Waters. I. Preparatory Steps, Filtration and Storage of Water Samples," *Fresenius Z. Anal. Chem.* 296:350–357 (1979).

113. Hart, B.T. and S.H.R. Davies. "Trace Metal Speciation in Three Victorian Lakes," *Aust. J. Mar. Freshwater Res.* 32:175–189 (1981).

114. Mc Donald, C. and H.J. Duncan. "Reproducibility of Elemental Impurity Levels in Millipore Filters," *Anal. Chim. Acta* 102:241–244 (1978).

115. Spencer, D.W. and F.T. Manheim. "Ash Content and Composition of Millipore HA Filters," Geological Survey Research Document 288–290 (1969).

116. Nurnberg, H.W., P. Valenta, L. Mart, B. Raspor, and L. Sipos. "Application of Polarography and Voltammetry to Marine and Aquatic Chemistry," *Z. Anal. Chem.* 282:357–367 (1976).

117. Salim, R. and B.G. Cooksey. "Adsorption of Lead on Container Surfaces," *J. Electroanalyt. Chem.* 106:251–262 (1980).

118. Marvin, K.T., R.R. Proctor, and R.A. Neal. "Some Effects of Filtration on the Determination of Copper in Fresh Water and Salt Water," *Limnol. Oceanogr.* 15:320–325 (1970).

119. Burrell, D.C. *Atomic Spectrometric Analysis of Heavy Metal Pollutants in Water.* (Ann Arbor, MI: Ann Arbor Science Publishers, 1974).

120. Gardiner, J. "The Chemistry of Cd in Natural Water. I. A Study of Cd Complex Formation Using the Cd Specific-Ion Electrode," *Water Res.* 8:23–30 (1974).

121. Batra, S. "Aqueous Solubility of Steroid Hormones. Explanation for the Discrepancy in the Published Data," *J. Pharm. Pharmacol.* 27:777–779 (1975).

122. Liu, S.T., C.F. Carney, and A.R. Hurwitz. "Adsorption as a Possible Limitation in Solubility Determination," *J. Pharm. Pharmacol.* 29:319 (1977).

123. Ghanem, A.H. "Adsorption of Drugs on Filter Membranes," *Egypt. J. Pharm. Sci.* 15:51–59 (1974).

124. Gardner, M.J. and D.T.E. Hunt. "Adsorption of Trace Metals During Filtration of Potable Water Samples with Particular References to the Determination of Filtrable Lead Concentration," *Analyst* 106:471–474 (1981).

125. Robbe, D., P. Marchandise, D. Baudet, and A. Magnin. "Filtration des solutions aqueuses en vue du dosage des cations métalliques," *Environ. Technol. Letters* 1:283–290 (1980).

126. Eaton, A.D. and V. Grant. "Sorption of Ammonium by Glass Frits and Filters: Implications for Analyses of Brackish and Fresh Water," *Limnol. Oceanogr.* 24:397–399 (1979).

127. Gardner, M.J. "Adsorption of Trace Metals from Solution during Filtration of Water Samples," Water Research Center Technical Report TR-172 (1982).

128. Mill, A.J.B. Ph.D. thesis, Imp. College Sci. Technol., London (1976).

129. Guy, R.D. and C.L. Chakrabarti. *Symp. Proc. Int. Conf. Heavy Metals Env.,* Vol. I. Oct. 27–31 (1975) Toronto, p. 275.

130. Hart, B.T. and H.R. Davies. "A Study of the Physico-chemical Forms of Trace Metals in Natural Waters and Waste Waters," *Aust. Water Resour. Counc. Tech. Rep.* 35 (1978).

131. Quinn, J.G. and P.A. Meyers. "Retention of Dissolved Organic Acids in Seawater by Various Filters," *Limnol. Oceanogr.* 16:129–131 (1971).

132. Karger, B.L., L.R. Snyder, and C. Horvath. *An Introduction to Separation Science.* (New York: Wiley-Interscience, 1973).

133. Essede, L.A. "Effect of Organic Anion Adsorption on Water Permeability of Microporous Membranes," *J. Colloid Interface Sci.* 100:414–422 (1984).

134. Mc Donoogh, R.M., C.J.D. Fell, and A.G. Fane. "Surface Charge and Permeability in the Filtration of Non-Flocculating Colloids," *J. Membr. Sci.* 21:285–294 (1984).

135. Fane, A.G., C.J.D. Fell, and A.G. Waters. "Ultrafiltration of Proteins Solutions through Partially Permeable Membranes — The Effect of Adsorption and Solution Environment," *J. Memb. Sci.* 16:211–224 (1983).

136. Errede, L.A. "Effect of Molecular Adsorption on Water Permeability of Microporous Membranes," *J. Membr. Sci.* 20:45–61 (1984).

137. Stumm, W. and J.J. Morgan. *Aquatic Chemistry.* (New York: John Wiley & Sons, 1981).

138. Liu, M.Y., H.M. Lindsay, D.A. Weitz, R.C. Ball, R. Klein, and P. Meakin. "Universality of Colloid Aggregation," *Nature* 339:360–362 (1989).

139. Buffle, J. Unpublished results.

140. Danielsson, L.G. "On the Use of Filters for Distinguishing between Dissolved and Particulate Fractions in Natural Waters," *Water Res.* 16:179–182 (1982).

141. Igawa, M., T. Yoshida, C. Ohtake, and T. Hayashita. "Clogging and Gel-Layer Formation during Membrane Filtration and its Effect on Analytical Data," *Bull. Chem. Soc. Jpn.* 60:3183–3188 (1987).

142. Florence, T.M. and G.E. Batley. "Chemical Speciation in Natural Waters," *CRC Crit. Rev. Analyt. Chem.* 9:219–296 (1980).

143. Bowen, V.T., J.S. Olsen, C.L. Osterberg, and J. Ravera. *Radioactivity in the Marine Environment.* (Washington, D.C.: National Academy of Sciences, 1971), chap. 8.

144. Jones, J.G. and B.M. Simon. "Increased Sensitivity in the Measurement of ATP in Fresh Water Samples with a Comment on the Adverse Effect of Membrane Filtration," *Freshwater Biol.* 7:253–260 (1977).

145. Kamide, K. and S. Manabe. "Mechanisms of Permselectivity of Porous Polymeric Membrane in Ultrafiltration Process," *Polym. J.* 13:459–79 (1981).

146. Jonsson, G. "Overview of Theories for Water and Solute Transport in UF/RO Membranes," *Desalination* 35:21–38 (1990).

147. Meares, P. "The Physical Chemistry of Transport and Separation by Membranes," in *Membrane Separation Processes,* Meares, P., Ed. (Amsterdam: Elsevier, 1976), chap. 1.

148. Matsuura, T. "Transport Processes in Membranes for Reverse Osmosis," *Chem. Eng. Tech.* 50:565–573 (1978).

149. Rangarajan, R., T. Matsuura, E.C. Goodhue, and S. Sourirajan. "Predictability of Membrane Performance for Mixed Solute Reverse Osmosis Systems. II. System Cellulose Acetate Membrane- 1:1 and 2:1 electrolytes-water," *Ind. Eng. Chem. Process Des. Dev.* 18:278–287 (1979).

150. Grace, H.P. "Structure and Performance of Filter Media," *AlChE J.* 2:307–336 (1956).

151. Hermans, P.H. and H.L. Bredée. "Mathematical Treatment of Constant-Pressure Filtration," *J. Soc. Chem. Ind.* 55:1T (1936).
152. Suki, A.J, A. G. Fane, and C.J.D. Fell. "Modeling Fouling Mechanisms in Protein Ultrafiltration," *J. Membr. Sci.* 27:181–193 (1986).
153. Fane, A.G. "Ultrafiltration of Suspensions," *J. Membr. Sci.* 20:249–259 (1984).
154. Zydney, A.L. and C.K. Colton. "A Concentration Polarisation Model for the Filtrate Flux in Cross-Flow Microfiltration of Particulate Suspensions," *Chem. Eng. Commun.* 47:1–21 (1986).
155. Suki, A., A.G. Fane, and C.J.D. Fell. "Flux Decline in Protein Ultrafiltration," *J. Membr. Sci.* 21:269–283 (1984).
156. Choe, T.B., P. Masse, and A. Verdier. "Membrane Fouling in the Ultrafiltration of Polyelectrolyte Solutions: Polyacrylic and Bovine Serum Albumine," *J. Membr. Sci.* 26:17–30 (1986).
157. Tettin, D.R. and M.R. Doshi. "Limiting Flux in Ultrafiltration of Macromolecular Solutions," *Chem. Eng. Commun.* 4:507–522 (1980).
158. Wigmans, J.G., S. Nakao, and C.A. Smolers. "Flux Limitation in Ultrafiltration: Osmotic Pressure Model and Gel Layer Model," *J. Membr. Sci.* 20:115–124 (1984).
159. Reihanian, H., C.R. Robertson, and A.S. Michaels. "Mechanisms of Polarization and Fouling of Ultrafiltration Membranes by Proteins," *J. Membr. Sci.* 16:237–258 (1983).
160. Matthiasson, E. "The Role of Macromolecular Adsorption in Fouling of Ultrafiltration Membranes," *J. Membr. Sci.* 16:23–26 (1983).
161. Weilenmann, U., C.R. O'Melia, and W. Stumm. "Particle Transport in Lakes: Models and Measurements," *Limnol. Oceanogr.* 34:1–18 (1989).
162. Perret, D. "Caractéristiques physico-chimiques et dynamique de transport des formes du fer dans un lac eutrophe," Ph.D. thesis No. 2395, University of Geneva (1989).
163. Leppard, G.G., R.R. De Vitre, D. Perret, and J. Buffle. "Colloidal Iron Oxyhydroxyphosphate: The Sizing and Morphology of an Amorphous Species in Relation to Partitioning Phenomena," *Sci. Total Environ.* 87/88:345–54 (1989).
164. Lee, S., Y. Aurelle, and H. Roques. "Concentration Polarisation, Membrane Fouling and Cleaning in Ultrafiltration of Soluble Oil," *J. Membr. Sci.* 19:23–38 (1984).
165. Choe, T.B., P. Masse, and A. Verdier. "Flux Decline in Batch Ultrafiltration: Concentration Polarization and Cake Formation," *J. Membr. Sci.* 26:1–15 (1986).
166. van Bruggen, J.T., J.D. Boyett, A.L. van Bueren, and W.R. Galey. "Solute Flux Coupling in a Homopore Membrane," *J. Gen. Physiol.* 63:639–656 (1974).
167. Porter, M.C. and A.S. Michaels. *Membrane Ultrafiltration.* (Chem. Tech., July, 440–445, 1971).
168. Do, D.D. and A.A. Elhassadi. "A Theory of Limiting Flux in a Stirred Batch Cell," *J. Membr. Sci.* 25:113–132 (1985).
169. Nguyen, Q.T. and J. Neel. "Characterization of Ultrafiltration Membranes. IV. Influence of the Deformation of Macromolecular Solutes on the Transport Through Ultrafiltration Membranes," *J. Membr. Sci.* 14:111–128 (1983).
170. Melling, J. "Application of Ultrafiltration — Modifying Factors," *Process. Biochem.* 9:7–10 (1974).

171. Lakshminarayanaiah, N. *Transport Phenomena in Membranes.* (New York: Academic Press, 1969).
172. Blair, P. *Reverse Osmosis and Synthetic Membranes,* S. Sourirajan, Ed. (National Research Council Canada 1977), chap. 9.
173. Conway, B.E. *Ionic Hydration in Chemistry and Biophysics.* (Amsterdam: Elsevier, 1981).
174. Nguyen, Q.T. and J. Neel. "Characterization of Ultrafiltration Membranes. III. Role of Solvent Media and Conformational Changes in Ultrafiltration of Synthetic Polymers," *J. Membr. Sci.* 14:97–109 (1983).
175. Hijnen, H.J.M., J. van Daalen, and J.A.M. Smit. "The Application of the Space-Charge Model to the Permeability Properties of Charged Microporous Membranes," *J. Colloid Interface Sci.* 107:525–539 (1985).
176. Kaibara, K., Y. Nagata, T. Kimotsuki, and H. Kimizuka. "Study of Ion Transport across an Amphoteric Ion-Exchange Membrane — General Transport Properties of a Simple Electrolyte," *J. Membr. Sci.* 29:37–47 (1986).
177. Tasaka, M. and N. Aoki. "Membrane Potentials and Electrolyte Permeation Velocities in Charged Membranes," *J. Phys. Chem.* 79:1307–1314 (1975).
178. Kamo, N., M. Oikawa, and Y. Kobatake. "Effective Fixed-Charge Density Governing Membrane Phenomena. V. A Reduced Expression of Permselectivity," *J. Phys. Chem.* 77:92–95 (1973).
179. Nguyen, Q.T., P. Apter, and J. Neel. "Characterization of Ultrafiltration Membranes. II. Mass Transport Measurements for Low and High Molecular Weight Synthetic Polymers in Water Solutions," *J. Membr. Sci.* 7:141–155 (1980).
180. Blatt, W.F., B.G. Hudson, S.M. Robinson, and E.M. Zipilivan. "Fractionation of Protein Solutions by Membrane Partition Chromatography," *Nature* 216:511–513 (1967).
181. Lopez-Leiva, M. "Ultrafiltration at Low Degree of Concentration Polarisation: Technical Possibilities," *Desalination* 35:115–128 (1980).
182. Kwak, J.C.T. and R.W.P. Nelson. "Ultrafiltration of Fulvic and Humic Acids; a Comparison of Stirred Cell and Hollow Fibre Techniques," *Geochim. Cosmochim. Acta* 41:993–996 (1977).
183. Chang, D.P.Y. and S.K. Friedländer. "Particle Collection from Aqueous Suspensions by Permeable Hollow Fibres," *Am. Chem. Soc., Div. Org. Coat. Plast. Chem.,* Paper 34:540–544 (1974).
184. Zeli, D.W. and W.N. Gill. "Convective Diffusion in Rotating Disk Systems with an Imperfect Semipermeable Interface," *AIChE J.* 14:715–719 (1968).
185. Dahlheimer, J.A., D.G. Thomas, and K.A. Krans. "Application of Woven Fibre Hoses to Hyperfiltration of Salts and Cross-Flow Filtration of Suspended Solids," *Ind. Eng. Chem. Process Design Dev.* 9:566 (1970).
186. Degueldre, C., B. Baeyens, W. Goerlich, H. Grimmer, M. Mohos, A. Portmann, J. Riga, and J. Verbist. "In Laboratory, On Site, In Situ Sampling and Characterization of Grimsel Colloids. Phase I.," (EIR (PSI) internal report TM-42-87-20 1987).
187. Bauser, H., H. Churiel, and E. Walitza. "Control of Concentration Polarization and Fouling of Membranes in Medical, Food and Biotechnical Applications," *J. Membr. Sci.* 27:195–202 (1986).
188. De Filippi, R.P. "Ultrafiltration," *Chem. Process. Eng.* 10:475–518 (1977).
189. Gregor, H.P. and Ch. D. Gregor. "Les applications des membranes synthetiques," *Recherche* (juin 1982).

190. Kunst, B. and S. Sourirajan. "Development of Cellulose Acetate Ultrafiltration Membranes," *J. Appl. Polym. Sci.* 18:3423 (1974).
191. Kutowy, O. and S. Sourirajan. "Cellulose Acetate Ultrafiltration Membranes," *J. Appl. Polym. Sci.* 19:1449 (1975).
192. Degueldre, C., G. Langworth, V. Moulin, and P. Vilks. "Grimsel Colloid Exercise," (PSI-Bericht No 39, 1989).

CHAPTER **6**

EVALUATION OF ELECTRON MICROSCOPE TECHNIQUES FOR THE DESCRIPTION OF AQUATIC COLLOIDS

Gary G. Leppard

Lakes Research Branch, National Water Research, Environment Canada — C.C.I.W., Burlington, Ontario, Canada

TABLE OF CONTENTS

1.   Introduction and Context for Assessing Progress.................233
     1.1   General Context 233
     1.2   Electron Microscopes....................................234
     1.3   Correlative Microscope Technology 235
     1.4   Preparatory Techniques..................................237
           1.4.1 An Overview of the "Art" of Sample
                  Preparation......................................237
           1.4.2 Ultrathin Sections 241
           1.4.3 Technology Transfer from Cytochemistry.........243
           1.4.4 Cryotechnology 244
           1.4.5 Whole Mount Preparations......................245

2.   The Most Frequently Encountered Aquatic Colloids 246
        2.1   Nonliving Organic Materials 246
                2.1.1   Fibrils (Fibrillar Extracellular Polymeric
                        Substances) 246
                2.1.2   Fulvic and Humic Substances.................... 251
                2.1.3   Organic Skeletal Materials and Protein-
                        Rich Cell Fragments 252
        2.2   Living Colloids.. 253
        2.3   Mineral Colloids and Mineral-Organic
                Associations 259
                2.3.1   Inorganic Associations with Organic High
                        Polymers.. 260
                2.3.2   Iron-Rich Colloids 261
                2.3.3   Manganese Oxyhydroxides....................... 266
                2.3.4   Colloidal Phosphorus 267
                2.3.5   Biogenic Calcium and Silicon 267
                2.3.6   Clay Minerals.................................. 268

3.   Where Are We Now? — A Summary 270
        3.1   The Behavior and Interactions of Aquatic
                Colloids 270
        3.2   How to Minimize the Artifact Problem.................. 270
        3.3   The State-of-the-Art 272

4.   Data Analysis and Interpretation 272

5.   Values and Parameters in the Literature 274
        5.1   The Point of View of a Biological Electron
                Microscopist.. 274
        5.2   Suggestions for Revising Values and Parameters 274

6.   Future Needs... 275

References.. 276

1. INTRODUCTION AND CONTEXT FOR ASSESSING PROGRESS
1.1 General Context

Colloids can be considered as "particles" having a least diameter in the range of about 0.001 to 1.0 μm.[1] The significance of the submicron size range is that (1) an appreciable fraction of the molecules of a colloid is located at the boundary region between particle and aquatic milieu, and that (2) a microscope investigation should employ at least one microscope whose resolution permits analyses of the smallest colloids and their surfaces. In addition to their small size, colloids have a "glue-like" or adhesive aspect which gives them complex properties in water. Considerations of these properties from diverse scientific points of view, including coagulation/flocculation and ion binding, are available in the literature.[1-8]

As is increasingly evident,[4,5,9,10-14] colloids play significant multifaceted roles in the structure and function of aquatic ecosystems and water treatment systems, both healthy and polluted. From the point of view of human needs and frailties, colloids can be major factors in modulating the quality of natural waters, playing roles both positive and negative. To better understand these roles, improved methods of characterization are a necessity.

Electron microscopy (EM) in its various forms is providing an essential technology for realistic descriptions of aquatic colloids of all kinds. This critical review will show that, as artifacts become more readily identified and minimized, EM is increasingly useful in the characterization of colloidal materials previously considered too artifact-sensitive for realistic analyses. The focus will be on the techniques of transmission electron microscopy (TEM) which contribute by providing data on size, shape, porosity, internal heterogeneity, crystallinity, and colloid-colloid associations. As well, TEM is being adapted to permit increasingly sophisticated microchemical analyses, including some for reactivity, and elemental composition analyses.

Among the quantitatively important kinds of colloid under scrutiny by recent technology transfer into the aquatic sciences are the following:

(1) Oxyhydroxides of iron and manganese
(2) Fibrils, or linear aggregates of biopolymers rich in polysaccharide
(3) Aggregated fulvic acids
(4) Clays
(5) Water-borne viruses
(6) Picoplankton, or living cellular organisms of colloidal size
(7) Refractory skeletal materials, both organic and mineral

All of these are treated here, as are some of their natural associations as revealed by minimal perturbation studies. Some of the smallest colloids yield little ultrastructural information on an individual basis, but TEM analyses of their mode of aggregation can be revealing.[15,16]

1.2 Electron Microscopes

The conventional transmission electron microscope,[17] TEM, is currently the most useful kind of microscope for morphological analyses of colloids and aggregates of colloids. The TEM is similar in purpose to the classical light microscope but with a much greater resolving power; it can provide clear well-defined images of objects which are ca. 1000 times smaller than the smallest objects resolved by a conventional light microscope. A TEM passes an electron beam through the object of study. An image of the object, mainly through differential electron scattering, is carried forward in the electron beam to a viewing and/or recording device. Because imaging depends on the penetrating power of electrons and because this penetrating power is low for the voltages used in a conventional instrument, an object for study in the specimen plane cannot be thick (less than 0.1 μm is preferred). As a consequence of this limitation, most objects are sectioned for examination and an ultrathin section is placed in the specimen plane. The technology for achieving this is outlined in a later section below.

In an up-to-date, well-funded TEM laboratory, there will be a TEM with a resolution better than 0.001 μm. Consequently, the choice of which kind of TEM to use for a particular type of sample should be based on factors such as (1) the variety and quality of accessory equipment, (2) the ease of operation, (3) the completeness of the laboratory with regard to specialized electron microscope techniques, and, last but not least (4) the level of skill displayed by the technical personnel.

The laboratory chosen should have accessories (see also Chapter 3 in this volume) which can be coupled to the TEM to permit analyses beyond the purely morphological. One very useful accessory is the increasingly sophisticated energy-dispersive spectroscopy or EDS.[18,19] It allows X-ray microanalysis of colloids, providing information on elemental composition for those elements of atomic number greater than 10; EDS is applicable to the microanalysis of individual colloids in the mid-size range and above. Two other useful accessories are the apparatus of electron diffraction[17,20] for examining crystalline colloids, and the apparatus associated with electron energy-loss spectroscopy or EELS.[21,22] Electron diffraction in association with a field emission scanning electron microscope or FESEM (see Crewe[23] for an introduction to the technology) can be especially useful in analyzing colloids at the lower limit of the size range. EELS is explored elsewhere in this volume.[24]

The conventional scanning electron microscope or SEM[25] can provide a useful correlative technology when used in conjunction with a TEM. A conventional SEM has a resolution which is not so good as that of a TEM, but it has a much greater depth of focus and requires less complication in specimen preparation. These latter positive features present some advantages relative to TEM (and its requirement for ultrathin specimens) especially in orienting the analyst within large heterogeneous aggregates of colloids where some of the colloids are many times larger than the thickness of an ultrathin section. This problem of the disposition in three dimensions of colloids within a large aggregate and the need for correlative microscopies to provide orientation is discussed more fully below. The capacity of the SEM to analyze extremely thick specimens is a result of the nature of its image formation which is different from that of TEM. In a SEM, a narrow beam of electrons is focused onto the surface of the specimen and is caused to scan it in a regular pattern of lines; the complete response at each instant is used for modulating the signal which governs the image on a viewing screen and/or camera.

There are various hybrids of TEM and SEM which are called scanning transmission electron microscopes or STEM.[26,27] Some are relatively inexpensive in comparison with a high quality TEM and can be fitted with EDS to provide a versatile instrument for the elemental analysis of colloids in the upper part of the size range.[28] Other accessories are contributing to the use of STEM for analyzing relatively thick sections (ca. 0.5 μm).[29]

While a conventional TEM is restricted to analyzing ultrathin sections, it is possible to use TEM to analyze colloids in thick sections through the use of its big brother, the high-voltage electron microscope or HVEM[30,31] whose electrons have extra penetrating power. A HVEM operated at 1000 keV can achieve the same resolution as a TEM operated at 100 keV when using sections 10 to 20 times the thickness permitted by TEM. Specimens of several micrometers thickness can be imaged on a routine basis, but since the images represent volumes, they require special recording and display methods. Because these special methods are being improved, HVEM has considerable potential as an instrument for the analysis of aquatic colloids, despite its cost.

An instrument of the future having fascinating possibilities is an electron microscope adapted to the visualization of colloids in water through the use of an environmental device (sample compartment) placed within the microscope column.[32-35]

1.3 Correlative Microscope Technology

To examine large colloids and aggregates of colloids, especially heterogeneous native aggregates whose size extends far into the range for conventionally defined "true particles" and whose activities lead to the formation of settling particles in aquatic ecosystems, it can be necessary to employ a battery of correlative microscope technologies so as to bridge the gap between the resolution of the unaided human eye (near 100 μm) and that of the conventional TEM (∼0.001 μm). The light microscope can provide a resolution of ca. 0.2 μm when used optimally, while permitting an examination of aggregates taken freshly from nature or from controlled experiments. Examinations of gross features can be done directly without any processing or with processing restricted to a simple chemical fixation followed by differential staining. This permits the localization of specific functional groups/families of macromolecules/selected biological materials[36,37] in contexts where the artifact contribution is known. Hayat[38] provides a related guide to the differential staining of biological macromolecules in TEM sections.

There are many useful variations on the theme represented by the standard light microscope which permit improvements/refinements in information yield. Some of the techniques involved have analogs in EM[39] and include phase, interference, darkfield and polarization microscopy[36] as well as the recently developed confocal laser microscopy[40-42] and high resolution digital imaging microscopy.[42,43] Attempts to extend structural analyses well into the submicron range are becoming increasingly more successful through the use of image processing technology.[44-46] Some of it is being developed specifically to study living aquatic microbes, with some of the effort being focused on small natural living aggregates of relevance to environmental stresses on surface waters.[47,48] Additionally, biologists are active in studying structural details of aquatic phototrophic organisms in the submicron size range[49] through the use of epifluorescence microscopy[50] on these "living colloids".

A systematic approach to bridging the gap between the near million-fold difference in resolution above (naked eye vs. TEM) becomes essential when one wishes to study (1) the fine details of the three-dimensional disposition of different colloid types and their associations in a natural aggregate; (2) the subunit structure/microheterogeneity/porosity of an individual colloid; or (3) the formation of heterogeneous aggregates in experiments. In theory, serial sectioning of embedded colloids would allow one to carry out such TEM morphological studies unaided by accessory microscopes of lower resolution. Practically, however, a blend of correlative technologies is often a necessity and always a blessing to the completion of a successful analysis.[51] The rationale is illustrated by a simplified example as follows; it is based on the need to orient oneself within a volume for the interpretation of the essentially planar images of ultrathin sections, and do it within a reasonably short time frame.

Let us assume that, to analyze a certain kind of aggregate effectively, one must take sections of 0.050 μm thickness through embedded aggregates of 5 μm diameter whose heterogeneity is such that one must examine at least 100 examples to draw a meaningful conclusion. Let us further assume that the purpose of the study is to relate details of aggregate morphology to physicochemical factors, thus requiring five different experiments in at least two replicates. Thus, in a relatively modest project, one finds a requirement for a photographic documentation of each of 100,000 different sections. This is clearly a formidable task and one which is attracting the attention of innovators who strive to make the task easier through TEM modifications, new accessory techniques and novel methods of image analysis.[52,53]

A more feasible research strategy, and one which can incorporate innovations as they become available, is outlined as follows:

(1) One can visualize the three-dimensional disposition of the larger components of an aggregate using a technology which allows one to look at the entire volume of each 5 μm aggregate, even though the resolution may be relatively low (e.g., by light microscopy)

(2) One can then look at finer components with a technology which provides a higher level of resolution while still permitting an appropriately varied selection of views relating to the entire volume of the aggregate (e.g., by SEM)

(3) Then, properly oriented with respect to volume, one can employ a selective sectioning approach[34] to provide essentially planar specimens for TEM analyses of all fine details of interest.

This strategy permits an approximate reconstruction of an aggregate which, with an appropriate selection of preparatory techniques for each microscope technology employed, will be close to realistic. One can repeat the work using a different selection of preparatory techniques in a multi-method approach and analyze technique-specific variations in detail for an assessment of artifact.[15] Also, one might wish to employ a TEM-based morphometric analysis,[55] a HVEM approach and some of the novel methods of image analysis mentioned earlier.

The assessment and minimization of artifact can be carried out in a systematic manner through the minimum perturbation preparatory technology currently in development for TEM[15,56] and for SEM.[57] Although the procedures for artifact minimization are increasingly more systematic, there is still a premium to be placed on technical skills. Thus the evaluation of electron microscope techniques for a realistic description/characterization of colloids (and their aggregates) must also be (1) an evaluation of preparatory techniques (especially TEM preparatory techniques transferred from the biomedical sciences); and (2) an evaluation of strategies for selecting the most effective blend of preparatory techniques and microscope accessories for a given research goal. The latter evaluation was covered in part in Section 1.2 and its coverage will be completed in later sections. An evaluation of preparatory techniques was the subject of a recent review[15] and thus will be treated briefly below.

1.4 Preparatory Techniques
1.4.1 An Overview of the "Art" of Sample Preparation

One cannot automatically assume that a specific colloid will be altered adversely by a perturbing mode of preparation. However, it is certainly wise to prepare an incompletely known specimen (or aggregated mixtures/flocs of known and unknown colloids) in a minimally perturbing way if one's goal is to initiate a new literature, that of the characterization of aquatic colloids. The fact that many aquatic colloids show instability in natural waters has been amply confirmed by physicochemical investigations.[6,58-60] It is also evident that the colloidal extracellular extensions of many microbes can be involved in aggregation/flocculation events, and that the morphology and/or aggregation-promoting behavior of the fibrillar colloids can be altered according to choice of preparatory technique.[12,15,61] Natural organic coatings, which form on particles spontaneously in surface waters, can modulate particle chemistry and behavior;[62-67] thus, interactions at the particle-water interface are of special interest when preparatory technique is of high enough quality to permit morphological analyses near the extreme lower size limit for colloids. The fact that the pursuit of such an extreme level of detail is feasible is attested by recent observations of fibrillar surface coatings[68,69] and on the pore structure of the colloid-water interface material of a hydrated iron oxyhydroxyphosphate.[70]

The considerable attention paid to specimen preparation in this critical review is the result of an appreciation, documented in part above, of the current lack of understanding of many limnological materials with regard to structural stability, especially stability changes related to degree of hydration. Because of this concern, there are two guiding principles which one must consider in transferring EM technology from a base in biomedical science to the analysis of aquatic colloids. These principles, which will be developed in a historical context below, before enunciation, give insight into the "art" of selecting the most appropriate preparatory techniques for a given goal.

The best techniques were developed on a trial-and-error basis, by and for biologists for facilitating the analysis of unstable organic-rich materials, such as parts of living cells. Some are readily adapted to the ultrastructural analysis of colloidal gels and colloidal aggregates rich in minerals, as well as to individual colloids, both inorganic and organic.[15] Originally, they were developed in response to a need to (1) stabilize complex biologicals for examination of specific features under the harsh dehydrating condition of the high vacuum required by the specimen chamber of a conventional TEM; and (2) reduce the thickness of many kinds of specimens so as to permit an optimal transmission of the electron beam at the accelerating voltages used routinely in the third quarter of this century. The need for stabilization led to the development of modern chemical fixatives[71] and techniques for physical fixation based on rapid freezing,[72-74] and also to sophisticated combinations of these two general approaches. The restrictions on thickness of specimen led to the development of embedding resins which retained the integrity of the three dimensional disposition of (at the time intracellular) colloids while permitting the cutting of ultrathin sections by ultramicrotomy.[38]

Large numbers of scientists from many disciplines participated with the biologists for decades in the development of EM preparatory technology and in refining each variation of it for artifact recognition, assessment, and minimization. As a consequence of their highly successful efforts, we have a literature on most of the colloids present in living cells and present as extracellular materials on the surface of living cells, as well as a literature on the artifacts created by the application of perturbing preparatory techniques. Additionally, there are scattered contributions in the literature of the earth sciences (see later section for selected contributions) which provide useful ultrastructural information on minerals and on refractory organics much altered from their biological precursors, even though the artifact problem is usually addressed incompletely. In contrast, many of the ontogenetically intermediate colloidal substances/materials in natural waters (ones neither freshly synthesized/exposed nor altered/degraded to a final refractory form) and their natural aquatic associations have no proper ultrastructural literature at all. Thus the situation with regard to the description/characterization of aquatic colloids is one of having to employ different levels of technological effort and skill for any given research, depending on the background knowledge already extant for the material(s) to be investigated. Those who cannot develop well the "art" of technique selection and blending will tend to use time and resources far in excess of the needs for their research.

The historical context above and the uneven way in which technology has developed for the characterization of "unstable" colloids leads to two useful principles to be applied simultaneously:

(A) The literature, especially the literature of *cell biology*, can serve as an *excellent guide* for suggestions on how best to analyze a given kind of sample and on how best to seek out and minimize artifacts.

(B) However, investigations on most natural aquatic colloids/aggregates, despite assistance from the literature, will have to be *complete researches in themselves*, including a state-of-the-art treatment of the artifact problem, regardless of increased cost and difficulty, for some time into the future.

The stringent requirements imposed on those who wish to produce *realistic* ultrastructural descriptions will follow mainly from the specific problems below related to colloid stability, problems which must be addressed now to provide a framework for future research:

(1) Fresh biological colloids, once removed from their normal cellular milieu, will undergo a variable series of chemical and physical changes at varying rates, as well as changes in their associations with solutes.[75]

(2) Surface active organics, both refractory and fresh, will tend to form coatings on newly arrived particles/colloids, coatings whose physical nature must become better known for an improved understanding of nutrient and contaminant dispersion in surface waters.[5,76,77]

(3) Important aquatic mineral colloids in their natural state have been incompletely documented with respect to both reactivity[75,78] and morphology.[70,79,80]

A note of caution is necessary for those aquatic scientists who choose to work in direct collaboration with cell biologists. Those ultrastructural cell biologists involved in perfecting the "art" of artifact minimization will often be found selecting (but deliberately and with good reason) a highly perturbing approach for preparing certain specimens.[73,81,82] This is because they know their specimen so well (from the literature and by experience) that they can predict (and verify later) that their colloid of interest will be visualized realistically, despite the harsh technique, while allowing them to reduce the time and cost of analysis. They may deliberately select a harshly perturbing technique for an additional reason, the fact that it may render more accessible to

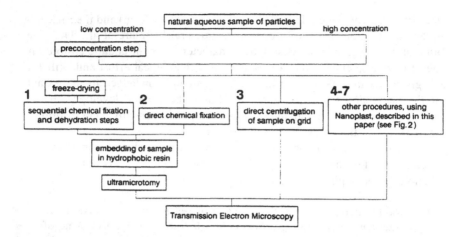

Figure 1. An overview of methods for the preparation of natural aquatic colloids/particles for visualization by TEM; Part 1. (From Perret, D. et al. *Water Res.* 25:1333–1343 (1991). With permission.)

measurement some feature which would otherwise be difficult to visualize. This artistic approach has an encouraging aspect because it provides for a possibility that short cuts may be permitted in the aquatic colloid analyses of the future. The drawback is that it may encourage one to take unwarranted and premature short cuts.

To complete this overview of the ''art'' of sample preparation, I present Figures 1 and 2. They illustrate an overview of methods for the preparation of natural aquatic colloids/particles for visualization by TEM. Schemes 1 and 2 of Figure 1 have been presented in detail in the literature.[28,83,84] Scheme 3 of Figure 1 has been discussed by the Nomizu group.[85-87] Schemes 4 through 7 of Figure 2 have recently been developed in a paper[56] in which a comparison

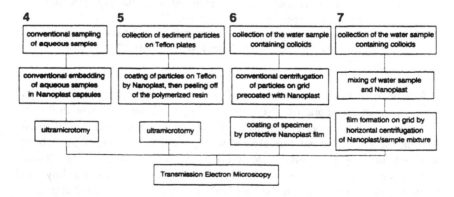

Figure 2. An overview of methods for the preparation of natural aquatic colloids/particles for visualization by TEM; Part 2. (From Perret, D. et al. *Water Res.* 25:1333–1343 (1991). With permission.)

of all seven schemes may be found. For optimal effectiveness, any scheme chosen should be applied to samples which were perturbed minimally prior to their receipt by the electron microscopist.

1.4.2 Ultrathin Sections

The basic problems of sample stabilization and thickness reduction were solved decades ago through (1) the use of chemical fixatives[71] which preserved colloids/aggregates in a natural state and (2) through gently replacing the water of the fixed specimen by molecules of resin monomer which could be polymerized to produce a hard block of resin-embedded specimen for ultrathin sectioning. The most widely used resins until recently were hydrophobic resins which imposed an extra step in processing; the water of fixed specimens had to be replaced by an organic solvent miscible with both water and resin monomers,[38] with the solvent being replaced in turn by the resin, thus increasing the level of extraction artifact and exacerbating any artifacts of dehydration.

The availability of a high-quality hydrophilic resin, Nanoplast FB 101,[88,89] appropriate to the embedding of aquatic colloids[15,56] has improved this latter situation and opened up new opportunities in the minimal perturbation approach. For extracellular organic colloids and colloids free in the aquatic milieu, both organic and inorganic, the need for chemical fixation is removed through the use of Nanoplast FB 101. Thus, embedding in such a hydrophilic resin can begin in the field immediately after the water sample has been taken.[56,80] For comparative analyses involving several embedding media, both hydrophobic and hydrophilic, with and without chemical fixation, one can consult published strategies.[15] The Nanoplast formulations of Bachhuber and Frosch[89] permit section thicknesses down to ca. 0.010 μm, a feature which optimizes the potential for high resolution morphological analyses.[90] This melamine resin has such a fine grain structure that it permits a *practical resolution* in ultrathin sections of ca. 0.001 μm, which is several times better than that of the most widely used epoxy formulations of the cell biologists, including that of Spurr.[91] The clarity of image produced by this improvement in practical resolution is helpful but taking full advantage of it is difficult because of physical effects in the specimen plane which interfere with the interpretation of morphology for details in the size range near 0.002 μm.[92] During the polymerization process, there is always a possibility of adverse interactions between the molecules of the embedding medium and the specimen; consequently, any artifacts produced by this interaction are continually analyzed and documented.[88,93]

Morphological analyses are not the only application of a melamine resin for aquatic colloids. Its ultrathin sections are stable enough to permit refined EDS analyses[94] in conjunction with either TEM or STEM. Furthermore, it is suitable for use with some cytochemical techniques (see next section) which permit identification and localization of chemical components.[95]

When sections are viewed by TEM, localized concentrations of heavy elements in the specimens appear darker than the background matrix of the embedding medium. This is because of a superior electron scattering power (greater contrast/electron opacity) relative to that of the light elements making up the resin. This effect has been utilized in the development of counterstains (solutions of heavy metals which have differential affinities for different substances) which can be applied to sections prior to inserting them into the specimen plane of a TEM.[38] This situation is analogous to the use of stains for differential coloring of sections to be viewed in the optical microscope;[37] through the use of TEM counterstains, different colloids rich in light elements receive an artificial increase in electron opacity for improved visualization (and also may become differentiated one from another by taking on different shades of grey according to counterstain uptake). Two of the most useful elements for differentiating biological colloids are uranium[96] and lead;[97] they are often used by cell biologists in conjunction with chemical fixatives which contribute other heavy metals to the image to refine further the level of differentiation. In addition to their general use, counterstains (and also stains employed with fixatives) can be designed so as to be specific to certain families of chemicals and then used with sections which have received no artificial inputs of heavy metal, an experimental tool not to be neglected in the analysis of organic colloids as shown below.

1.4.3 Technology Transfer from Cytochemistry

Technology transfer from cytochemistry should provide many advances in the characterization of organic particles. The potential of this biomedical technology is great, especially when applied to sections in conjunction with the improved correlative optical techniques noted earlier and with EDS.

Cytochemistry is the identification and localization of chemical components (of the cell), with a view to relating functional changes to chemical changes in a morphological context.[36,71,98] Individual techniques can be either chemical or physical and can be applied to either sections or entire particles/colloids mounted on a transparent support (whole mounts). The physical analytical techniques can be either optical or electron-optical[99] while physical preparatory techniques tend to revolve around rapid freezing.[71,72,100] The chemical localization techniques can be conventional, immunochemical, or related to *in situ* enzymatic activity[38,71] and they can be applied to specimens prepared either chemically or physically. All the TEM-based cytochemical technology has its origins in methods which are related to the science of histochemistry, the localization of chemical components on a tissue level. Thus it is based on methods for optical microscopy[101-104] which are adapted for TEM analyses as needed in an ongoing process. In histochemistry, the ideal requirements for quantification of a localized substance are as follows:

(1) The substance to be measured must be kept *in situ*.
(2) The fluids into which the specimens must be passed must neither extract the substance nor damage its chemical reactivity to the subsequent iden-tifying reaction.

(3) The reaction used to identify the localized substance should involve a reaction with all of the substance of interest.

(4) The newly "labeled" substance should be readily quantifiable.

There are some stains in general use, both counterstains and stains applied directly to wet specimens, which serve as "markers" for some families of organic polymers likely to be found in at least some surface waters. Examples are the use of lanthanum as a marker for mucopolysaccharides and glycoproteins; the use of ruthenium as a marker for polyanions such as acid polysaccharides rich in uronic acid residues; the use of silver as a marker for mucopolysaccharides and proteins rich in cystine. The choice of formulation and mode of use of stains based on these metals is determined by a knowledge of the cytochemical literature and some knowledge of the approximate composition of the sample. The lanthanum and ruthenium formulations are applied to the wet sample prior to embedding[38] whereas the silver formulations are applied to sections. Ruthenium stains employing the mineral dye ruthenium red[105,106] have been especially useful in the description of environmental colloids having a polyanionic character.

Immunocytochemistry or immunoelectron microscopy, a blend of cytochemistry and immunology,[81,107,108] has shown some potential for the characterization of aquatic colloids. Colloids composed of or enriched in sugar polymers containing certain sequences of monomers can act as antigenic determinants. Such colloids permit an immunologist to make antibodies which can be modified to accept a heavy metal component, such as gold, and then used as a molecule-specific stain. A variation on this theme is to couple an enzyme to the antibody and then use the enzyme's activity to generate a localized deposit of heavy metal at the site of the antibody-antigen union.[109]

The immunocytochemical approach to marking specific polymers is proceeding well at the histochemical level for some gelling phycocolloids produced by marine algae.[110-113] The progress of this work in moving from fluorescent markers (for *optical* microscope observations) to heavy metal markers is awaited with interest. For living colloids in the submicron size range, principally with medically important microbes, the jump to the highest resolution techniques employing heavy metal markers has been a successful ongoing process extending beyond the scope of this critical review.

Enzyme cytochemistry[71,109,114,115] can become a tool for assessing local aquatic impacts of protein-rich cell parts recently derived from living cells, especially parts which contain active phosphatases capable of altering the nutrient chemistry of aquatic microniches.[77] Techniques for identifying and localizing enzymes in colloid aggregates are based on the incubation of the sample with an appropriate substrate in a specially designed artificial medium. For example, in one method for phosphatase, phosphoric esters of glycerol are used as the substrate. As a result of the selection of chemicals in the incubation medium, the phosphate ion liberated by hydrolysis is converted into an insoluble metal compound at the site of enzyme action. The buildup of metal at that site can be controlled so that it is large enough to be an obvious marker but small enough to allow a localization in terms of the finest units of morphology. The incubation is done directly with the wet sample, usually after application of a chemical fixative. After the embedding, the three-dimensional distribution of the marker metal is analyzed in ultrathin sections with respect to overall morphology. This technology is not without its artifacts and transferring it to the analysis of organic and organomineral colloids from aquatic ecosystems will not be accomplished without difficulties. However, it has been in existence for five decades so there should be no paucity of literature to assist one in making the technology transfer. For living colloids, especially bacteria,[116] a direct literature is already extant.

Despite the exciting potential of technology transfer from the biomedical sciences, one should not neglect the fact that chemistry itself has evolved a use of microscopy which parallels that found in some branches of cytochemistry.[117] The field of chemical microscopy, sometimes called microchemistry, is adept at using various microscopes for analyzing rigid colloids, crystals, high polymers, and particle behavior. In an analogous manner, physics has evolved a microscope technology for pursuing these same subjects. Some of their technology has contributed to our understanding of the colloidal structural polysaccharides of higher plants, algae, fungi, insects, and aquatic invertebrates,[118,119] especially those likely to enter surface waters as refractory debris particles in significant quantities.[75,120]

1.4.4 Cryotechnology

Physical fixation by rapid freezing was mentioned in earlier sections as a sometimes useful alternative[73,81,95,100] to chemical fixation for the TEM or SEM preparation of colloidal materials. While useful in the hands of skilled cell biologists investigating well-known biological materials, techniques of rapid freezing place one at the risk of inducing major artifacts of dehydration, such as the extreme shrinkage which can occur to *loose aggregates* of colloidal organic fibrils.[15] Among the cryotechniques is an extraordinary exception of great potential in all kinds of ultrastructure research. This exception is the time-consuming and costly freeze-etch technique.[74,121,122]

Freeze-etching consists of freezing a sample rapidly enough to vitrify it, mechanically generating a fracture plane through it and then making a metallic replica of the fracture surface (usually following a pre-set level of etching), all the while maintaining the vitrified colloid-rich sample below the recrystallization temperature. The etching consists of a controlled sublimation of bulk water from the fracture surface, so as to place individual colloids in relief. The product of the technique for viewing is an ultrathin replica, created by vaporizing a metal-rich material at an angle onto the fracture face, while maintaining this face below the recrystallization temperature. This replica, when placed in the specimen plane of the TEM, yields a topographical image of a colloid or colloid aggregate, unperturbed by chemical agents or the physical separation phenomena associated with rapid freezing at a rate inferior to 10,000 K.s^{-1}.[122] This metallic replica will reveal detail with a resolution as good as that of an ultrathin section made with a hydrophobic resin, and will be more faithful to reality. Because the replica is topographical rather than planar and because the fracture plane may change levels within the sample, image analysis can present complications.[123] Despite its overall complexity and high cost, however, the freeze-etch technique is an ideal alternative confirmatory technique, unrelated to the major standard preparatory techniques normally devoted to particle analysis. Used as part of a multi-method approach in conjunction with ultrathin section analyses, it can permit conclusive decisions to be made about the shape and size and porosity of hydrated colloids when other combinations of particle analysis techniques have been found inconclusive.

1.4.5 Whole Mount Preparations

A whole mount preparation is one in which an entire colloid (or aggregate) is visualized by being placed in its entirety in the specimen plane of a microscope. For the SEM, this presents no difficulty, but one is limited to a view of the particle surface; one can fracture a particle to see inside but again one is limited to a view of a surface, the internal one exposed by the fracture. For the TEM, the particle is usually placed on top of an electron transparent support film made of plastic.[38] The electron opacity of the particle puts an upper limit on the size (thickness) of particle which can be examined effectively by transmitted electrons, thus limiting the analyst to particles in the low end of the colloidal size range. Despite this constraint and problems associated with the usually perturbing approaches to whole mount preparation, some progress has been made, most recently by the laboratory of Nomizu.[85-87] It is noteworthy that studies of the aggregation behavior of soil fulvic acids received an impetus from whole mount preparations.[124,125]

An alternative to the traditional TEM whole mount has been explored in detail recently (Figure 2). It is the Nanoplast film technique,[126] which has exciting potential;[56] with it, the smallest native colloids can be added to a fresh Nanoplast resin preparation in such a way as to form a support film having the colloids embedded within. The support film can be made as thin as is needed to allow for a practical TEM resolution similar to that achieved with ultrathin sections. Such a technique for realistic descriptions of aquatic colloids has to be considered promising from the point of view of minimal perturbation. This is because:

(1) It can be applied in the field directly to a sample freshly taken.
(2) It avoids the application of chemical fixatives to the sample.
(3) It avoids air-drying artifacts and the dehydration artifacts which can occur with traditional methods of whole mount preparation.
(4) It can be applied as a practical finishing step in situations which permit the ideal multi-method approach, as delineated by Buffle et al.[78]

2. THE MOST FREQUENTLY ENCOUNTERED AQUATIC COLLOIDS

The most frequently encountered aquatic colloids thus far described represent a compromise between what actually is common and the specialized interests of the investigator. It is likely that some common colloidal materials have been ignored.

2.1 Nonliving Organic Materials

Although the most common organic families of molecules in natural waters are known[75,127] and the more refractory components of the organisms at the base of the food web are known,[120,128,129] there is a lack of detailed knowledge about the extent to which specific molecules relate to given types of natural waters.[75,127,130] Concomitantly, while some organic colloidal materials can be tentatively identified on morphological grounds by TEM, there can be difficulty in relating morphological details at high resolution to specific arrangements of (and species of) molecules, with exceptions being found among some crystalline organics[131,132] and certain cell parts if visualized at or before the earliest stages of breakdown. For a compendium of cell parts, some general references serve as a good starting point.[128,133,134] Structure, function, and chemistry meet at a resolution near 0.001 μm and some aquatic colloids have shown themselves to be inherently interesting from ultrastructural, ecological, and biogeochemical points of view. When considering the potential significance of colloidal activity in surface waters two decades ago, Breger[135] showed profound insight in the title of his paper used as a closing address to a symposium on aquatic organic matter, "What you *don't* know can hurt you: organic colloids and natural waters". Since that time, a strong economic interest has also developed.[12]

2.1.1 Fibrils (Fibrillar Extracellular Polymeric Substances)

Judging from the volume and variety of literature published, the most interesting of the aquatic organic colloids appears to be the almost ubiquitous extracellular "fibril". This elongate organic colloid rich in high polymers[12,54] is readily recognized in TEM images by its distinctive ribbon-like aspect and greatly increased electron opacity following heavy metal staining.[136] Individual fibrils, whether branched or not, have a diameter typically in the range of 0.002 to 0.010 μm; the most common examples are composed at least in part of polysaccharides whose monomeric composition tends to be rich in uronic acid moieties (sugar acids with a projecting carboxyl group). Examples of some morphological varieties, taken from lakewaters, are shown in Figure 3. It is unfortunate that studies to relate fibril morphology to fibril chemistry have yielded little information.

Potential impacts of fibrils on aquatic ecosystems are outlined in Figure 4 whose context can be found in Reference 137. There is evidence that all the phenomena shown do occur, but quantification of their significance has not yet been achieved (with the exception that fibril roles in biofilm formation/

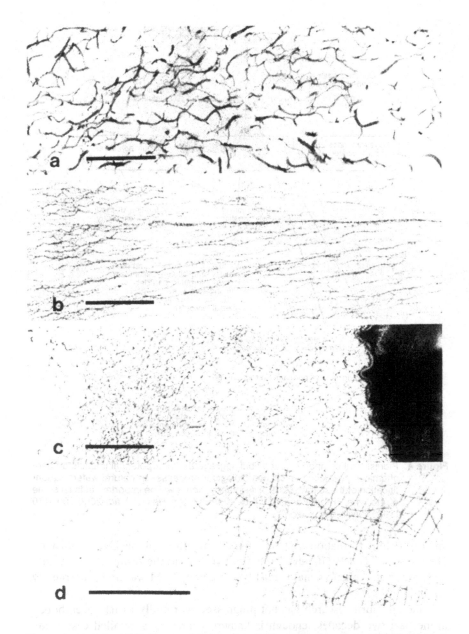

Figure 3. Varieties of fibrils sampled from lakes. Figure 3a shows some of the thickest fibrils documented to date. It and Figures 3b through d all show fibrils in counterstained ultrathin sections. Each micrograph was reprinted with permission from Massalski and Leppard[155] where details of preparation may be found. *The bar and all subsequent bars represent 0.45 μm,* which represents the pore size of the traditional filter used by limnologists/oceanographers to separate particles from solutes. Note that fibril length cannot be measured in ultrathin sections; a fibril has enough curvature to move in and out of the section.

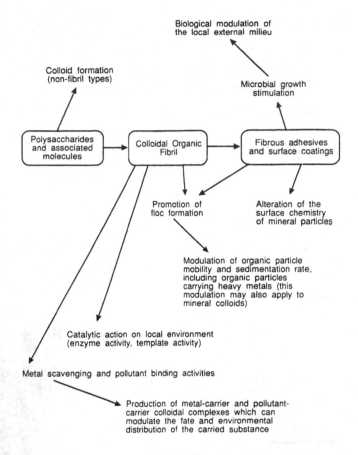

Figure 4. Fibrils, fibril components and fibril aggregates; their potential impacts on geo-chemical, physicochemical and biological processes in natural waters. Quantification of these impacts is in its infancy; some will be important at least some of the time. (From Leppard, G. G. *Water Pollut. Res. J. Can.* 20(2):100–110 (1985). With permission.)

microbe colony formation are vitally important, a topic to be developed later). The value of TEM for the description of fibrils (and the resultant insight into fibril-associated phenomena) is readily evaluated; TEM was and is absolutely essential.

Although fibril research has not progressed as rapidly as had been hoped in the past two decades, enough is known to present a detailed case study pertinent to this critical review. To begin, it is given that fibrils represent a family (or families) of morphologically and cytochemically similar colloids which share many common properties. The most evident of these are a biological derivation (from algae, bacteria, and plant roots), a limited size range and the presence of carbohydrate moieties in polymerized form. The literature which reveals their similarities is extremely scattered but guides to much of it can be found in two older references.[12,77] Despite generalized similarities, however, it must be kept in mind that fibrils do differ in details such that differences in function and reactivity might be considerable.

Fibrils were noted on the surfaces of a variety of algae and bacteria in the 1960s.[54] During this decade, they were also documented as a component of:

(1) Slime layers in streams, in association with microbiota[138]
(2) Organic flocs taken from contaminated waters[139,140]
(3) Highly acid mine waters[141]

Because of the fear of artifact, many of the conventional TEM observations were confirmed by freeze-etch analyses in the same publication.

In the 1970s, fibrils were found to be a component of the rhizoplane, the interface between the outer surface of a plant root and the mineral particles of the soil solution. It can be a zone of high metabolic activity, especially when microbes are present, and it influences the chemistry of the adjacent soil zone.[142] The reality of rhizoplane fibrils was demonstrated by both structural and cytochemical analyses of ultrathin sections[143-145] and by freeze-etching;[146] they were shown through the use of axenic root culture (no bacteria present) to be produced by root cells,[146] although in nature one would expect also a contribution of fibrils to the rhizoplane by soil bacteria. At this time, fibrils were also identified as the likely agents for contact cation exchanges between mineral colloids and plant cells,[147,148] a role which finds an analogy in lake water.[54] Thus came the link between fibrils and nutrition.

Also in the 1970s, fibrils were shown to be

(1) A polymeric bridging structure within many microbial biofilms[149,150] regardless of the biological speciation within the biofilm,[12] a bridging structure which could either cross-link microbes or encapsulate them as shown in Figure 5
(2) A natural adhesive promoting pelagic associations of algae and bacteria in lakes,[151] including their associations with suspended abiotic particles
(3) A functional component of activated sludge flocs, although the evidence tended to be circumstantial[152,153]
(4) A secretion product of higher plant cells grown as suspension cultures in mineral media, and used for a brief time as a source of sufficient quantities of fresh fibrils for wet chemical analyses[154]
(5) A common component of many Canadian lakewaters, being detected in lakes of various trophic states and sizes at all levels of the water column from surface microfilm to bottom sediments[54,136,155]

One can reasonably assume[69] that fibrils comprised the *fibers* shown by the relatively low resolution SEM images of the 1970s to be a major component of the fine debris of lakewaters. Since those fibers appeared to bind together the individual components of heterogeneous debris particles, then one can infer that the fibrils within them helped to mediate particle aggregation in lakes. Convincing SEM studies of these fibers can be found in the innovative ecological researches of Paerl.[156-159]

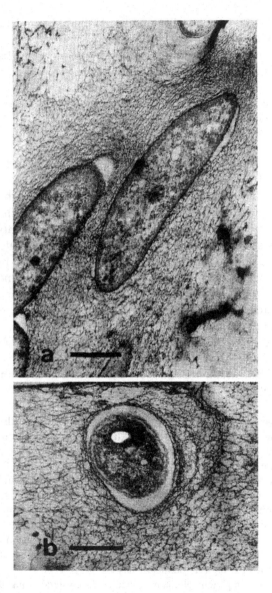

Figure 5. Microbial biofilms. These and all subsequent micrographs show counter-stained ultrathin sections *unless stated otherwise*. Figure 5a shows a fibrillar biofilm matrix in which the fibrils form direct cross-bridges between adjacent bacteria. Figure 5b, reprinted with permission from Leppard,[68] shows a bacterium residing in a cavity whose boundary is made of oriented fibrils in a locally differentiated portion of the fibrillar matrix.

In the 1980s, some attention was placed on quantifying fibrils in lakewater;[84,160] levels up to 7 mg.dm^{-3} were recorded. Research on water treatment systems then became focused more intensively on fibril roles in flocculation.[161-165] In keeping with the increased emphasis on environmental pollution, especially with regard to recalcitrant organic contaminants,[166] some research emphasis is being placed on possible roles of aquatic flocs as natural decontaminators of surface waters. The focus has been placed on the action of the flocs per se, however, rather than on the interactions of the fibril component with organic pollutants.

2.1.2 Fulvic and Humic Substances

Another group of aquatic organics of interest from ultrastructural, ecological and geochemical points of view is the group of fulvic and humic substances.[75,167-169] They have become increasingly interesting because of researches in the past decade to relate their chemistry, structure, and behavior to morphological parameters.[83,124,125,170] Their great abundance, refractory nature, and interactions with organic and inorganic pollutants[167,171] demand a greater understanding of their colloidal behavior. Their nature as colloids, when aggregated, is increasingly amenable to analysis by TEM.

In 1980, Ghosh and Schnitzer[170] produced a model for the macromolecular structure of soil fulvic acids which showed that their colloidal structure was a dynamic feature controlled by three environmental parameters:

(1) Sample concentration
(2) pH of the system
(3) Ionic strength of the medium

At a sample concentration of 100 mg.dm^{-3} (other parameters controlled), Stevenson and Schnitzer[124] found five common structural entities, including three classes of giant aggregates amenable to study by TEM. These findings, based on whole mount preparations, compared favorably with those obtained later in analyses of ultrathin sections of soil derived water fulvic acids.[83] The large colloidal aggregates were composed of granules in the ca. 0.002 μm size range, organized into dynamic structures as attested by the fact that they could pass ultrafiltration membranes even when the aggregate size was much larger than the pore size.[83] The presence in soils, lakes, and groundwaters of moderately large colloid aggregates (0.05 to 0.20 μm scatterers) of humic acids has been confirmed by photon correlation spectroscopy.[172] The meaning of these observations is

(1) Fulvic and humic substances can form a continuum of aggregated particles with widely varying size, in rather fast equilibrium.
(2) Ultrafiltration is not always a simple process when applied to waters rich in humic substances and it could become an unreliable one if induced aggregate formation is not controlled.

Further confirmation of the size and shape of humic substance colloids should be done by freeze-etching[74,121] and by Nanoplast embedding techniques[56] in association with experiments on the mechanics of aggregate formation. Soil fulvic acids already present a context for such confirmation and behavioral studies.[83] Some aggregated humic substances are illustrated in the ultrathin section views of Figure 6.

2.1.3 Organic Skeletal Materials and Protein-Rich Cell Fragments

Many organic colloids of direct biological origin are readily identified on morphological grounds; these are (sometimes unique) combinations of size and shape and electron opacity coupled with specific indicator details such as surface sculpturing or a geometric arrangement of subunits. Characterization on this basis is especially informative for a variety of refractory skeletal

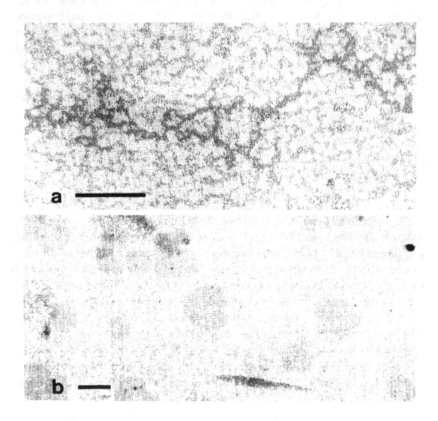

Figure 6. Lacustrine humic substances. For Figure 6a, the humic substances were concentrated before being embedded, so as to aggregate them. Note the weak fibrous aspect which is commonly seen. Figure 6b shows undegraded fulvic acid, concentrated prior to embedding to provoke colloid formation. This fulvic acid, whose smallest granules are ca. 0.002 μm in diameter, was isolated from metal contaminated water; no heavy metal counterstain was necessary for this preparation.

materials, algal cell walls, and protein-rich cell wall appendages of small organisms. At the earliest stages of degradation, some protein-rich cell parts (e.g., flagellae and complex organelles) of decomposing organisms are readily classified provided that the fragment's least dimension is at the upper level of the colloidal size range. Subclassification can be carried out in principle on the basis of reactivity (e.g., the presence of hardy enzymes) using the technology of cytochemistry. For an understanding of the potential of such characterizations, one need simply consult the literatures of cell and ultra-structural biology for their characterizations and descriptions permitting the identification of cell parts. One can use some of the general references quoted herein as a guide to this literature, most of which was highly developed some time ago.[128,133,134] Figures 7 through 10 illustrate some organic and organo-mineral skeletal materials and protein-rich cell parts and fragments recorded by the author in his limnological investigations. A very extensive phase of cataloguing is necessary to assess the relative importance of the nonliving organic materials, a subject which is in its infancy.

2.2 Living Colloids

Living colloids come in two classes, cells and viruses. Aquatic cells of colloidal dimensions come in two fundamental kinds;[36] these are (1) the prokaryotes such as bacteria and the so-called blue-green algae and (2) the eukaryotes, such as the true algae and all other cell types. A full treatment of living colloids extends beyond the scope of this critical review. However, several exciting subject areas of this aquatic research topic have blossomed recently and can be reviewed briefly. The advances concern the viruses (Figure 11) and the picoplankton (Figure 12).

Despite the role of viruses as agents of human disease via water supplies,[173,174] and despite several decades of development of particle counting techniques for viruses,[175] the technology for quantifying viruses in aquatic environments was not fully exploited until recently.[176] Using a new method for quantitative enumeration, the group of Bergh[176] found up to 2.5×10^8 virus-like particles per milliliter in natural waters. Their counts showed viruses to be present in numbers 10^3 to 10^7 times higher than was previously reported. Considering the virus species which attack and infect bacteria, those with a head size of 0.060 μm predominated, smaller than what had been anticipated.

If the viruses are active, the implications of such great numbers are far-reaching indeed. Viral infection of cells at the base of the food web could be an important factor in the ecological control of such cells/organisms, and in turn of their effects on water quality. After this discovery, additional TEM technology was devoted immediately to a search for morphological correlates of viral action on other living colloids in lakewater.[177,178]

Recent atlases of viral ultrastructure show TEM views of both whole mounts and ultrathin sections of common viruses in general.[179,180] Works specialized in TEM examinations and identifications of plant and insect viruses are available[181,182] as are reviews of literature pertinent to algal viruses.[177,178] A critical appraisal of viral taxonomy is found in Matthews.[183]

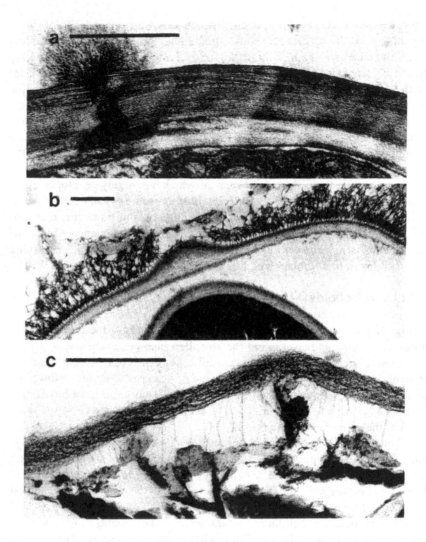

Figure 7. Microcrystalline organic cell walls which can degrade to submicron size fragments. The wall of a healthy algal cell prepared according to Reference 54 is shown in Figure 7a, while Figure 7b shows a recently discarded wall prepared in the same way. Figure 7c shows a degraded wall fragment aggregated with minerals and fibrils. (From Massalski, A. and G. G. Leppard. *J. Fish. Res. Board. Can.* 36:906–921 (1979). With permission.)

In the context of "small is important", the picoplankton, or aquatic cellular organisms in the size range of 0.2 to 2.0 μm, are now known to be much more important to the biological processes of surface waters than had been believed until quite recently.[49,184] Understanding their physiological roles in the "metabolism" of surface waters and their relations to viral predators are now important goals of the biological aquatic sciences.[185] Moving from physiology and disease to microbial ecology, another goal is concerned with the

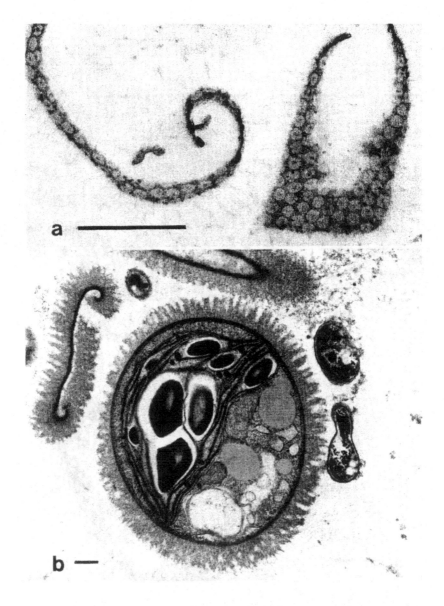

Figure 8. Extracellular patterned wall layers which can become detached from cells. Figure 8a shows patterned fragments of wall from a lacustrine microbe, reprinted with permission from Massalski and Leppard.[155] Figure 8b reveals an analogous situation for a eukaryote alga.

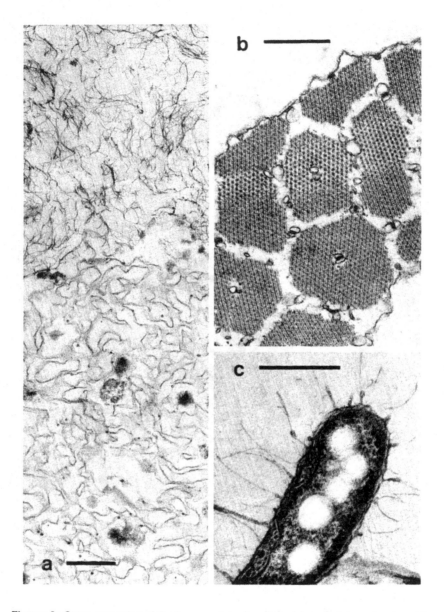

Figure 9. Some examples of the large variety of protein-rich materials encountered in survey work on surface waters. Figures 9a through c present, in order, some biological membranes, a relatively undegraded bit of muscle and some differentiated extensions of bacteria. In a quantitative sense, these are minor aquatic components. Figure 9c is reprinted with permission from Massalski and Leppard;[151] its colloids of interest are described in detail in Reference 134.

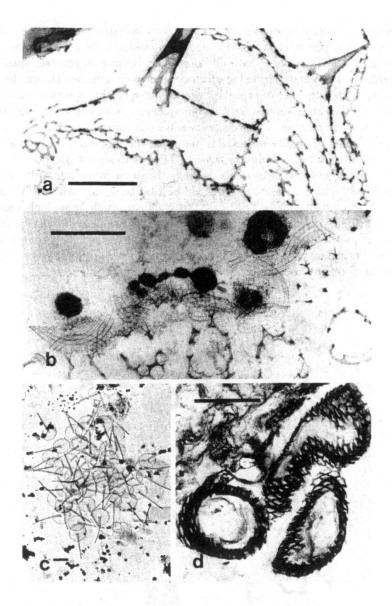

Figure 10. Partially degraded extracellular skeletal structures. Figure 10a is reprinted with permission from Massalski and Leppard.[155] Figures 10b,d were prepared according to Reference 54. Figure 10c is a whole mount preparation.

secretion of colloidal adhesives (often fibrils) by bacterial picoplankton, adhesives which permit organisms to attach themselves to debris and mineral particles.[186,187] The importance of this attachment is great; in general, aquatic bacteria are more active in biogeochemical processes after attachment to a surface.[188] The sequence of processes whereby attachment influences metabolism, and the subsequent ecological consequences, is an important focus of aquatic microbe research. The mechanics of the attachment process, including the positioning of stabilizing colloids and cross-linking bridges involved in biofilm formation, have been amenable to TEM analyses for decades.[189,190]

The speciation of the algal picoplankton in the submicron range was begun a little more than a decade ago.[191] It can be continued only with the aid of TEM. Concomitant with the TEM-based speciation research, limnologists and ecotoxicologists anticipate using TEM as a tool for health and toxicological assessments[192] by correlating changes in the chemistry and morphology of sensitive picoplankton with specific environmental stresses. Some picoplankton with distinctive morphological features[193] are shown in Figure 12.

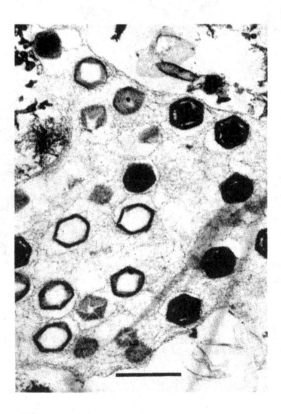

Figure 11. Viruses aggregated within the fibrillar matrix of a slime particle. This virus aggregate, prepared according to Reference 54, was sampled from the water column of a lake. Included are several incomplete viruses.

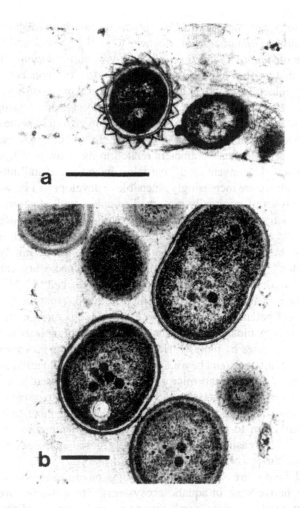

Figure 12. Pelagic picoplankton. Figure 12a shows a cell with a distinctive type of wall, reprinted with permission from Leppard et al.[193] Figure 12b reveals different aspects of the cell contents of a prokaryote picoplankton by showing different section planes through a given species prepared according to Reference 84.

2.3 Mineral Colloids and Mineral-Organic Associations

Mineral colloids have been studied extensively and well by microscopy for decades.[194] The general literature on mineral colloids is large and readily accessible.[195,196] However, its EM literature is essentially about the minerals per se, featuring "cleaned" minerals free of naturally associated materials or particles of a pure mineral which had no aquatic context. The experience of this author in examinations of surface water colloids is that the minerals are typically eroded/coated/irregularly fractured/aggregated with dissimilar colloids to a considerable extent, such that their classical features are not always evident in samples which have been perturbed minimally. While the focus of this critical review is on realistic descriptions of both individual colloids and natural colloid associations, it must adhere to its theme of descriptions/characterizations of *aquatic colloids in their native state in aquatic ecosystems*. This means that, with a few noteworthy exceptions, our concerns herein are restricted to mineral colloid associations with organic coatings, inorganic coatings, microbiota, extracellular enzymes, colloidal ion exchangers (such as fibrils), and other mineral colloids. Where pertinent to the theme of "native state", individual colloids will be considered. In this context, mention will be made of morphological parameters related to the growth of hydrated colloids into hydrated "conventional" particles, including crystallinity changes.

The topics above are increasingly amenable to development by way of TEM and EDS analyses of minimally perturbed samples, with electron diffraction and cytochemistry as useful adjuncts. In response to recent transfers of high technology, the aquatic sciences are creating a literature on such topics. The situation with respect to unperturbed mineral colloids has definitely advanced for both water[28,94] and sediment samples.[80,197] The necessary concepts and technology for continued progress are increasingly refined[15,56,78,107,121,198] and there is no lack of interesting experiments to do and hypotheses to test.

Two topics of potential interest worthy of mention are the genesis of mineral colloids in water by biota,[198-200] and the synthesis and organization of mineral-rich skeletal structures by biota.[201] However, they take us far enough into the biological literature to extend beyond our scope; studies of natural associations between mineral colloids and microbes are readily accessed in the bacteriological and phycological literatures.[198,199] With regard to mixed inorganic/organic flocs, there are good experimental systems available to analyze the mineral contribution to mutual flocculation between minerals and cells/organics.[161,202-204] Increasingly amenable to TEM-based analyses are molecular interactions at the mineral-microbe interface.[198,205]

Discussed below are some specific studies on mineral colloids in their approximate native state in aquatic ecosystems. These studies were chosen because they involve various combinations of improved sampling, improved preservation, selective staining and the use of EDS or electron diffraction on a "per colloid" basis. The remarks will focus on a few mineral colloids (containing Fe, Mn, P, Ca, Si, and Al) which play important biogeochemical roles in aquatic ecosystems.[206]

2.3.1 Inorganic Associations with Organic High Polymers

In 1988, Stone[207] made the following statement in reference to the interactions of mineral particles with surface active aquatic compounds of high molecular weight. "Whether such organic compounds are spread over entire mineral surfaces or are found in patches on mineral surfaces is not known. The nature of this surface coverage can be expected to have an important impact on the availability of mineral surface sites for chemical reaction." In making this statement, he is providing a comment on the limited capacity of conventional chemistry to detect the detailed distribution of a coating material at a mineral surface. There is, however, a microscopical/limnological technology which does not have this limitation, at least for the case of organic fibrils (Figures 3 and 4). The possibility of elucidating the disposition of fibrils on various particles in mixed aggregates, using TEM, was demonstrated some time ago,[68,69] although a systematic use of the technology on the major types of mineral particles has not yet been attempted. The mode of attachment of fibrils to an inorganic surface was demonstrated two decades ago[190] when the nature of the binder was identified as acid polysaccharide.[189] Considering that organic fibrils are rich in acid polysaccharide,[54] the facts above are relevant to the finding of Davis[208] that acidic functional groups on natural organic matter are important in complex formation at the mineral surface. TEM clearly has potential to assist the efforts of environmental analytical chemists in the analysis of high polymer organic coatings.

The relations between metals/mineralization in an aquatic milieu and active biopolymers (extracellular enzymes and/or secreted colloidal ion exchangers) as studied by TEM[77,209] is a specialized field which is growing in sophistication.[198] Considering the enormous variety of organo-metallic species in natural waters[75,210] and their various reactivities with respect to biota,[211,212] TEM should play a larger role in describing the relations between metals, biopolymers, and mineral formation.

In conjunction with metal-organic interactions, inorganic coatings on minerals is a subject of interest to many scientific disciplines. In aquatic environments, the abundant hydrous iron and manganese oxides can act as scavengers of (and eventual sinks for) heavy metals[213] including toxic ones. In a pollution context, such incorporation of inorganic substances by mineral surfaces is a scientific field connected to socio-political issues. In this context, some novel attempts to analyze iron and manganese oxyhydroxide colloids have begun using TEM and STEM-EDS[80] applied to diagenetic colloids collected *in situ* from sediments and embedded directly in Nanoplast for ultrathin sectioning.[214,215] While the genesis of inorganic coatings is an important research area on its own, such genesis in the presence of organic coating agents should become a major research topic within the area.

2.3.2 Iron-Rich Colloids

Iron-rich particles in freshwaters have become much better understood in recent years.[216] Concomitantly, the growth of hydrated mineral particles from molecules is increasingly amenable to microscopical analyses. While investigating the iron-rich materials of lakewaters, Laxen and Chandler[217] found stable iron-rich particles in the submicron range. In 1984, Tipping and Ohnstad[218] were able to assess the colloid stability of such particles. This put a focus on describing the genesis of iron particles in lakes[78] and the speciation of such particles.[28,219] As a result, through the use of minimally perturbing techniques and a multi-method approach, the formation of a mineral colloid and its aggregation "growth" into "true" particles was analyzed in detail for an amorphous iron-rich material found in the redox transition boundary layer of a lake.[28,70,78,79,94] In this study, near-spherical iron-rich colloids (globules) in the range of 0.05 to 0.31 μm were shown to be composed of subunits in the nanometer size range. Accessory techniques[220] showed the iron component to be close to one third Fe(II). A provoked flocculation of colloid-rich lake water showed the iron-rich globules as participants in the formation of loose aggregates which were well into the size range for true particles (by sticking to each other and to other colloids, both organic and inorganic). Some examples of the globules are shown in the micrographs of Figure 13, accompanied by a representative spectrum. The relationships between globules and their aggregates is diagrammed in Figure 14; globule subunit structure was recently investigated with regard to iron partitioning phenomena in relation to filter fractionation.[70]

EDS showed individual iron-rich globules to contain also the elements Ca and P (identified as PO_4 by laser mass spectrometry microprobe analysis) in the mean molar ratios Ca:Fe = 0.19 and P:Fe = 0.25. As a generalization, individual globules came in three morphological varieties, one of which was rare and another of which was strongly associated with other mineral particles as a mixed colloid. A bulk approach to analysis showed many small aggregates to be Fe/P/Ca/Si/Al and others to be Fe/P/Ca/Si. However, the Si/Al and Si components of the mixed colloids involved could be shown to be inadvertent associations.[94] Individual Fe/P/Ca globules, with or without adhering clays or silicates, in turn could be part of larger heterogeneous aggregates containing recognizable patches of organics, clays, calcium-rich colloids, and silicon-rich structures (some of which were identifiable by their morphology as bits of the silica frustules, or mineral walls, of diatom algae). The globules, either alone or in aggregates, had a surprisingly restricted size range; however, the lower limit was a function of anomalous filter behavior.[70,75] The cutoff filter, a standard filter of 0.45 μm used by limnologists, would retain globules down to a diameter a little less than 0.05 μm but would pass globules smaller than 0.04 μm at the filter flow rates used at that time. This phenomenon was investigated in detail so as to use flow-rate comparative studies as a tool to analyze the behavior of iron colloids,[70,75] a subject treated in detail elsewhere in this volume.[9] To summarize, in globule-rich samples, colloidal iron of average dimensions much smaller than 0.04 μm was found in experiments not based on capture by a 0.45 μm filter; these appeared to be singlets and multiples of the near 0.002 μm granules which were the substructural units of globules and the basis for the internal porosity of globules. In this context, it is interesting to note that Schneider and Schwyn[221] have proposed for iron hydroxides a hexameric basic "building block" which has a diameter near 0.002 μm.

Morphological analysis of aggregates revealed a feature often seen in studies of distinctive colloids; despite the tight association of globules in many aggregates, a globule-globule association was readily recognized as such. Thus, *the effect of perturbation was to increase the size of the aggregate but not the size of individual globules.*

Figure 13 illustrates part of a combined morphological-EDS analysis of amorphous iron colloids and makes evident the "separation effect" gained through the use of ultrathin sections to localize and identify individual colloids within an aggregate. In this case the section thickness is less than the thickness of most of the colloids of interest, and, of course, colloids above and below

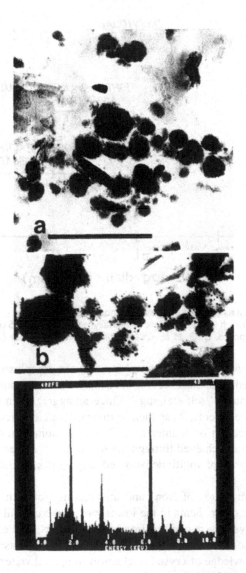

Figure 13. Amorphous iron globules from the oxic-anoxic interface of a small eutrophic lake. Figures 13a, b owe their electron opacity entirely to the native heavy elements within them, and are reprinted with permission from Leppard et al.[28] A typical spectrum from the globules in Figure 13a is presented in Figure 13c. The principal Fe peak illustrated is the Kα peak centered near 6.4 keV. P is near 2.0 and Ca is near 3.7 keV.

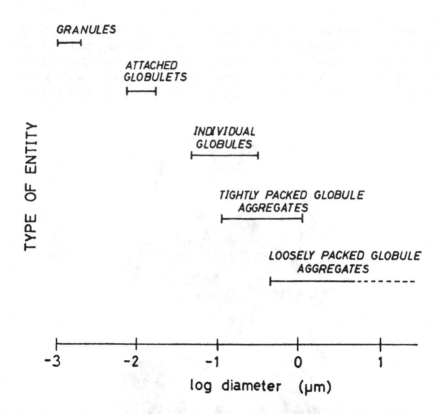

Figure 14. Size classes of iron-rich globules, globule aggregates and globule parts found in an iron-rich lakewater fraction originally isolated on a 0.45 μm filter. (Adapted from Leppard, G. G. et al. *Arch. Hydrobiol.* 113:405–424 (1988).

those of interest have a minimal contribution to the complexity of the image (by being in preceding and succeeding sections). Those colloids rich in heavy elements are of course "self-staining". Once an aggregate in section view is mapped for heavy elements, it can then be mapped for biologicals (sometimes) through the application of counterstains.[28] The combination of spectral and morphological detail achieved through the use of ultrathin sections *cannot* be achieved in analyses of multimicron-sized aggregates visualized as whole mounts.

Crystalline compounds of iron, and amorphous iron-rich materials containing crystalline components in the lower part of the colloid size range, are amenable to analyses by a combination of TEM, EDS, and electron diffraction. Such analyses are currently being pushed to their limits[222] and promise to extend our knowledge of crystal nucleation in natural waters. They should also help to refine our understanding of the "amorphous" state for iron compounds, a subject of growing interest.[223] Sectioned iron compounds from aquatic environments have permitted a morphological resolution of ca. 0.001 μm in Nanoplast[70,222] and permit high grade spectra from colloids as small as 0.04 μm diameter during routine use of TEM-EDS. Diffraction patterns potentially useful for "fingerprinting" can be obtained from iron-rich crystals with a diameter below 0.005 μm.[222] This last figure may seem surprisingly good to some, since electron diffraction parallels X-ray diffraction, but one must remember that electrons interact about 10^6 times more strongly with matter than do X-rays (and that the identification of near-nanometer crystals in an ultrathin section is not difficult).

Even with variable mixtures of several different iron-rich compounds in a complex organic matrix, one has a possibility to selectively identify specific crystalline iron compounds in samples whose size is extremely small in relation to the sample size requirements of conventional wet chemistry. This capacity of TEM-based technology for research on microcrystals within complex mixtures certainly merits further exploration and has already been well utilized in studies of iron-rich magnetosomes, the crystalline colloids of magnetite found within the cytoplasm of some aquatic bacteria.[224,225] These intracellular crystals of cuboidal-to-octahedral shape have diameters mainly in the range of 0.040 to 0.050 μm and are arranged in chains. Evidence from cell remagnetization studies indicates that individual bacteria possessing these chains have properties of single domain ferromagnets. TEM has contributed greatly to confirming that the individual magnetosomes contain iron in the form of magnetite.

In 1983, the crystal habit and magnetic domain structure of individual magnetosomes was analyzed[226] using a field emission electron microscope.[227] Each colloid was determined to be a single crystal with a hexagonal prism shape truncated by {111} planes; the lattice spacings agreed with those of magnetite. A subsequent study in 1984[228] by high resolution TEM with electron diffraction on the *growth and development* of magnetosomes revealed that the 1983 story[226] was not complete. In the later study, direct evidence was presented for both crystalline and non-crystalline phases within individual magnetosomes. It led to an interesting hypothesis on the mechanism of biogenic magnetite formation, one whereby magnetite crystallization involves hydrated iron (III) oxides as non-crystalline precursors.

The isolation and detailed characterization of native iron colloids is intimately tied to an improved understanding of minimal sample perturbation. Mention was made earlier of colloids of 0.05 μm diameter (Fe/P/Ca globules) being captured at the *upper* surface of a cut-off filter whose pore size was *nine times larger* than the colloid. The basis for this is found in colloid "stickiness" phenomena which allow some variables in the filtration process (mainly flow rate) to promote aggregate formation at the filter surface.[9,70,79] This filtration-induced creation of "true particles" was investigated further in relation to the (increasingly misleading) dogma that the 0.45 μm cut-off filter of limnologists separates particles from solutes, regardless of colloid instability problems (and in a conceptual framework often oblivious to the existence of colloids). This "particle creation" artifact is treated in depth in another chapter of this volume.[9] A brief and to-the-point commentary on the current use of 0.45 μm filters by limnologists/oceanographers (applicable also to the 0.7 μm filters of water quality analysts) was published in 1990.[229] Well-presented arguments for considering colloid fractions separately from solute and true particle fractions can be found in the literature since the 1970s.[230] When this concept finally takes hold in aquatic science laboratories, progress in research on iron colloids, and colloids in general, should leap forward. As a final comment on the need for considering colloids as entities presenting special problems in characterization and definition, look at the magnification marker in the figures showing photomicrographs. It shows 0.45 μm as a bar which is larger than the colloids of interest, thus defining them as solutes by conventional limnological/oceanographic thinking.

2.3.3 Manganese Oxyhydroxides

Research on manganese colloids in sediments is being done in parallel with investigations of iron-rich colloids in the same sediments.[80,215] One of the goals is to ascertain just what is meant by the term "ferromanganese colloid" when it is applied to a natural colloid "system". For example, is the system composed of colloids rich in *Mn plus Fe*, or is it a two-component system consisting of Mn colloids coexisting with Fe colloids? There has been some technological success at achieving the means to make such distinctions.[214,222] In concert with this effort, there has been some success in demonstrating a partitioning of cations between Mn and Fe in natural aquatic colloid systems.[80]

Manganese oxyhydroxides are ubiquitous constituents of aquatic ecosystems and, within the sediments, they are believed to impact on the cycling of toxic trace elements. Despite their obvious importance, little has been learned of their native colloid structure and physicochemical properties.[215] This lack of information is a result of (1) their dilution in the sediment matrix coupled with (2) the lack of a minimally perturbing separation technology for isolating them until 1989.[80,215] At present, one can effect an *in situ* collection of diagenetic Mn colloids, spatially separated with respect to Fe colloids and readily dissociated from much of the organic matrix, through the use of sheets of an inert material inserted vertically into sediment,[215] and left there for varying periods of time. The preparation for TEM can be initiated directly after removal of a sheet, via Nanoplast embedding applied to the Mn-rich film adhering to the sheet.[80,214] The narrow, minimally perturbed film is large enough to be split into subsamples for multi-method analyses involving optical microscopy, SEM, TEM, EDS, and many of the conventional techniques of wet chemistry.

The genesis of Mn colloids can involve an active participation by bacteria.[200] Through the use of bacterial cultures capable of making colloidal Mn from added manganese sulfate, Ghiorse and Hirsch[209] made the following observations of relevance to aquatic colloid studies. The bacterial cells produced fibrils extending from their surface into the aquatic milieu; colloidal Mn oxide particles formed in association with the fibrils, apparently through enzyme action. In other cases bacteria coat themselves by a layer of the oxide, and then behave as spherical Mn oxide particles.[242]

2.3.4 Colloidal Phosphorus

Phosphorus in the form of phosphate is a major nutrient, often a limiting one, which can be a contaminant when too abundant in surface waters.[206] Changes in the availability of P to biota can cause dramatic alterations in aquatic ecosystems. Nearly two decades ago, Lean[231] demonstrated that a colloidal phosphorus material, rich in an organic component, might be significant in nutritional exchanges between plankton and lakewater. Later, Francko and Heath[232] distinguished two kinds of complex phosphorus compounds with regard to nutrient phosphate release in lakes; those sensitive to enzymatic hydrolysis and those sensitive to photodegradation. In 1984, Persson[233] described a promising physical technology for the separation and characterization of phosphorus-rich colloids from lakes and rivers. Recently, Ridal and Moore[234] conducted a re-examination of the measurement of dissolved organic P in seawater. They showed the colloidal fraction to comprise 20 to 50% of the total "dissolved" organic phosphorus. The moment is propitious to develop a mode of analysis of colloidal P based on TEM, EDS, and correlative technologies. With this idea in mind, one must remember that the Fe/P/Ca globules described earlier, the nucleic acid-rich viruses and some nucleic acid-rich picoplankton, all qualify as colloidal P (and even in many laboratories as "dissolved" P when they pass 0.45 μm filters).

2.3.5 Biogenic Calcium and Silicon

The biogenic nature of much of the abundant kinds of aquatic particulates rich in calcium[206,235] or silicon,[133,206] and the concomitant variations in ultrastructure, complicate morphological analyses of submicron-sized versions of them. However, the importance of Ca and Si activity to the modulation of biological processes in aquatic ecosystems[201,206] merits a greater research effort on their aquatic colloidal species. Descriptions of the most abundant particulate species of Ca and of Si, including some biogenic kinds, can be found in mineral atlases for earth scientists;[194] some are distinctive morphologically but their distinctive aspects tend to disappear for eroded fragments in the lower part of the colloidal size range. They are readily analyzed by EDS, however, because both elements have peaks in EDS spectra which are distinguished readily from each other and from the most abundant "heavy" elements whose particulate species are sampled from surface waters. Their natural associations with mineral colloids of distinctive appearance are currently under analysis by minimal perturbation techniques.[94]

Colloidal $CaCO_3$, calcite, presents a mineral surface in natural waters which is effective in adsorbing many organic compounds. Photosynthetic cells, through removal of CO_2 from their immediate aquatic milieu, can precipitate calcite at the cell-water interface and initiate the formation of calcium-rich particles whose fresh calcite surfaces are readily exposed to secreted organic molecules. Some roles of organics in determining the nature of mineral species formed at the surface of coated calcite are reviewed in Morse.[235] Analyses by TEM and EDS of these phenomena can contribute to their understanding.

Silicon, in terms of the compound silica, is relatively unreactive chemically in many surface waters. However, Si is important in the cycles of (often abundant) diatom algae which assimilate it for the synthesis of their mineral cell walls (frustules).[133,206] Silicon occurs in soluble forms (silicic acids) and in particulate forms as well as adsorbing to/complexing with other substances. In the author's experience, silicon-rich materials sampled from freshwaters are difficult to recognize in the lower part of the true particle size range when identification is based purely on morphology; colloidal fragments derived from them are even more difficult to recognize. There are two exceptions to this generalization: (1) some of the clay minerals[194,196] as shown in the next section; (2) the spectacular frustules of many varieties whose colloid-sized fragments are readily identified by their regular geometric aspect (provided that the fragments are not too small). Figure 15 illustrates frustules in different stages of fragmentation/degradation, including one approaching colloidal dimensions and accompanied by its EDS spectrum. In the case of the living example shown in cross-section with its cell structure intact, one can see evidence, at the frustule-water interface, of the high polymer organics which are usually associated with frustule mineral. While diatoms are the most noted examples of single-celled organisms which make silicon-rich structures, they are not alone. Of relevance to this critical review, there are examples of aquatic bacteria which can precipitate silicon-rich materials; some examples are summarized briefly in Beveridge.[198]

2.3.6 Clay Minerals

For several decades, there has been a literature available on TEM descriptions of mineral structure per se and much of this has featured clays.[194] This literature contributed to an understanding of both the ultrastructure and activity of clay colloids.[196] However, the literature was not concerned with minimal perturbation techniques, having little need for them in descriptions of rigid materials and having only recently developed a need to visualize better the coatings on minerals, and other colloid-colloid associations. Simultaneous with this TEM contribution to mineralogy, there developed a biological literature focused on clay-microbe-fibril associations. The researches were concerned with the effects of the soil environment on the morphology of submicron biota,[236] and with the spatial relations between clays, microbes, and high polymer organics,[145] including cytochemical analyses of ultrathin sections.[143]

Submicroscopic studies of soils per se[237] and of clays,[197,238] however, have begun only recently to receive the detailed attention that they merit. A few advances are especially exciting in the context of this critical review. Using a hydrophobic embedding resin and placing emphasis on correcting problems of sample perturbation, the Bennett group[197] has made considerable information gains through TEM analysis of clay sediments in ultrathin sections. They were able to show, for sediments of unconsolidated high-porosity marine clays visualized at high resolution, that the pore profiles of these sediments

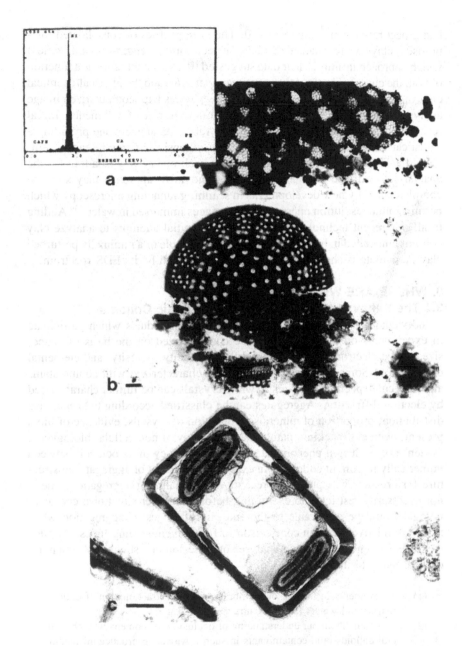

Figure 15. The fragmentation of the silicon-rich cell walls of diatom algae. A whole mount of a fragment of frustule with its EDS spectrum is presented in Figure 15a, while Figure 15b shows a frustule in an early stage of fragmentation. Compare the whole mount of the dead cell of Figure 15b with that of the sectioned "live" cell of Figure 15c. This latter figure is sectioned so as to show the overlap of the two frustule walls when encasing the algal cell contents. Organic polymers are visible at the outer frustule surface in Figure 15c but the section reveals very little of the geometric pore structure.

had aspect ratios which approach 1.0. The pore profiles of consolidated low-porosity clays were characterized by aspect ratios (length-to-width ratios) which approach infinity. Their data suggested that sediment fabric is a function of both the characteristics of the constituent particles and the physical/chemical environments of deposition. They had some success with computerized image analyses carried out on SEM and TEM micrographs of sediments. Initial results appeared promising for quantifying fabric parameters and providing a statistical basis for fabric descriptions. An extension of this research[32] employed a TEM environmental device[35] in a preliminary way as a tool for the analysis of hydrated clay colloids. This innovative approach may soon be complemented by new developments in scanning tunneling microscopy which permit atomic resolution microscopy of surfaces immersed in water.[239] Adding to all this recent technology transfer are the initial attempts to analyze clay colloids embedded in hydrophilic resins. An example of a minimally perturbed clay particulate is shown in Figure 16 accompanied by its EDS spectrum.

3. WHERE ARE WE NOW? — A SUMMARY
3.1 The Behavior and Interactions of Aquatic Colloids
Individual colloids within an aggregate and individuals which participate in experiments on aggregation can be characterized on the basis of shape, size, native electron opacity, internal heterogeneity, porosity, and elemental composition. Some organics can be further characterized with counterstains and molecule-specific "markers", while crystals can be further characterized by electron diffraction. Aggregates can be classified according to shape, size distribution, proportion of minerals, proportion of crystals, evidence of biota present, degree of packing, nature and frequency of non-cellular biologicals, evidence of occlusion phenomena, and the frequency of association between numerically important colloids. An expanded treatment of aggregate substructure for organics was published recently.[15] In analyzing aggregation, one is not necessarily restricted to examining before-and-after situations; cryotechnology should be applicable to the analysis of stages of aggregation when these are relatively fast in comparison to fixation/embedding times. Applied spin-offs from high-resolution analyses of adhesion or "stickiness" phenomena are likely to include:

(1) Improvements in the use of membrane filters in fractionation of colloid-rich natural waters (and, potentially, wastewaters)
(2) Improvements in our understanding of occlusion phenomena whereby natural colloids bind contaminants in such a way as to produce misleading chemical analyses for the contaminants.

3.2 How to Minimize the Artifact Problem
To maximize the amount of information which one can obtain from an aquatic environmental sample, it is necessary to analyze the sample in a state

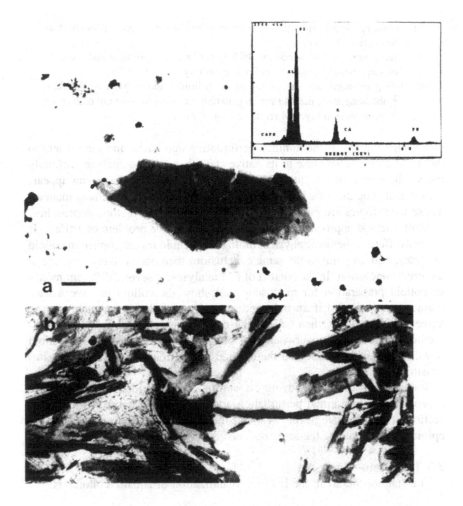

Figure 16. Clay particles. Figure 16a illustrates the detail that one can document in a whole mount of a suspended clay particle near the colloid-true particle size overlap. The inset shows its EDS spectrum with obvious peaks for Al, Si, and K. Figure 16b shows an aggregate of submicron particles, including clay colloids, and is reprinted with permission from Massalski and Leppard.[155]

as close to the native state as is possible. To achieve this goal, one must *handle and process the sample as little as possible.*

The minimal perturbation approach consists of:

(1) Taking the sample in the most gentle manner possible
(2) Isolating the material of interest in the most gentle way possible, and doing so in a manner so as to keep the number of processing steps to a minimum
(3) Avoiding, if possible, concentrating the sample — otherwise let the water body concentrate the sample for you, as one can do with lakes by conducting a search beforehand for the stratum most rich in the colloid of interest

(4) Avoiding sample storage — otherwise store it for a duration as short as possible (<1 day)
(5) Preparation for EM which involves as few steps as possible and a choice of steps which minimizes colloid instability artifacts
(6) Using extraordinary means to ensure colloid stability prior to the final embedding step, means such as isolating oxygen-sensitive colloids at the oxygen tension level where the colloids are found

Through the use of the minimal perturbation approach, one can present to the TEM a sample so close to its native state that one can analyze profitably the smallest units of structure at high resolution. While artifacts can appear, they can also be identified, assessed, and minimized in a systematic manner. These latter topics are pursued through the use of multi-method approaches.

Multi-method approaches allow one to surround the problem of artifact. If several different chemical/physical/biological techniques are applied in colloid analyses, and all point to the same conclusion, then one is likely to arrive at a correct conclusion. In the context of EM analyses, if several different modes of colloid preparation for microscopy all show the colloid to have a least diameter of "x" and if an independent verification by freeze-etching is in agreement with "x", then "x" is the correct answer. This use of a multi-method approach can overwhelm all objections to the analyses of colloid instability which are based on the attitude that extreme difficulty equals impossibility.

Some successes in employing an optimal combination of multi-method approaches with minimal perturbation approaches were described in earlier sections. General strategies are available in the literature[15,56,78] and the author's opinion of the state-of-the-art is outlined below.

3.3 The State-of-the-Art

The state-of-the-art in the realistic visualization of aquatic colloids is

(1) The application of the Nanoplast film technique (see Figure 2) for preparing small colloids and small aggregates of them directly after sampling[56]
(2) The application of the Nanoplast embedding technique for preparing large colloids and large aggregates of them directly after sampling[70,80]
(3) Independent verification by freeze-etch technology of any contentious finding
(4) The most gentle sampling technique available, used without sample storage and employing extraordinary precautions when necessary[70,78]

4. DATA ANALYSIS AND INTERPRETATION

The nature of the potentially available data is as follows for unit colloids:

(1) Shapes
(2) Sizes and size distributions
(3) Porosity
(4) Native electron opacity levels, an indication of heavy element levels

(5) Elemental composition on a "per colloid" basis

(6) Crystallinity identification/fingerprinting on a "per colloid" basis

(7) Acquired electron opacity, an indication of selective affinity for heavy elements

(8) Internal differentiations in addition to porosity, for large complex colloids

The nature of the potentially available data is as follows for aggregates:

(1) Shapes

(2) Sizes and size distributions

(3) Degree of packing

(4) Relative frequency of colloid types

(5) Distribution of colloid types with respect to the surface of an aggregate

(6) Natural associations between colloid types

For experimental work, a classification scheme based on static images can be used in principle to interpret dynamic processes. Time sequence experiments on colloid aggregation and aggregate ageing are feasible with cryotechnology. It is already evident that identifying individual colloids within a compact aggregate is not necessarily a serious problem for image analysis techniques. For example, work on the perturbation of globular Fe/P/Ca colloids, in the presence of other colloid types in the same sample, allowed one to visualize changes in aggregate morphology while permitting conclusive identification of individual Fe/P/Ca globules, even disintegrating ones.[28,70]

For describing large complex shapes, in the case of individual colloids, one can employ the techniques of morphometry developed in the past for biologists and now being expanded for soil and clay scientists as they turn to the use of ultrathin sections. Automated analyses[240,241] may be useful for some kinds of samples provided that sample preparation is compatible with a minimal perturbation approach. For large complex aggregates, HVEM and the use of multiple correlative microscope technologies present a difficult but feasible aid to data collection. To analyze the data of complex samples, systematic computerized approaches are evolving rapidly to assist with three-dimensional reconstructions, EDS spectra, and electron diffraction patterns. Also, a recently purchased electron microscope will arrive with a battery of useful computer apparatus and software.

Despite the aid given by computerization, and the promise of a more highly evolved technology, photomicrography will remain the backbone of aquatic colloid descriptive studies well into the future. Since *recognizing what to photograph* (within the myriad of images displayed by the microscope's viewing screen) is a function of skill and experience, there will persist an "art" of particle analysis well into the future. With it will follow the great losses in time required to bring the skills of a laboratory up to the level permitting a competent practice of the art. There is an urgent need to break these skills down into their individual components and create systematic procedures based on them. In keeping with an urgent need to become more systematic, the colloid investigator must increasingly relate his interpretations to earlier works which included uncontrolled and unassessed artifacts. The urgency is heightened by the fact that data on environmental samples containing unknown distortions from colloid instability artifact is data employed to assess pollution problems which impact on public health and highly valued natural resources.

5. VALUES AND PARAMETERS IN THE LITERATURE
5.1 The Point of View of a Biological Electron Microscopist

The ultrastructure literature of cell biology is rich in information on the colloids within living cells. Chemically, these include proteins, nucleic acids, polysaccharides, lipids, and mineral inclusions. Structurally, these include granules, membranes, fibrils, tubules, and certain hybrids of these. Most of these substances and structural entities are readily degraded in surface waters. There is also a wealth of information on the colloidal components of extracellular structures, including refractory organic cell walls, mineralized cell walls, various coatings on small organisms, secreted scales, a plethora of types of skeletons of small plankton, and the extremely fine layered walls of prokaryotes. The principal refractory molecules of concern to fresh waters are cellulose, lignin, chitin, pectin, and perhaps the protein-polysaccharide and protein-lipid hybrids of prokaryote cell walls. In oceanic waters, the molecules derived from extracellular structures would include the principal algal polysaccharides. In addition, there are the tannins released by many kinds of plant and algal cells. While the cell biologists will take shortcuts with the minimal perturbation approach, they arrive at sound conclusions about colloid systems when they confine themselves to their specialty. In the case of those colloids, both organic and inorganic, whose genesis as an extracellular material is directed metabolically, the picture painted by cell biologists is likely to be a good one. Included in these comments on their literature are the closely related ultrastructure literatures of specialists in bacteria, protozoa, algae, fungi, and the lower animals.

5.2 Suggestions for Revising Values and Parameters

The literature of limnology and oceanography is seriously distorted in its considerations of colloidal phenomena. As remarked upon earlier, this is a result of their out-of-date working assumption that all aquatic materials can be defined as either particles or solutes (no colloids) by causing natural water to flow through a standard cut-off filter of, usually, 0.45 μm pore size (often without assessing flow-rate effects on colloid stability and adjusting the rate accordingly). There is great potential for an acceptance of colloids and an enlightened view of colloid behavior making a valuable contribution to the limnology and oceanography of the future. TEM could play a lead role in making this contribution, both with descriptive work and monitoring work on water fractionation procedures.

The literature of soil science is finally evolving in the manner of the literature of cell biology. Hopefully, it will be guided so as to avoid unnecessary contentious issues. Through the use of ultrathin sections it is leaping forward and perhaps now is the time to introduce Nanoplast techniques and freeze-etching.

Given the profound changes in technology produced in the past decade, one must question the value of past researches which impinge on our understanding of aquatic colloids. The mineralogists who contributed a vast literature on mineral colloids have placed most of their emphasis on cleaned minerals whereas it is increasingly evident that mineral surfaces in an unclean (coated) state play major roles in the biogeochemical processes of aquatic ecosystems. The cell biologists and their cousins in the biomedical fields did many things well; yet, despite their prodigious efforts, the picoplankton were ignored until the late 1970s and the abundance of aquatic viruses was not realized until the late 1980s. With regard to the soil scientists, I have yet to see my first view in their literature of a minimally perturbed, dehydration-sensitive, clay colloid photographed at high resolution.

It would be wise to use the literature of the past as a guide, particularly the more highly evolved versions of it, but *only as a general guide* to form hypotheses. Every observation of importance to the success of a new research should be checked. Even publications which led to obviously correct conclusions may be in error with regard to specific details. Priorities for revising published values should be set by the various specialists as opportunities arise to make a scientific contribution (with the exception of the high priority research on water fractionation already suggested for limnology/oceanography which should begin immediately).

6. FUTURE NEEDS

Science needs to create a proper literature on the structure, composition, activity, and behavior of aquatic colloids in relation to the structure, function, and quality of aquatic ecosystems. Specific attention in this literature should be placed on the cell-mineral interface and on those aggregation processes in the water column which lead to sedimentation and to occlusion phenomena.

Technical advances are needed, and appear to be already forthcoming, for reducing the time necessary to survey sectioned samples. Time is currently the single biggest factor inhibiting research on aquatic colloids. Reducing the skill components of analysis and interpretation to a system of standard routine procedures is needed to change the artistic component of electron microscopy to a more systematic one. Success in this endeavor will in turn reduce the constraints to progress imposed by the excessive time requirements of the past. With further regard to standardization of methods, the approach of "tuning" the minimal perturbation technology to the specific properties of the natural water sample under investigation must evolve in sophistication.

Immediately, there should begin a reassessment of past "particle analysis" data and derivative information currently utilized by environmental managers and modelers in the service (and potential disservice) of public health and environmental conservation. There are reasons why some pollutants are mod-

eled as solutes by some groups and as particle-bound substances by others
— and the reasons do not reflect well on aquatic scientists. This reassessment
should be followed up by a positive research effort to correct the basic problems. In this regard, the inappropriateness of the two-fraction scenario in
water fractionation schemes must be addressed; natural waters do contain
colloids in addition to solutes and true particles.[9,229]

Last, but not least, the effect of native colloids on the bioavailability of
toxic substances and nutrients should be pursued with greater vigor; TEM
can be a great aid to such research.

REFERENCES

1. Vold, M. J. and R. D. Vold. *Colloid Chemistry, The Science of Large Molecules, Small Particles, and Surfaces*. (New York: Reinhold, 1964), pp. 1–
   118.
2. Van Riemsdijk, W. H. and L. K. Koopal. "Ion Binding by Natural Heterogeneous Particles," in *Environmental Particles*, J. Buffle and H. P. van Leeuwen, Eds., IUPAC Environmental Chemistry Series, Vol. 1 (Chelsea, MI:
   Lewis Publishers, in press), chap. 12.
3. Gregory, J. "Fundamentals of Flocculation," *CRC Crit. Rev. Environ. Control*
   19:185–230 (1989).
4. Morel, F. M. M. and P. M. Gschwend. "The Role of Colloids in the Partitioning
   of Solutes in Natural Waters," in *Aquatic Surface Chemistry — Chemical
   Processes at the Particle-Water Interface*, W. Stumm, Ed. (New York: John
   Wiley & Sons, 1987), pp. 405–422.
5. O'Melia, C. R. "The Influence of Coagulation and Sedimentation on the Fate
   of Particles, Associated Pollutants, and Nutrients in Lakes," in *Chemical Processes in Lakes*, W. Stumm, Ed. (New York: Wiley-Interscience, 1985), pp.
   207–224.
6. Ali, W., C. R. O'Melia, and J. K. Edzwald. "Colloidal Stability of Particles
   in Lakes: Measurement and Significance," *Wat. Sci. Technol.* 17:701–712
   (1985).
7. Stumm, W. and J. J. Morgan. *Aquatic Chemistry*, 2nd ed. (New York: Wiley-
   Interscience, 1981), pp. 1–780.
8. Tanaka, T. "Gels," *Sci. Am.* 244(1):124–138 (1981).
9. Buffle, J., D. Perret, and M. Newman, "The Use of Filtration and Ultrafiltration
   for Size Fractionation of Aquatic Particles, Colloids and Macromolecules," in
   Environmental Particles, J. Buffle and H. P. van Leeuwen, Eds., IUPAC
   Environmental Chemistry Series, Vol. 1 (Chelsea, MI: Lewis Publishers, in
   press), chap. 5.
10. van Leeuwen, H. P. "Dynamic Aspects of Metal Speciation in Aquatic Colloidal
    Systems," in *Environmental Particles*, J. Buffle and H. P. van Leeuwen, Eds.,
    IUPAC Environmental Chemistry Series, Vol. 1 (Chelsea, MI: Lewis Publish-
    ers, in press), chap. 13.
11. Sigleo, A. C. and J. C. Means. "Organic and Inorganic Components in Es-
    tuarine Colloids: Implications for Sorption and Transport of Pollutants," *Rev.
    Environ. Contam. Toxicol.* 112:123–147 (1990).

12. Geesey, G. G. "Microbial Exopolymers: Ecological and Economic Considerations," *ASM News* 48:9–14 (1982).

13. Means, J. C. and R. Wijayaratne. "Role of Natural Colloids in the Transport of Hydrophobic Pollutants," *Science* 215:968–970 (1982).

14. Lawler, D. F., C. R. O'Melia, and J. E. Tobiason. "Integral Water Treatment Plant Design — From Particle Size to Plant Performance," in *Particulates in Water — Characterization, Fate, Effects, and Removal,* M. C. Kavanaugh and J. O. Leckie, Eds. (Washington, D.C.: American Chemical Society, 1980), pp. 354–387.

15. Leppard, G. G., B. K. Burnison, and J. Buffle. "Transmission Electron Microscopy of the Natural Organic Matter of Surface Waters," *Anal. Chim. Acta* 232:107–121 (1990).

16. Lin, M. Y., H. M. Lindsay, D. A. Weitz, R. C. Ball, R. Klein, and P. Meakin. "Universality in Colloid Aggregation," *Nature* 339:360–362 (1989).

17. Meek, G. A. *Practical Electron Microscopy for Biologists,* 2nd ed. (London: John Wiley & Sons, 1976), pp. 1–528.

18. Goldstein, J. I., D. E. Newbury, P. Echlin, D. C. Joy, C. Fiori, and E. Lifshin. *Scanning Electron Microscopy and X-ray Microanalysis — A Text for Biologists, Materials Scientists, and Geologists.* (New York: Plenum Press, 1981), pp. 1–673.

19. Chandler, J. A. *X-ray Microanalysis in the Electron Microscope.* (Amsterdam: North-Holland, 1977), pp. 317–547.

20. Beeston, B. E. P., R. W. Horne, and R. Markham. *Electron Diffraction and Optical Diffraction Techniques.* (Amsterdam: Elsevier, 1974), pp. 1–260.

21. Egerton, R. F. *Electron Energy-Loss Spectroscopy in the Electron Microscope.* (New York: Plenum Press, 1986), pp. 1–410.

22. Ottensmeyer, F. P. "Electron Spectroscopic Imaging: Parallel Energy Filtering and Microanalysis in the Fixed-Beam Electron Microscope," *J. Ultrastruct. Res.* 88:121–134 (1984).

23. Crewe, A. V. "High Resolution Scanning Microscopy of Biological Specimens," in *New Developments in Electron Microscopy,* H. E. Huxley and A. Klug, Eds. (London: The Royal Society, 1971), pp. 61–70.

24. Xhoffer, C., L. Wouters, P. Artaxo, A. vanPut, and R. van Grieken. "Characterization of Individual Environmental Particles by Different Beam Techniques," in *Environmental Particles,* J. Buffle and H. P. van Leeuwen, Eds., IUPAC Environmental Chemistry Series, Vol. 1 (Chelsea, MI: Lewis Publishers, in press), chap. 3.

25. Hearle, J. W. S., J. T. Sparrow, and P. M. Cross. *The Use of the Scanning Electron Microscope.* (Oxford: Pergamon Press, 1972), pp. 1–278.

26. Murr, L. E. *Electron and Ion Microscopy and Microanalysis — Principles and Applications.* (New York: Marcel Dekker, 1982), pp. 1–793.

27. Brown, L. M. "Scanning Transmission Electron Microscopy: Microanalysis for the Microelectronic Age," *J. Phys.* F11:1–26 (1981).

28. Leppard, G. G., J. Buffle, R. R. De Vitre, and D. Perret. "The Ultrastructure and Physical Characteristics of a Distinctive Colloidal Iron Particulate Isolated from a Small Eutrophic Lake," *Arch. Hydrobiol.* 113:405–424 (1988).

29. Colliex, C., C. Mory, A. L. Olins, D. E. Olins, and M. Tence. "Energy Filtered STEM Imaging of Thick Biological Sections," *J. Microsc.* 153:1–21 (1989).

30. Turner, J. N. and C. W. Allen. "Introduction to High-Voltage Electron Microscopy," *EMSA Bull.* 20:104–105 (1990).

31. Humphreys, C. "High Voltage Electron Microscopy," in *Principles and Techniques of Electron Microscopy — Biological Applications*, Vol. 6, M. A. Hayat, Ed. (New York: Van Nostrand Reinhold, 1976), pp. 1–39.

32. Fischer, K. M. and R. H. Bennett. "Environmental Cell for the Transmission Electron Microscope — Applications in Marine Science," *Eos* 71:133 (1990).

33. Danilatos, G. D. "Foundations of Environmental Scanning Electron Microscopy," *Adv. Electron. Electron Phys.* 71:109–250 (1988).

34. Fukami, A., K. Fukushima, A. Ishikawa, and K. Ohi. "New Side-Entry Environmental Cell Equipment for Dynamic Observation," in *Proc. of the 45th Annu. Meet. Electron Microscopy Society of America*, G. W. Bailey, Ed. (San Francisco: San Francisco Press, 1987), pp. 142–143.

35. Allinson, D. L. "Environmental Devices in Electron Microscopy," in *Principles and Techniques of Electron Microscopy — Biological Applications*, Vol. 5, M. A. Hayat, Ed. (New York: Van Nostrand Reinhold, 1975), pp. 62–113.

36. De Robertis, E. D. P. and E. M. F. De Robertis. *Cell and Molecular Biology*, 7th ed. (Philadelphia: Saunders College/HRW, 1980), pp. 1–673.

37. Lillie, R. D. *H. J. Conn's Biological Stains — A Handbook on the Nature and Uses of the Dyes Employed in the Biological Laboratory*, 8th ed. (Baltimore: Williams and Wilkins, 1969), pp. 1–498.

38. Hayat, M. A. *Basic Techniques for Transmission Electron Microscopy*. (Orlando: Academic Press, 1986), pp. 1–411.

39. Hayat, M. A., Ed. *Principles and Techniques of Electron Microscopy — Biological Applications*, Vol. 3. (New York: Van Nostrand Reinhold, 1973), pp. 1–321.

40. Arndt-Jovin, D. J., M. Robert-Nicoud, and T. M. Jovin. "Probing DNA Structure and Function with a Multi-Wavelength Fluorescence Confocal Laser Microscope," *J. Microsc.* 157:61–72 (1990).

41. Pawley, J. B., Ed. *Handbook of Biological Confocal Microscopy*, (rev. ed.). (New York: Plenum Press, 1990), pp. 1–246.

42. Jovin, T. M. and D. J. Arndt-Jovin. "Luminescence Digital Imaging Microscopy," *Annu. Rev. Biophys. Biophys. Chem.* 18:271–308 (1989).

43. Arndt-Jovin, D. J., M. Robert-Nicoud, S. J. Kaufman, and T. M. Jovin. "Fluorescence Digital Imaging Microscopy in Cell Biology," *Science* 230:247–256 (1985).

44. Russ, J. C. *Computer-Assisted Microscopy — The Measurement and Analysis of Images*. (New York: Plenum Press, 1990), pp. 1–453.

45. Delgado, R. M., M. J. Fink, and R. M. Brown. "Imaging of Submicron Objects with the Light Microscope," *J. Microsc.* 154:129–141 (1989).

46. Misell, D. L. *Image Analysis, Enhancement and Interpretation*. (Amsterdam: Elsevier, 1978), pp. 1–306.

47. Mayfield, C. I. and M. Munawar. "Microcomputer-Based Measurement of Algal Fluorescence as a Potential Indicator of Environmental Contamination," *Bull. Environ. Contam. Toxicol.* 41:261–266 (1988).

48. Mayfield, C. I. "A Simple Microcomputer-Based Video Analysis System and Potential Applications to Microbiology," *J. Microbiol. Methods* 3:61–67 (1985).

49. Stockner, J. G. "Phototrophic Picoplankton: An Overview from Marine and Freshwater Ecosystems," *Limnol. Oceanogr.* 33:765–775 (1988).

50. Pick, F. R. and D. A. Caron. "Picoplankton and Nanoplankton Biomass in Lake Ontario: Relative Contribution of Phototrophic and Heterotrophic Communities," *Can. J. Fish. Aquat. Sci.* 44:2164–2172 (1987).

51. Barer, R. "Microscopes, Microscopy, and Microbiology," *Annu. Rev. Microbiol.* 28:371–389 (1974).

52. Bron, C., P. Gremillet, D. Launay, M. Jourlin, H. P. Gautschi, T. Bachi, and J. Schupbach. "Three-Dimensional Electron Microscopy of Entire Cells," *J. Microsc.* 157:115–126 (1990).

53. Peachey, L. D. and J. P. Heath. "Reconstruction from Stereo and Multiple Tilt Electron Microscope Images of Thick Sections of Embedded Biological Specimens Using Computer Graphic Methods," *J. Microsc.* 153:193–204 (1989).

54. Leppard, G. G., A. Massalski, and D. R. S. Lean. "Electron-Opaque Microscopic Fibrils in Lakes; their Demonstration, their Biological Derivation and their Potential Significance in the Redistribution of Cations," *Protoplasma* 92:289–309 (1977).

55. Weibel, E. R. and R. P. Bolender. "Stereological Techniques for Electron Microscopic Morphometry," in *Principles and Techniques of Electron Microscopy — Biological Applications*, Vol. 3, M. A. Hayat, Ed. (New York: Van Nostrand Reinhold, 1973), pp. 237–296.

56. Perret, D., G. G. Leppard, M. Muller, N. Belzile, R. De Vitre and J. Buffle. "Electron Microscopy of Aquatic Colloids: Non-Perturbing Preparation of Specimens in the Field," *Water Res.* 25:1333–1343 (1991).

57. Inoue, T. and H. Koike. "High-Resolution Low-temperature Scanning Electron Microscopy for Observing Intracellular Structures of Quick Frozen Biological Specimens," *J. Microsc.* 156:137–147 (1989).

58. Weilenmann, U., C. R. O'Melia, and W. Stumm. "Particle Transport in Lakes: Models and Measurements," *Limnol. Oceanogr.* 34:1–18 (1989).

59. Stumm, W., Ed. *Aquatic Surface Chemistry — Chemical Processes at the Particle-Water Interface*. (New York: John Wiley & Sons, 1987), pp. 1–520.

60. O'Melia, C. R. "Aquasols: The Behavior of Small Particles in Aquatic Systems," *Environ. Sci. Technol.* 14:1052–1060 (1980).

61. Cagle, G. D. "Fine Structure and Distribution of Extracellular Polymer Surrounding Selected Aerobic Bacteria," *Can. J. Microbiol.* 21:395–408 (1975).

62. Gibbs, R. J. "Effect of Natural Organic Coatings on the Coagulation of Particles," *Environ. Sci. Technol.* 17:237–240 (1983).

63. Baccini, P., E. Grieder, R. Stierli, and S. Goldberg. "The Influence of Natural Organic Matter on the Adsorption Properties of Mineral Particles in Lake Water," *Schweiz. Z. Hydrol.* 44:99–116 (1982).

64. Theng, B. K. G. "Clay-Polymer Interactions: Summary and Perspectives," *Clays Clay Miner.* 30:1–10 (1982).

65. Tipping, E. "The Adsorption of Aquatic Humic Substances by Iron Oxides," *Geochim. Cosmochim. Acta* 45:191–199 (1981).

66. Hunter, K. A. "Microelectrophoretic Properties of Natural Surface-Active Organic Matter in Coastal Seawater," *Limnol. Oceanogr.* 25:807–822 (1980).

67. Neihof, R. and G. Loeb. "Dissolved Organic Matter in Seawater and the Electric Charge of Immersed Surfaces," *J. Mar. Res.* 32:5–12 (1974).

68. Leppard, G. G. "Organic Coatings on Suspended Particles in Lake Water," *Arch. Hydrobiol.* 102:265–269 (1984).

69. Leppard, G. G. "The Ultrastructure of Lacustrine Sedimenting Materials in the Colloidal Size Range," *Arch. Hydrobiol.* 101:521–530 (1984).

70. Leppard, G. G., R. R. De Vitre, D. Perret, and J. Buffle. "Colloidal Iron Oxyhydroxy-Phosphate: The Sizing and Morphology of an Amorphous Species in Relation to Partitioning Phenomena," *Sci. Total Environ.* 87/88:345–354 (1989).

71. Hayat, M. A. *Fixation for Electron Microscopy.* (New York: Academic Press, 1981), pp. 1–501.

72. Robards, A. W. and U. B. Sleytr. *Low Temperature Methods in Biological Electron Microscopy.* (Amsterdam: Elsevier, 1985), pp. 1–551.

73. Nermut, M. V. "Freeze-Drying for Electron Microscopy," in *Principles and Techniques of Electron Microscopy — Biological Applications,* Vol. 7, M. A. Hayat, Ed. (New York: Van Nostrand Reinhold, 1977), pp. 79–117.

74. Benedetti, E. L. and P. Favard, Eds. *Freeze-Etching Techniques and Applications.* (Paris: Société Française de Microscopie Electronique, 1973), pp. 1–274.

75. Buffle, J. *Complexation Reactions in Aquatic Systems: An Analytical Approach* (Chichester: Ellis Horwood, 1988), pp. 1–692.

76. Davis, J. A. "Complexation of Trace Metals by Adsorbed Natural Organic Matter," *Geochim. Cosmochim. Acta* 48:679–691 (1984).

77. Leppard, G. G. "Relationships Between Fibrils, Colloids, Chemical Speciation, and the Bioavailability of Trace Heavy Metals in Surface Waters — A Review," National Water Research Institute Contrib. No. 84–45, Burlington, Ontario, Canada (1984), pp. 1–53.

78. Buffle, J., R. R. De Vitre, D. Perret, and G. G. Leppard. "Combining Field Measurements for Speciation in Non-Perturbable Water Samples — Application to the Iron and Sulfide Cycles in a Eutrophic Lake," in *Metal Speciation: Theory, Analysis and Application,* J. R. Kramer and H. E. Allen, Eds. (Chelsea, MI: Lewis Publishers, 1988), pp. 99–124.

79. Perret, D., R. De Vitre, G. G. Leppard, and J. Buffle. "Characterizing Autochthonous Iron Particles and Colloids — The Need for Better Particle Analysis Methods," in *Large Lakes: Ecological Structure and Function,* M. M. Tilzer and C. Serruya, Eds. (Heidelberg: Springer-Verlag/Science Tech, 1990) pp. 224–244.

80. De Vitre, R., N. Belzile, G. G. Leppard, and A. Tessier. "Diagenetic Manganese and Iron Oxyhydroxide Particles Collected from a Canadian Lake: Morphology and Chemical Composition," in *Heavy Metals in the Environment,* Vol. 1, J. P. Vernet, Ed. (Edinburgh: CEP Consultants, 1989), pp. 217–220.

81. Nermut, M. V. "Strategy and Tactics in Electron Microscopy of Cell Surfaces," *Electron Microsc. Rev.* 2:171–196 (1989).

82. Haschemeyer, R. H. and R. J. Myers. "Negative Staining," in *Principles and Techniques of Electron Microscopy — Biological Applications,* Vol. 2, (New York: Van Nostrand Reinhold, 1972), pp. 99–147.

83. Leppard, G. G., J. Buffle, and R. Baudat. "A Description of the Aggregation Properties of Aquatic Pedogenic Fulvic Acids — Combining Physico-Chemical Data and Microscopical Observations." *Water Res.* 20:185–196 (1986).

84. Burnison, B. K. and G. G. Leppard. "Isolation of Colloidal Fibrils from Lake Water by Physical Separation Techniques," *Can. J. Fish. Aquat. Sci.* 40:373–381 (1983).

85. Nomizu, T., K. Goto, and A. Mizuike. "Electron Microscopy of Nanometer Particles in Freshwater," *Anal. Chem.* 60:2653–2656 (1988).

86. Nomizu, T., T. Nozue, and A. Mizuike. "Electron Microscopy of Submicron Particles in Natural Waters: Morphology and Elemental Analysis of Particles in Fresh Waters," *Mikrochim. Acta* 2:99–106 (1987).

87. Nomizu, T. and A. Mizuike. "Electron Microscopy of Submicron Particles in Natural Waters: Specimen Preparation by Centrifugation," *Mikrochim. Acta* 1:65–72 (1986).

88. Frosch, D. and C. Westphal. "Melamine Resins and Their Application in Electron Microscopy," *Electron Microsc. Rev.* 2:231–255 (1989).

89. Bachhuber, K. and D. Frosch. "Melamine Resins, a New Class of Water-Soluble Embedding Media for Electron Microscopy," *J. Microsc.* 130:1–9 (1983).

90. Frosch, D. and C. Westphal. "Choosing the Appropriate Section Thickness in the Melamine Embedding Technique," *J. Microsc.* 137:177–183 (1985).

91. Spurr, A. R. "A Low-Viscosity Epoxy Resin Embedding Medium for Electron Microscopy," *J. Ultrastruct. Res.* 26:31–43 (1969).

92. Cowley, J. M. "The Principles of High Resolution Electron Microscopy," in *Principles and Techniques of Electron Microscopy — Biological Applications,* Vol. 6, M. A. Hayat, Ed. (New York: Van Nostrand Reinhold, 1976), pp. 40–84.

93. Causton, B. E. "Does the Embedding Chemistry Interact with Tissues?" in *The Science of Biological Specimen Preparations for Microscopy and Microanalysis 1985,* M. Muller, R. P. Becker, A. Boyde, and J. J. Wolosewick, Eds. (Chicago: SEM Inc./AMF O'Hare, 1985), pp. 209–214.

94. Buffle, J., R. R. De Vitre, D. Perret, and G. G. Leppard. "Physico-Chemical Characteristics of a Colloidal Iron Phosphate Species Formed at the Oxic-Anoxic Interface of a Eutrophic Lake," *Geochim. Cosmochim. Acta* 53:399–408 (1989).

95. Frosch, D., C. Westphal, and H. Bohme. "Improved Preservation of Glycogen in Unfixed Cyanobacteria Embedded at $-82°C$ in Nanoplast," *J. Histochem. Cytochem.* 35:119–121 (1987).

96. Watson, M. L. "Staining of Tissue Sections for Electron Microscopy with Heavy Metals," *J. Biophys. Biochem. Cytol.* 4:475–478 (1958).

97. Reynolds, E. S. "The Use of Lead Citrate at High pH as an Electron-Opaque Stain in Electron Microscopy," *J. Cell Biol.* 17:208–212 (1963).

98. Wied, G. L. *Introduction to Quantitative Cytochemistry.* (New York: Academic Press, 1966), pp. 1–623.

99. Slayter, E. M. *Optical Methods in Biology.* (New York: Wiley-Interscience, 1970) pp. 1–757.

100. Simard, R. "Cryoultramicrotomy," in *Principles and Techniques of Electron Microscopy — Biological Applications,* Vol. 6, M. A. Hayat, Ed. (New York: Van Nostrand Reinhold, 1976), pp. 290–311.

101. Bancroft, J. D. and A. Stevens, Eds. *Theory and Practice of Histological Techniques,* 3rd ed. (Edinburgh: Churchill Livingstone, 1990), pp. 1–726.

102. Gahan, P. B. *Plant Histochemistry and Cytochemistry, An Introduction.* (London: Academic Press, 1984), pp. 1–301.

103. Pearse, A. G. E. *Histochemistry, Theoretical and Applied,* Vol. 2, 3rd ed. (Edinburgh: Churchill Livingstone, 1972), pp. 761–1518.

104. Pearse, A. G. E. *Histochemistry, Theoretical and Applied,* Vol. 1, 3rd ed. (London: J. & A. Churchill, 1968), pp. 1–759.

105. Hanke, D. E. and D. H. Northcote. "Molecular Visualization of Pectin and DNA by Ruthenium Red," *Biopolymers* 14:1–17 (1975).

106. Luft, J. H. "Ruthenium Red and Ruthenium Violet. I. Chemistry, Purification, Methods of Use for Electron Microscopy, and Mechanism of Action," *Anat. Rec.* 171:347–368 (1971).

107. Bullock, G. R. and P. Petrusz, Eds. *Techniques in Immunocytochemistry,* Vol. 1. (London: Academic Press, 1986), pp. 1–306.

108. Williams, M. A. *Autoradiography and Immunocytochemistry.* (Amsterdam: Elsevier, 1985), pp. 1–218.

109. Shnitka, T. K. and A. M. Seligman. "Ultrastructural Localization of Enzymes," *Annu. Rev. Biochem.* 40:375–396 (1971).

110. Zablackis, E., V. Vreeland, B. Doboszewski, and W. M. Laetsch. "Localization of Kappa Carrageenan in Cell Walls of *Eucheuma alvarezii* var. *tambalang* with *in situ* Hybridization Probes," in *Algal Biotechnology,* T. Stadler, J. Mollion, M.-C. Verdus, Y. Karamanos, H. Morvan, and D. Christiaen, Eds. (London: Elsevier Applied Science, 1988), pp. 441–449.

111. Vreeland, V., E. Zablackis, and W. M. Laetsch. "Monoclonal Antibodies to Carrageenan," in *Algal Biotechnology,* T. Stadler, J. Mollion, M.-C. Verdus, Y. Karamanos, H. Morvan, and D. Christiaen, Eds. (London: Elsevier Applied Science, 1988), pp. 431–439.

112. Vreeland, V., E. Zablackis, B. Doboszewski, and W. M. Laetsch. "Molecular Markers for Marine Algal Polysaccharides," *Hydrobiologia* 151/152:155–160 (1987).

113. Vreeland, V., M. Slomich, and W. M. Laetsch. "Monoclonal Antibodies as Molecular Probes for Cell Wall Antigens of the Brown Alga, *Fucus,*" *Planta* 162:506–517 (1984).

114. Roodyn, D. B., Ed. *Enzyme Cytology.* (London: Academic Press, 1967), pp. 1–587.

115. Sabatini, D. D., K. Bensch, and R. J. Barrnett. "Cytochemistry and Electron Microscopy, the Preservation of Cellular Ultrastructure and Enzymatic Activity by Aldehyde Fixation," *J. Cell Biol.* 17:19–58 (1963).

116. Costerton, J. W., J. M. Ingram, and K.-J. Cheng. "Structure and Function of the Cell Envelope of Gram-Negative Bacteria," *Bacteriol. Rev.* 38:87–110 (1974).

117. Mason, C. W. *Handbook of Chemical Microscopy,* Vol. 1, 4th ed. (New York: Wiley-Interscience, 1983), pp. 1–505.

118. Preston, R. D. *The Physical Biology of Plant Cell Walls.* (London: Chapman & Hall, 1974), pp. 1–491.

119. Walton, A. G. and J. Blackwell. *Biopolymers.* (New York: Academic Press, 1973), pp. 1–604.

120. Seki, H. *Organic Materials in Aquatic Ecosystems.* (Boca Raton, FL: CRC Press, 1982), pp. 1–201.

121. Steinbrecht, R. A. and K. Zierold, Eds. *Cryotechniques in Biological Electron Microscopy.* (Berlin: Springer Verlag, 1987), pp. 1–305.

122. Moor, H. "Recent Progress in the Freeze-Etching Technique," *Phil. Trans. R. Soc. Lond. B* 261:121–131 (1971).

123. Steere, R. L. "Preparation of High-resolution Freeze-Etch, Freeze-Fracture, Frozen-Surface, and Freeze-Dried Replicas in a Single Freeze-Etch Module, and the Use of Stereo Electron Microscopy to Obtain Maximum Information from Them," in *Freeze-Etching Techniques and Applications,* E. L. Benedetti and P. Favard, Eds. (Paris: Société Française de Microscopie Electronique, 1973), pp. 223–255.

124. Stevenson, I. L. and M. Schnitzer. "Transmission Electron Microscopy of Extracted Fulvic and Humic Acids," *Soil Sci.* 133:179–185 (1982).

125. Ghosh, K. and M. Schnitzer. "A Scanning Electron Microscope Study of Effects of Adding Neutral Electrolytes to Solutions of Humic Substances," *Geoderma* 28:53–56 (1982).

126. Yaffee, M. "Visualization of Untreated Bio-Macromolecular Structures: Development of Extremely Thin Nanoplast Embedding Films," Diploma Dissertation, Swiss Federal Institute of Technology, Zurich, Switzerland (1988).

127. Thurman, E. M. *Organic Geochemistry of Natural Waters.* (Dordrecht, Netherlands: Martinus Nijhoff/Dr. W. Junk, 1986), pp. 1–497.

128. Beveridge, T. J. "Ultrastructure, Chemistry, and Function of the Bacterial Wall," *Int. Rev. Cytol.* 72:229–317 (1981).

129. Siegel, B. Z. and S. M. Siegel. "The Chemical Composition of Algal Cell Walls," *CRC Crit. Rev. Microbiol.* 3:1–26 (1973).

130. Hunter, K. A. and P. S. Liss. "Organic Sea Surface Films," in *Marine Organic Chemistry — Evolution, Composition, Interactions and Chemistry of Organic Matter in Seawater*, E. K. Duursma and R. Dawson, Eds. (Amsterdam: Elsevier Scientific, 1981), pp. 259–298.

131. Brown, R. M., W. Herth, W. W. Franke, and D. Romanovicz. "The Role of the Golgi Apparatus in the Biosynthesis and Secretion of a Cellulosic Glycoprotein in *Pleurochrysis*: A Model System for the Synthesis of Structural Polysaccharides," in *Biogenesis of Plant Cell Wall Polysaccharides*, F. Loewus, Ed. (New York: Academic Press, 1973), pp. 207–257.

132. Colvin, J. R. "The Structure and Biosynthesis of Cellulose," *CRC Crit. Rev. Macromol. Sci.* 1:47–81 (1972).

133. Dodge, J. D. *The Fine Structure of Algal Cells.* (London: Academic Press, 1973), pp. 1–261.

134. Lima-de-Faria, A., Ed. *Handbook of Molecular Cytology* (Amsterdam: North-Holland, 1969), pp. 1–1508.

135. Breger, I. A. "What You Don't Know Can Hurt You: Organic Colloids and Natural Waters," in *Organic Matter in Natural Waters*, D. W. Hood, Ed. (College, Alaska: Inst. Marine Sci. Occasional Publ. No. 1, 1970), pp. 563–574.

136. Leppard, G. G. and B. K. Burnison. "Bioavailability, Trace Element Associations with Colloids and an Emerging Interest in Colloidal Organic Fibrils," in *Trace Element Speciation in Surface Waters and its Ecological Implications*, G. G. Leppard, Ed. (New York: Plenum Press, 1983), pp. 105–122.

137. Leppard, G. G. "Transmission Electron Microscopy Applied to Water Fractionation Studies — A New Look at DOC," *Water Pollut. Res. J. Can.* 20(2):100–110 (1985).

138. Jones, H. C., I. L. Roth, and W. M. Sanders. "Electron Microscope Study of a Slime Layer," *J. Bacteriol.* 99:316–325 (1969).

139. Friedman, B. A., P. R. Dugan, R. M. Pfister, and C. C. Remsen. "Structure of Exocellular Polymers and their Relationship to Bacterial Flocculation," *J. Bacteriol.* 98:1328–1334 (1969).

140. Friedman, B. A. and P. R. Dugan. "Concentration and Accumulation of Metallic Ions by the Bacterium *Zoogloea*," *Dev. Ind. Microbiol.* 9:381–388 (1968).

141. Dugan, P. R., C. B. MacMillan, and R. M. Pfister. "Aerobic Heterotrophic Bacteria Indigenous to pH 2.8 Acid Mine Water: Predominant Slime-Producing Bacteria in Acid Streamers," *J. Bacteriol.* 101:982–988 (1970).

142. Bowen, G. D. and A. D. Rovira. "The Influence of Micro-Organisms on Growth and Metabolism of Plant Roots," in *Root Growth*, W. J. Whittington, Ed. (New York: Plenum Press, 1969), pp. 170–201.

143. Foster, R. C. "Polysaccharides in Soil Fabrics," *Science* 214:665–667 (1981).

144. Guckert, A., H. Breisch, and O. Reisinger. "Interface Sol-Racine. I. Etude au Microscope Electronique des Relations Mucigel-Argile-Microorganismes," *Soil Biol. Biochem.* 7:241–250 (1975).

145. Leppard, G. G. "Rhizoplane Fibrils in Wheat: Demonstration and Derivation," *Science* 185:1066–1067 (1974).

146. Leppard, G. G. and S. Ramamoorthy. "The Aggregation of Wheat Rhizoplane Fibrils and the Accumulation of Soil-Bound Cations," *Can. J. Bot.* 53:1729–1735 (1975).

147. Ramamoorthy, S. and G. G. Leppard. "Fibrillar Pectin and Contact Cation Exchange at the Root Surface," *J. Theor. Biol.* 66:527–540 (1977).

148. Lagerwerff, J. V. "The Contact-Exchange Theory Amended," *Plant Soil* 13:253–264 (1960).

149. Leppard, G. G. "The Fibrillar Matrix Component of Lacustrine Biofilms," *Water Res.* 20:697–702 (1986).

150. Geesey, G. G., W. T. Richardson, H. G. Yeomans, R. T. Irvin, and J. W. Costerton. "Microscopic Examination of Natural Sessile Bacterial Populations from an Alpine Stream," *Can. J. Microbiol.* 23:1733–1736 (1977).

151. Massalski, A. and G. G. Leppard. "Morphological Examination of Fibrillar Colloids Associated with Algae and Bacteria in Lakes," *J. Fish. Res. Board Can.* 36:922–938 (1979).

152. Brown, M. J., and J. N. Lester. "Metal Removal in Activated Sludge: The Role of Bacterial Extracellular Polymers," *Water Res.* 13:817–837 (1979).

153. Steiner, A. E., D. A. McLaren, and C. F. Forster. "The Nature of Activated Sludge Flocs," *Water Res.* 10:25–30 (1976).

154. Colvin, J. R. and G. G. Leppard. "Fibrillar, Modified Polygalacturonic Acid in, on, and between Plant Cell Walls," in *Biogenesis of Plant Cell Wall Polysaccharides*, F. Loewus, Ed. (New York: Academic Press, 1973), pp. 315–331.

155. Massalski, A. and G. G. Leppard. "Survey of Some Canadian Lakes for the Presence of Ultrastructurally Discrete Particles in the Colloidal Size Range," *J. Fish. Res. Board Can.* 36:906–921 (1979).

156. Paerl, H. W., R. D. Thomson, and C. R. Goldman. "The Ecological Significance of Detritus Formation During a Diatom Bloom in Lake Tahoe, California-Nevada," *Verh. Int. Verein. Limnol.* 19:826–834 (1975).

157. Paerl, H. W. "Microbial Attachment to Particles in Marine and Freshwater Ecosystems," *Microbial Ecol.* 2:73–83 (1975).

158. Paerl, H. W. "Bacterial Uptake of Dissolved Organic Matter in Relation to Detrital Aggregation in Marine and Freshwater Systems," *Limnol. Oceanogr.* 19:966–972 (1974).

159. Paerl, H. W. "Detritus in Lake Tahoe: Structural Modification by Attached Microflora," *Science* 180:496–498 (1973).

160. Burnison, B. K. and G. G. Leppard. "Ethanol Fractionation of Lacustrine Colloidal Fibrils," *Can. J. Fish. Aquat. Sci.* 41:385–388 (1984).

161. Bernhardt, H., O. Hoyer, H. Schell, and B. Lusse. "Reaction Mechanisms Involved in the Influence of Algogenic Organic Matter on Flocculation," *Z. Wasser-Abwasser-Forsch.* 18:18–30 (1985).

162. Lusse, B., O. Hoyer, and C. J. Soeder. "Mass Cultivation of Planktonic Freshwater Algae for the Production of Extracellular Organic Matter (EOM)," *Z. Wasser-Abwasser-Forsch.* 18:67–75 (1985).

163. Hoyer, O., B. Lusse, and H. Bernhardt. "Isolation and Characterization of Extracellular Organic Matter (EOM) from Algae," *Z. Wasser-Abwasser-Forsch.* 18:76–90 (1985).

164. Brown, M. J. and J. N. Lester. "Role of Bacterial Extracellular Polymers in Metal Uptake in Pure Bacterial Culture and Activated Sludge. I. Effects of Metal Concentration," *Water Res.* 16:1539–1548 (1982).

165. Brown, M. J. and J. N. Lester. "Role of Bacterial Extracellular Polymers in Metal Uptake in Pure Bacterial Culture and Activated Sludge. II. Effects of Mean Cell Retention Time," *Water Res.* 16:1549–1560 (1982).

166. Alexander, M. "Environmental and Microbiological Problems Arising from Recalcitrant Molecules," *Microbial Ecol.* 2:17–27 (1975).

167. Choudhry, G. G. *Humic Substances — Structural, Photophysical, Photochemical and Free Radical Aspects and Interactions with Environmental Chemicals.* (New York: Gordon & Breach Science, 1984), pp. 1–185.

168. Gjessing, E. T. *Physical and Chemical Characteristics of Aquatic Humus.* (Ann Arbor, MI: Ann Arbor Science, 1976), pp. 1–120.

169. Schnitzer, M. and S. U. Khan. *Humic Substances in the Environment.* (New York: Marcel Dekker, 1972), pp. 1–327.

170. Ghosh, K. and M. Schnitzer. "Macromolecular Structures of Humic Substances," *Soil Sci.* 129:266–276 (1980).

171. Suffet, I. H. and P. MacCarthy, Eds. *Aquatic Humic Substances — Influence on Fate and Treatment of Pollutants.* (Washington, D.C.: American Chemical Soc., 1989), pp. 1–864.

172. Caceci, M. S. and A. Billon. "Evidence for Large Organic Scatterers (50—200 nm diameter) in Humic Acid Samples," *Org. Geochem.* 15:335–350 (1990).

173. Rose, J. B. "Emerging Issues for the Microbiology of Drinking Water," *WATER/ Eng. Man.*, July, pp. 23–29 (1990).

174. Payment, P. and R. Armon. "Virus Removal by Drinking Water Treatment Processes," *CRC Crit. Rev. Environ. Control* 19:15–31 (1989).

175. Miller, M. F. "Particle Counting of Viruses," in *Principles and Techniques of Electron Microscopy — Biological Applications*, Vol. 4, M. A. Hayat, Ed. (New York: Van Nostrand Reinhold, 1974), pp. 89–128.

176. Bergh, O., K. Y. Borsheim, G. Bratbak, and M. Heldal. "High Abundance of Viruses Found in Aquatic Environments," *Nature* 340:467–468 (1989).

177. Proctor, L. M. and J. A. Fuhrman. "Viral Mortality of Marine Bacteria and Cyanobacteria," *Nature* 343:60–62 (1990).

178. Klut, M. E. and J. G. Stockner. "Virus-Like Particles in an Ultra-Oligotrophic Lake on Vancouver Island, British Columbia," *Can. J. Fish. Aquat. Sci.* 47:725–730 (1990).

179. Palmer, E. L. and M. L. Martin. *Electron Microscopy in Viral Diagnosis.* (Boca Raton, FL: CRC Press, 1988), pp. 1–194.

180. Doane, F. W. and N. Anderson. *Electron Microscopy in Diagnostic Virology: A Practical Guide and Atlas.* (New York: Cambridge University Press, 1987), pp. 1–192.

181. Horne, R. W. "The Development and Application of Electron Microscopy to the Structure of Isolated Plant Viruses," in *Molecular Plant Virology*, Vol. 1, J. W. Davies, Ed. (Boca Raton, FL: CRC Press, 1985), pp. 1–41.

182. Maramorosch, K., Ed. *The Atlas of Insect and Plant Viruses — Including Mycoplasmaviruses and Viroids*. (New York: Academic Press, 1977), pp. 1–478.

183. Matthews, R. E. F., Ed. *A Critical Appraisal of Viral Taxonomy*. (Boca Raton, FL: CRC Press, 1983), pp. 1–256.

184. Stockner, J. G. and N. J. Antia. "Algal Picoplankton from Marine and Freshwater Ecosystems: A Multidisciplinary Perspective," *Can. J. Fish. Aquat. Sci.* 43:2472–2503 (1986).

185. Sherr, E. B. "And Now, Small is Plentiful," *Science* 340:429 (1989).

186. Lewin. R. "Microbial Adhesion is a Sticky Problem," *Science* 224:375–377 (1984).

187. Sutherland, I. W. "Microbial Exopolysaccharides — Their Role in Microbial Adhesion in Aqueous Systems," *CRC Crit. Rev. Microbiol.* 10:173–201 (1983).

188. Van Loosdrecht, M. C. M., J. Lycklema, W. Norde, and A. J. B. Zehnder. "Influence of Interfaces on Microbial Activity," *Microbiol. Rev.* 54:75–87 (1990).

189. Fletcher, M. and G. D. Floodgate. "An Electron-Microscopic Demonstration of an Acidic Polysaccharide Involved in the Adhesion of a Marine Bacterium to Solid Surfaces," *J. Gen. Microbiol.* 74:325–334 (1973).

190. Marshall, K. C., R. Stout, and R. Mitchell. "Mechanism of the Initial Events in the Sorption of Marine Bacteria to Surfaces," *J. Gen. Microbiol.* 68:337–348 (1971).

191. Johnson, P. W. and J. M. Sieburth. "Chroococcoid Cyanobacteria in the Sea: A Ubiquitous and Diverse Phototrophic Biomass," *Limnol. Oceanogr.* 24:928–935 (1979).

192. Munawar, M., I. F. Munawar, and G. G. Leppard. "Early Warning Assays: An Overview of Toxicity Testing with Phytoplankton in the North American Great Lakes," *Hydrobiologia* 188/189:237–246 (1989).

193. Leppard, G. G., D. Urciuoli, and F. R. Pick. "Characterization of Cyanobacterial Picoplankton in Lake Ontario by Transmission Electron Microscopy," *Can. J. Fish. Aquat. Sci.* 44:2173–2177 (1987).

194. Beutelspacher, H. and H. W. van der Marel. *Atlas of Electron Microscopy of Clay Minerals and their Admixtures: A Picture Atlas*. (Amsterdam: Elsevier, 1968), pp. 1–333.

195. Yariv, S. and H. Cross. *Geochemistry of Colloid Systems for Earth Scientists*. (Berlin: Springer-Verlag, 1979), pp. 1–450.

196. Swartzen-Allen, S. L. and E. Matijevic. "Surface and Colloid Chemistry of Clays," *Chem. Rev.* 74:385–400 (1974).

197. Bennett, R. H., K. M. Fischer, D. L. Lavoie, W. R. Bryant, and R. Rezak. "Porometry and Fabric of Marine Clay and Carbonate Sediments: Determinants of Permeability," *Mar. Geol.* 89:127–152 (1989).

198. Beveridge, T. J. "Role of Cellular Design in Bacterial Metal Accumulation and Mineralization," *Annu. Rev. Microbiol.* 43:147–171 (1989).

199. Poole, R. K. and G. M. Gadd, Eds. *Metal — Microbe Interactions*, Spec. Publ. Vol. 26, Society for General Microbiology. (New York: Oxford University Press, 1989) pp. 1–146.

200. Ghiorse, W. C. "Biology of Iron- and Manganese-Depositing Bacteria," *Annu. Rev. Microbiol.* 38:515–550 (1984).

201. Lowenstam, H. A. and S. Weiner. *On Biomineralization*. (New York: Oxford Univeristy Press, 1989), pp. 1–324.

202. Oliver, R. L., R. H. Thomas, C. S. Reynolds, and A. E. Walsby. "The Sedimentation of Buoyant *Microcystis* Colonies Caused by Precipitation with an Iron-Containing Colloid," *Proc. R. Soc. Lond. B* 223:511–528 (1985).

203. Sukenik, A., W. Schroder, J. Lauer, G. Shelef, and C. J. Soeder. "Coprecipitation of Microalgal Biomass with Calcium and Phosphate Ions," *Water Res.* 19:127–129 (1985).

204. Avnimelech, Y., B. W. Troeger, and L. W. Reed. "Mutual Flocculation of Algae and Clay: Evidence and Implications," *Science* 216:63–65 (1982).

205. Mayers, I. T. and T. J. Beveridge. "The Sorption of Metals to *Bacillus subtilis* Walls from Dilute Solutions and Simulated Hamilton Harbour (Lake Ontario) Water," *Can. J. Microbiol.* 35:764–770 (1989).

206. Wetzel, R. G. *Limnology.* (Philadelphia: W. B. Saunders, 1975), pp. 1–743.

207. Stone, A. T. "Introduction to Interactions of Organic Compounds with Mineral Surfaces," in *Metal Speciation: Theory, Analysis and Application,* J. R. Kramer and H. E. Allen, Eds. (Chelsea, MI: Lewis Publishers, 1988), pp. 69–80.

208. Davis, J. A. "Adsorption of Natural Dissolved Organic Matter at the Oxide/Water Interface," *Geochim. Cosmochim. Acta* 46:2381–2393 (1982).

209. Ghiorse, W. C. and P. Hirsch. "An Ultrastructural Study of Iron and Manganese Deposition Associated with Extracellular Polymers of *Pedomicrobium*-Like Budding Bacteria," *Arch. Microbiol.* 123:213–226 (1979).

210. Mantoura, R. F. C. "Organo-Metallic Interactions in Natural Waters," in *Marine Organic Chemistry — Evolution, Composition, Interactions and Chemistry of Organic Matter in Seawater,* E. K. Duursma and R. Dawson, Eds. (Amsterdam: Elsevier Scientific, 1981), pp. 179–223.

211. Leppard, G. G., Ed. *Trace Element Speciation in Surface Waters and its Ecological Implications.* (New York: Plenum Press, 1983), pp. 1–320.

212. Florence, T. M. and G. E. Batley. "Chemical Speciation in Natural Waters," *CRC Crit. Rev. Anal. Chem.* 9:219–296 (1980).

213. Singh, S. K. and V. Subramanian. "Hydrous Fe and Mn Oxides — Scavengers of Heavy Metals in the Aquatic Environment," *CRC Crit. Rev. Environ. Control* 14:33–90 (1984).

214. Belzile, N., R. R. De Vitre, G. G. Leppard, D. Fortin, and A. Tessier. "Physicochemical Characteristics of Natural Iron and Manganese Oxyhydroxides Collected from Canadian Lakes," in preparation.

215. Belzile, N., R. De Vitre, and A. Tessier. "*In situ* Collection of Diagenetic Iron and Manganese Oxyhydroxides from Natural Sediments," *Nature* 340:376–377 (1989).

216. Davison, W. and R. R. De Vitre, "Iron Particles in Freshwaters," in *Environmental Particles,* J. Buffle and H. P. van Leeuwen, Eds., IUPAC Environmental Chemistry Series, Vol. 1. (Chelsea, MI: Lewis Publishers, in press), chap. 8.

217. Laxen, D. P. H. and I. M. Chandler. "Size Distribution of Iron and Manganese Species in Freshwaters," *Geochim. Cosmochim. Acta* 47:731–741 (1983).

218. Tipping, E. and M. Ohnstad. "Colloid Stability of Iron Oxide Particles from a Freshwater Lake," *Nature* 308:266–268 (1984).

219. Tipping, E., D. W. Thompson, and C. Woof. "Iron Oxide Particulates Formed by the Oxygenation of Natural and Model Lakewaters Containing Fe(II)," *Arch. Hydrobiol.* 115:59–70 (1989).

220. De Vitre, R. R., J. Buffle, D. Perret, and R. Baudat. "A Study of Iron and Manganese Transformations at the $O_2/S(-II)$ Transition Layer in a Eutrophic Lake (Lake Bret, Switzerland): A Multi-Method Approach," *Geochim. Cosmochim. Acta* 52:1601–1613 (1988).

221. Schneider, W. and B. Schwyn. "The Hydrolysis of Iron in Synthetic, Biological, and Aquatic Media," in *Aquatic Surface Chemistry — Chemical Processes at the Particle-Water Interface,* W. Stumm, Ed. (New York: John Wiley & Sons, 1987), pp. 167–194.

222. Fortin, D. and G. G. Leppard. Unpublished results (1991).

223. Combes, J. M., A. Manceau, G. Calas, and J. Y. Bottero. "Formation of Ferric Oxides from Aqueous Solutions: A Polyhedral Approach by X-ray Absorption Spectroscopy. I. Hydrolysis and Formation of Ferric Gels," *Geochim. Cosmochim. Acta* 53:583–594 (1989).

224. Blakemore, R. P. "Magnetotactic Bacteria," *Annu. Rev. Microbiol.* 36:217–238 (1982).

225. Frankel, R. B., R. P. Blakemore, and R. S. Wolfe. "Magnetite in Freshwater Magnetotactic Bacteria," *Science* 203:1355–1356 (1979).

226. Matsuda, T., J. Endo, N. Osakabe, and A. Tonomura. "Morphology and Structure of Biogenic Magnetite Particles," *Nature* 302:411–412 (1983).

227. Tonomura, A., T. Matsuda, J. Endo, H. Todokoro, and T. Komoda. "Development of a Field Emission Electron Microscope," *J. Electron Microsc.* 28:1–11 (1979).

228. Mann, S., R. B. Frankel, and R. P. Blakemore. "Structure, Morphology and Crystal Growth of Bacterial Magnetite," *Nature* 310:405–407 (1984).

229. Buffle, J., G. G. Leppard, R. R. De Vitre, and D. Perret. "Submicron Sized Aquatic Compounds: From Artefacts to Ecologically Meaningful Data," *Proc. Am. Chem. Soc.* 30(1):337–340 (1990).

230. Sharp, J. H. "Size Classes of Organic Carbon in Seawater," *Limnol. Oceanogr.* 18:441–447 (1973).

231. Lean, D. R. S. "Phosphorus Dynamics in Lake Water," *Science* 179:678–680 (1973).

232. Francko, D. A. and R. T. Heath. "Functionally Distinct Classes of Complex Phosphorus Compounds in Lake Water," *Limnol. Oceanogr.* 24:463–473 (1979).

233. Persson, G. "Characterization of Particulate and Colloidal Phosphorus Forms in Water by Continuous Flow Density Gradient Centrifugation," *Verh. Int. Verein. Limnol.* 22:149–154 (1984).

234. Ridal, J. J. and R. M. Moore. "A Re-Examination of the Measurement of Dissolved Organic Phosphorus in Seawater," *Mar. Chem.* 29:19–31 (1990).

235. Morse, J. W. "The Surface Chemistry of Calcium Carbonate Minerals in Natural Waters: An Overview," *Mar. Chem.* 20:91–112 (1986).

236. Bae, H. C., E. H. Cota-Robles, and L. E. Casida. "Microflora of Soil as Viewed by Transmission Electron Microscopy," *Appl. Microbiol.* 23:637–648 (1972).

237. Bisdom, E. B. A. and J. Ducloux, Eds. *Submicroscopic Studies of Soils,* Geoderma (Spec. Issue) 30:1–356 (1983).

238. Bennett, R. H. and M. H. Hulbert. *Clay Microstructure* (Boston: International Human Resource Development Corp., 1986), pp. 1–161.

239. Sonnenfeld, R. and P. K. Hansma. "Atomic-Resolution Microscopy in Water," *Science* 232:211–213 (1986).

240. Bernard, P. C., R. E. van Grieken, and L. Brugmann. "Geochemistry of Suspended Matter from the Baltic Sea. I. Results of Individual Particle Characterization by Automated Electron Microprobe," *Mar. Chem.* 26:155–177 (1989).
241. Bernard, P. C., R. E. van Grieken, and D. Eisma. "Classification of Estuarine Particles Using Automated Electron Microprobe Analysis and Multivariate Techniques," *Environ. Sci. Technol.* 20:467–473 (1986).
242. R. R. De Vitre. "Multimethod Characterization of the Forms of Iron, Manganese and Sulfur in a Eutrophic Lake (Bret, VD)," Ph.D. thesis n° 2224, University of Geneva, Switzerland (1986).

CHAPTER 7

CHARACTERIZATION OF PARTICLE SURFACE CHARGE

Garrison Sposito

Department of Soil Science, University of California, Berkeley, California

TABLE OF CONTENTS

1.   Introduction 292

2.   Surface Charge: Definitions 293
        2.1   Surface Chemical Speciation 293
        2.2   Components of Surface Charge 294
        2.3   Points of Zero Charge 296

3.   Determination of the Net Total Particle Surface Charge
        Density 300

4.   Determination of Intrinsic Surface Charge Density 302

5.   Determination of Net Structural Surface Charge 303
      5.1   Nonspecific Methods 303
      5.2   Alkylammonium Cation Exchange 303

6.   Determination of the Net Proton Surface Charge 305
      6.1   Electrometric Titration 305
      6.2   Side-Reaction Interferences.............................307
             6.2.1   Aqueous Phase.................................307
             6.2.2   Solid Phase 308
      6.3   Electrometric Interferences 308

7.   Conclusions..309

Acknowledgments...310

Glossary..310

References..312

1. INTRODUCTION

Solid particles that appear in the natural environment as a result of weathering phenomena and human activity play a critical role in determining the fate of both airborne and waterborne pollutants.[1,2] In recent years, significant advances have been made in the experimental characterization of these particles and their reactions with aqueous solutions.[3-5] Of particular interest are charged particles, whose surface reactions are essential to the biogeochemical cycles of trace elements and the pathways of detoxification of these elements when present in aqueous environments at hazardous concentrations.[1,2]

Fundamental concepts and nomenclature regarding surface reactions on charged solid particles have been discussed under IUPAC auspices by Everett,[6] whose work should be consulted for the basic definitions of *adsorption*, *adsorbent*, *adsorbate*, *adsorptive*, *surface excess*, *surface charge density*, and other well-established terminology in particle surface chemistry. The present chapter builds on these fundamental definitions to describe concepts and methods necessary to the experimental characterization of solid particle surface charge. The approach taken is informed by recent advances in the aqueous surface chemistry of natural particles.[4,5,7] These new results have shown conclusively that particle charge and surface chemical speciation are connected intimately, and that measurements of particle surface charge components can be interpreted without reliance on detailed mathematical models of the interfacial region.

2. SURFACE CHARGE: DEFINITIONS
2.1 Surface Chemical Speciation

Solid particle surfaces develop electrical charge in two principal ways: either permanently, from isomorphic substitutions of component ions in the bulk structure of the solid, or conditionally, from the reactions of surface functional groups with adsorptive ions in aqueous solution. A *surface functional group* is a chemically reactive molecular unit bound into the structure of an adsorbent at its periphery, such that the reactive portion of the functional group is exposed to an aqueous solution contacting the adsorbent.[7] Surface functional groups occur on both organic and inorganic adsorbents (e.g., surface hydroxyls on both colloidal humus and metal oxide minerals). After reaction with an adsorptive ion in aqueous solution (which then becomes an adsorbate), they can form *adsorption complexes,* which are immobilized molecular entities comprising the adsorbate and the surface functional group to which it is bound closely.[6] A further classification of adsorption complexes can be made into inner-sphere and outer-sphere surface complexes.[3,9] An *inner-sphere surface complex* has no water molecule interposed between the surface functional group and the small ion or molecule it binds, whereas an *outer-sphere surface complex* has at least one such interposed water molecule. Outer-sphere surface complexes thus contain at least partially solvated adsorbate ions or molecules.

These two concepts are illustrated in Figure 1 for the idealized surface of a 2:1 layer type clay mineral, such as montmorillonite. (For a description of the structures and surface characteristics of clay minerals and hydrous oxides, see, e.g., Sposito[3] or Brown et al.[10]) In this example, the surface functional group is the siloxane ditrigonal cavity formed by oxygen ions at the bases of six, corner-sharing silica tetrahedra in the mineral structure.[3,10] This cavity has a diameter of about 0.26 nm and can mediate the permanent negative charge resulting from cation substitutions nearby in the clay mineral (e.g., Al^{3+} for Si^{4+}, or Mg^{2+} for Al^{3+}). When charged, it then can form inner-sphere or outer-sphere complexes with aqueous cations, as shown in Figure 1, for both external basal-plane and interlayer siloxane surfaces.

Ions in surface complexes are to be distinguished from those in the *diffuse layer*[6] (also illustrated in Figure 1) because the former species remain immobilized on the particle surface over time scales that are long when compared, e.g., with the ca. 10 ps required for a diffusive step by a solvated ion in aqueous solution. For example, the well known outer-sphere surface complex formed in the interlayer of montmorillonite by Ca^{2+} or Mg^{2+} (cf. the left side of Figure 1) has been shown to be immobile on the time scale of

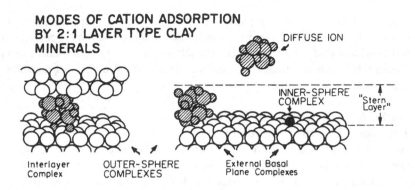

Figure 1. Three modes of adsorption for small aqueous ions, illustrated for a layer type aluminosilicate adsorbent (e.g., smectite).

0.1 to 100 ps probed by scattering neutrons from the protons in the six water molecules solvating the bivalent cation.[3] This clear separation of residence time scales will not be as sharp for outer-sphere complexes on external surfaces vs. diffuse-layer ions. Moreover, if the diffuse-layer ions are in very close proximity to the particle surface (i.e., *counter-ion condensation*[11,12] occurs), then it becomes more difficult to contrast their behavior with that of ions in outer-sphere complexes, since both kinds of surface species are solvated.

The three types of surface chemical species — inner-sphere complex, outer-sphere complex, and diffuse-layer — represent three modes of adsorption of small aqueous ions that contribute to the formation of the *electrochemical double layer*[6] on particle surfaces. No inference of special "planes" containing adsorbed ions is required by these speciation concepts, nor is any detailed molecular structure implied, other than the general notions of surface complexes and dissociated ions. It is sometimes convenient, although not necessary, to group surface complexes into a *Stern layer*[6] to distinguish them from diffuse-layer ions (see Figure 1). This geometric distinction among surface species, however, should not be taken to mean that diffuse-layer ions must necessarily approach a particle surface less closely than Stern-layer ions.

2.2 Components of Surface Charge

The *net permanent structural surface charge density*, denoted σ_o and measured in coulombs per square meter, is created by isomorphic substitutions in minerals. These substitutions occur in both primary and secondary minerals, but they produce significant surface charge only in the 2:1 layer type aluminosilicates.[3,10,13] In these minerals, $\sigma_o < 0$ invariably because of cation substitutions. The *net proton surface charge density*, denoted σ_H, and measured in coulombs per square meter, is proportional to the difference between the moles of protons and the moles of hydroxide ions complexed by surface functional groups:[14]

$$\sigma_H = F(q_H - q_{OH})/a_s \tag{1}$$

where q_i is the *specific adsorbed charge* (moles of charge per kilogram) of ion i complexed by surface groups (i.e., the product of the valence of ion i and its specific surface excess [in moles per kilogram] attributed to adsorption complexes), F is the Faraday constant, and a_s is specific surface area (square meters per kilogram). Conceptually, diffuse-layer protons are not included in the definition of σ_H. The most important surface functional groups that complex protons are hydroxyl groups on colloidal humus, metal oxides, and 1:1 layer type aluminosilicates (e.g., kaolinite). Values of σ_H can be negative, zero, or positive, depending on pH, ionic strength, etc.

Besides σ_H, particle surfaces can bear an *inner-sphere complex surface charge density*, σ_{IS} and an *outer-sphere complex surface charge density*, σ_{OS} both measured in coulombs per square meter. Contributing to σ_{IS} is the net total charge of the ions, other than H^+ or OH^-, which are bound into inner-sphere surface complexes. Similarly, σ_{OS} receives contributions from the net total charge of the ions, other than H^+ or OH^-, that are bound into outer-sphere surface complexes. These two components of surface charge density do not include H^+ or OH^- because of the traditional emphasis given to these two latter ions as components of surface-reactive solid phases and aqueous solutions.[14] Other ions (e.g., Ca^{2+}, K^+, SO_4^{2-}, or $H_2PO_4^-$) that form surface complexes are less ubiquitous in mineral particle structures than H^+ or OH^- and are less often so central to the chemical behavior of aqueous systems.

It is useful in applications to group the four component surface charge densities into the *intrinsic surface charge density*,[3]

$$\sigma_{in} \equiv \sigma_o + \sigma_H \tag{2}$$

and the *Stern layer surface charge density*,[5]

$$\sigma_S \equiv \sigma_{IS} + \sigma_{OS} \tag{3}$$

The intrinsic surface charge density reflects particle charge developed either from isomorphic substitutions and surface complex formation involving H^+ or OH^-. The Stern layer surface charge density reflects particle charge developed from counterions immobilized on an adsorbent surface. The *net total particle surface charge density* then can be defined mathematically by the equation.[3]

$$\sigma_p \equiv \sigma_{in} + \sigma_S = \sigma_o + \sigma_H + \sigma_{IS} + \sigma_{OS} \tag{4}$$

It should be emphasized that the definitions in Equations 2 through 4 are not unique and that other groupings of surface charge components (e.g., the net total adsorbed ion charge) are useful as well. In general, both σ_p and its components can result from a variety of inorganic and organic surface species in a myriad of molecular configurations. Implicit in Equation 4, however, is the concept that a charged particle and its surface complexes represent an identifiable molecular unit in aqueous systems.

Although particles may bear electrical charge, aqueous suspensions of particles are always electrically neutral. Thus σ_p in Equation 4 must be balanced, when it is non-zero, by another kind of surface charge. This balancing charge arises from the ions in the *diffuse layer*,[6] which move about freely in aqueous solution while remaining near enough to particle surfaces to create an effective surface charge density, σ_d, that balances σ_p. On the molecular scale, this effective surface charge density can be apportioned to diffuse-layer ions according to the equation:[6]

$$\sigma_d \equiv \frac{F}{ma_s} \sum_i z_i \int_V [c_i(x) - c_i(\infty)] \, dV \tag{5}$$

where z_i is the valence of diffuse-layer ion i, $c_i(x)$ is its concentration at point x in aqueous solution, and $c_i(\infty)$ is its concentration in solution far enough from any particle surface to avoid adsorption in a diffuse layer. The integral in Equation 5 is over the entire volume V of aqueous solution contacting the mass m of solid adsorbent whose specific surface area is a_s. Thus Equation 5 represents the surface excess charge of the ions in aqueous solution: if $c_i(x) = c_i(\infty)$ uniformly, there is no contribution of ion i to σ_d. Note that Equation 5 applies to all ions in solution, including H^+ and OH^-, and defines the *diffuse-layer surface charge density* (in coulombs per square meter) required to balance σ_p in order to maintain electrical neutrality:

$$\sigma_p + \sigma_d = 0 \tag{6}$$

Equation 6, which expresses the *balance of surface charge* (cf. Everett[6]), can be applied both to an individual particle in suspension and to an entire suspension. It serves as a general conservation law for the characterization of particle surface charge.

2.3 Points of Zero Charge

Speaking generically, *points of zero charge* are pH values at which one or more of the surface charge components in Equation 4 vanishes at fixed temperature, pressure, and aqueous solution composition. Five points of zero charge are summarized in Table 1 and Figure 2. A standard nomenclature for points of zero charge has not been established. For example, p.z.n.p.c. is often called the "zero point of charge"[14] and, in much of the surface chemistry literature concerning natural particles, p.z.s.e. has been termed the "point of zero charge", as has p.z.n.c. Irrespective of this unfortunate variability in terminology, agreement does exist on the general importance of points of zero charge to particle surface characterization. It should be borne in mind, however, that pH is not the only chemical property whose variation can lead to a vanishing component of surface charge density.

Table 1. Points of Zero Charge

Symbol	Name	Defining Condition[a]
p.z.c.	Point of zero charge[6]	$\sigma_p = 0$
p.z.n.p.c.	Point of zero net proton charge[6]	$\sigma_H = 0$
p.z.n.c.	Point of zero net charge[15]	$\sigma_{in} = 0$
i.e.p	Isoelectric point[6]	$u = 0$
p.z.s.e.	Point of zero salt effect[15]	$\partial \sigma_H / \partial I = 0$

[a] See text for the definitions of the symbols used.

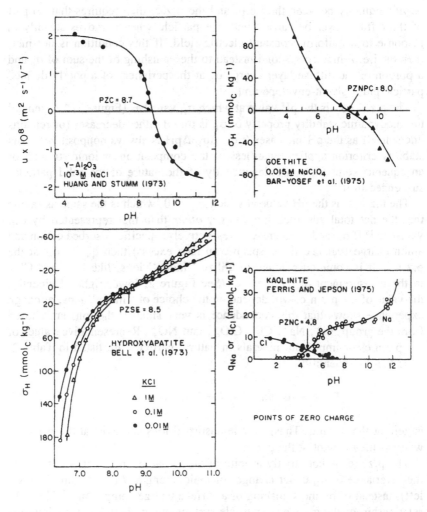

Figure 2. Experimental illustrations of four "points of zero charge". (From Sposito, G. The Surface Chemistry of Soils. (New York: Oxford University Press, 1984). With permission.)

The p.z.c. is the pH value at which the net total particle charge vanishes: $\sigma_p = 0$.[6] Thus, by Equation 6, at the p.z.c. there is no net particle surface charge neutralized by ions in the diffuse layer and all adsorbed ions are immobilized in surface complexes. In principle, the p.z.c. can be measured by ascertaining the pH value at which colloids do not respond to an applied electric field or, more generally, by determining the pH value at which charge balance exists in an aqueous solution in which particles are suspended. The first method actually determines the pH value at which the particle *electro-phoretic mobility u* vanishes (see Figure 2, upper left), termed the *isoelectric point*.[6] Equality between the i.e.p. and the p.z.c. then requires that no part of the diffuse layer be carried with the particle when it moves steadily in response to a uniform, constant electric field. If this condition is not met, then the i.e.p. may correspond instead to the vanishing of the sum of σ_p and a portion of the diffuse-layer charge σ_d at the periphery of a poorly defined particle-plus-solvent-envelope unit.

The p.z.n.p.c. is the pH value at which σ_H vanishes (Figure 2).[8] A general thermodynamic stability property of σ_H is that it either decreases (or remains unchanged) as the pH increases (i.e., $\Delta\sigma_H/\Delta pH$ is always nonpositive). This stability criterion applies regardless of the composition or ionic strength of an aqueous solution, and independently of the nature of the solid particles suspended in it.

The p.z.n.c. is the pH value at which $\sigma_{in} = 0$, which is the same as saying that the net total adsorbed ion charge, *other* than that represented by σ_H, vanishes.[15] If q_+ and q_- represent, respectively, specific adsorbed cation and anion charge (valence times specific surface excess) then $q_+ = q_-$ at the p.z.n.c. It is common practice to utilize "index" ions, like Na^+ and Cl^-, in the measurement of the p.z.n.c. (See Figure 2, lower right). Evidently, the value of the p.z.n.c. will depend on the choice of "index" ions, although experience shows that this dependence is very small if the ions are chosen from the group: Li^+, Na^+, Cl^-, ClO_4^-, and NO_3^-. Representative values of the p.z.n.c for important minerals in natural particles are listed in Table 2. Note that the quantity:

$$F(q_+ - q_-)/a_s = \sigma_{IS} + \sigma_{OS} + \sigma_d = -\sigma_{in} \tag{7}$$

is zero at the p.z.n.c. Thus, mobile adsorbed ions can exist at the p.z.n.c., whereas they cannot at the p.z.c.

The p.z.s.e. is not strictly a point of zero charge, in that it is defined by the invariance of σ_H under changes of ionic strength, I (see Figure 2, lower left), instead of by the vanishing of a surface charge component.[15] Thus, the relationship of the p.z.s.e. to particle surface charge is indirect, and usually it is necessary to appeal to a detailed molecular model of the particle/aqueous solution interface in order to interpret the p.z.s.e.[15-17]

A set of general statements about points of zero charge can be proved using only the law of conservation of surface charge and the criterion, $\partial\sigma_H/\partial pH <$

Table 2. Some Representative Values of p.z.n.c. for Specimen Minerals

Mineral	p.z.n.c.	Mineral	p.z.n.c.
Quartz (α-SiO$_2$)	2.0–3.0	Goethite (α-FeOOH)	7.0–8.0
Birnessite (α-MnO$_2$)	1.5–2.5	Hematite (α-Fe$_2$O$_3$)	8.0–8.5
Kaolinite (Si$_4$Al$_4$O$_{10}$(OH)$_8$)	4.0–5.0	Gibbsite (Al(OH)$_3$)	8.0–9.0

Source: Sposito[7]

0. These statements, the *PZC Theorems*, do not require molecular details of chemical speciation at the particle/aqueous solution interface and so may be applied to validate mathematical surface speciation models (e.g., modified Gouy-Chapman theory or site-binding models) or to examine experimental surface speciation data for internal consistency. Proofs of the theorems are given by Sposito.[7]

Theorem 1

Let q_+ be the specific adsorbed cation charge (excluding H^+ in surface complexes) and let q_- be the specific adsorbed anion charge (excluding OH^- in surface complexes). If $(\partial\sigma_H/\partial pH) < 0$, then:

$$\sigma_o = -\frac{F}{a_s}(q_+ - q_-) \quad \text{at} \quad pH = \text{p.z.n.p.c.} \tag{8}$$

and

$$\sigma_o \begin{array}{c} > \\ = \\ < \end{array} 0 \text{ if p.z.n.c.} \begin{array}{c} > \\ = \\ < \end{array} \text{p.z.n.p.c.} \tag{9}$$

Theorem 2

The p.z.c. will equal the p.z.n.c. if and only if $(\sigma_{IS} + \sigma_{OS}) = 0$ when pH = p.z.n.c.

Theorem 3

If $(\sigma_{IS} + \sigma_{OS})$ decreases (resp. increases), then the p.z.c. decreases (resp. increases).

Theorem 4

If $\partial(q_+ - q_-)/\partial I = 0$ at the p.z.n.c., then p.z.s.e. = p.z.n.c. If $\sigma_o = 0$ as well, then p.z.n.c. = p.z.n.p.c. also. If $(\sigma_{IS} + \sigma_{OS}) = 0$ as well, then p.z.s.e. = p.z.c. also.

Theorem 1 often is applied to measurements of adsorbed "index" ion charge to calculate σ_o.[3] Equation 9 shows that it also can be applied to measurements of the p.z.n.c. and p.z.n.p.c. to determine the sign of σ_o[3] or, if $\sigma_o = 0$, to determine the p.z.n.p.c. by measuring the p.z.n.c.[7] For clay minerals, one expects p.z.n.c. $<$ p.z.n.p.c. ($\sigma_o < 0$), whereas for metal oxides p.z.n.c. $\approx$ p.z.n.p.c. ($\sigma_o \approx 0$).

Theorem 2 is an identifying characteristic of a system in which all adsorbed ions (except H^+ and OH^-) are in the diffuse layer, hence it will be a property of *any* diffuse-layer model, such as modified Gouy-Chapman theory. If adsorbed ions are in surface complexes as well as in the diffuse layer, then p.z.c. = p.z.n.c. requires a precise electrical neutrality among the complexes alone at the p.z.n.c., a condition that could be realized when "index" ions like Na^+ and Cl^- saturate a particle surface to which they are attracted primarily electrostatically with approximately equal affinities.

Theorem 3 demonstrates, *solely* on the basis of charge balance and the stability condition $\partial\sigma_H/\partial pH < 0$,[7] that the formation of surface complexes by *any* mechanism will shift the p.z.c. Surface complexation of anions shifts it downward, whereas surface complexation of cations shifts it upward. Theorem 3 thus can be applied to determine whether a given adsorptive ion forms surface complexes, but it cannot provide details of the mechanism by which the complexes form. Conversely, the fact that a molecular model of adsorption is consistent with Theorem 3 does not constitute evidence for the correctness of the detailed mechanistic hypotheses in the model. One can conclude only that models inconsistent with Theorem 3 are also inconsistent with charge balance and $\partial\sigma_H/\partial pH < 0$. Charge balance as expressed in Equations 2 and 7, along with the stability condition, $\partial\sigma_H/\partial pH < 0$, can be used also to show that p.z.n.p.c. is shifted upward (resp. downward) with increasing cation (resp. anion) adsorption by any mechanism.

Theorem 4 addresses the issue of the relation between p.z.s.e. and other points of zero charge. If the net total adsorbed ion charge (other than that from adsorbed H^+ and OH^-) is invariant under changes in ionic strength at the p.z.n.c, then p.z.s.e. = p.z.n.c. If there is no structural particle charge (e.g., most hydrous oxides, kaolins, and colloidal humus), then p.z.s.e. = p.z.n.p.c. as well. Equality between p.z.s.e. and p.z.c. requires the further condition that electrical neutrality is satisfied among the surface complexes alone. The condition of invariance of the net total (nonprotonic) adsorbed ion charge under changes in ionic strength is essential to these relationships. It makes clear the point that there is no *necessary* condition of vanishing surface charge to be associated with the "crossover point" determined through measurements of σ_H carried out at different background ionic strengths.

3. DETERMINATION OF THE NET TOTAL PARTICLE SURFACE CHARGE DENSITY

The steady migration of charged particles through a dilute suspension to which a uniform, constant electric field is applied has long been investigated as a method to infer the behavior of the net total particle surface charge density (for reviews see Hunter[17] or Harsh and Xu[18]). *Microscopic electrophoresis,*[6] which involves observation of the migration of individual charged particles, is a useful method, even for suspensions of heterogeneous particles, like soil colloids, because of the availability of instrumentation to determine either the average particle velocity by imaging techniques or the distribution of particle velocities by the Doppler shift of scattered light.[17,19] Once values of the migration velocity are determined, the corresponding electrophoretic mobilities in a suspension of charged particles are calculated as the ratio of the (steady) velocity to the strength of the (uniform, constant) applied electric field, with the sign of the ratio taken positive if migration is from a region of high electric potential to a region of low potential.[6] The electrophoretic mobility, u, is thus expressed in the units of square meters per volt per second. Typical values of u for natural colloids lie in the range 10^{-8} to 10^{-7} m^2 V^{-1} s^{-1}.[3,17]

A qualitative connection between u and σ_p can be inferred from the variation of the former parameter with electrolyte composition or concentration, pH, or other chemical variables. As implied in Figure 2 for γ-Al_2O_3, the sign of u is expected to be the same as that of σ_p, and the two parameters are expected to vanish together at the p.z.c. These qualitative concepts can be refined, in principle, by introducing the molecular concept of *electrokinetic potential*, or *zeta potential*, ζ, which is defined to be the inner potential difference across that portion of the electrochemical double layer which does *not* migrate with a particle moving in an applied electric field.[6] The concept of ζ has been discussed at length by Dukhin and Derjaguin,[20] who point out that the precise relation between u and ζ depends sensitively on the detailed molecular structure of the electrochemical double layer around a migrating, charged particle, including not only the adsorbed ion configuration, but also that of the solvent molecules. In the simplest case, one assumes that a solvent boundary layer remains attached to the moving particle. Outside the boundary layer, the solvent is assumed to have its bulk liquid characteristics and the electrochemical double layer is assumed to comprise only a diffuse ion swarm that has the same configuration regardless of whether the particle is under the influence of an applied electric field.[17,20] Under these circumstances, ζ can be identified with an inner potential calculated according to some molecular model of the equilibrium diffuse layer and evaluated at the slip surface ("plane of shear") of the immobile boundary layer.[3,17,20] A variety of recent models of ion configurations in the diffuse layer exists for this purpose,[20,24] each with its own set of hypotheses about particle and boundary layer shape, diffuse-layer properties, and solvent characteristics. Once a model is selected and applied to compute values of ζ from measurements of u, an appropriate form of the Gauss law then can be invoked to calculate a corresponding value of the surface charge density at the slip surface.[17] The connection between this charge density and σ_p, of course, still requires molecular details on how the slip surface is conceived relative to the outer periphery of the Stern layer.[20] A popular (but unproven) assumption has been to consider the two as congruent.[17] It should be evident, even from this brief discussion, that the quantitative relationship between u and σ_p is highly model-dependent. Therefore, it is essential to indicate in all cases wherein mobilities are converted to surface charge densities the model expressions and hypotheses used to develop this relationship.[20]

The ambiguity (i.e., sensitive model dependence) associated with the inference of σ_p from measurements of the electrophoretic mobility suggests that the development of methods based on direct quantitation of the component terms on the right side of Equation 4 would be worthwhile. A method of this type has been used by Charlet et al.[25] to determine σ_p for carbonate solids using a flow-through titration reactor. In their study, σ_o was assumed to be zero, and the ionic strength in the particle suspension, controlled by NaCl, was assumed to be high enough to ignore contributions to σ_d other than those from Na^+ and Cl^-. Under these assumptions, a calculation of the net total charge concentration in the aqueous phase of the suspension both just before and after its reaction with the solid phase permitted an inference of changes in σ_p from the changes in net total charge concentration for the aqueous ionic species assumed to form surface complexes (H^+, OH^-, HCO_3^-, etc.). The changes in σ_p so computed were converted to absolute values by assuming that $\sigma_p = 0$ under the equilibrium conditions established just before the solid adsorbent was titrated.[25]

4. DETERMINATION OF INTRINSIC SURFACE CHARGE DENSITY

A classical technique for measuring the intrinsic surface charge density on a particle is the *Schofield method*.[26-28] In this method, particles are reacted with an electrolyte solution (e.g., NaCl or CsCl) at a given pH value and ionic strength; the specific surface excess of the cation and the anion adsorbed from the electrolyte is determined (see, e.g., Sposito[9] for a description of the method of measurement); and the value of σ_{in} is calculated with Equation 7. Often the factor F/a_s is deleted from Equation 7 to leave σ_{in} in the convenient units of moles of charge per kilogram when the specific surface area is not readily measured.

The surface-chemical interpretation of σ_{in} as measured by the Schofield method depends on the type and concentration of electrolyte used. If the cation in the reacting electrolyte neutralizes precisely the exposed functional group charge associated with isomorphic substitutions and dissociated hydroxyls, and the anion neutralizes only exposed protonated functional groups, then q_+ and q_- will have "optimal" magnitudes, for a given pH value, and the measured σ_{in} will be truly an intrinsic surface charge density. On the other hand, if the cation in the probe electrolyte is not able to displace all of the native adsorbed cations (e.g., those in inner-sphere surface complexes), or if the anion cannot displace all of the native anions bound to protonated functional groups, then σ_{in} will differ from its "optimal" value.

Thus, the intrinsic surface charge density, viewed operationally, can exhibit different values for the same particle at a given pH value. The "optimal" σ_{in} represents the difference between the largest quantities of positive and negative adsorbed ion charge achievable by reaction with exposed surface functional groups. Conditional values of σ_{in} fall into the broad spectrum of possible differences between the amounts of adsorbed positive and negative charge that can be brought to particle surfaces by cations and anions of varying chemical characteristics. If they are positive, these conditional values of q_+ or q_- usually are termed *cation* or *anion exchange capacities*, respectively.[3] They reflect only the reactivities of the chosen probe ions with particle surfaces under prescribed chemical conditions. If these conditions simulate those of interest in natural colloidal systems, however, the resulting σ_{in} may be of significant practical utility.

5. DETERMINATION OF NET STRUCTURAL SURFACE CHARGE
5.1 Nonspecific Methods

Surface-reactive particles comprising layer-type aluminosilicates (clay minerals) are almost the only ones that bear net structural surface charge among the natural colloids.[3] Organic particles never exhibit structural charge and metal oxides in nature show very small σ_o values.[3,9] In principle σ_o can be determined by doing a charge balance calculation with exhaustive chemical composition data for a particle sample, including its surface species.[3,29,30] In practice, this method is difficult to apply unless the particle sample comprises a single mineral comparable to a geological specimen.

Theorem 1 of the PZC Theorems shows that measurements of the net total adsorbed ion charge can be used to calculate σ_o if the net proton surface charge density, σ_H, is zero, or at least negligible when compared to σ_o. Procedures for these kinds of measurements with common electrolytes have been described by Rhoades[31] in the context of cation exchange capacity determinations for soil particles. Fundamentally, an "index" strong electrolyte (e.g., NaCl or $BaCl_2$) is reacted with the particles repeatedly to replace all adsorbed ionic species by the ions of the index electrolyte. These ions then are extracted by another strong electrolyte [e.g., $Mg(NO_3)_2$] and quantitated in order to calculate their surface excess. The surface excess is converted to a specific adsorbed ion charge and σ_o is calculated with Equation 8, either with or without the factor F/a_s.

Rhoades[31] has discussed several operational sources of error in the "index" electrolyte method of measuring σ_o: (1) incomplete replacement of the adsorbed species indigenous to the particles; (2) dissolution or precipitation of solid phases containing adsorptive ions; (3) loss of either the adsorbed "index" ions or the colloid, or excess retention of the "index" electrolyte after the replacement step; and (4) incomplete extraction of adsorbed "index" ions or dissolution-precipitation reactions induced by the second strong electrolyte. These sources of error can be minimized by careful choices of both the "index" and extracting electrolyte, and by careful experimental protocol. Ideally, the electrolyte choice will be predicated on some prior information about the natural surface speciation of the particles.

5.2 Alkylammonium Cation Exchange

The structural charge on 2:1 layer type aluminosilicates is localized to some extent in or near the cavities formed by the roughly hexagonal rings of oxygen

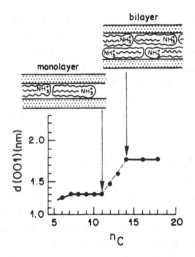

Figure 3. Basal-plane [d(001)] spacings of a smectite intercalated by n-alkylammonium cations, showing the monolayer-bilayer transition.

ions that make up their siloxane surfaces.[3] These cavities are known to form inner-sphere complexes with n-alkylammonium cations ($C_nH_{2n+1}NH_3^+$) and this phenomenon has been applied to develop a method to measure σ_o.[32] In outline, the method consists of reacting Na-saturated particles with a series of n-alkylamine hydrochloride solutions at 65°C, washing the reacted particles with ethanol to remove excess, non-adsorbed electrolyte, and determining the basal plane [d(001)] spacing by X-ray diffraction after drying under vacuum. The basal plane spacing is then plotted against n_c, the number of carbon atoms in the n-alkylammonium cation used in an adsorption experiment, with the entire series of experiments included in the range $1 \leq n_c \leq 20$. When a statistical monolayer of complexed alkylammonium cations lies in the interlayers of a clay mineral, the basal plane spacing is 1.36 nm and, when a bilayer is present, it is 1.77 nm (Figure 3). A simple procedure for preparing n-alkylamine hydrochlorides in crystalline form that is in widespread use was published by Ruehlicke and Kohler,[33] who also recommended a technique for preparing surface complexes of a given n-alkylammonium cation with 2:1 layer type clay minerals, such as smectite, vermiculite, or illite, after they have been Na-saturated.

When n-alkylammonium cations form the ideal 1.36-nm monolayer structure, they lie flat at between opposing siloxane surfaces and each cation is assumed to cover a Van der Waals packing area equal to $(0.057n_c + 0.14)$ nm². Since the cation is univalent, this packing area is associated with one surface electron charge. As the value of n_c increases, the area required by an adsorbed cation increases. At some n_c value, the area required becomes larger than the area per electron charge on a siloxane surface and the monolayer structure is no longer stable. The 1.77-nm bilayer structure then becomes favored because the area requirement of the adsorbed organic cation then can be met independently on each opposing basal plane in the interlayer region (Figure 3). At the monolayer-bilayer transition, the packing area per adsorbed cation just equals the area of one unit cell on a siloxane surface, a, divided by the moles of structural charge per unit cell, x. Therefore, at the transition point,[32]

$$0.057n_c + 0.14 = \frac{a}{x} \tag{10}$$

and

$$x = \frac{a}{0.057n_c + 0.14} \tag{11}$$

where x is termed the *layer charge* of the clay mineral.[1] With the values of x characteristic of 2:1 clay minerals ($0.5 < x < 2.0$) and a typical value of $0.46 \ nm^2$ per unit cell for a,[3] the monolayer-bilayer transition is expected for $4 \leq n_c \leq 14$ in smectite and $2 \leq n_c \leq 4$ in vermiculite. The layer charge of illite ($x > 1.4$) is usually too large for the monolayer-bilayer transition to be observed at any n_c. Often, as in Figure 3, the transition does not take place sharply at a single value of n_c, but instead ranges over several values.[34,35] This gradual transition is interpreted to mean that the layer charge does not have a unique value; i.e., there is layer charge heterogeneity. In this case, the layer charge distribution can be estimated with the help of the theory of X-ray diffraction from randomly interstratified layer silicates.[32]

The alkylammonium cation exchange method has been applied widely both to specimen clay minerals and to those extracted from heterogeneous weathering environments (for reviews, see Lagaly[32] and Laird et al.[30,36]). Lagaly et al.[35] have noted that ambiguities in the interpretation of data like those in Figure 3 can arise from variability in the detailed configurations of monolayers or bilayers, variability in the values of x among different unit cells, and variability in adsorbent particle size. Laird et al.[30] have suggested that variability in the conformation of adsorbed alkylammonium cations can produce deviations of up to 30% between values of the layer charge calculated with Equation 11 and those calculated with chemical composition data. Laird et al.[36] also demonstrated that alkylammonium cations can replace K^+ bound in inner-sphere complexes on weathered mica, with the result that diagnostic X-ray diffraction patterns are difficult to reproduce when the method is applied to a mixture of 2:1 layer type silicates. Senkayi et al.[37] also found these mixtures yield poorly defined monolayer-bilayer transitions.

6. DETERMINATION OF THE NET PROTON SURFACE CHARGE
6.1 Electrometric Titration

The importance of the proton as a constituent of both surface-reactive solids and natural waters has led to many published measurements of the net proton surface charge density, σ_H (see reviews by Huang,[38] Stumm and Morgan,[14] James and Parks,[39] and Sposito[3]). Typically σ_H is determined for aqueous particle suspensions by electrometric titration as a function of pH under chosen conditions of ionic strength, background electrolyte concentration and composition, and particle suspension density, at fixed temperature and pressure. A common approach involves the use of a glass electrode and a double-junction reference electrode in the titration cell:

glass electrode	suspension of solid particles in background electrolyte solution	background electrolyte solution	liquid junctions	reference electrode

The emf of the electrode assembly is measured while known volumes of either strong acid or strong base are added to the suspension. These data, in turn, are converted to proton *concentrations* with the help of a calibration curve prepared from similar titration data obtained without the suspended particles in the cell. Values of the principal experimental parameter, the *apparent net proton surface excess*, δn_H, measured in moles, are calculated with the equation:[3,14,38]

$$\delta n_H = (c_A - c_B + [OH^-] - [H^+]) \, V \tag{12}$$

where c_A is the concentration of strong acid added, c_B is the concentration of strong base added, $[OH^-]$ and $[H^+]$ are free hydroxide ion and proton concentrations, respectively, and V is the suspension volume. The *apparent net proton surface charge density* is then:[14]

$$\delta\sigma_{H,titr} = \delta n_H F/ma_s \tag{13}$$

where m is the mass of particles in the suspension volume. The experimental parameter $\delta\sigma_{H,titr}$ has the units of coulombs per square meter, but these units often are replaced by moles per square meter (surface excess concentration) after deletion of F, or by moles per kilogram (specific surface excess) after deletion of F/a_s from Equation 13.

As written in Equation 12, δn_H is the net proton surface excess (surface excess of H^+ less that of OH^-) *relative* to the net proton surface excess that existed *before* either strong acid or base was added to the particle suspension. The initial proton surface excess usually is not known experimentally,[16] so δn_H must be renormalized by adding (algebraically) to it some reference value that is accessible to experimental determination. For example, at very low pH, it might be possible to set δn_H equal to a maximum positive value. In practice, unless a well-defined plateau appears in δn_H as pH decreases and the number of protonatable, complexing surface functional groups is known, this kind of renormalization procedure is not possible. Alternatively, it might be possible to determine the p.z.n.p.c. using one of the PZC Theorems. In this case, an "absolute" value of the net proton surface charge density, σ_H, would be calculated with the expression (cf. Charlet and Sposito[40]):

$$\sigma_H = \delta\sigma_{H,titr} - \delta\sigma_{H,titr} \ (pH = p.z.n.p.c.) \tag{14}$$

where the second term on the right side is the apparent net proton surface charge density at the p.z.n.p.c.

It is an unfortunate matter of common practice that $\delta\sigma_{H,titr}$ in Equation 13 is often equated with σ_H, which is correct only if the net proton surface charge density happens to be zero when an electrometric titration is begun. Equally problematic is the procedure of measuring $\delta\sigma_{H,titr}$ as a function of pH at two or more different background electrolyte ionic strengths, then setting it equal to zero at the p.z.s.e., i.e., the point of intersection of the $\delta\sigma_{H,titr}$ vs. pH curves.[16,38] If the value of $\delta\sigma_{H,titr}$ at the p.z.n.p.c. depends on ionic strength, it follows that the true point of intersection of the σ_H vs. pH curves will not be the same as that of the $\delta\sigma_{H,titr}$ vs. pH curves. Even if it were the same, equating the p.z.s.e. with the p.z.n.p.c. requires either special assumptions about the speciation of particle surfaces[16,38] or additional measurements of the components of particle surface charge.[7] At present, the only unambiguous operational method to identify the p.z.n.p.c. applies to particles whose net structural charge density is zero. In this case ($\sigma_o = 0$), as shown in Theorem 1 of the PZC Theorems,[7,40] the p.z.n.p.c. ($\sigma_H = 0$) is equal to the p.z.n.c. ($\sigma_{in} = 0$), which can be measured independently of a titration experiment by the Schofield method.

6.2 Side-Reaction Interferences

6.2.1 Aqueous Phase

Implicit in the use of δn_H to calculate $\delta\sigma_{H,titr}$ is the critical assumption that the only protons consumed in a suspension upon addition of strong acid or base are those that have reacted with solid particles to form surface complexes. This assumption cannot be true, among other things, unless δn_H has been corrected for side-reactions of the added protons or hydroxide ions with dissolved chemical species. A correction of δn_H for aqueous-phase side re-actions can be performed operationally following a procedure suggested by Huang.[38] A suspension of the particles to be titrated is equilibrated, then centrifuged (or filtered) to provide a clear supernatant solution, which is then titrated in the same way as the particle suspension. Each value of δn_H for the solution is calculated with Equation 12 and subtracted (algebraically) from the δn_H value for the suspension at the same proton concentration. This method requires the titration equilibration times for the suspension and its supernatant solution to be the same and depends on the insignificance of processes that can alter the aqueous phase composition during the titration equilibration period (e.g., particle dissolution or gas diffusion), a problem more serious for discontinuous titrations than for rapid, continuous titrations.[38] Changes in aqueous phase composition can be monitored during a titration by multielemental analysis (e.g., with inductively coupled plasma emission spectrometry) of the supernatant solution in sequentially withdrawn aliquots of the suspension. If solution compositional changes are found to be significant, aliquot supernatant solutions can be titrated individually to provide a sequential correction to δn_H.

6.2.2 Solid Phase

Side-reaction interferences directly involving the solid particles being titrated are potentially more important than those involving solely the aqueous phase. Most obvious are particle dissolution-precipitation reactions, H^+ or OH^- adsorption in the diffuse layer, and flocculation-dispersion (or conformation change) processes. Parker et al.,[15] Huang,[38] and Bales and Morgan[41] have studied the first problem for suspensions of soil, metal oxide, and aluminosilicate particles, respectively. They point out that proton reactions with dissolving natural particles can contribute spuriously to δn_H, and that this contribution will depend on the interplay between pH, dissolution kinetics, and the titration equilibration time. A similar statement can be made about pH-induced precipitation. Since σ_H conventionally should include only protons or hydroxide ions bound in surface complexes, contributions to δn_H from H^+ or OH^- adsorbed in the diffuse layer will produce interpretive errors in σ_H based on Equation 12. It is conceivable also that surface functional groups involved in strong, inner-sphere surface complexes with metal ions could resist protonation during the titration equilibration period, such that δn_H is underestimated if the number of these complexes is large before a titration begins (depending on the surface chemical history of the particles). These and related problems with solid-phase side-reactions led Parker et al.[15] to recommend that data from titration measurements of σ_H for heterogeneous soil particles should be "regarded with considerable caution." Charlet and Sposito[40] noted that the range of pH over which σ_H can be measured accurately often will be bracketed between the highest pH value at which proton-induced particle dissolution (or disaggregation) is significant and the lowest pH value at which hydroxide-induced disaggregation (or dissolution) of composite heterogeneous particles occurs. None of the solid-phase side-reaction problems is as serious for synthetic, well-characterized, homogeneous particles (e.g., synthetic metal oxides, silica, or clay minerals) as it is for natural, heterogeneous particles.

6.3 Electrometric Interferences

Prototypical electrometric titration measurements of σ_H with a glass electrode have been described by Breeuwsma and Lyklema,[42] Yates and Healy,[43] Stumm and Morgan,[14] Bales and Morgan,[41] Lövgren et al.,[44] and Charlet et al.[25] Two fundamental questions that arise in connection with the electrometric measurement of σ_H are (1) Does the electrode assembly respond significantly to dissolved or solid constituents in the titrated suspension that were not in the aqueous system used to calibrate the glass electrode? (2) Does the glass electrode respond to diffuse-layer protons as well as "bulk" solution protons?

In respect to question (1), the answer is a qualified "no", if the glass electrode is calibrated *in terms of* $[H^+]$ by Gran titration[45-47] of a solution whose ionic strength is controlled by a "swamping" background electrolyte (e.g., KCl) that also controls the ionic strength in the suspension, and if the reference electrode is immersed only in an overlying supernatant solution of the suspension during titration.[44] With this method, liquid junction potentials created by particles, by dissolved species not in the calibrating solution, or by the added protons or hydroxide ions in the suspension should be minimized and the *suspension effect*[5,48,49] should be obviated. If pH buffer solutions are used instead to calibrate a glass electrode in the presence of liquid junctions, or even if a reversible electrode assembly is used with account taken of the actual ionic strength of the supernatant solution, there is still a need to convert pH or emf values to $[H^+]$ by calculation, which cannot be done unambiguously.[50] Moreover, under the best circumstances, electrometric pH cannot be measured more *accurately* than ± 0.05,[50] with a corresponding, amplified uncertainty in the calculated $[H^+]$, especially at pH ≤ 4. On the other hand, the use of a "swamping" background electrolyte, like KCl, may produce σ_H values that bear little relevance to the net proton surface charge density developed by the particles of interest in a natural aqueous environment.

In respect to question (2), there appears to be no unambiguous answer because the question itself is ill-posed. Since diffuse-layer protons are relatively mobile aqueous species, it is reasonable to suppose that they can contribute to the emf developed by a glass electrode/reference electrode pair and, therefore, that $[H^+]$ in Equation 12, when determined electrometrically, will include a contribution from diffuse-layer protons. If this is true, δn_H will be linearly related to the particle charge created by surface-complexed protons, as required by the definition of σ_H. But how is this conclusion to be reconciled with a common observation, viz., that no difference in emf is found by moving a glass electrode from a suspension to its overlying supernatant solution while keeping the reference electrode fixed?[51,52] This result could be interpreted as evidence either that diffuse-layer protons are typically of negligible concentration or that a glass electrode does not in fact respond to diffuse-layer protons, present only in the suspension. That both of these interpretations are inappropriate can be appreciated after noting that electrochemical equilibrium of the proton between any two phases must result in zero emf across a pair of glass electrodes inserted in the phases.[3] An emf measurement alone cannot be used to resolve a question about the molecular distribution of a chemical species.

7. CONCLUSIONS

Particle surface charge can be characterized conceptually by generic definitions of surface chemical species and by the law of surface charge conservation. The structures of solid adsorbents and their reactions with aqueous species lead to the definition of five principal components of surface charge density: structural, net proton, inner-sphere complex, outer-sphere complex, and diffuse-layer, the sum of which must equal zero to fulfill surface charge balance. This latter condition and the stability criterion, $\partial\sigma_H/\partial pH < 0$, where σ_H is net proton surface charge density, can be applied in various ways to derive rigorous results concerning "points of zero charge", pH values at which surface charge components vanish (or show invariance properties). These results, the PZC Theorems, suggest that the point of zero net charge (p.z.n.c.) is the most accessible experimentally and the most versatile conceptually.

The measurement of the five component surface charge densities is a difficult issue on which research must continue, especially for heterogeneous natural particles. Most experimental methods involve the reaction of charged particles with aqueous species whose ultimate surface speciation is presumed known and, therefore, whose adsorption reactions can be used to help quantitate one or more components of the particle surface charge. Difficulties with the interpretation of the results of existing methods arise because of side-reactions of the probe adsorptive or the adsorbent, uncertainties in the surface speciation of the adsorbate, and ambiguities in the method used to measure the surface excess. At present, no method of quantifying natural particle surface charge has universal applicability.

ACKNOWLEDGMENTS

The undertaking of this review was supported in part by U.S. National Science Foundation grant no. EAR-8915291. Gratitude is expressed to Dr. S. J. Anderson, Dr. J. Buffle, Dr. L. Charlet, Dr. L. K. Koopal, and Dr. H. P. van Leeuwen for critical reviews that led to significant improvement in the technical content of this chapter. Thanks to Ms. Joan Van Horn for her excellent typing of the manuscript, to Mr. Frank Murillo for drawing Figures 1 and 3, and to Ms. Linda Bobbitt for drawing Figure 2.

GLOSSARY

N.B. Key terminology not found in this list has been defined previously by Everett.[6]

> *Apparent net proton surface charge density*: the net proton surface charge density calculated directly from electrometric proton titration data for an adsorbent, without knowledge of its point of zero net proton charge or the existence of interfering side-reactions
>
> *Apparent net proton surface excess*: a net proton surface excess calculated directly from electrometric proton titration data for an adsorbent, without knowledge of its point of zero net proton charge or the existence of interfering side-reactions
>
> *Counterion condensation*: the close association of diffuse-layer counterions with an adsorbent surface, such that a non-zero fraction of the counterions remains within a finite distance of the adsorbent

surface even when the concentration of background electrolyte solution approaches zero (infinite-dilution limit)

Diffuse-layer surface charge density: the net surface charge density created by adsorbed ions in a diffuse layer

Inner-sphere complex surface charge density: the net charge, per unit area of adsorbent, of ions bound in inner-sphere surface complexes, excepting protons and hydroxide ions

Inner-sphere surface complex: an adsorption complex in which there is no water molecule interposed between the surface functional group and the ion or molecule it binds

Intrinsic surface charge density: the sum of net permanent and net proton surface charge densities

Ion exchange capacity: a conditional specific adsorbed cation or anion charge

Layer charge: the net permanent surface charge density of a layer type aluminosilicate, expressed in units of moles of charge per mole of unit cells

Net permanent surface charge density: surface charge density created by isomorphic substitutions of ions in the structure of an adsorbent

Net proton surface charge density: surface charge density created by the formation of adsorption complexes involving protons or hydroxide ions as the adsorbate; mathematically, it is the difference between the surface excess of protons and that of hydroxide ions in adsorption complexes

Net total particle surface charge density: the sum of intrinsic surface charge density and Stern layer surface charge density

Outer-sphere complex surface charge density: the net charge, per unit area of adsorbent, of ions bound in outer-sphere surface complexes, excepting protons and hydroxide ions

Outer-sphere surface complex: an adsorption complex in which at least one water molecule is interposed between the surface functional group and the ion or molecule it binds

PZC theorems: mathematical relationships among different points of zero charge based solely on the law of charge conservation and on the general variation of net proton surface charge density with pH

Point of zero charge (generic): values of pH at which one or more surface charge density components equal zero at a given temperature, pressure, and aqueous solution composition

Point of zero net charge: the pH value at which the intrinsic surface charge density is zero

Point of zero net proton charge: the pH value at which the net proton surface charge density is zero

Point of zero salt effect: the pH value at which the net proton surface charge density is invariant under changes of ionic strength

Schofield method: an experimental procedure for measuring intrinsic
surface charge density by adsorption of the ions in a chosen "index"
electrolyte solution
Specific adsorbed ion charge: the product of the valence of an ion
and its surface excess, divided by the mass of the adsorbent
Surface functional group: a chemically reactive molecular unit bound
into the structure of an adsorbent at its periphery

REFERENCES

1. Tessier, A. "Sorption of Trace Elements in Natural Waters," in *Characteri-zation of Environmental Particles*, J. Buffle and H. P. van Leeuwen, Eds., IUPAC Environmental Analytical Chemistry Series, Vol. I (Chelsea, MI: Lewis Publishers, 1991), chap. 11.
2. Honeyman, B. D. and P. H. Santschi. "The Role of Particles and Colloids in the Transport of Radionuclides and Trace Metals in the Oceans," in *Charac-terization of Environmental Particles*, J. Buffle and H. P. van Leeuwen, Eds., IUPAC Environmental Analytical Chemistry Series, Vol. I (Chelsea, MI: Lewis Publishers, 1991), chap. 10.
3. Sposito, G. *The Surface Chemistry of Soils*. (New York: Oxford University Press, 1984).
4. Stumm, W., Ed. *Aquatic Surface Chemistry*. (New York: John Wiley & Sons, 1987).
5. Hochella, Jr., M. F. and A. F. White, Eds. *Mineral-Water Interface Geo-chemistry*. (Washington, D.C.: Mineralogical Society of America, 1990).
6. Everett, D. H. "Definitions, Terminology and Symbols in Colloid and Surface Chemistry," *Pure Applied Chem.* 31:578–638 (1972).
7. Sposito, G. "Surface Reactions in Natural Aqueous Colloidal Systems," *Chimia* 43:169–176 (1989).
8. Sposito, G. "The Operational Definition of the Zero Point of Charge in Soils," *Soil Sci. Soc. Am. J.* 45:292–297 (1981).
9. Sposito, G. *The Chemistry of Soils*. (New York: Oxford University Press, 1989).
10. Brown, G., A. C. D. Newman, J. H. Rayner, and A. H. Weir. "The Structures and Chemistry of Soil Clay Minerals," in *The Chemistry of Soil Constituents*, D. J. Greenland and M. H. B. Hayes, Eds. (Chichester, U.K.: John Wiley & Sons, 1978), pp. 29–178.
11. Manning, G. S. "Counterion Binding in Polyelectrolyte Theory," *Acc. Chem. Res.* 12:443–449 (1979).
12. Zimm, B. H. and M. Le Bret. "Counter-ion Condensation and System Di-mensionality," *J. Biomol. Struct. Dyn.* 1:461–471 (1983).
13. Greenland, D. J. and C. J. B. Mott. "Surfaces of Soil Particles," in *The Chemistry of Soil Constituents*, D. J. Greenland and M. H. B Hayes, Eds. (Chichester, U.K.: John Wiley & Sons, 1978), pp. 321–353.
14. Stumm, W. and J. J. Morgan. *Aquatic Chemistry*, 2nd ed. (New York: John Wiley & Sons, 1981).
15. Parker, J. C., L. W. Zelazny, S. Sampath, and W. G. Harris. "Critical Eval-uation of the Extension of Zero Point of Charge (ZPC) Theory to Soil Systems," *Soil Sci. Soc. Am. J.* 43:668–673 (1979).

16. Lyklema, J. "Points of Zero Charge in the Presence of Specific Adsorption," *J. Colloid Interface Sci.* 99:109–117 (1984).

17. Hunter, R. J. *Zeta Potential in Colloid Science* (London: Academic Press, 1981).

18. Harsh, J. B. and S. Xu. "Microelectrophoresis Applied to the Surface Chemistry of Clay Minerals," *Advan. Soil Sci.* 14:131–165 (1990).

19. Rees, T. F. "A Review of Light-Scattering Techniques for the Study of Colloids in Natural Waters," *J. Contaminant Hydrol.* 1:425–439 (1987).

20. Dukhin, S. S. and B. V. Derjaguin. "Electrokinetic Phenomena," *Surface Colloid Sci.* 7:1–356 (1974).

21. O'Brien, R. W. and R. J. Hunter. "The Electrophoretic Mobility of Large Colloidal Particles," *Can. J. Chem.* 59:1878–1887 (1981).

22. Anderson, J. L. "Effect of Nonuniform Zeta Potential on Particle Movement in Electric Fields," *J. Colloid Interface Sci.* 105:45–54 (1985).

23. O'Brien, R. W. and D. N. Ward. "The Electrophoresis of a Spheroid with a Thin Double Layer," *J. Colloid Interface Sci.* 121:402–413 (1988).

24. Fair, M. C. and J. L. Anderson. "Electrophoresis of Nonuniformly Charged Ellipsoidal Particles," *J. Colloid Interface Sci.* 127:388–400 (1989).

25. Charlet, L., P. Wersin, and W. Stumm. "Surface Charge of $MnCO_3$ and $FeCO_3$," *Geochim. Cosmochim. Acta* 54:2329–2336 (1990).

26. Schofield, R. K. "Effect of pH on Electric Charges Carried by Clay Particles," *J. Soil Sci.* 1:1–8 (1949).

27. van Raij, B. and M. Peech. "Electrochemical Properties of Some Oxisols and Alfisols of the Tropics," *Soil Sci. Soc. Am. J.* 36:587–593 (1972).

28. Greenland, D. J. "Determination of pH Dependent Charges of Clays Using Caesium Chloride and X-ray Fluorescence Spectrography," *Trans. 10th Int. Congr. Soil Sci. (Moscow)* 2:278–285 (1974).

29. Greenland, D. J. and M. H. B. Hayes, Eds. *The Chemistry of Soil Constituents.* (Chichester, U.K.: John Wiley & Sons, 1978), pp. 54–58.

30. Laird, D. A., A. D. Scott, and T. E. Fenton. "Evaluation of the Alkylammonium Method of Determining Layer Charge," *Clays Clay Miner.* 37:41–46 (1989).

31. Rhoades, J. "Cation Exchange Capacity," in *Methods of Soil Analysis,* 2nd ed., A. L. Page, R. H. Miller, and D. R. Keeney, Eds. (Madison, WI: American Society of Agronomy, 1982), pp. 149–157.

32. Lagaly, G. "Characterization of Clays by Organic Compounds," *Clay Miner.* 16:1–21 (1981).

33. Ruehlicke, G. and E. E. Kohler. "A Simplified Procedure for Determining Layer Charge by the N-Alkylammonium Method," *Clay Miner.* 16:305–307 (1981).

34. Stul, M. S. and W. J. Mortier. "The Heterogeneity of the Charge Density in Montmorillonites," *Clays Clay Miner.* 22:391–396 (1974).

35. Lagaly, G., M. Fernández González, and A. Weiss. "Problems in Layer-Charge Determination of Montmorillonites," *Clay Miner.* 11:173–187 (1976).

36. Laird, D. A., A. D. Scott, and T. E. Fenton, "Interpretation of Alkylammonium Characterization of Soil Clays," *Soil Sci. Soc. Am. J.* 51:1659–1663 (1987).

37. Senkayi, A. L., J. B. Dixon, L. R. Hossner, and L. A. Kippenberger. "Layer Charge Evaluation of Expandable Soil Clays by an Alkylammonium Method," *Soil Sci. Soc. Am. J.* 49:1054–1060 (1985).

38. Huang, C.-P. "The Surface Acidity of Hydrous Solids," in *Adsorption of Inorganics at Solid-Liquid Interfaces*, M. A. Anderson and A. J. Rubin, Eds. (Ann Arbor, MI: Ann Arbor Science, 1981), pp. 183–217.

39. James, R. O. and G. A. Parks. "Characterization of Aqueous Colloids by their Electrical Double-Layer and Intrinsic Surface Chemical Properties," *Surf. Colloid Sci.* 12:119–216 (1982).

40. Charlet, L. and G. Sposito. "Monovalent Ion Adsorption by an Oxisol," *Soil Sci. Soc. Am. J.* 51:1155–1160 (1987).

41. Bales, R. C. and J. J. Morgan. "Surface Charge and Adsorption Properties of Chrysotile Asbestos in Natural Waters," *Environ. Sci. Technol.* 19:1213–1219 (1985).

42. Breeuwsma, A. and J. Lyklema. "Interfacial Electrochemistry of Haematite (α-Fe_2O_3)," *Disc. Faraday Soc.* 52:324–333 (1971).

43. Yates, D. E. and T. W. Healy, "Titanium Dioxide — Electrolyte Interface. II. Surface Charge (Titration) Studies." *J. Chem. Soc. Faraday I* 76:9–18 (1980).

44. Lövgren, L., S. Sjöberg, and P. W. Schindler. "Acid/Base Reactions and Al(III) Complexation at the Surface of Goethite," *Geochim. Cosmochim. Acta* 54:1301–1306 (1990).

45. Gran, G. "Determination of the Equivalence Point in Potentiometric Titrations. II," *Analyst* 77:661–671 (1952).

46. Ingman, F. and E. Still. "Graphic Method for the Determination of Titration End-Points," *Talanta* 13:1431–1442 (1966).

47. Pehrsson, L., F. Ingman, and A. Johansson. "Acid-Base Titrations by Stepwise Additions of Equal Volumes of Titrant with Special Reference to Automatic Titrations," *Talanta* 23:769–788 (1976).

48. Babcock, K. L. and R. Overstreet. "On the Use of Calomel Half Cells to Measure Donnan Potentials," *Science* 117:686–687 (1953).

49. Sposito. G. *The Thermodynamics of Soil Solutions.* (Oxford, U.K.: Clarendon Press. 1981), pp. 123–124.

50. Bates, R. G. "The Modern Meaning of pH," *Crit. Rev. Anal. Chem.* 10:247–278 (1981).

51. Jenny, H., T. R. Nielsen, N. T. Coleman, and D. E. Williams. "Concerning the Measurement of pH, Ion Activities, and Membrane Potentials in Colloidal Systems," *Science* 112:164–167 (1950).

52. Bates, R. G. *Determination of pH*, 2nd ed. (New York: John Wiley & Sons, 1973), pp. 323–324.

CHAPTER 8

Iron Particles in Freshwater

William Davison

Institute of Environmental and Biological Sciences, Lancaster University, Lancaster, United Kingdom

and

Richard De Vitre

Department of Inorganic, Analytical, and Applied Chemistry, Sciences II, Geneva, Switzerland

TABLE OF CONTENTS

1.   Introduction 316

2.   Iron Particles Formed by Weathering 317

3.   Fe(III) Oxyhydroxide Particles Formed in Water Bodies 320
      3.1   Origin, Formation, and Fate 320
      3.2   Oxidation/Hydrolysis Reactions 328
             3.2.1 Rate of Oxidation 328

315

            3.2.2   Structure and Nature of Oxidation
                    Products 330
      3.3   Reduction/Dissolution Reactions 335
            3.3.1   Theoretical Aspects of Dissolution
                    Reactions 335
            3.3.2   Laboratory Studies..............................336
                    3.3.2.1   Nonreductive Dissolution 336
                    3.3.2.2   Reductive Dissolution 338
            3.3.3   Reductive Dissolution Under Natural
                    Conditions 339

4.   Other Iron Particles Formed in Freshwaters.....................340

5.   Sampling of Iron Particulates 344
      5.1   Sampling of Iron Particles in the Water Column 344
      5.2   Sampling of Iron Particles from Sediments..............346

References...347

1. INTRODUCTION

Most freshwaters receive a plentiful supply of iron, as it is one of the most abundant elements in the earth's crust. Its perceived importance is attributable to its early use as a flocculant in water treatment, and to its biogeochemical interactions, which have so intrigued scientists that it must be a contender for the most studied inorganic element in freshwaters. Fe(III), which is stable in the presence of oxygen, is hydrolyzed to insoluble oxyhydroxides, whereas Fe(II), which is stable in the absence of oxygen, is soluble and relatively free from complexation. This ready interchange between these two oxidation states of iron leads to the production and removal of particles in various freshwater environments. These processes of interconversion, which are highly dependent on pH, are influenced or controlled by microorganisms. This link with the biota is compounded by the interaction of both Fe(III) and Fe(II) with phosphate, and hence the mediation of the iron cycle in the biological productivity of a water. Moreover, iron is an essential element for most living organisms and is involved in many important metabolic processes. Naturally occurring iron particles are by no means simple. They are often present in association with bacteria and other mineral particles, and their surface properties are affected by the adsorption of humic substances.

The basic biological and chemical interactions of iron in freshwaters have been known for many years.[1,2] In the past decade, however, new discoveries and advances have been made. Many studies have contributed to understanding the chemical and microbiological mechanisms of reduction and oxidation, and to elucidating the role of light and organic material on these processes. There is now a better appreciation of sediment-water interactions and in-lake processes, but our knowledge of particles themselves has perhaps increased the most. A range of techniques, including electron microscopy, ultrafiltration, and polarography, have been used to show that in natural freshwaters there is a complete spectrum of sizes of iron particles from molecular sizes upwards, and surface chemical studies have elucidated the role of iron oxyhydroxides in controlling the concentration of various trace components.

This review is concerned with the formation, characteristics, and role of iron particles in freshwaters. It examines recent developments in the context of the firm foundations of the past and shows how understanding of iron particles and their interactions is at the forefront of developments in aquatic science.

2. IRON PARTICLES FORMED BY WEATHERING

Iron occurs in a wide variety of rocks, as a major component of discrete minerals such as pyrite, FeS_2, as a substitution element in alumino-silicates, and as a minor constituent of sedimentary rocks such as sandstone and limestone. Consequently it has a high mean crustal abundance of 4.1% and accounts for 4.8% of the suspended load in rivers.[3] Weathering processes, which dissolve or abrade rocks by physical, chemical, or biological actions, are responsible for the transfer of iron to rivers. The ratios of iron concentrations in river waters to its concentrations in rocks are similar to those for aluminum, and generally lower than for other elements. This relative immobility of iron can be attributed to its extreme insolubility in oxidizing conditions at pH >4. Weathering processes undoubtedly involve the dissolution of Fe(II),[4] particularly in local reducing environments, such as in soils, but once this iron is transported into an oxidized region it is rapidly oxidized in neutral solutions to insoluble oxyhydroxides. Humic material in soils is responsible for stabilizing oxidized iron as small colloidal particles which may contribute to the dissolved fraction, as operationally defined by filtration, usually 0.45 μm; the transfer of this fraction to running water does not occur readily and is poorly understood. Physical erosion processes are mainly responsible for the transfer of iron particles to streams and rivers. Consideration of the flux of iron in a wide variety of rivers indicates that the ratio of particulate to dissolved flux is about 500:1.[3] This ratio is arbitrary[9] as the dissolved fraction contains colloidal particles, sufficiently small to pass through a filter of typically 0.45 μm.

The concentration of iron in rivers is usually in the range of 0.4 to 2 μmol dm^{-3}, but much lower values of, for example 0.04 μmol dm^{-3}, have been recorded.[5] Higher values are usually associated with local pollution, distur-

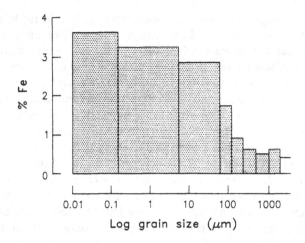

Figure 1. Size fractions of iron in sediments from the Saddle River. (Compiled from data in Wilbur, W.G. and J.V. Hunter, *Water Res. Bull.* 15:790–800 (1979).)

bance, or floods. Quarrying and mining activities generally increase erosional components. However, if pyrite is present its exposure to oxygen leads to the production of sulfuric acid which can result in waters with a pH as low as 2.[6] These waters are then very rich in iron with concentrations approaching 20 mmol dm^{-3} in exceptional circumstances.[7]

Iron is generally associated with particles of small grain size (Figure 1). In rivers with a high suspended load this distribution can be rationalized by considering that larger quartz particles have a low iron content, whereas smaller clay particles contain iron within their lattice, as for chlorite, and as coatings of oxyhydroxides. When river flow, and consequently the suspended load, is low there are few large particles. Detailed analysis by spectrophotometric and atomic absorption methods of the iron fraction in a slow-moving stream revealed that most iron was present as oxyhydroxides. Sequential filtration showed that the iron was mainly in the size range of 0.1 to 3 μm (Figure 2).[8] If filtration was performed immediately in the field there was an appreciable fraction associated with particles less than 0.1 μm, but this was substantially removed and the rest of the distribution modified if filtration was delayed for a few hours. These results indicate that the distribution of iron between particle sizes is not at equilibrium, but rather reflects the dynamics of supply and removal within the river. For a more detailed discussion of filtration, see Chapter 5, in this volume.[9] The proportion of iron in various ultrafiltrate fractions in river samples is variable, depending on the time of sampling and the particular river, indicating possible problems in the reproducibility of sampling or filtration, or the absence of any systematic distribution of colloidal particles.[10] Electron microscopic techniques, ultrafiltration, and use of ion exchange resins have shown that "dissolved" iron (<1 μm) is predominantly present as negatively charged colloids which may or may not be associated with humic substances.[11] Additionally there may be a small fraction (<10%) of positively charged colloids. Many substances are capable of stabilizing colloidal iron particles at a concentration of 1 mg dm^{-3}, including humic (ca. 1 mg dm^{-3}) and tannic acids (2 mg dm^{-3}), surfactants, and the inorganic ions silicate (ca. 500 mg dm^{-3}) and phosphate (0.4 mg dm^{-3}).[12] Divalent cations such as Ca^{2+} destabilize the colloids. Large particles (5–30 μm) rich in iron were thought to comprise clay minerals with a surface coating of iron oxyhydroxides, to which humic substances were attached.[13] A more mineralogical approach, which used electron microscopy to study systematically particles collected from an estuary, showed that iron was largely present in "chlorite/smectite" and 'biotite' phases.[14] X-ray diffraction revealed that iron-rich chlorite was present. The proportion of 'biotite' systematically increased from 0 to 30% as the grain size decreased from 1–2 μm, through 0.5–1 μm and 0.2–0.5 μm to <0.2 μm, fitting in well with other observations of iron increasing with decreasing particle size (Figure 1). 'Biotite' in this case was thought to be glauconite, which is an iron-rich illites-mectite mineral.

The distribution of heavy metals between various phases of suspended solids in rivers has been investigated using schemes based on chemical reactivity (see Reference 15 for a review of these procedures). Tessier et al.[16] sequentially treated the samples as follows: (i) 1 M $MgCl_2$ at pH 7 (exchangeable metal); (ii) 1 M acetate, pH 5 (metals bound to carbonates); (iii) 0.04 M NH_2OH at 96°C (metals bound to Fe-Mn oxides); (iv) 30% H_2O_2, pH 2, 85°C (metals bound to organic matter); (v) fusion with lithium metaborate (residual metals). For two rivers draining mixed catchments, with quartz, plagioclase, and chlorite dominating the suspended solids, the iron was found to be mainly associated with fraction (v), 61 and 73%, and fraction (iii), 32 and 22%. The remainder was in the organic (4.8 and 3.8%) and carbonate (2.1 and 1.3%) fractions. These results indicated that about one third of the iron was present as reactive, presumably amorphous oxyhydroxides and two thirds incorporated in the crystal lattices of clay minerals and silicates or present as refractory iron oxide.

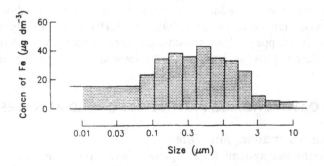

Figure 2. Size fractions of iron from a slow moving river, Black Beck, determined by serial filtration in the field. (Adapted from Laxen, D.P.H. and I.M. Chandler, *Geochim. Cosmochim. Acta* 47:731–741 (1983).)

Iron particles in rivers are removed by settling to form the river bed where they are periodically re-suspended according to their size and the hydraulic velocity. By this process the particles are graded, coarser ones remaining in the river but finer ones being progressively transported downstream. When rivers encounter lakes the quiet water ensures a more permanent removal of all particle sizes. The other major loss is to estuaries where a complicated cycle of deposition and resuspension is controlled by tidal cycles and salinity gradients. When sediment material from a wide variety of rivers was subjected to a fractionation scheme similar to that used by Tessier et al.,[16] iron was always mainly in the residual fraction, the remainder largely comprising oxyhydroxides.[3] For some rivers there could be an appreciable organic fraction. Although the oxyhydroxides represent only a small fraction their reactivity makes them particularly important.

In most lakes almost all the iron particles which enter are removed to the sediment. Large mineral particles (>1 μm) sink quite quickly and therefore their instantaneous concentration in the water column is low. Very small particles (colloids) theoretically sink so slowly that they would not be removed if they were completely inert. However, processes of aggregation and association with biota and other particles undoubtedly occur, and so even colloidal particles are ultimately transported to the sediment even though their residence time in the water is sufficiently long for them to dominate the instantaneous concentration.[17]

Sequential extraction schemes have been used to show that a large fraction of iron within lake sediments is inert and bound within mineral lattices, the remainder, as for suspended material, being mostly in the form of oxyhydroxides.[18] In a sediment of a productive lake 90% of the variation of Fe with depth from 0 to 60 cm could be correlated with magnesium and aluminum present in clay minerals,[19] but a combination of wet chemistry and Mossbauer spectroscopy[20] showed that only 50% of the iron was actually in clay minerals. As the concentration of sediment components reflects fluxes of material to the lake, the implication is that the remaining 40% of the iron had been transported to the sediment in association with clay minerals. The other 10% was taken to be the iron which is supplied to the sediment by in-lake recycling involving oxidation and reduction process. Some of the oxyhydroxides associated with clay minerals are reduced to Fe(II) within highly reducing sediments, but it appears that a considerable proportion of the oxyhydroxides are not reduced, possibly because of the formation of a protective sulfide coating.[20,21]

3. Fe(III) OXYHYDROXIDE PARTICLES FORMED IN WATER BODIES
3.1 Origin, Formation, and Fate

Most freshwater systems have regions where oxygen can become sufficiently depleted to provide a reducing environment in which Fe(II) is stable. In sediments the supply of oxygen is limited by molecular diffusion which is usually slower than the rate of oxidation of labile organic material. Consequently, electron acceptors other than oxygen are used in the oxidation process. Nitrate and manganese oxyhydroxides are usually reduced most readily, followed by iron oxyhydroxides. This simple thermodynamic ranking is not always adhered to strictly, as each reduction reaction depends on local concentrations and is usually controlled by microorganisms.[6] In very productive aquatic systems where there is an ample supply of readily oxidizable organic material, the oxygen is completely consumed very close (<1 mm) to the sediment water interface.[22,23] However, in unproductive systems such as oligotrophic lakes and many rivers the redox boundary is much deeper (centimeters) within the sediment.[24] Any freshwater sediment which is a fine grained mud rather than a sand or gravel is likely to be reducing, even though the supply of organic material may be very low. Sediments are heterogeneous and so, even within an ostensibly oxidizing zone, there may be reducing microniches associated with discrete particles rich in organic material.

In certain circumstances the water overlying the sediment may be devoid of oxygen. This is particularly so in lakes where thermal stratification can lead to relative isolation of the colder bottom layer of water. If there is sufficient decomposition of organic material these waters become devoid of oxygen, either permanently as in meromictic systems or seasonally as in monomictic and dimictic systems.[25] Exceptionally, in lakes and rivers where productivity is sufficiently high that they are classified as hypereutrophic, the whole of the water column may become devoid of oxygen.[26] There appear to be no systematic studies of iron chemistry in these latter systems, or in less pronounced situations where oxygen minima occur in mid-water regions of deep lakes, associated with the decomposition of settling organic material as it sinks through the water column.[25]

Wherever the redox boundary resides within a freshwater system, similar processes govern the transport of iron and its oxidation and reduction. A generalized conceptual model for the transport of iron at a redox boundary has been proposed[27] (Figure 3). The simple model assumes that there is a well-defined boundary between oxidizing and reducing conditions, where there is a ready interconversion between particulate oxidized forms ($Fe^{III}OOH$) and the dissolved reduced form [Fe(II)]. Oxidized particulate material is supplied continuously to this boundary which is situated in a region where random, diffusion-style transport processes operate. In water columns oxidized material is supplied by settling particles; in sediments the process of sediment accumulation provides fresh material. Both dissolved species and particles are subject to eddy diffusion in water columns. Random transport of dissolved species in sediments is by molecular diffusion which is sometimes augmented by bioirrigation, but particles in sediments are relatively little affected by random transport unless bioturbation is important.

The iron oxyhydroxides which pass through the oxidizing region are reduced to soluble Fe(II) when they encounter the reducing zone. There is thus a point source of Fe(II) at the redox boundary. Fe(II) diffuses from this point source

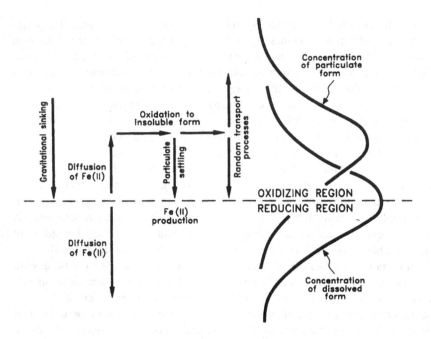

Figure 3. Generation of concentration profiles of dissolved and particulate iron in the vicinity of a redox boundary. (Adapted from Davison, W. in *Chemical Processes in Lakes,* W. Stumm. Ed. (New York: Wiley-Interscience, 1985.)

up into the oxidizing region, or down into the reducing zone. The upwardly diffusing Fe(II) is re-oxidized to oxyhydroxides as it encounters oxygen or perhaps other oxidants such as nitrate or manganese oxides. There is then a point source of particulate material which, in turn, is subject to random transport processes and may diffuse upwards or downwards. Additionally, larger particles may sink due to the effects of gravity. As a consequence of these processes, concentration profiles develop which, if random transport operates uniformly throughout the region of the redox boundary, approximate to a Gaussian shape.

For such a model to represent a steady state, removal processes must operate. Within reducing sediments Fe(II) may be removed by the formation of authigenic minerals such as siderite, vivianite, or iron sulfides.[18,28-30] Particulate material which is mixed into surface waters is removed by being flushed from the system. Sediments also act as a sink for particulate material because it is only the very reactive oxyhydroxides of iron which are reduced at the redox boundary. Most of the particulate iron entering freshwaters is bound within minerals and not available for reduction.

Truly steady-state conditions rarely apply in lakes and rivers, but a pseudo-steady state may sometimes be approached. Under certain circumstances lakes may permanently stratify. The bottom waters, or hypolimnion, become enriched in dissolved components which provide a density gradient that helps to prevent complete mixing.[31,32] Oxygen is completely removed from these bottom waters, allowing Fe(II) to accumulate. An iron cycle is established about the redox boundary in the mid-water region where upwardly diffusing Fe(II) is oxidized. The iron particles which are produced then sink back into the reducing zone where they complete the cycle which Campbell and Torgensen[32] have called the "ferrous wheel". In one well-studied Canadian lake,[32] where the deep waters had a mean renewal time of 2.5 years, Fe(II) reached concentrations of 4.2 mmol dm^{-3}. Ninety percent of the Fe(II) diffusing upwards was returned to the hypolimnion via the redox cycle, resulting in a mean residence time for iron in the lake of 15 years. The remaining 10% was lost from the lake by flushing. To preserve steady state this loss must be compensated by a supply of fresh, reducible iron entering the lake. The additional iron stokes the "ferrous wheel" by either dissolving at the redox boundary or at the sediment water interface.

When sediments are overlain by well-oxygenated water they can also provide pseudo-steady-state conditions. If iron is reduced a few millimeters or centimeters below the interface it can diffuse upwards away from its point source and as it encounters more oxidizing conditions it can be oxidized. At pH 7 and 10°C, Fe(II) has a half-life of about 4 h in well-oxygenated (0.3 mmol dm^{-3} O_2) soft water.[33] If only molecular diffusion operates in the sediment Fe(II) can be expected to travel on average about 0.5 cm in a plentiful supply of oxidants before it is re-oxidized. Using traditional sampling techniques of sediment slicing or dialysis cells, such gradients of dissolved iron are too steep to measure, but a new technique of diffusive equilibration in a thin film of gel (DET), capable of sub-millimeter resolution, has shown gradients over a few millimeters.[34] Belzile et al.[35] have recently developed a procedure for isolating authigenic solid phases at the required resolution (see Section 5.2). They showed that high iron concentrations occurred at depths greater than 2 mm below the sediment surface, and that they were overlain by a layer enriched in manganese.

Surface sediments overlain by oxygenated waters are frequently rich in iron and manganese.[36] Sometimes this appears as a red-brown surface layer and occasionally as a crust. Thus, the generalized view of transport at a redox boundary (Figure 3) provides a mechanism for the localized enrichment of iron and manganese, which can result in manganese nodules.[37] Most lacustrine deposits are richer in iron compared with manganese and occur as crusts. They are generally found in lakes where sedimentation rates are low, waters are well oxygenated and the sediments have low concentrations of organic matter.[38] Discrete nodules occur in more productive systems and may be linked with a supply of iron and manganese from the water.[39]

Concentrations of Fe(II) in pore waters at depths greater than a few centimeters are thought to be controlled by authigenic mineral formation processes rather than a simple diffusion model. In some systems, concentrations increase systematically with depth,[29] and in others they show no particular trend.[40] Many minerals, including siderite, amorphous iron sulfide, and vivianite are capable of being formed, as the pore waters may be supersaturated with respect to these phases. Wersin et al.[41] have modeled the systematic increase in Fe(II) with depth in the sediments of Lake Greifen. They showed that apparent supersaturation was consistent with a very slow rate of mineral precipitation. Moreover, very little of the total iron in the sediment was required to dissolve to maintain a steady-state concentration of Fe(II).

In circumneutral lakes where there is an ample supply of organic material, it is thought[33] that iron is reduced so close to the sediment-water interface that it can diffuse out of the sediment before it is re-oxidized. As the rate of vertical eddy diffusion in isothermal water columns is typically 3 to 5 orders of magnitude greater than molecular diffusion,[42] the Fe(II) is effectively well mixed in the overlying water and diluted to such an extent that a concentration gradient is not observed. Rapid oxidation in oxygenated waters ensures that there is no build-up of the reduced species. However, when a lake is stratified and the bottom waters are still well oxygenated the rate of vertical mixing is sufficiently low for an accumulation of total iron to be observed in the bottom waters, mainly in the 1 m overlying the sediment, and for Fe(II) to be detectable in the bottom 10 cm (Figure 4). Because of its much slower rate of oxidation, Mn(II) can be measured throughout a substantial part of the water column under these conditions. That iron is released from sediments overlain by oxygenated water has been confirmed using sediment traps placed near the bottom of a lake.[43] They showed elevated concentrations of iron which corresponded to the sinking of oxyhydroxides freshly formed from Fe(II)

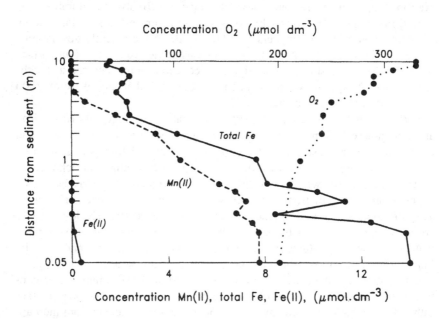

Figure 4. Concentration of polarographically measured Mn(II) and Fe(II), total iron and dissolved oxygen near the bottom of Esthwaite Water, at the onset of stratification, 22 April, 1981.

recently released from the sediment water interface. Use of traps at different sites ruled out the possibility of the elevated concentrations being due to physical resuspension.

The above discussion has considered the *in situ* formation of iron particles when the redox boundary is fixed at one site where a pseudo-steady state is assumed to operate. In practice, in most freshwaters the conditions which determine the location of the redox boundary, such as the hydrodynamic regime, the supply of readily reducible organic material, and its rate of decomposition, are continually changing. When such dynamic conditions prevail, numerical modeling of concentrations and fluxes using steady-state assumptions is no longer valid. Moreover, the redox boundary is no longer fixed in one location. In a seasonally anoxic lake there is a well-documented migration of the redox boundary.[27,33,44,45] Distributions of Fe(II) and particulate iron associated with this migration are illustrated schematically in Figure 5 for productive, circumneutral, dimictic lakes. During the winter months, when the waters are isothermal and well oxygenated at all depths, the oxic/anoxic boundary extends from a few millimeters to a few centimeters within the sediment, typically 5 to 10 mm. Coincident with the onset of stratification in the spring, there can be an increased supply of organic material which, along with higher temperatures, can force the boundary to migrate very near the sediment water interface. Some Fe(II) can then diffuse into the water column where it is oxidized to Fe(III), resulting in a slight elevation of particulate iron in the bottom waters. By early summer the oxic/anoxic boundary has migrated into the water column. Although there is only a trace amount of Fe(II) in the waters immediately overlying the sediment, there is a much greater concentration of Fe(III) particles which extend to, and penetrate, the oxic/anoxic boundary. By late summer the bottom waters are dominated by accumulating Fe(II) and the particulate iron forms a pronounced peak corresponding to a decline in the concentration of Fe(II). Examination of actual data for this time of the year shows that the peak of particulate iron occurs below the oxic/anoxic boundary (Figure 6). As oxygen is often not detected for several meters above the peak of particulate iron,[46] there is an insufficient supply to account for the oxidation of Fe(II) directly. It has been suggested that manganese oxides or nitrate may serve as oxidants.[47] There may be a coupled manganese cycle operating just above the iron cycle in the water column, with oxygen serving as the oxidant for Mn(II) (Figure 7). Iron particles may also be removed below the peak by aggregation and subsequent more rapid sinking. However, estimates of the residence time of iron particles in the particulate peak are about 10 days,[8,45] indicating that aggregation is slow, in keeping with laboratory studies.[48]

The peak of particulate iron in the water column sometimes immediately overlies an accumulation of sulfide in the bottom waters[47] (Figure 5). In this case the iron peak may be due to sulfide reducing Fe(III) to Fe(II).

The above description of the migration of the redox boundary is very much simplified, as it focuses on a single vertical dimension. The reality is of course

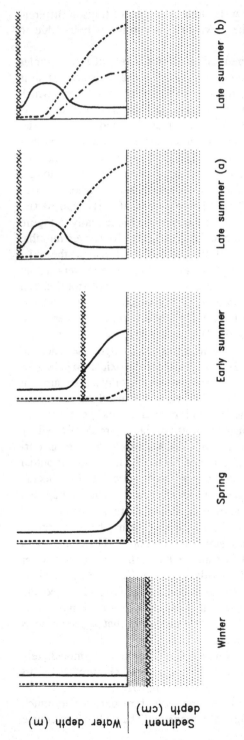

Figure 5. Schematic representation of the concentration of particulate iron (—) and Fe(II) (——) in the water column of a lake as the oxic/anoxic boundary (xxxxxx) undergoes seasonal migration. The sediment is represented by light hatching, elevated concentrations of particulate iron in the sediment are represented by darker hatching. An alternate late summer situation is shown where sulfide (——) may be responsible for forming the Fe(III) peak by reduction.

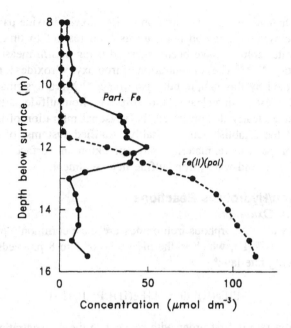

Figure 6. Polarographically measured Fe(II) and particulate iron = Fe$_{total}$ − Fe(II) in Esthwaite Water on 11 August, 1977. There was no detectable oxygen below a depth of 8 m.

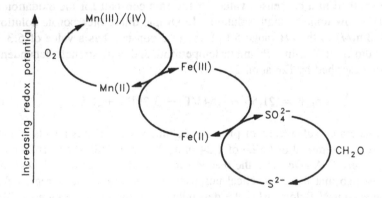

Figure 7. Possible interactions of the iron cycle in a redox gradient as may be found in a vertical water column.

three dimensional. However, relatively rapid horizontal water movements ensure that horizontal concentration gradients are usually negligible. This does not mean that contributions from lateral sediments can be neglected, but rather that they can be regarded as providing instantaneous inputs. Such an approach has been used in formulating mathematical models to describe the development of vertical concentration profiles with time.[42]

Detailed changes in the concentration of iron oxyhydroxide particles, iron sulfide particles, and plankton populations with respect to time and depth within the water column have been observed using *in situ* measurements of light attenuation.[45,49,50] The concentration of iron oxyhydroxides is particularly well represented by this simple bulk parameter.[45] More systematic work is still required to establish a clear relationship with iron sulfide particles. Measurements have clearly demonstrated the seasonal migration of iron oxyhydroxides and the establishment, in stably stratified systems, of three compartments for particulate material: phytoplankton in surface waters; iron oxyhydroxides in mid-water; iron sulfide near sediments.

3.2 Oxidation/Hydrolysis Reactions

3.2.1 Rate of Oxidation

In natural waters amorphous iron oxides are most commonly produced by the oxidation of Fe(II), which in the pH range of 5 to 8 proceeds according to the following rate law:[6]

$$- d[Fe(II)]/dt = k[Fe(II)] \, Po_2[OH^-]^2 \qquad (1)$$

The reaction rate is first order with respect to the concentration of Fe(II) and the partial pressure of oxygen, but its second-order dependence on hydroxyl ion makes it very sensitive to pH. Recent measurements under carefully controlled conditions, combined with a critical appraisal of previous data,[51-53] have resulted in a consensus value for the rate constant for the oxidation of Fe(II) in low ionic strength solutions. In simple dilute bicarbonate solutions $(I = 2 \, mM)$ in the pH range 5 to 8 the rate constant has a value of $6.3 \times 10^{-17} \, dm^3.mol^{-3}. \, min^{-1}$,[52] and its temperature and ionic strength dependence can be described by Equation 2.

$$\log k = 21.56 - 1545/T - 3.29I^{1/2} + 1.52I \qquad (2)$$

In air-saturated soft waters of pH 8, 7, 6, and 5 at 10°C this results in half-lives for the removal of Fe(II) of 2.34 min, 234 min, 390 h, and 1625 days, respectively.[33] Above pH 8 the reaction order with respect to hydroxyl ions increases so that for all practical purposes the oxidation can be regarded as instantaneous.[52] Below pH 5 the dependence on the concentration of OH^- decreases. In solutions more acid than pH 4 the rate of oxidation is virtually independent of pH.[6,54] Although the oxidation is catalyzed by trace metals, phosphate, fluoride, and particles, including freshly formed iron oxides (autocatalysis), the concentration thresholds for catalysis are sufficiently high that the effect is negligible in unpolluted, neutral freshwaters (Table 1),[53] but it may be significant in porewaters where there are effectively high concentrations of solid phase. Autocatalysis is believed to proceed by adsorbed Fe(II) being rapidly oxidized.[55,56] A mechanism also exists for accelerating the rate of oxidation photochemically by the production of hydrogen peroxide,[57] as it

Table 1. The Effect of Other Substances on the Rate of Oxidation of Fe(II). ? Indicates Conflicting Results, + Acceleration and − Inhibition

Substance	Conc. Range (mmol dm^{-3})	Effect
HCO$_3^-$	0.3–5.0	?
Si(OH)$_4$	0.1–1.2	?
Zeolite	?	+
Cl$^-$	0.3	0
Cl$^-$	100–500	—
SO$_4^{2-}$	0.1	0
SO$_4^{2-}$	30–165	—
NO$_3^-$	100	—
Br$^-$	100	—
I$^-$	100	—
PO$_4^{3-}$	0.01	0
PO$_4^{3-}$	0.04–400	+
F$^-$	20–100	+
Cu^{2+}	>0.0003	+
Co^{2+}	0.04	+
Mn^{2+}	0.04	+
Fe(III)	<0.01	0
Fe(III)	0.2–1	+
Mn(IV)	0.0002	?
Tetramine	?	—
EDTA	?	—
Tannic acid	0.002–0.1	—
Histidine	0.1	0
Histidine	0.6	—
Gallic acid	0.1	—
Pyrogallol	0.1	—
Glutamic acid	0.1	—
Tartaric acid	0.1	—
Glutamine	0.1	—
Vanillic acid	0.1	0
Phenol	0.1	0
Resorcinol	0.1	0
Syringic acid	0.1	0
Vanillin	0.1	0
Humic acid[a]	1–3	0
Humic acid[a]	12–145	—

[a] Concentration given in mg dm^{-3}.

Data taken from Davison, W. and G. Seed. *Geochim. Cosmochim. Acta* 47:67–79 (1983).

is well known that hydrogen peroxide rapidly oxidizes Fe(II) over a wide range of pH.[58,59] However, in natural waters light also induces the reduction of amorphous iron oxides by catalyzing the reduction by organic matter, and any hydrogen peroxide which is simultaneously produced merely affects the steady-state concentration of Fe(II) through a back reaction.[57] Thus the observed net effect of light on iron in freshwaters is always reduction, rather than oxidation.[60]

It is well established that microorganisms can accelerate the oxidation of Fe(II) in natural waters. Measurements of the rate of oxidation of iron in acid streams (pH 2.4 to 3.8) revealed that the rate was five to eight orders of magnitude faster than predicted by the chemical reaction rate expression.[61,62] In an acid lake (pH 5.4) it was about 300 times faster.[63] The bacteria *Thiobacillus ferrooxidans* is thought to be responsible for these dramatic increases in rate. Direct measurements of the rate of oxidation of iron in circumneutral (pH 6.5 to 7.4) lakewater have shown that the oxidation proceeds at the same rate as in comparable synthetic solutions,[53,64] suggesting that the reaction under these conditions is not mediated by microorganisms.

Some organic compounds, including naturally occurring humic material, can slow down the oxidation of iron,[53] but if the humic material is only present at natural concentrations it has little effect on the rate at least when the concentration of iron is sufficiently high (>10 μmol dm^{-3}) to allow measurement of its oxidation rate (Table 1). It is possible that the oxidation may be inhibited by natural organics when Fe(II) is present at much lower concentrations.[52] This perturbation of the rate of oxidation is quite distinct from the ability of humic material, through complexation, to stabilize Fe(II) and to reduce Fe(III).[65] Inhibition of the rate of Fe(II) oxidation by inorganic ions such as sulfate and chloride has been attributed to the formation of ion pairs which are not easily oxidized.[51] Luther[66] considers that this effect may be explained by molecular orbital theory with chloride out-competing hydroxyl ions for Fe(II) coordination.

3.2.2 Structure and Nature of Oxidation Products

According to thermodynamics, well-oxygenated waters contain only Fe(III), which at neutral pH is completely hydrolyzed. The concentration of the dominant monomeric species, Fe(OH)$_3$°, at pH 7 is not known precisely, but probably lies in the range of 5×10^{-10}[67] to 5×12^{-12} mol dm^{-3},[68] so it is below the limit of detection of most analytical methods. Even at pH 4 the total concentration of dissolved species, mainly Fe(OH)$_2$$^+$ and FeOH^{2+}, only reaches 10^{-7} mol dm^{-3}.[68]

Although a wide variety of polymeric hydrolysis products of iron may be formed, the initial formation steps have some common features. Irrespective of whether hydrolysis products of iron (III) form due to a change in pH or to the oxidation of Fe(II), the same mononuclear precursor species, such as Fe(OH)$_3$, are thought to be involved.[69] Dimers and trimers have been reported to be formed in the initial reaction,[70] but this quickly leads to the formation of higher molecular weight species. At low pH (<3) polymeric material of distinct composition, depending on the anions present, may be obtained,[71] but such species are transient because precipitates are produced as the solutions age. The wide variety of iron solid phases which form depend on many factors, including the pH and chemical composition of the solution and the extent of aging processes. In the pH range 5 to 7 there are two main products of iron oxidation. In the absence of silica, γFeOOH, known as lepidocrocite, is produced.[64,72] It has a characteristic X-ray diffraction pattern, but may have several morphologies, including rafts, needles, and plates.[73] When Fe(II) is allowed to oxidize in neutral solutions whose major ions simulate lake water, with or without humic substances, the lepidocrocite product takes the form of crumpled sheets.[64] When silicate is present a more amorphous product known as ferrihydrite[72] is formed. It is characterized by its X-ray diffraction pattern and the very small size of its primary particles which confer a high specific surface area.[74] In both synthetic and natural lake waters containing silica the oxidation products comprise a network of aggregated, fused or cross-linked primary particles, 1 to 5 nm in diameter.[44,74,77] Aging experiments in strongly alkaline solutions have shown that geothite is formed from both ferrihydrite[74] and lepidocrocite.[75]

The above discussion refers primarily to laboratory studies of oxidation products, but material collected from the peak of freshly produced particles in the vicinity of the redox boundary in stratified lakes has also been well characterized. Examination of particles from a soft water lake[76] by electron microscopy showed that they mainly comprised spherical or ellipsoidal particles with mean diameters in the range of 0.05 to 0.5 μm. They could be largely classified as ferrihydrite and consisted of 30 to 40% iron by weight and small amounts of P, N, Mn, Si, S, Ca, and Mg. Organic matter could account for as much as 36% (by weight), and humic substances accounted for about one third of the organic matter. Calcium, phosphate, and humic substances could be removed by changing the pH, without changing the appearance, suggesting an open structure. Silicate, however, could not be removed without changing the structure, consistent with laboratory observations for the formation of ferrihydrite and lepidocrocite. Particles formed by allowing filtered lake water to oxidize are only a few nanometers in diameter, indicating that natural particulate material may be essential for the formation of relatively large particles,[64] although there are no indications to suggest that living organisms are involved.

Iron particles recently formed by oxidation have only been extensively characterized in one hard water and one soft water lake. Electron microscopy showed that in the hard water example they were amorphous, nearly spherical globules (2 to 300 nm) which appeared to exist either as individual entities, or associated in either tightly (0.1 to 1 μm) or loosely (0.5 to 10 μm) packed aggregates.[77] Substantial concentrations of Ca and P were present in the particles (Table 2) and the iron consisted of similar amounts of Fe(II) and Fe(III).[47] The constancy of Fe/P/Ca and Fe(II)/Fe(III) ratios with both depth and time in the water column suggests that there is a well-defined chemical entity.[44]

Humic substances control the surface chemistry of iron particles freshly formed in lakes. Tipping[78] suggests that only part of the humic molecule interacts with the oxide surface in a ligand exchange reaction with surface co-ordinated OH^- or H_2O groups. The functional groups of the non-adsorbed parts of the molecule extend away from the surface and determine the surface

Table 2. Characterization of Iron Particles Found in Freshwater Environments

Nature of Iron Compound	Composition	Method of Characterization	Environment	Ref.
Ferrihydrite; reddish brown mud, rounded aggregates, > 10 μm, primary units, 10 nm	Fe 22–51% C 0.5–17%	Electron microscopy, X-ray and electron diffraction, IR spectroscopy, thermal analysis and chemical analysis	Deposits in soil water drainage ditches	74
Iron oxyhydroxide; spherical/cylindrical particles, < 0.5 μm, very variable	Variable Ca, ≤ Fe. Si, P, S consistently present	Microprobe electron microscopy	Surface waters, soft water lake	17
Iron oxyhydroxide; spherical/ellipsoidal particles, 0.05–0.5 μm diameter, poorly crystalline ferrihydrite	Fe $P_{0.055–0.14}$ $Ca_{0.012–0.02}$ high carbon content, appreciable Si and Mn	Microprobe and transmission electron microscopy, X-ray diffraction, wet chemistry	Redox transition layer in anoxic soft water lake	76 64
Iron oxyhydroxide; roughly spherical globules, 0.04–0.4 μm diameter	$Fe(II)_{0.5}$ $Fe(III)_{0.5}$ $P_{0.25}$ $Ca_{0.19}$	Microprobe electron microscopy, polarography, colorimetry, atomic absorption spectroscopy	Redox transition layer in anoxic hard water lake	44 77
Nontronite; ~1 μm flakes aggregated to 0.05–2 mm pellets	Fe 21% Si 20%	Electron and optical microscopy, X-ray diffraction, differential thermal analysis, wet chemistry	Freshwater sediment	109
Limonite; yellow-brown amorphous iron oxyhydroxide	Fe 35% Si 9%	Electron and optical microscopy, X-ray diffraction, Mossbauer spectroscopy, wet chemistry	Freshwater sediment	109
Hematite; detected by Mossbauer spectroscopy	Not determined	Mossbauer spectroscopy	Polluted freshwater sediment	110
Goethite	Fe 40–56%	Differential thermal analysis, X-ray diffraction, wet chemistry	Bog ores	111, 112
Wustite; detected by Mossbauer spectroscopy	Not determined	Mossbauer spectroscopy	Polluted freshwater sediment	110
Vivianite; small nodules with roughened surface, ca 60 μm	Fe 29%, PO_4 36% Mg 3.3%, Mn 2% Ca 0.1%	Chemical analysis and optical spectroscopy	Freshwater sediment	113
Vivianite; white spots turning blue on exposure to air.	Contained only Fe and P	Optical microscopy, microprobe electron microscopy	Freshwater sediment	114

Mineral/description	Composition	Method	Sample	Ref.
Vivianite brown-orange or blue concretions	Fe $P_{0.46}$ $Mn_{0.26}$ $Si_{0.17}$ $Al_{0.08}$ Ca,K,Mg ~0.5%	Optical and microprobe electron spectroscopy, Mossbauer spectroscopy	Freshwater sediment	115
Vivianite; banded and fibrous with a spherulitic texture	Not determined	Optical microscopy and x-ray diffraction	Freshwater sediment	109
Siderite, detected by X-ray diffraction and Mossbauer spectroscopy	Not determined	X-ray diffraction and Mossbauer spectroscopy	Freshwater sediment	18
Siderite, detected by Mossbauer spectroscopy	Not determined	Mossbauer spectroscopy	Polluted freshwater sediment	110
Siderite; distinct grains 2–10 μm long comprised of smaller units	Not determined	Electron microscopy and X-ray diffraction	Freshwater sediment	114
Acid volatile sulfide	Probably amorphous FeS, mackinawite, pyrrhotite and greigite	Wet chemistry	Numerous freshwater sediments	116, 117
Iron sulphide; black amorphous particles some evidence for surface coating of FeS on oxy-hydroxide	Fe $S_{0.4-0.7}$	Mossbauer spectroscopy, wet chemistry, X-ray diffraction	Anoxic waters soft water lake	118, 21
Mackinawite, 0.2 mm thick coats on detritus	Fe 53–59%, S 35–40% SiO_2 0.8%, Al_2O_3 0.2–0.8%	X-ray diffraction, microprobe electron and optical spectroscopy	Deposits from a thermal spring	119
Greigite; small, black, irregular grains < 1 mm–3 mm, magnetic	Not determined	X-ray diffraction	Freshwater sediment	120
Pyrite; 0.2 mm thick coats on detritus	Not determined	X-ray diffraction and optical spectroscopy	Deposits from a thermal spring	119
Pyrite; detected by Mossbauer spectroscopy	Not determined	Mossbauer spectroscopy	Freshwater sediment	122
Pyrite; detected by Mossbauer spectroscopy	Not determined	Mossbauer spectroscopy	Freshwater sediment	121
Pyrite; framboids composed of octahedral crystallites and kaolinite sheath	Not given	Electron microscopy, X-ray diffraction	Sedimenting material from a lake	123

Table 2. (continued) Characterization of Iron Particles Found in Freshwater Environments

Nature of Iron Compound	Composition	Method of Characterization	Environment	Ref.
Pyrite; determined by sequential reaction of iron	Not determined, probably pyrite + marcasite	Wet chemistry	Soft water sediments	116
Pyrite; determined by sequential reaction of sulfur	Not determined, probably pyrite + marcasite	Wet chemistry	Polluted, soft water sediments	124
Framboidal pyrite; comprised of 5–70 μm spherules, also crystalline grains	Not given	Electron microscopy, X-ray diffraction	Freshwater sediment	125
Pyrite; framboids and single crystals	Not given	Electron microscopy	Freshwater sediment	126

charge. The negative charge which results from the presence of humic substances encourages the adsorption of divalent cations, by electrostatic attraction.

3.3 Reduction/Dissolution Reactions

As discussed in Section 3.1 redox cycling of iron occurs at the oxic-anoxic boundaries found in aquatic systems. Dissolution of iron(III) oxyhydroxide particles formed by oxidative hydrolysis of Fe(II) may theoretically occur via non-reductive or reductive reaction pathways.[79] Reductive dissolution mechanisms are more important in oceans, lakes, and rivers because of pH buffering, the low relative concentration of complexing agents, and the frequent occurrence of reducing conditions, whereas nonreductive dissolution may be significant in soils. The role of bacteria and phytoplankton may be important,[80,81] since they have been shown to participate either directly or indirectly in the redox cycling of iron. Biota may release redox active substances such as organic acids (formate, acetate, oxalate, pyruvate, etc.), reduced sulfur species (H_2S, HS, polysulfides, sulfites, etc.), phenols, and humic and fulvic type acids, maintain redox microenvironments,[82,83] colonize particles, and induce aggregation or particle microfracturing. The discussion here is restricted to chemical aspects of the dissolution of iron (III) oxyhydroxides.

3.3.1 Theoretical Aspects of Dissolution Reactions

It is generally accepted that the rate of dissolution of many minerals under natural conditions, and in particular iron oxides, is surface controlled.[79,84,85] The overall reaction pathway may be schematically represented by the following two-step sequence, although each step is composed of a series of chemical interactions such as surface complexation (inner or outer sphere), protonation or deprotonation, charge transfer, ligand exchange, steric rearrangement, etc. Step 1 involves the formation of a surface species between the hydrated mineral's surface sites ($\equiv$Fe-ØH for an iron oxide) and solution reactants such as H^+, OH^-, ligands and/or reductants; in the case of reductive dissolution, formation of the surface species involves the surface complexation of a reductant followed by a charge transfer between the reductant and the surface site.

$$\equiv\text{Fe-OH} + \text{reactants (H}^+, \text{OH}^-, \text{L, Red)} \xrightarrow{\text{fast}} \text{surface complex}$$

$$\xrightarrow[\text{fast}]{(+ \text{ charge transfer})} \text{(reduced) surface complex}$$

The second step (below) is generally slower and involves the detachment of the metal species from the crystal lattice and its release to the solution.

$$\text{(reduced) surface complex} \xrightarrow[\text{slow}]{\text{metal detachment}} \text{Fe}_{\text{aq}} +$$

$$\text{oxidized reactant}$$

Because the first step is generally fast,[86-88] the second step is often the rate-limiting one, and the rate of dissolution may therefore be expected to be proportional to the concentration of the surface species.

Consequently, if one assumes (a) that a steady-state amount of surface phase is formed during the dissolution i.e., surface area and mole fraction of active sites relative to total sites are both constant, (b) that the system is far from equilibrium implying that the back reaction may be neglected, and (c) that the concentration of solution reactants is constant, then it should be expected that a constant rate of dissolution, dN/dt, should be observed. In well-controlled systems this has been found to be the case.[89] However, surface heterogeneity due to mineral grinding, the presence of ultrafine particles, the diffusion of molecular water into grain dislocations, and the precipitation of secondary phases may all affect the dissolution process and lead to nonlinear dissolution rates.[90,91]

3.3.2 Laboratory Studies

3.3.2.1 Nonreductive Dissolution. The dissolution kinetics of Fe(III) particles such as goethite, hematite, and ferrihydrites without a change in the Fe oxidation state may be modeled using the concepts developed for the dissolution of other oxides.[92] Because of the slight solubility of Fe(III) above pH 4, nonreductive dissolution occurs primarily in the presence of ligands which effectively increase the solubility of Fe(III) by forming dissolved Fe(III) complexes. As previously mentioned, the critical rate-determining step is, in most cases, the detachment of the surface metal species. This requires a polarization and weakening of the Fe-oxygen bonds at the surface of the lattice structure, which may be brought about by protonation and/or by surface complexation of a ligand as shown in Figure 8.[a,b] Dissolution rate constants for different iron oxide phases have been experimentally measured and were found to be proportional to the surface concentrations of protons and/or ligands.[93] Zinder et al.[93] found that the dissolution kinetics could be expressed for a proton promoted dissolution as follows:

$$dN/dt = k_H(C_{H,S})^n \tag{3}$$

and for a ligand promoted dissolution

$$dN/dt = k_L(C_{H,S})^n (C_{ML,S}) \tag{4}$$

where dN/dt are the respective dissolution rates (moles m^{-2} s^{-1}), k_H and k_L are rate constants, $C_{H,S}$ and $C_{ML,S}$ are the surface concentrations (moles m^{-2}) of protons and of a metal-ligand surface chelate. For a proton promoted dissolution, n is the number of protons required to weaken the metal oxygen bond, whereas for a ligand promoted dissolution n is an integer often equal to zero. At constant pH, under nonreductive conditions, ligands such as oxalate were found to increase the dissolution rate constant of goethite by a factor of 30.

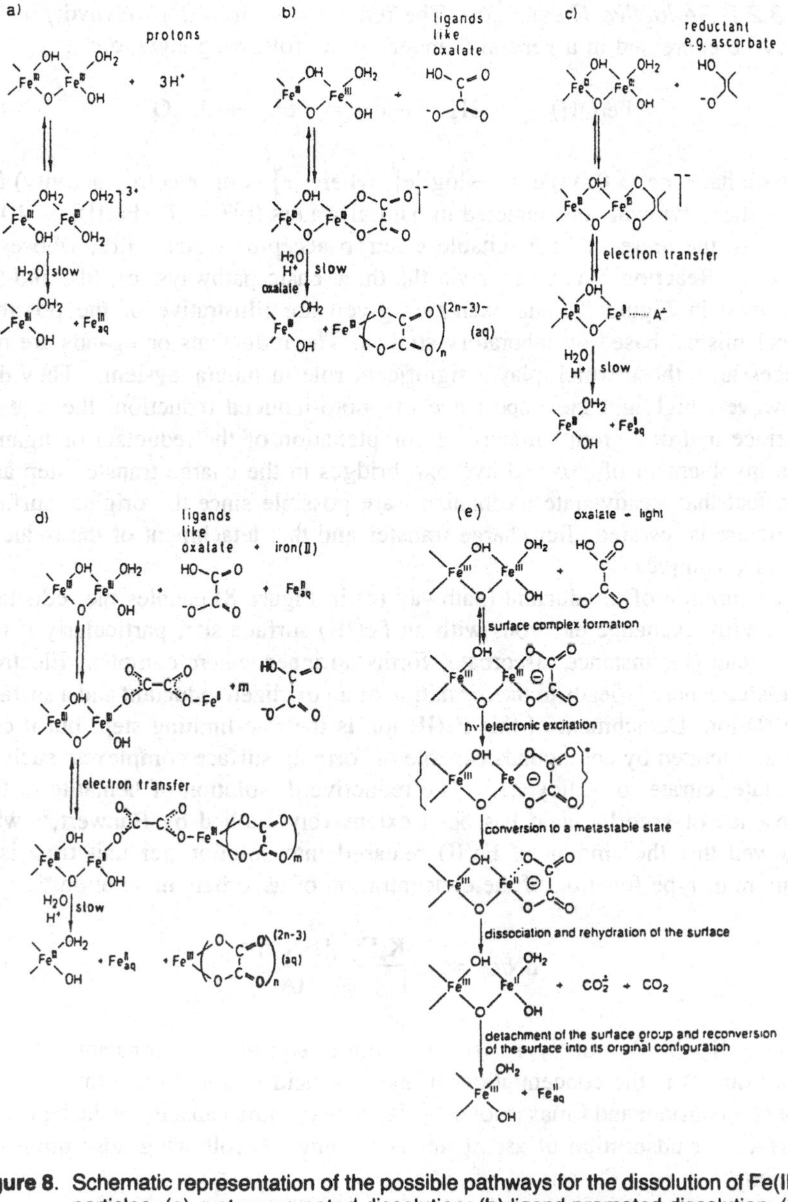

Figure 8. Schematic representation of the possible pathways for the dissolution of Fe(III) particles: (a) proton promoted dissolution; (b) ligand promoted dissolution; (c) reductive dissolution; (d) Fe(II) catalyzed ligand promoted dissolution; (e) photocatalyzed reductive dissolution. (Modified from Hering, J.G. and W. Stumm. in *Reviews in Mineralogy,* Vol. 23, M.F. Hochella and A.F. White, Eds. (Washington, D.C.: Mineralogical Society of America, 1990.)

3.3.2.2 Reductive Dissolution. The reduction of iron(III) (oxyhydr)oxides may be expressed in a general manner by the following equation:

$$Fe(OH)_{3(s)} + 3H^+ + e^- = Fe^{2+} + 3H_2O \tag{5}$$

which has a pe $\cong 0.0$ (pe $= -\log \{e\}$, where $\{e\}$ is the electron activity) for conditions typically encountered in natural waters (pH $= 7$, $[Fe(II)] = 10^{-5}$ M).[6] In the presence of a suitable electron acceptor system (i.e., whose pe < 0.0), Reaction 5 may occur via the three basic pathways (c), (d), and (e) outlined in Figure 8. The examples given are illustrative of the possible mechanisms, based on laboratory studies. The reductants or ligands are not necessarily those which play a significant role in natural systems. They do, however, highlight the importance of photo-induced reduction, the role of surface hydroxy groups in surface complexation of the reductant or ligand, the involvement of oxo and hydroxo bridges in the charge transfer step and the fact that steady-state mechanisms are possible since the original surface structure is restored after charge transfer and the detachment of the reduced surface complex.

Adsorption of a reductant (pathway (c) in Figure 8) enables the reductant to readily exchange electrons with an Fe(III) surface site, particularly if the reductant (for instance, ascorbate) forms an inner sphere complex. Electron transfer generally leads to the formation of an oxidized reductant and a surface Fe(II) ion. Detachment of this Fe(II) ion is the rate-limiting step, but it can be accelerated by compounds capable of forming surface complexes, such as oxalate, citrate, or salicylate.[94] The reductive dissolution of hematite in the presence of ascorbic acid has been extensively studied by Banwert,[95] who showed that the amount of Fe(II) released into solution per unit time is a Langmuir-type function of the concentration of ascorbate in solution:[95]

$$dN/dt = k \frac{K_a \Gamma_{max}[HA^-]}{1 + K_a[HA^-]} \tag{6}$$

where K_a (dm^3 mol^{-1}) is the adsorption equilibrium constant, $[HA^-]$ (mol.dm^{-3}) is the concentration of ascorbic acid in solution, k (m^2.s^{-1}), is the rate constant and Γmax (mol m^{-2}) is the maximum capacity of the hematite surface for adsorption of ascorbate. Assuming the following adsorption reaction:

$$\equiv FeOH + HA^- \overset{K_a}{\rightleftarrows} \equiv FeA^- + H_2O$$

the Langmuir quotient in Equation 6 gives the concentration (mol m^{-2}) of surface bound ascorbate:

$$[FeA^-] = \frac{K_a \Gamma_{max}[HA^-]}{1 + K_a[HA^-]} \tag{7}$$

Consequently, it may be readily seen that the rate of dissolution is directly proportional to the concentration of surface bound ascorbate.

$$dN/dt = k[FeA^-] \tag{8}$$

Photo-induced reductive dissolution reactions (pathway e in Figure 8) may also be important in environments such as shallow sediments and in surface waters.[96] Oxalate, for instance, which has been studied as a model reactant does not reduce iron (oxyhydroxy)oxides in the absence of light although this reaction is thermodynamically favorable. Charge transfer requires prior formation of an excited state by absorption of a photon, followed by a ligand-metal charge transfer. The mechanism is probably similar to that proposed by Parker and Hatchard[97] for the photolysis of dissolved trioxalato ferric iron, which involves further redox reactions of the oxalato radical after dissociation from the surface. These reactions have been omitted from Figure 8 for simplicity. Rehydration and detachment of the surface Fe(II) ion leads to the reformation of the original surface configuration. A rate law for photo-induced reductive dissolution has been derived[63,98] and leads to the autocatalytic dissolution mechanism mentioned below in the absence of a sink for the released Fe(II).

Iron (III) (oxyhydroxy)oxides may also be reduced via a Fe(II) catalytic mechanism (pathway d in Figure 8) in the presence of a bridging ligand, through which an inner sphere electron transfers between an adsorbed Fe(II) and a surface Fe(III) site. This mechanism has been recently demonstrated by Suter et al.[99] and Siffert[98] using oxalate as the bridging ligand. Fe(II) is not consumed during the reaction as it is continually regenerated and released into the solution. The rate of dissolution has been reported to follow a Langmuir-type function of the concentration of dissolved Fe(II) at a given oxalate concentration.[98]

$$dN/dt = d[Fe(II)]/dt = k_3[C_2O_4^{2-}]^2 \frac{K_a\Gamma_{max}[Fe(II)]}{1 + K_a[Fe(II)]} \tag{9}$$

3.3.3 Reductive Dissolution Under Natural Conditions

Dissolution of Fe(III) (oxyhydroxy)oxides formed either in the water column or in the sediments may proceed via any of the three above-mentioned mechanisms, as well as by biotically mediated pathways. The model compounds used in the above-mentioned laboratory studies may not play a significant role in environmental systems. In such systems, most dissolved organics and S(−II) compounds resulting from the decomposition of natural organic matter are, thermodynamically, possible reductants of Fe(III), however, in many cases photo-activation and/or bio-mediation (perhaps via exudates) is probably necessary for the reaction to proceed. Photo-induced dissolution and the release of bioavailable Fe(II) may be of significance in surface waters,[57,100] and Fe(III) reductive reactions at the surface of phytoplankton may also occur.[101] Reduction by dissolved sulfide is certainly important.[102] It can proceed in the dark at pH 6 to 8 (De Vitre, unpublished results) and therefore could play a significant role in deep sediments or when the oxic-anoxic boundary is below the photic zone. Finally, reductive dissolution by combinations of (i) organic reductants and ligands and (ii) inorganic reduced ions (Fe(II), for instance) and organic ligands (pathway d in Figure 8) may also be quite important in natural systems. However, it should be underlined that the nature of the most important reactions have still to be identified, and much further work is necessary in order to gain a better insight into reductive dissolution in real systems.

4. OTHER IRON PARTICLES FORMED IN FRESHWATERS

There is little doubt that the major iron particles formed in oxygenated lake waters are oxyhydroxides. Iron is also taken up into growing phytoplankton; depending on the species and the background concentration of iron in the water, it constitutes 0.01 to 0.13% of the dry weight.[103-105] Even in very productive lakes, the phytoplankton only accounts for a fraction of the iron present in surface waters.[106] Moreover, in sedimenting material, where the rapidly sinking iron within clay minerals is more abundant than oxyhydroxides, the ratio of iron to either carbon or phosphorus is much greater than would be expected if the iron was wholly associated with biota.[107,108]

Iron particles formed within the peak of particulate iron found in lakes with anoxic bottom waters (Figures 3 and 5) are by no means simple oxyhydroxides. Buffle et al.,[44] who found that in a hard water lake the ratio of Fe:Ca:P in the particles was fairly constant at 1:0.19:0.25, suggested that a well-defined chemical entity might be present. In a soft water lake the particles contained a smaller proportion of calcium and the ratio of 1:0.012-0.020:0.055-0.140 was more variable.[76]

It is in the anoxic waters of lakes and sediments that most attention on authigenic mineral formation has been focused. Fe(II) is relatively resistant to hydrolysis and so the possibility of forming insoluble Fe(II) carbonates, sulfides, and phosphates arises. Table 2 lists reported occurrences of such minerals.

Calculations of ion activity products have shown that the anoxic bottom waters of lakes may become supersaturated with respect to siderite, $FeCO_3$,[118,127-129] which has a $pK_{so} = 10.2$ to 10.7.[130] Direct analysis of the solid phase failed to provide evidence of ferrous carbonate[118] in a soft water lake, but the stoichiometric excess of Fe(II) to S(−II) in the particles of a hard water lake led to the suggestion that siderite might be present.[127] Similarly the pore waters of lake sediments may be saturated with respect to siderite.[18,40,130,131] Emerson[130] and Nembrini et al.[18] considered that siderite does not form in the presence of dissolved sulfide, probably because the formation of FeS is kinetically favored. However, by using chemical extraction procedures followed by Mossbauer spectroscopy, Nembrini et al.[18] were able to show that siderite was formed when pore water sulfide concentrations were low. Manning et al.[110] have also identified siderite in freshwater sediments using Mossbauer spectroscopy and Anthony[132] reported siderite identified by X-ray diffraction.

Although anoxic bottom waters have been reported to be saturated with respect to vivianite, $Fe_3PO_4.8H_2O$,[133] which has a solubility product, pK_{SO}, in the range of 33.5 to 36,[130] the solid phase has not been isolated and identified. The particles identified by Buffle et al.[44] in the vicinity of the redox boundary, which contained calcium and phosphorus, as mentioned above, may be related to the variety of iron and phosphorus phases identified in sediments. These include vivianite,[132] ludlamite, $(Fe, Mn, Mg)_3 (PO_4)_2.H_2O$, and phosphoferrite, $(Mn, Fe)_3(PO_4)_2.3H_2O$.[38] Nriagu and Dell[133] have also calculated that anapaite, $Ca_3Fe(PO_4)_3.4H_2O$, formation should be favored at lacustrine sediment-water interfaces, and that lipscombite, $Fe_3(PO_4)_2(OH)_2$, should form within soft water sediments. Measurements of pore waters have shown that anoxic sediments are often supersaturated with respect to vivianite.[18,130,133] Although some attempts to identify this phase have been successful,[132] others have failed.[18,41] Berner[134] suggests that vivianite is only formed in mildly reducing environments where sulfide is absent, while Wersin et al.[41] consider that the rate of formation of such phases is sufficiently slow to prevent them accumulating to measurable concentrations.

There can be no doubt that iron sulfides readily form in anoxic environments. In anoxic lake water, black particles are clearly observed,[117,118] and they can be isolated by filtration. When ion activity products (IAP) appropriate to Equation 10 have been calculated at various depths within water columns,

$$K_s = aFe^{2+} \times aHS^-/(H^+) \qquad (10)$$

a constant value has usually resulted.[30] For a wide variety of waters the IAP falls in a narrow range (pIAP = 2.6 to 3.22) which agrees with the measured solubility product for amorphous ferrous sulfide of $pK_{SO} = 2.95 \pm 0.1$. Calculation of an ion activity product from the simple measurement of Fe(II), S(−II), and pH is valid because Fe(II) is substantially uncomplexed near to pH 7,[135] the pH at which most anoxic freshwaters are buffered by proton exchanging redox reactions. Recently the formation of the soluble species Fe_2S_2 has been reported[136] when concentrations of Fe(II) and S(−II) approach saturation. However, this species usually accounts for a negligible proportion of the iron or sulfide and therefore does not invalidate simple calculations of ion activity products.[30]

Characterization of the iron sulfide particles is as yet poor, due largely to the difficulties of performing measurements on samples which are very sensitive to oxygen. It has been observed that when anoxic lake water is freshly collected into a sealed bottle it may be clear or have a light grey appearance, but on standing for a few hours the solution visibly darkens (Davison et al., unpublished observations in different laboratories). With further aging (24 h) distinct black particles appear and they settle to form a layer on the bottom of the bottle. It appears that the particles are initially of a very small size (20 to 200 nm), but that they subsequently aggregate.[127] By making measurements on site within minutes of sampling and comparing them with measurements made a few hours later in the laboratory De Vitre et al.[47] have shown that, even when samples are collected into sealed glass bottles, changes to their chemistry can occur. Sulfide is degassed, forming very small bubbles within the bottle, the concentration of the dissolved Fe_2S_2 species decreases and the concentration of Fe(II) increases. There appears to be a re-equilibration as larger particles of insoluble FeS form, perhaps from a solution which was previously slightly supersaturated.

Analyses of black iron sulfide particles collected from a soft water lake using a combination of wet chemistry and Mossbauer spectroscopy,[21] have suggested that they may comprise iron oxyhydroxides with a surface coating of sulfide. This observation implies that iron sulfide particles in lake water are formed by the reaction of sulfide with sinking iron oxyhydroxides, rather than by the combination of Fe(II) and S($-$II) to form first Fe_2S_2 and then FeS when saturation is exceeded. In reality both mechanisms probably operate, depending on the local conditions. The molar ratio of Fe(II), as measured spectrophotometrically or polarographically, to S($-$II) measured spectrophotometrically, in particles (>0.8 μm) collected from a softwater lake on three separate occasions was 2.52, 2.14, and 0.69.[21] These particles were obtained by filtration 2 h after collection and so they may have included colloidal particles which had aggregated while in the sealed glass bottle. When the concentration of Fe(II) in colloidal particles collected from a hard water lake was plotted against the concentration of S($-$II) in the same colloidal particles, a fairly constant slope of 2.1 $\pm$ 0.25 emerged.[127] These two sets of results agree well when plotted on the same graph, the colloidal data spanning a range of concentrations, whereas the data for particles being restricted to very low concentrations (Figure 9). Buffle et al.[127] have hypothesized that the stoichiometric excess of iron may be due to the presence of siderite. Although this explanation is more likely for a hard water, Davison and Heaney[118] failed to find any carbonate in their samples from a soft water lake where an excess of iron was observed. Measurements of amorphous iron sulfide particles from a soft water lake using a proton microprobe with 1 μm resolution have revealed the presence of two distinct elemental compositions, one largely comprising iron and oxygen, and the other iron, sulfur, calcium, and phosphorus (Davison et al., unpublished results). Black iron sulphide particles appear, then, to be a heterogeneous mixture of very small units of various oxidation states and composition.

Framboidal pyrite particles have been reported in the water column of a productive lake,[123] as well as in marine water columns.[137,138] Nuhfer and Pavlovic[123] performed the X-ray diffraction measurements of the framboids after the samples had been stored for an unspecified time in formalin, so the possibility that pyrite had formed during this storage time cannot be ruled out. It is, however, more likely that pyrite may form in appropriate microniches within the water column. The surfaces of clay minerals may provide suitable sites, as Nuhfer and Pavlovic[123] reported that pyritic framboids were associated with kaolinite. Alternatively, pyrite may be formed within bacterial cells.[139]

There is ample evidence for the presence of iron sulfide minerals in sediments.[30,139] Bulk chemical measurements have distinguished between acid volatile sulfur, which includes amorphous FeS and mackinawite, and pyritic iron or sulfur, which embraces greigite and pyrite. Although this categorization is not perfect,[140,141] it clearly shows that acid volatile sulfur usually dominates in the highly reducing sediments of productive lakes.[116] Pyrite is usually at low concentrations in freshwater sediments, and, based on field data, it appears to form directly in the vicinity of a redox boundary from Fe(II), S(−II) and possibly elemental sulfur.[116,142]

Reports of the occurrence of iron sulfide minerals in sediments have been compiled by Jones and Bowser[38] and Morse et al.[139] In addition to amorphous FeS, mackinawite, greigite, and euhedral and framboidal pyrite have been found. It has long been proposed[143] that pyrite may form diagenetically and that there may be a sequence of transformations such as $FeS_{am} \rightarrow$ mackinawite $\rightarrow$ greigite $\rightarrow$ pyrite. Although such processes undoubtedly operate in marine systems, it has been suggested that the relatively low concentrations of sulfur in freshwater systems may lead to stabilization of monosulfides.[144] Pyrite is known to form directly in salt marshes when the solution is periodically partially oxidizing and undersaturated with respect to monosulfides, but supersaturated with respect to pyrite.[145] Such a direct mechanism of formation

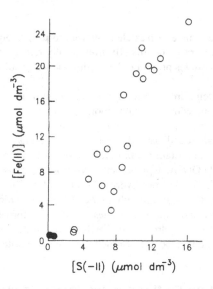

Figure 9. Correlation between colloidal Fe (II) and S(−II) in the anoxic layer of Lake Bret, ○, and of particulate Fe (II) and S(−II) in the anoxic layer of Esthwaite Water, ●. Data taken from References 21 and 127.

may also operate in freshwater sediments as pyrite has been found in apparently partially reducing environments.[126,142] However, measurements of the ion activity products for iron sulfides in the porewaters of lacustrine sediments have shown that pIAP systematically increases with increasing depth within the sediment.[117,126,130] The trend is from a pIAP of about 3, characteristic of amorphous FeS, in surface sediment, to a pIAP near to 4.4, characteristic of greigite, at depths of typically 10 or more cm.[30] It is tempting to interpret this progression as evidence for diagenetic transformations of iron sulfides,[117] but it has also been suggested that the solution composition may be determined by a kinetic balance between supply of $Fe(II)$ and $S(-II)$ and formation of the solid.[30] The presence of rapidly forming FeS_{am} at the sediment surface could reflect the greater flux of $Fe(II)$ and $S(-II)$ near to their most active site of generation, the sediment-water interface.

5. SAMPLING OF IRON PARTICULATES

Analytical measurements in natural waters must take into account the complexity, diversity, and the reactivity of the analyte, in particular during the sampling step. For instance, traditionally, the distinction between dissolved and particulate species (often defined by a 0.45 μm membrane filter) is misleading since a continuous distribution of particles is often found between ca. 10 nm and 0.45 nm and above (Figure 2).[146] In the same manner, no clear distinction between organic and inorganic compounds can be made since many "inorganic particles" such as $CaCO_3$ or Fe and Mn oxyhydroxides are coated with organics, which may alter their behavior in a natural environment. The principal difficulties involved in field studies of iron particles are the following:

(1) There is a wide range of particle sizes (nm to μm) in aquatic systems.
(2) Particles which are predominantly iron usually represent only 0.1 to 2% of the total particles present either within the water column or within the sediments.
(3) In natural systems iron particles are continually fluxed through physical (sedimentation, eddy diffusion, bioturbation) and chemical (reduction, oxidation) processes and are therefore inherently metastable.
(4) Iron particles are found under different crystalline and/or amorphous chemical phases. For instance iron oxides may exist as: hematite (α-Fe_2O_3), goethite (α-FeOOH), lepidocrocite (γ-FeOOH), maghemite (γ-Fe_2O_3), magnetite (Fe_3O_4) and ferrihydrite ($Fe_2O_3 \cdot nH_2O$). The latter phase is the group name for X-ray amorphous Fe(III) phases which are thought to originate from mononuclear species $Fe^{III}(OH)_i^{3-i}$ formed during the authigenic oxidative hydrolysis of Fe(II). Ferrihydrites are metastable and may, for instance, be transformed into lepidocrocite and hematite in sediments.[147]

5.1 Sampling of Iron Particles in the Water Column

The various techniques for collecting iron (III) particles from lakes, such as sediment traps and centrifugation or filtration, each have specific limitations. Centrifugation suffers from a lack of size selectivity for particles. Furthermore, particle coagulation and aggregation in the centrifugation tubes are important problems[148,149] which are extremely difficult to minimize for such amorphous natural particles. Finally, chemical modifications in the sample makeup, in particular in the centrifugate (due to changes in partial pressure of O_2, CO_2, and H_2S), may also induce artefacts. The use of sediment traps has been extensively reviewed in the literature.[150,151] When using them to collect iron particulates from the water column of lakes their particular features must be recognized.

(1) Because sediment traps must be deployed for at least several days during which time they collect particles from various depths within the water column, they do not give real time data, but rather information integrated with respect to time and space (depth). Furthermore, they cannot be used to collect nonaggregated colloidal sized particles (<1 μm) which do not sediment fast enough.

(2) Traps are nonselective and will therefore collect all types of particles, not only the iron rich ones. This is often a problem in the study of authigenic iron(III) (oxyhydroxy)oxides since they represent a minor ($\sim$1%) fraction of the total flux of sedimenting particles such as biotic debris, calcite, clays, and allochthonous Fe particulates. Since, the latter two contain a relatively large amount of Fe, this may lead to considerable uncertainty[43] in the measurement of the downward flux of authigenic particulate iron.

(3) Finally, since traps must be left *in situ* for several days both biological and chemical transformations may occur during the sampling time and lead to artefactual data.[151] For instance, if conditions within the trap become reducing due to biotic activity, Fe(III) particles may be reduced to dissolved Fe(II) and diffuse out of the sediment trap. Although various biocides may be used to prevent such problems, they inevitably modify the chemical environment.

Filtration on 0.45 μm filters has often been used to collect Fe particles from natural systems and to distinguish operationally between dissolved and particulate fractions, despite a growing body of evidence supporting the importance of colloidal sized Fe particles.[64,76,77,152] The main limitations of filtration may be classified into three categories (see Reference 9 for details):

(a) Contamination of the sample by trace components present in the membrane or losses of a trace compound by adsorption onto the membrane[153]

(b) Artefacts induced by long sample storage prior to filtration or long filtration times if very small flow rates are used[154]

(c) Particle coagulation at, or adsorption onto, the membrane surface[155]

Contamination is rarely a problem for studying iron particles since they are generally present at concentrations greater than μmol dm^{-3}. Artifacts due to sample modification before or during filtration can generally be minimized by performing the filtration in the field[8] or directly *in situ*.[156] However, artefacts induced by coagulation and/or adsorption at the membrane surface are more difficult to avoid. They can, however, be minimized by using a low flow rate, stirring the solution during the filtration and by using a minimum V_o/V_f ratio, where V_o and V_f are the initial and final volumes in the filtration cell.[9]

5.2 Sampling of Iron Particles from Sediments

Direct sampling of iron particles from the sediment matrix is difficult because they are generally very minor constituents (<1 to 3%) and are often compacted or aggregated with other particles and are therefore lost in a complex matrix (see also Chapter 11 of this book). A study of the literature reveals that in nearly all cases, sedimentary iron particles have been studied on whole sediment samples by spectroscopic techniques such as Mossbauer or X-ray fluorescence spectroscopy or by electron microscopy coupled to an EDS probe. Recently, however, a technique has been developed that enables the *in situ* collection of diagenetic iron and manganese oxyhydroxides.[35] It involves inserting Teflon sheets into a sediment for a few weeks depending on the upward flux of iron.

Once removed from the sediments, depending on the studied lake, spatially resolved deposits of Mn and Fe oxyhydroxides can be found on the Teflon surface. An example is shown in Figure 10 for a circumneutral lake. The use of a scanning electron microscope coupled to an energy dispersive spectroscopy probe revealed three distinct zones: (a) a Mn rich layer, (b) an intermediate transition zone ca. 500 μm thick, and (c) an Fe rich zone. The only other peaks found were Ca and Si which were minor and had a random distribution across the intersection of the Fe and Mn zones, suggesting slight contamination by other inorganic and organic sediment constituents.

This novel approach should enable considerable progress to be made in our knowledge of the properties and physicochemical structure of diagenetic iron(III) (oxyhydroxy)oxides. In particular, progress in the following areas should be attainable: *in situ* determination of conditional adsorption constants of trace elements on Fe and Mn oxyhydroxides, determination of the oxidation states of adsorbed species Cr(III,VI), As(III/V), etc., and the physicochemical study of diagenetic Fe and Mn oxyhydroxides.

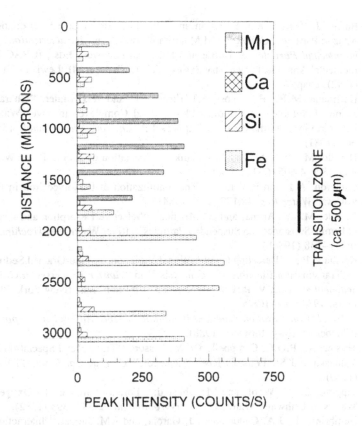

Figure 10. Energy dispersive spectroscopy analysis across iron and manganese oxy-hydroxide deposited on a Teflon collector inserted in a lake sediment.

REFERENCES

1. Mortimer, C.H. "The Exchange of Dissolved Substances between Mud and Water in Lakes: I and II," *J. Ecol.* 29:280–329 (1941).

2. Mortimer, C.H. "The Exchange of Dissolved Substances between Mud and Water in Lakes: III and IV," *J. Ecol.* 30:147–201 (1942).

3. Salomons, W. and V. Forstner. *Metals in the Hydrocycle.* (Berlin: Springer-Verlag, 1984).

4. Sulzberger, B., J.L. Schnoor, R. Giovanoli, J.G. Hering, and J. Zobrist. "Biogeochemistry of Iron in an Acidic Lake," *Aquat. Sci.* 52:56–74 (1990).

5. Gibbs, R.J. "Water Chemistry of the Amazon River," *Geochim. Cosmochim. Acta* 36:1061–1066 (1972).

6. Stumm, W. and J.J. Morgan. *Aquatic Chemistry,* 2nd ed. (New York: Wiley-Interscience, 1981).

7. Wittmann, G.T.W. and V. Forstner. "Metal Enrichment of Sediments in Inland Waters — the Jukskei and Hennops River Drainage Systems," *Water,* 2:67–72 (1976).

8. Laxen, D.P.H. and I.M. Chandler. "Size Distribution of Iron and Manganese Species in Freshwaters," *Geochim. Cosmochim. Acta* 47:731–741 (1983).

9. Buffle, J. "The Use of Filtration and Ultrafiltration for Size Fractionation of Aquatic Particles, Colloids, and Macromolecules," in *Characterization of Environmental Particles*, J. Buffle and H.P. van Leeuwen, Eds., IUPAC Environmental Analytical Chemistry Series, Vol. I, Chelsea, MI: Lewis Publishers, (1992), chap. 5.

10. Hoffmann, M.R., E.C. Yost, S.J. Eisenreich, and W.J. Maier. "Characterization of Soluble and Colloidal-Phase Metal Complexes in River Water by Ultrafiltration. A Mass Balance Approach," *Environ. Sci. Technol.* 15:655–661 (1981).

11. Hiraide, M., M. Ishi, and A. Mizuike. "Speciation of Iron in River Water," *Anal. Sci.* 4:605–609 (1988a).

12. Cameron, A.J. and P.S. Liss. "The Stabilization of Dissolved Iron in Freshwaters," *Water Res.* 18:179–185 (1984).

13. Hiraide, M., Y. Arima, and A. Mizuika. "Selective Desorption and Analysis of Humic Substances on Suspended Particles in River Water," *Mikrochim. Acta* II:231–238 (1988b).

14. Kershaw, P.J. "Detecting Compositional Variations in Fine-Grained Sediments by Transmission Electron Microanalysis," in *Transfer Processes in Cohesive Sediment Systems*, W.R. Parker and D.J.J. Kinsman, Eds. (New York: Plenum Press, 1984), 87–108.

15. Buffle, J. *Complexation Reactions in Aquatic Systems: An Analytical Approach.* (Chichester: Ellis Horwood, 1988).

16. Tessier, A., P.G.C. Campbell, and M. Bisson. "Trace Metal Speciation in the Yamaska and St. Francois River (Quebec)," *Can. J. Earth Sci.* 17:90–105 (1980).

17. Tipping, E., C. Woof, and M. Ohnstad. "Forms of Iron in the Oxygenated Waters of Esthwaite Water, UK," *Hydrobiologia* 92:383–393 (1982).

18. Nembrini, G., J.A. Capobianco, J. Garcia, and J.M. Jaquet. "Interaction between Interstitial Water and Sediment in Two Cores of Lac Léman, Switzerland," *Hydrobiologia* 92:363–375 (1982).

19. Hilton, J., W. Davison, and U. Ochsenbein. "A Mathematical Model for Analysis of Sediment Core Data: Implications for Enrichment Factor Calculations and Trace Metal Transport Mechanisms," *Chem. Geol.* 48:281–291 (1985).

20. Hilton, J., G.J. Long, J.S. Chapman, and J.P. Lishman. "Iron Mineralogy of Sediments. A Mossbauer Study," *Geochim. Cosmochim. Acta* 50:2147–2151 (1986).

21. Davison, W. and D.P.E. Dickson. "Mossbauer Spectroscopic and Chemical Studies of Particulate Iron Material from a Seasonally Anoxic Lake," *Chem. Geol.* 42:177–187 (1984).

22. Carlton, R.G. and R.G.B. Wetzel. "A Box Corer for Studying Metabolism of Epipelic Microorganisms in Sediment under In Situ Conditions," *Limnol. Oceanogr.* 30:422–426 (1985).

23. Lundesen, J.K. and B.B. Jorgensen. "Microstructure of Diffusion Boundary Layers and the Oxygen Uptake of the Sea Floor," *Nature* 345:604–607 (1990).

24. Jones, J.G., M.J.L.G. Orlandi, and B. Simon. "A Microbiological Study of Sediments from the Cumbrian Lakes," *J. Gen. Microbiol.* 115:37–48 (1979).

25. Hutchinson, G.E. *A Treatise on Limnology: Geography, Physics and Chemistry*, Vol. 1. (New York: John Wiley & Sons, 1957).

26. Wetzel, R.G. *Limnology,* 2nd ed. (New York: CBS College Publishing, 1983).

27. Davison, W. "Conceptual Models for Transport at a Redox Boundary," in *Chemical Processes in Lakes,* W. Stumm, Ed. (New York: Wiley-Interscience, 1985) pp 31–53.

28. Boesen, C. and D. Postma. "Pyrite Formation in Anoxic Environments of the Baltic," *Am. J. Sci.* 288:575–603 (1988).

29. Emerson, S. and G. Widmer. "Early Diagenesis in Anaerobic Lake Sediments. II. Thermodynamic and Kinetic Factors Controlling the Formation of Iron Phosphate," *Geochim. Cosmochim. Acta* 42:1307–1316 (1978).

30. Davison, W. "The Solubility of Iron Sulphides in Synthetic Solutions and Natural Waters," *Aquat. Sci.* in press (1991a).

31. Kjensmo, J. "The Development and Some Main Features of "Iron-Meromictic" Soft Water Lakes," *Arch. Hydrobiol. Suppl.* 32:137–312 (1967).

32. Campbell, P. and T. Torgersen. "Maintenance of Iron Meromixis by Iron Redeposition in a Rapidly Flushed Monimolimnion," *Can. J. Fish. Aquat. Sci.* 8:1303–1313 (1980).

33. Davison, W. "Oxidative and Reducing Environments in Lakes: Iron and Manganese," in *Lakes II,* A. Lerman, Ed. (New York: Springer-Verlag, 1992).

34. Davison, W., G.W. Grime, J.A.W. Morgan, and K. Clarke. "Microstructure of Dissolved Iron in Sediment Porewaters Using a New Technique of Diffusive Equilibration in a Thin Film (DET)," *Nature* 352:323–325 (1991).

35. Belzile, N., R.R. DeVitre, and A. Tessier. "In Situ Collection of Diagenetic Iron and Manganese Oxyhydroxides from Natural Sediments," *Nature* 340:376–377 (1989).

36. Gorham, E. and D.J. Swaine. "The Influence of Oxidizing and Reducing Conditions Upon the Distribution of Some Elements in Lake Sediments," *Limnol. Oceanogr.* 10:268–279 (1965).

37. Callender, E. and C.J. Bowser. "Freshwater Ferromangenese Deposits," in *Handbook of Strata-Bound and Stratiform Ore Deposits,* Vol. 7, K.H. Wolf, Ed. (New York: Elsevier, 1976).

38. Jones, B.F. and C.J. Bowser. "The Mineralogy and Related Chemistry of Lake Sediments," in *Lakes: Chemistry, Geology, Physics,* A. Lerman, Ed. (New York: Springer-Verlag, 1978).

39. Lungren, D.G. and W. Dean. "Biogeochemistry of Iron," in *Biogeochemical Cycling of Mineral-Forming Elements,* P.A. Trudinger and D.J. Swaine, Eds. (Amsterdam: Elsevier, 1979).

40. Matisoff, G., A.H. Lindsay, S. Matis, and F.M. Foster. "Trace Metal Mineral Equilibria in Lake Erie Sediments," *J. Great Lakes Res.* 6:353–366 (1980).

41. Wersin, P., P. Hohener, R. Giovanoli, and W. Stumm. "A Kinetic Model for Iron Diagenesis in a Lake Sediment," *Chem. Geol.* 84:210–211 (1990).

42. Imboden, D.M. and R.P. Schwarzenbach. "Spatial and Temporal Distribution of Chemical Substances in Lakes: Modelling Concepts," in *Chemical Processes in Lakes,* W. Stumm, Ed. (New York: Wiley-Interscience, 1985), pp 1–30.

43. Davison, W., C. Woof, and E. Rigg. "The Dynamics of Iron and Manganese in a Seasonally Anoxic Lake; Direct Measurement of Fluxes Using Sediment Traps, *Limnol. Oceanogr.* 27:987–1003 (1982).

44. Buffle, J., R.R. DeVitre, D. Perret, and G.G. Leppard. "Physico-Chemical Characteristics of a Colloidal Iron Phosphate Species Formed at the Oxic-Anoxic Interface of a Eutrophic Lake," *Geochim. Cosmochim. Acta* 53:399–408 (1989).

45. Davison, W., S.I. Heaney, J.F. Talling, and E. Rigg. "Seasonal Transformations and Movements of Iron in a Productive English Lake with Deep-Water Anoxia," *Schweiz. Z. Hydrol.* 42:196–224 (1980).

46. Yagi, A. and I. Shimodaira. "Seasonal Changes of Iron and Manganese in Lake Fukami-ike. Occurrence of Turbid Manganese Layer," *Jpn. J. Limnol.* 47:279–289 (1986).

47. DeVitre, R.R., J. Buffle, D. Perret, and R. Baudat. "A Study of Iron and Manganese Transformations at the $O_2/S(-II)$ Transition Layer in a Eutrophic Lake (Lake Bret, Switzerland): A Multi-Method Approach," *Geochim. Cosmochim. Acta* 52:1606–1613 (1988).

48. Tipping, E. and M. Ohnstad. "Colloid Stability of Iron Oxide Particles from a Freshwater Lake," *Nature* 308:266–268 (1984).

49. Talling, J.F., "The Development of Attenuance Depth-Profiling to Follow the Changing Distribution of Phytoplankton and Other Particulate Material in a Productive English Lake," *Arch. Hydrobiol.* 93:1–20 (1981).

50. Heaney, S.I., W.J.P. Smyly, and J.F. Talling. "Interactions of Physical, Chemical and Biological Processes in Depth and Time within a Productive English Lake During Summer Stratification," *Int. Rev. Hydrobiol.* 71:441–494 (1986).

51. Millero, F.J. "The Effect of Ionic Interactions on the Oxidation of Metals in Natural Waters," *Geochim. Cosmochim. Acta* 49:547–533 (1985).

52. Millero, F.J., S. Sotolongo, and M. Izaguirre. "The Oxidation Kinetics of Fe(II) in Seawater," *Geochim. Cosmochim. Acta* 51:793–801 (1987).

53. Davison, W. and G. Seed. "The Kinetics of the Oxidation of Ferrous Iron in Synthetic and Natural Waters," *Geochim. Cosmochim. Acta* 47:67–79 (1983).

54. Lowson, R.T. "Aqueous Oxidation of Pyrite by Molecular Oxygen," *Chem. Rev.* 82:461–497 (1982).

55. Sung, W. and J.J. Morgan. "Kinetics and Product of Ferrous Iron Oxygenation in Aqueous Systems," *Environ. Sci. Technol.* 14:561–568 (1980).

56. Wehrli, B. "Redox Reactions of Metal Ions at Mineral Surfaces," in *Aquatic Chemical Kinetics*, W. Stumm, Ed. (New York: Wiley-Interscience, 1990).

57. Waite, T.D. and F.M.M. Morel. "Photoreductive Dissolution of Colloidal Iron Oxides in Natural Waters," *Environ. Sci. Technol.* 18:860–868 (1984).

58. Walling, C. "Fenton's Reagent Revisited," *Acc. Chem. Res.* 8:125–131 (1975).

59. Moffet, J.W. and R.G. Zika. "Reaction Kinetics of Hydrogen Peroxide with Copper and Iron in Seawater," *Environ. Sci. Technol.* 21:804–810 (1987).

60. Colliene, R.H. "Photoreduction of Iron in the Epilimnion of Acidic Lakes," *Limnol. Oceanogr.* 28:83–100 (1983).

61. McKnight, D.M., B.A. Kimball, and K.E. Bencala. "Iron Photoreduction and Oxidation in an Acidic Mountain Stream," *Science* 240:637–640 (1988).

62. Nordstrom, D.K. "The Rate of Ferrous Iron Oxidation in a Stream Receiving Acid Mine Effluent," *U.S. Geol. Surv. Water-Supply Pap.* 2270:112–119 (1985).

63. Sulzberger, B. "Photoredox Reactions at Hydrous Metal Oxide Surfaces; a Surface Coordination Chemistry Approach," in *Aquatic Chemical Kinetics*, W. Stumm, Ed. (New York: Wiley-Interscience, 1990).

64. Tipping, E., D.W. Thompson, and C. Woof. "Iron Oxide Particulates Formed by the Oxygenation of Natural and Model Lakewaters Containing Fe(II)," *Arch. Hydrobiol.* 115/1:59–70 (1989).

65. Miles, C.J. and P.L. Brezonik. Oxygen Consumption in Humic-Colored Waters by a Photochemical Ferrous-Ferric Catalytic Cycle," *Environ. Sci. Technol.* 15:1089–1095 (1981).

66. Luther, G.W. "The Frontier-Molecular Orbital Theory Approach in Geochemical Processes," in *Aquatic Chemical Kinetics*, W. Stumm, Ed. (New York: Wiley-Interscience, 1990).

67. Jones, B.F., V.C. Kennedy, and G.W. Zellweger. "Comparison of Observed and Calculated Concentrations of Dissolved Al and Fe in Stream Water," *Water Resour. Res.* 10:791–793 (1974).

68. Baes, C.F. and R.E. Mesmer. *The Hydrolysis of Cations*. (New York: Wiley Interscience, 1976).

69. Schneider, W. and B. Schwyn. "The Hydrolysis of Iron in Synthetic, Biological and Aquatic Media," in *Aquatic Surface Chemistry*, W. Stumm, Ed. (New York: Wiley-Interscience, 1987), pp. 167–196.

70. Sylva, R.R. "The Hydrolysis of Iron (III)," *Rev. Pure Appl. Chem.* 22:115 (1972).

71. Spiro, T.C., S.E. Allerton, J. Renner, A. Terzis, R. Bils, and P. Saltmann. "The Hydrolytic Polymerization of Iron III," *J. Am. Chem. Soc.* 88:2721–2726 (1966).

72. Schwertmann, U. and H. Thalmann. "The Influence of [FeII], [Si], and pH on the Formation of Lepidocrocite and Ferrihydrite during Oxidation of Aqueous $FeCl_2$ Solutions," *Clay Miner.* 11:189–200 (1976).

73. Robinson, R.B., T. Demirel, and E.R. Baumann. "Identity and Character of Iron Precipitates," *J. Environ. Eng. Div.* 10:1211–1227 (1981).

74. Schwertmann, U. and W.R. Fischer. "Natural 'Amorphous' Ferric Hydroxide," *Geoderma* 10:237–247 (1973).

75. Schwertmann, U. and R.M. Taylor. "The Transformation of Lepidocrocite to Geothite," *Clays Clay Miner.* 20:151–158 (1972).

76. Tipping, E., C. Woof, and D. Cooke. "Iron Oxide from a Seasonally Anoxic Lake," *Geochim. Cosmochim. Acta* 45:1411–1419 (1981).

77. Leppard, G.G., J. Buffle, R.R. DeVitre, and D. Perret. "The Ultrastructure and Physical Characteristics of a Distinctive Colloidal Iron Particulate Isolated from a small Eutrophic Lake," *Arch. Hydrobiol.* 113:405–424 (1988).

78. Tipping, E. "Humic Substances and the Surface Properties of Iron Oxides in Freshwaters," in *Transfer Processes in Cohesive Sediment Systems*, W.R. Parker and D.J.J. Kinsman, Eds. (New York: Plenum Press, 1984).

79. Stumm, W. and G. Furrer. "The Dissolution of Oxides and Aluminium Silicates; Examples of Surface Coordination Controlled Kinetics," in *Aquatic Surface Chemistry*, W. Stumm, Ed. (New York: Wiley-Interscience, 1987), pp. 197–219.

80. Lovely, D.R. and E.J.P. Phillips. "Novel Mode of Microbial Energy Metabolism: Organic Carbon Oxidation Coupled to Dissimilatory Reduction of Iron and Manganese," *Appl. Environ. Microbiol.* 54:1472–1480 (1988).

81. Arnold, R.G., T.J. DiChristina, and M.R. Hoffmann. "Dissimilative Fe(III) Reduction by Pseudomonas sp. 200-Inhibitor Studies," *Appl. Environ. Microbiol.* 52:281–294 (1988).

82. Carpenter, E.J. and C.C. Price, IV. "Marine Oscillatoria (Trichodesmium): Explanation for Aerobic Nitrogen Fixation without Heterocysts," *Science* 191:1278–1280 (1976).

83. Richardson, L.L., C. Aguilar, and K.H. Nealson. "Manganese Oxidation in pH and O_2 Microenvironments Produced by Phytoplankton, *Limnol. Oceanogr.* 33:352–363 (1988).

84. Petrovic, R., R.A. Berner, and M.B. Goldhaber. "Rate Control in Dissolution of Alkali Feldspars. I. Study of Residual Feldspar Grains by X-ray Photoelectron Spectroscopy," *Geochim. Cosmochim. Acta* 40:537–548 (1976).

85. Berner, R.A. and G.R. Holdren, Jr. "Mechanism of Feldspar Weathering. II. Observations of Feldspars from Soils," *Geochim. Cosmochim. Acta* 43:1173–1186 (1979).

86. Wehrli, B., S. Ibric, and W. Stumm. "Adsorption Kinetics of Vanadyl(IV) and Chromium(III) to Aluminium Oxide: Evidence for a Two Step Mechanism," *Colloids Surf.* in press (1990).

87. Hachiya, K., M. Sasaki, T. Ikeda, N. Mikami, and T. Yasunga. Static and Kinetic of Adsorption-Desorption. II. Kinetic Study by Means of Pressure Jump Technique," *J. Phys. Chem.* 88:27–31 (1984).

88. Hayes, K.F. and J.O. Leckie. "Mechanism of Lead Ion Adsorption at the Goethite Water Interface," in *Geochemical Processes at Mineral Surfaces*, J.A. Davis and K.F. Hayes, Eds. (Washington, D.C.: American Chemical Society, 1986), pp. 114–141.

89. Sulzberger, B., D. Suter, C. Siffert, S. Banwert, and W. Stumm. Dissolution of Fe(III)(hydr)oxides in Natural Waters; Laboratory Assessment on the Kinetics Controlled by Surface Coordination," *Mar. Chem.* 28:127–144 (1989).

90. Schott, J. and J.-C. Petit. "New Evidence for the Mechanisms of Dissolution of Silicate Minerals," in *Aquatic Surface Chemistry*, W. Stumm, Ed. (New York: Wiley-Interscience, 1987).

91. Blum, A.E. and A.C. Lasaga. "Monte Carlo Simulations of Surface Reaction Rate Laws," in *Aquatic Surface Chemistry*, W. Stumm, Ed. (New York: Wiley-Interscience, 1987).

92. Furrer, G. and W. Stumm. "The Coordination Chemistry of Weathering. I. Dissolution Kinetics of $-Al_2O_3$ and BeO," *Geochim. Cosmochim. Acta* 50:1847–1860 (1986).

93. Zinder, B., G. Furrer, and W. Stumm. "The Coordination Chemistry of Weathering. II. Dissolution of Fe(III) Oxides," *Geochim. Cosmochim. Acta* 50:1861–1869 (1986).

94. Banwart, S., S. Davies, and W. Stumm. "The Role of Oxalate in Accelerating the Reductive Dissolution of Hematite (α-Fe_2O_3) by Ascorbate," *Colloids Surf.* in press (1990).

95. Banwart, S. "The Reductive Dissolution of Hematite (α-Fe_2O_3) by Ascorbate," Ph.D. thesis, No. 8934, ETH, 1989.

96. Schnoor, J.L., B. Sulzberger, R. Giovanoli, L. Sigg, W. Stumm, and J. Zobrist. "Fate of Iron and Aluminium in Lake Cristallina, Switzerland," in preparation (1990).

97. Parker, C.A. and C.G. Hatchard. "Photodecomposition of Complex Oxalate. Some Preliminary Experiments by Flash Photolysis," *J. Phys. Chem.* 63:22–26 (1959).

98. Siffert, C. "L'effect de la lumière sur la dissolution des oxydes de fer(III) dans les milieux aqueoux," Ph.D. thesis No. 8852, ETH, 1989.

99. Suter, D., C. Siffert, B. Sulzberger, and W. Stumm. "Catalytic Dissolution of Iron(III) Hydroxides by Oxalic Acid in the Presence of Fe(II)," *Naturwissenschaften*, 75:571–573 (1988).

100. Sunda, W.G., S.A. Huntsman, and G.R. Harvey. "Photoreduction of Manganese Oxides in Seawater and its Geochemical Biological Implications," *Nature* 301:234–236 (1983).

101. Price, N. and F.M.M. Morel. "Role of Extracellular Enzymatic Reactions in Natural Waters," in *Aquatic Chemical Kinetics*, W. Stumm, Ed. (New York: Wiley-Interscience, 1990).

102. Pysik, A.J. and S.E. Sommer. "Sedimentary Iron Monosulfides: Kinetics and Mechanism of Formation," *Geochim. Cosmochim. Acta* 45:687–698 (1981).

103. Trollope, D.R. and B. Evans. "Concentrations of Copper, Iron, Lead, Nickel and Zinc in Freshwater Algal Blooms," *Environ. Pollut.* 11:109–116 (1976).

104. Gerloff, G.C. and F. Skoog. "Availability of Iron and Manganese in Southern Wisconsin Lakes for the Growth of *Microcystis aeruginosa*," *Ecology* 38:551–556 (1957).

105. Udel'nova, T.M., E.N. Kondral'eva, and E.A. Boichenko. "Iron and Manganese Content in Various Photosynthesizing Microorganisms," *Mikrobiologiya* 37:197–200 (1968).

106. Sholkovitz, E.R. and D. Copland. "The Chemistry of Suspended Matter in Esthwaite Water, a Biologically Productive Lake with Seasonally Anoxic Hypolimnion," *Geochim. Cosmochim. Acta* 46:393–410 (1982).

107. Sigg, L. "Metal Transfer Mechanisms in Lakes: the Role of Settling Particles," in *Chemical Processes in Lakes*, W. Stumm, Ed. (New York: Wiley-Interscience, 1985).

108. Hamilton-Taylor, J., M. Willis, and C.S. Reynolds. "Depositional Fluxes of Metals and Phytoplankton in Windermere as Measured by Sediment Traps," *Limnol. Oceanogr.* 29:695–710 (1984).

109. Muller, G. and V. Forstner. "Recent Iron Ore Formation in Lake Malawi, Africa," *Miner. Deposita (Berlin)* 8:278–290 (1973).

110. Manning, P.G., W. Jones, and T. Birchall. "Mossbauer Spectral Studies of Iron-Enriched Sediments from Hamilton Harbor, Ontario," *Can. Mineral.* 18:291–299 (1980).

111. Ljunggren, P. "Geochemistry and Radioactivity of Some Mn and Fe Bog Ores," *Geol. Foren. Forhandl. (Stockholm)* 77:33–44 (1955a).

112. Ljunggren, P. "Differential Thermal Analysis and X-ray Examination of Fe and Mn Bog Ores," *Geol. Foren. Forhandl. (Stockholm)* 77:135–147 (1955b).

113. Nriagu, J.O. and C.I. Dell. "Diagenetic Formation of Iron Phosphates in Recent Lake Sediments," *Am. Mineral.* 59:934–946 (1974).

114. Postma, D. "Formation of Siderite and Vivianite and the Pore Water Composition of a Recent Bog Sediment in Denmark," *Chem. Geol.* 31:225–244 (1981).

115. Nembrini, G., J.A. Capobianco, A. Williams, and M. Viel. "A Mossbauer and Chemical Study of the Formation of Vivianite in Sediments of Lago Maggiore (Italy)," *Geochim. Cosmochim. Acta* 47:1459–1464 (1983).

116. Davison, W., J.P. Lishman, and J. Hilton. "Formation of Pyrite in Freshwater Sediments: Implications for C/S Ratios," *Geochim. Cosmochim. Acta* 49:1615–1620 (1985).

117. Cook, R.B. "Distribution of Ferrous Iron and Sulphide in an Anoxic Hypolimnion," *Can. J. Fish Aquat. Sci.* 41:286–293 (1984).

118. Davison, W. and S.I. Heaney. "Ferrous Iron-Sulphide Interactions in Anoxic Hypolimnetic Waters," *Limnol. Oceanogr.* 23:1194–1200 (1978).

119. Browne, P.R.L. and C.P. Wood. "Mackinawite and Pyrite in a Hot Spring Deposite, Mohaka River, New Zealand," *N. Jb. Miner. Mk.* H10:468–475 (1974).

120. Dell, C.I. "An Occurrence of Greigite in Lake Superior Sediments," *Am. Miner.* 57:1303–1304 (1972).

121. Manning, P.G. and L.A. Ash. "Mossbauer Spectral Studies of Pyrite, Ferric and High-Spin Ferrous Distributions in Sulphide-Rich Sediments from Moira Lake, Ontario," *Can. Mineral.* 17:111–115 (1979).

122. Manning, P.G., J.D.H. Williams, M.N. Charleton, L.A. Ask, and T. Birchall. "Mossbauer Spectral Studies of the Diagenesis of Iron in a Sulphide-Rich Sediment Core," *Nature* 280:134–136 (1979).

123. Nuhfer, E.B. and A.S. Pavlovic. "Association of Kaolinite with Pyrite Framboids," *J. Sed. Petrol.* 49:321–324 (1979).

124. Nriagu, J.O. and Y.K. Soon. "Distribution and Isotope Composition of Sulphur in Lake Sediments of Northern Ontario," *Geochim. Cosmochim. Acta* 49:823–834 (1985).

125. Dell, C.I. "Pyrite Concretions in Sediments from South Bay, Lake Huron," *Can. J. Earth Sci.* 12:1077–1083 (1975).

126. Psenner, R. "Die entstehung von pyrit in rezenten sedimenten des Piuburger Sees," *Schweiz. Z. Hydrol.* 45:219–232 (1983).

127. Buffle, J., O. Zali, J. Zumstein, and R. DeVitre. "Analytical Methods for the Direct Determination of Inorganic and Organic Species: Seasonal Changes of Iron, Sulphur, and Pedogenic and Aquogenic Organic Constituents in the Eutrophic Lake Bret, Switzerland," *Sci. Total Environ.* 64:41–59 (1987).

128. Mayer, L.M., F.P. Liotta, and S.A. Norton. "Hypolimnetic Redox and Phosphorus Cycling in Hypereutrophic Lake Sebasticook, Maine," *Water Res.* 16:1189–1196 (1982).

129. Verdouw, H. and E.M.J. Dekkers. "Iron and Manganese and Lake Vechten (The Netherlands); Dynamics and Role in the Cycle of Reducing Power," *Arch. Hydrobiol.* 89:509–532 (1980).

130. Emerson, S. "Early Diagenesis in Anaerobic Lake Sediments: Chemical Equilibria in Interstitial Waters," *Geochim. Cosmochim. Acta* 40:925–934 (1976).

131. Carignan, R. and J.O. Nriagu. "Trace Metal Deposition and Mobility in the Sediments of Two Lakes Near Sudbury, Ontario," *Geochem. Cosmochim. Acta* 49:1753–1764 (1985).

132. Anthony, R.S. "Iron-rich Rhythmically Laminated Sediments in Lake of Clouds, Northwestern Minn.," *Limnol. Oceanogr.* 22:45–54 (1977).

133. Anderson, N.J. and B. Rippey. "Diagenesis of Magnetic Minerals in the Recent Sediments of a Eutrophic Lake," *Limnol. Oceanogr.* 33:1476–1492 (1988).

134. Berner, R.A. "A new geochemical classification of sedimentary environments," *J. Sed. Petrol.* 51:359–365 (1981).

135. Davison, W. "Soluble inorganic ferrous complexes in natural waters," *Geochim. Cosmochim. Acta* 43:1693–1696 (1979).

136. Buffle, J., R.R. DeVitre, D. Perret, and G.G. Leppard. "Combining field measurements for speciation in non perturbable water samples," in *Metal Speciation: Theory, Analysis and Application,* J.R. Kramer and H.E. Allen, Eds. (Chelsea, M.I.: Lewis Publishers, 1988).

137. Ross, D.A. and E.T. Degens. "Recent sediment of Black Sea," in *The Black Sea: Geology, Chemistry and Biology,* E.T. Degens and D.A. Ross, Eds. AAPG Mem., 20, pp. 183–199 (1974).

138. Jacobs, L., S. Emersen, and J. Skei. "Partitioning and transport of metals across the O_2/H_2S interface in a permanently anoxic basin: Framvaren Fjord, Norway," *Geochim. Cosmochim. Acta* 49:1433–1444 (1985).

139. Morse, J.W., F.J. Millero, J.C. Cornwell, and D. Rickard. "The chemistry of hydrogen sulphide and iron sulphide systems in natural waters," *Earth Sci. Rev.* 24:1–42 (1987).

140. Canfield, D.E., R. Raiswell, J.T. Westrich, C.M. Reaves, and R.A. Berner. "The use of chromium reduction in the analysis of reduced inorganic sulphur in sediments and shales," *Chem. Geol.* 54:149–155 (1986).

141. Cornwell, J.C. and J.W. Morse. "The Characterization of Iron Sulphide Minerals in Anoxic Marine Sediments," *Mar. Chem.* 22:193–206 (1987).

142. Davison, W. "Interactions of Iron, Carbon and Sulphur in Marine and Lacustrine Sediments," in *Lacustrine Petroleum Source Rocks*, A.J. Fleet, K. Kelts, and M.R. Talbot, Eds. Geol. Soc. Spec. Publ. No. 40, pp. 131–137 (1988).

143. Berner, R.A. "Sedimentary Pyrite Formation." *Am. J. Sci.* 268:1–23 (1970).

144. Berner, R.A., T. Baldwin, and G.R. Holdren. "Authigenic Iron Sulphides as Paleosalinity Indicators," *J. Sed. Petrol.* 49:1345–1350 (1979).

145. Howarth, W. "Pyrite: its Rapid Formation in a Salt Marsh and its Importance to Ecosystem Metabolism," *Science* 203:49–51 (1979).

146. Perret, D., R.R. De Vitre, G.G. Leppard, and J. Buffle. "Characterizing Autochthonous Iron Particles and Colloids — the Need for Better Particle Analysis Methods," in *Large Lakes*, M. Tilzer and C. Serruya, Eds. (Berlin: Springer-Verlag, 1990), pp. 224–244.

147. Schwertmann, U. and W.R. Fischer. "Zur Bildung von α-FeOOH und α-Fe$_2$O$_3$ aus amorphem Eisen(III) hydroxid," *Z. Anorg. Allg. Chem.* 346:137–142 (1966).

148. Kavanaugh, M.C., U. Zimmermann, and A. Vagenknecht. "Determination of Particle Size Distributions in Natural Waters: Use of a Zeiss Micro-Videomat Image Analyser," *Schweiz. Z. Hydrol.* 39:86–98 (1977).

149. Salim, R. and and B.G. Cooksey. "The Effect of Centrifugation on the Suspended Particles of River Waters," *Water Res.* 15:835–839 (1981).

150. Simpson, W.R. "Particulate Matter in the Oceans: Sampling Methods, Concentration, Size Distribution and Particle Dynamics," *Oceanogr. Mar. Biol. Annu. Rev.* 20:119–172 (1982).

151. Bloesh, J. and N.M. Burns. "A Critical Review of the Sedimentation Trap Technique," *Schweiz. Z. Hydrol.* 42:15–55 (1980).

152. Nomizu, T., K. Goto, and A. Mizuike. "Electron Microscopy of Nanometer Particles in Freshwater," *Anal. Chem.* 60:2653–2656 (1988).

153. Truitt, R.E. and J.H. Weber. "Trace Metal Ion Filtration Losses at pH 5 and 7," *Anal. Chem.* 51:2057–2059 (1979).

154. Laxen, D.P.H. and I.M. Chandler. "Comparison of Filtration Techniques for Size Distribution in Freshwaters," *Anal. Chem.* 54:1350–1355 (1982).

155. Suki, A., A.G. Fane, and C.J.D. Fell. "Flux Decline in Protein Ultrafiltration," *J. Membr. Sci.* 21:269–283 (1984).

156. De Vitre, R.R., F.J. Bujard, C. Bernard, and J. Buffle. "A Novel in situ Cascade Ultrafiltration Unit Specifically Designed for Field Studies of Anoxic Waters," *Int. J. Env. Anal. Chem.* 31:145–163 (1987).

157. Wilbur, W.G. and J.V. Hunter. "The Impact of Urbanization on the Distribution of Heavy Metals in the Bottom Sediments of the Saddle River," *Water Res. Bull.* 15:790–800 (1979).

158. Hering, J.G. and W. Stumm. "Oxidative and Reductive Dissolution of Minerals," in *Reviews in Mineralogy*, Vol. 23, M.F. Hochella and A.F. White, Eds. (Washington, D.C.: Mineralogical Society of America, 1990), chap. II.

CHAPTER 9

Characterization of Oceanic Biogenic Particles

Clarice M. Yentsch

Bigelow Laboratory for Ocean Sciences, McKown Point, West Boothbay Harbor, Maine

and

Charles S. Yentsch

Boston University Remote Sensing Center, Department of Geography, Boston, Massachusetts

TABLE OF CONTENTS

1.   Introduction — The Time and Space Distribution of
      Ocean Particles 358

2.   Background on Optical Budgets 360

3.   Methods Used 361
         3.1   Light Absorption 362
         3.2   Light Scattering 365
         3.3   Impedance Volume......................................365
         3.4   Fluorescence Flow Cytometry 366

4.   Examples of Biogenic Particle Measurements in the
         Oceans by Flow Cytometry 368

5.   Biochemical Factors Regulating Cell Size 371

6.   What are the Majority of Particles?............................374

7.   Conclusions...374

Acknowledgments..374

References...374

1. INTRODUCTION — THE TIME AND SPACE DISTRIBUTION OF OCEAN PARTICLES

In the past, oceanographic observations have been wed to the research vessel and its umbilical wire/cords on which sensors or collectors are suspended. This approach has determined experimental design, and more importantly interpretation. Against the background of the enormity of the oceans and its manifold physical, chemical, optical, and biological processes, we can honestly ask whether or not these ship-based research techniques have provided valid information for predicting organic particles in the ocean environment. This remains an open question, but satellite data are providing new clues about time and spatial distribution, some of which are germane to particle distribution.

Remote sensing from space, (satellite altitudes ~500 miles) permits testing of the validity of ocean processes based on sea observations. The Coastal Zone Color Scanner (CZCS), which infers patches of photosynthetic particles from reflectance/water color, has demonstrated the enormity of the challenge. Consequently, biological oceanographers must move towards the utilization of instrumentation and methods that provide rapid sampling over both time and space. At sea, ocean biologists need instrumentation that at least matches the capability of data sampling and handling of their chemical and physical colleagues, and this instrumentation must encompass measurement of all particle sizes.

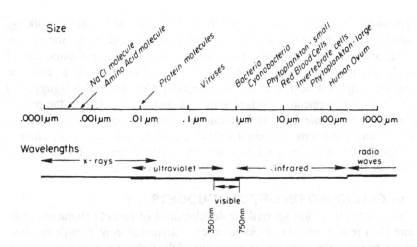

Figure 1. Relative sizes of various biogenic particles and molecules in proportion to the electronmagnetic spectrum wavelengths. Particles of interest span visible wavelengths and near infrared sizes. (Adapted from Yentsch, M. and Pomponi, S. *Annu. Rev. Cytol.* 105:183–243 (1986).)

Most of the microscopic organisms in the sea exist as single cells. In this chapter the focus is on a limited size group including bacteria, cyanobacteria, prochlorophytes, and other phytoplankton cells, and other particles spanning 0.2 to 200 μm in diameter (Figure 1). The emphasis is on chlorophyll-containing microautotrophs (nutrition derived from self) often referred to as microalgae or phytoplankton, detritus (decaying organic matter), single-celled microheterotrophs (nutrition derived from others), and suspended sediments.

Single-celled organisms, with their complicated intrinsic responses, are dictated by physiology and biochemistry bounded by genetic limits. These living particles must respond to ever-changing extrinsic geophysical variables. This fact is pivotal to our understanding of processes of global climatology and geochemistry. The understanding of these processes can only be achieved by observations of life patterns (patches of particles) in time and space coupled with knowledge of the wide diversity of responses of various cell types.

A new class of instruments important to the study of these microscopic ocean particles has evolved which is compatible with sampling frequencies and resolution of the automated instruments used in chemical and physical oceanography. These instruments are in large part optical, that is, exploit optical properties of the water mass and/or its light attenuating components. The result has been that ocean optics have coalesced with major ecological concepts of biological oceanography.

One of the amazing things about living microscopic particles is their diversity in size and shape. Most theorists account for the density in terms of the adaptive measures needed for a cell to survive in the seawater medium, such as sinking/buoyancy and nutrient absorption. A common argument involves the physiological values concerned with surface area to cell volume relationship, in relation to the fact that the cell volume and its interior workings must be serviced by ion transport through a cell membrane. As the volume of the cell increases, there is a point where surface area absorption cannot supply required substances. At this point the cell must divide to arrive at some more optimal surface area to volume ratio. Cell shape also is suggested as a means of augmenting surface area to volume. In particular the prolate spheroids and cylindrical shapes commonly observed appear to be ideal for low nutrient conditions and retard sinking out of the euphotic zone. There is a great deal of controversy on these subjects; and the reasons for the diverse nature of populations of living cells is poorly understood.

2. BACKGROUND ON OPTICAL BUDGETS

No other activity has spurred the development of particle characterization more than remote sensing of ocean color. Interpretation of change in ocean color depends upon accurate measurement of the light upwelled from the sea. This is not a trivial measurement, since we know that most of the light entering the sea surface (95%) is retained. But, what does this tell us of particles in the sea? First, the fate of most photons is to be absorbed by water itself, by colored particles, or by dissolved organics (Figure 2, i through iii). The value of 5% upwelled from the sea is, therefore, that which is not absorbed but scattered or fluoresced by organic and inorganic particles.

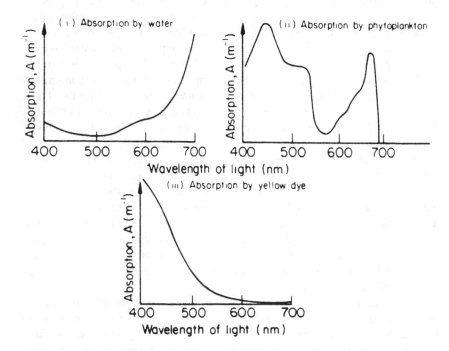

Figure 2. Absorption spectra of water (i), phytoplancton (ii), and fulvic compounds ($\approx$ dissolved organic matter) (iii).

All known substances which absorb light in the sea are wavelength specific (Figure 2). Whereas water strongly absorbs in the red regions of the visible spectrum, phytoplankton and dissolved organics absorb strongly in the blue region of the spectrum. A major part of dissolved yellow organics is terrestrial (fulvus) and humics which are at very low concentration in the open ocean. Consequently, the open ocean is relatively free of this substance.

Scattering of light by water occurs at the molecular level after the initial refraction at the sea surface. In addition, there is Raman scattering from water molecules which is somewhat analogous to fluorescence, that is, there is a spectral shift of light emitted such that the light scattered is always of greater wavelength than the illumination source.

One important feature in scattering of submicron particles in the oceans is that the wavelength of light is approaching the particle size. Accordingly, the variation of scattering with particle size and wavelength approaches the region of largest oscillation of the effective scattering area coefficient as described by Van de Hulst.[1] For very small particles, there is a great deal of Rayleigh scattering at the shorter wavelengths while at longer wavelengths there is relatively less scattering.

In the ocean, water absorption coupled with absorption by photosynthetic pigments dominates light absorption in the medium. In the evolutionary course of photosynthesis (3.5 billion years as compared to the 4.6 billion year history of planet Earth), organisms developed antennae for capturing sunlight which is the primary source of energy for the marine ecosystem. These antennae appear as pigmented proteins which absorb and fluoresce light in a manner similar to colored dyes and in the process drive photosynthesis. The pigmented proteins fill the visible window of the sunlight's radiation (Figure 3). In addition, pigments are found in the ultraviolet region of the spectrum. These ultraviolet pigments have a photoprotective role. Photosynthetic bacterial pigments extend into the infrared. Although no microalga or photosynthetic bacterium species has all of the pigments, most have a pigment suite, with some accessory pigments in addition to chlorophyll. Microalgae can vary their pigment complement and most are well adapted to their niche of growth.

There is no obvious relationship between algal particle size and pigment complement. The pigments shown in Figure 3 are found in cells as small as 0.2 μm and as large as sea weeds. Not only do the particles of photosynthetic autotrophs absorb light, they also scatter and reflect light. There exists a selectivity in both wavelength change and the amounts of scattering. In the case of the latter, the amount backscattered from some species is less than 10% of the incident light, whereas in others it can be far more than 10%.

3. METHODS USED

Light absorption, light scattering, impedance volume, and fluorescence flow cytometry are the methods of choice for quantitation of oceanic biogenic particles. A summary of limitations and promises of each is given in Table 1.

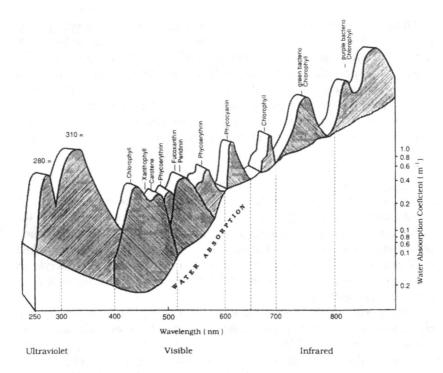

Figure 3. Absorption characteristics of pigments present in various bacterial and microalgal particles, superimposed on water absorption. Spectral range from the ultraviolet (250 nm) to the infrared (900 nm). Absorbance peaks at 280 and 319 nm represent ultraviolet photoprotective pigments.

3.1 Light Absorption

Light absorption is a standard method for pigmented particle quantitation.[2-4] It is frequently used in water quality assessment of lakes, ponds, streams, and oceans. The crudest *in situ* bulk measurements can be estimated with a Secchi disc, a standard-sized white disc submerged into the water and lowered to a depth where it is no longer visible. Strictly speaking, this is a reflectance measurement, but such has been found to be proportional to absorption. Satellite and aircraft remote sensing colorimetry is based on this principle. Absorption of natural samples is frequently too small to be measurable without preconcentration. For sea and laboratory observations then, 0.5 to 2.0 liters of sample are filtered through a 2.5 cm Whatman GF/F glass-fiber filter. The wet filter with the particulate matter is held upright in a Lucite holder in the spectrophotometer cuvette compartment, and an identical wet blank filter is used in the reference light path (Figure 4A). The diffuse attenuation spectrum as a function of wavelength is scanned between 750 and 350 nm with a dual beam spectrophotometer. This permits following the relative change of absorbance with conditions (e.g., depth or season) but not to estimate the number of phytoplankton cells in the sample. For this purpose, the absorbance of the glass-fiber filters must be converted into that of the equivalent suspension of

Table 1. Limitations and Advantages of Various Particle Measurements

	Instrument of choice	Platform for measurement	Major limitation	Major advantage	Ref.
Light absorption	Dual beam spectrophotometer	Shipboard laboratory	Best when coupled with individual particle measurements	Derive spectrum includes total photosynthetic pigments	2,3,5
	Spectroradiometer	Shipboard, in situ, winch, and data conducting wire	Best when coupled with individual particle measurements	Derive spectrum includes total photosynthetic pigments	4
Light scattering beam attenuation	Transmissometer	Shipboard, in situ, winch, and data conducting wire	Best when corresponding in situ particle and chlorophyll data are available	By-passes ambient light characteristics; inexpensive; rapid 1 m/min; continuous profile with water depth	1,7–10,20
Impedance volume	Coulter particle sizing and counting device	Shipboard laboratory	Best when coupled to in vivo fluorescence simultaneous detection as in flow cytometry	Measurement on individual particle basis; rapid 1000/s	10–13,31
Fluorescence flow cytometry	Multiple parameters of cell size, light scatter and three colors of fluorescence	Shipboard laboratory	Sample volume extremely small (less than 1 ml); 0.3–150 μm size range	Individual particle measurement can yield characterization and concentration of pigment groups present in phytoplankton; rapid to 20,000 particles per second	6,15–18,36
Synoptic global chlorophyll	Coastal Zone Color (CZCS) sensor	Satellite	Interference from seston and dissolved organics; data reduction costly	Broad scale; 1000 km; resolution 10 km	4,8,9,32

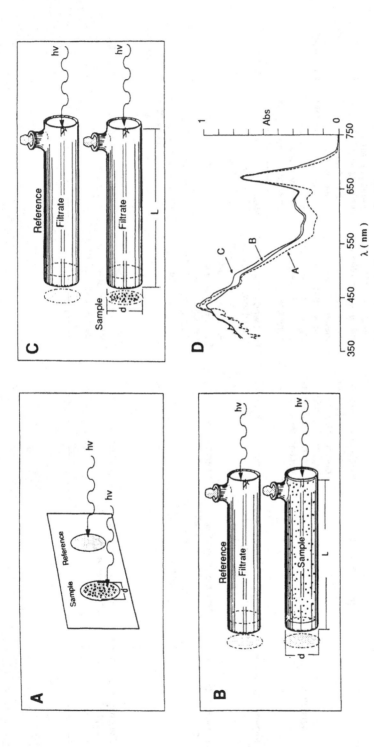

Figure 4. Cuvette calibration for the filter technique for absorption spectra. (A) Filter technique used at sea. (B) Cells in a 10-cm cuvette, 1.9-cm i.d., with blank GFF filters for diffusers. (C) Cells on filter at exit of cuvette; blank filter is diffuser reference. (D) Absorption spectra of panels A–C for diatom *Phaeodactylum tricornutum*. Filtered media used in reference cuvette. (From Yentsch, C.S. and Phinney, D.A., *Limnol. Oceanogr.* 34:1694–1705 (1989). With permission.)

phytoplankton measured traditionally with an optical light path of 10 cm. This is done by the following empirical calibration method.[5] A wetted, blank glass-fiber filter is placed over the end of two quartz 10 cm optical cuvettes. Volumes of mixed phytoplankton cultures are added to the sample cuvette while filtered growth medium occupied the reference cuvette (Figure 4B). The diffuse attenuation spectrum of the suspension between 750 and 350 nm is recorded. The sample cuvette contents is filtered onto a glass-fiber filter, the spectrum is measured with filtered growth medium in the sample cuvette and the sample filter placed on the end of the cuvette (Figure 4C). Finally, the cuvettes are removed and the spectrum recorded in the manner of measurements made at sea (Figure 4A). Because the internal diameter (d) of the cuvette and the area of the filter with particles is constant and care is taken to ensure even distribution of particulates on the filter, the three spectra measured in this manner are identical for samples with high adsorption (Figure 4D). Furthermore the absorbance due to phytoplankton may be related to the amount of its extracted chlorophyll on a per cell basis.[6] This is referred to as the specific absorption coefficient, a^*.

3.2 Light Scattering

Light is scattered in all directions when photons intersect a particle. Light scattering systems are capable of detecting particles as small as 0.1 μm when a laser is used as the illumination source. Laser-induced light scattering is presently the only system which can provide reliable, real-time information on particles smaller than 1 μm in liquids.[7] Yet there are many reliable light scattering measurement devices employing light-emitting diodes (LEDs) tungsten, xenon, and mercury lamps. Advantages of light scattering include automatic measurement, good repeatability, rapid production of hard copies of data, and low requirements for time and cost per sample. Disadvantages include the influence of particle refractive index and particle shape on determination of particle sizes, the importance of maintaining a constant flow rate through the sensor, and the requirement for moderate skill in operating the instrument.

A transmissometer is the most widely used light scattering instrument used at sea. This is strictly a known path length for the light to pass. The light source is at one end of the light path; the detector is at the other end of the light path. Non-spherical particles and particle surfaces (e.g., silica shells, calcium carbonate plates/liths complicate full interpretation of light scattering measurements).[8,9] Forward angle light scatter is said to be an indication of particle size, but is not a measure of particle size.

3.3 Impedance Volume

A non-optical measurement found highly useful for oceanography is so-called particle impedance volume. Electrical resistance measurement, commonly referred to as "Coulter counting", has been extensively used in the medical field, where it was originally developed for blood cell counting. A long history of use in oceanography now exists.[10-15]

Coulter particle counters measure particle size using electrical zone sensing. In operation, a thin stream of particle-bearing water flows through a small cylindrical aperture. Electrical current is passed through the sample stream or electrolyte, and the current monitored. When a particle is present in the sensing zone, electrical current is decreased. The drop in current is directly related to particle volume, so excellent size resolution is possible. The operating range of the instrument is from 2 to 50% of the diameter of the aperture; apertures are available in diameters ranging from 20 μm to several millimeters. Thus the measurement range of the instrument in particle diameters is from 0.4 to 800 μm. A recent version is the Coulter Multisizer which offers the advantage of zooming in on any size range of interest, rezooming in on some fraction of that size range, and so on. The resolution of 0.3 μm equivalent spherical diameter (ESD) is claimed, but the experience of the authors suggests that particle resolution below 0.6 μm ESD in seawater is not reliable.

3.4 Fluorescence Flow Cytometry

A major advance in particle analysis is the technique called flow cytometry (Figure 5). This technique provides state of the art technology which enables the sizing of particles (0.3 to 150 μm range) and measurement of other optical properties, simultaneously, on exactly the same particle. This is accomplished by the measurement of individual particles in a laminar-flow focused flow stream. The specific measurements include particle size using the Coulter principle, particle size by forward angle light scatter and by 90° light scatter, and up to three colors of fluorescence (e.g., chlorophyll autofluorescence).[16-18] A further capability is the physical separation of particles (sorting) based on the specific characteristics of the particle. Sorting is accomplished by the sample stream being separated into droplets caused by vibrations from a piezoelectric crystal. Individual cells are contained in droplets which are surrounded by saline sheath fluid. The droplets pass through a charging collar. Droplets with cells of interest (as determined by setting a sort logic) are electrostatically charged and deflected left or right when passing through charged plates. Two subpopulations can be sorted simultaneously.

Advantages of flow cytometric analysis include the speed of analysis (in excess of 10,000 particles per second), the multiparametric measurement for the same particle, extremely high precision and sensitivity, and the ability to physically sort out from the flow stream particles which meet specific size and/or fluorescence criteria. One disadvantage of this approach is that the instrumentation can be expensive, and with some equipment requires a high level of operator expertise and vigilance.

Fluorescence is a major, highly sensitive method of detection not only because of autofluorescent pigments and many useful stains, but because fluorescent labels have been linked to numerous antibody and molecular probes.[19] Figure 6 depicts detection capabilities: Coulter volume, light obscuration, and light scatter detection devices will count and measure each type of particle (S = suspended sediment; D = detritus or floc; H = heterotroph; A = autotroph or microalgae). Fluorescence detection eliminates the measurement of non-fluorescing particles, or particles which fluoresce at wavelengths shorter or larger than the target wavelengths. If cells are aligned horizontally in the detection region of any instrument, no discrimination of individual particles can be made. If these are aligned vertically and pass one-by-one through the detection region as is the case with flow cytometry, then each particle is resolved. This serves as a useful advantage in the characterization of oceanic biogenic particles.

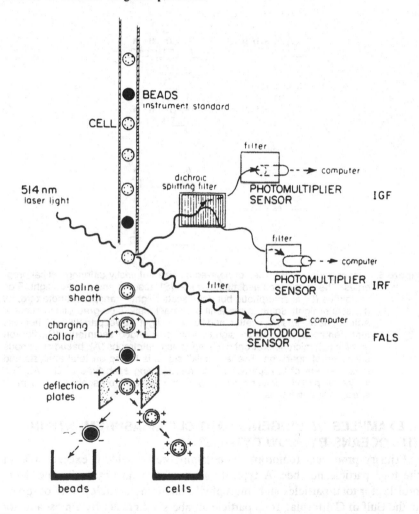

Figure 5. Schematic representation of a flow cytometer. Fate of particles and resulting signals obtained via flow cytometric analysis. Flow cytometric two-color fluorescence analysis, forward angle light scatter, and subsequent cell sorting. IGF is integrated "green" fluorescence; IRF is integrated "red" fluorescence; and FALS is forward angle light scatter. (From Yentsch, C.M. and Pomponi, S. *Annu. Rev. Cytol.* 105:183–243 (1986). With permission.)

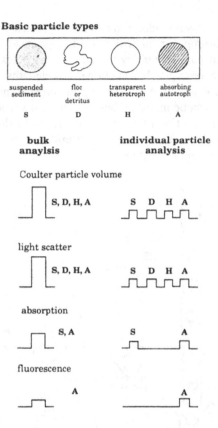

Figure 6. Upper panel — Particles of near-equal size but vastly differing optical properties. The suspended sediment particle (S) absorbs and scatters light. Floc
or detritus (D) is amorphous but does scatter light. Transparent heterotrophs
(H) scatter light. Some forms emit a characteristic blue-green fluorescence.
Autotrophs (A) absorb, scatter, and fluoresce light. Lower panels — Schematic
representation of detection signals from suspended sediments (S), floc on
detritus (D), microheterotrophs (H), and microautotrophs (A), based on various
principles of operation. The left "bulk" signal is based on total analysis and
measurement of four particles, one representing each type (S,D,H,A). The
individual particle analysis signal is at the right. Fluorescence is the most
specific for autotrophs.

4. EXAMPLES OF BIOGENIC PARTICLE MEASUREMENTS IN THE OCEANS BY FLOW CYTOMETRY

Primary producers (chlorophyll-containing cells) seldom exceed 50% of
the total particle number. A typical case is shown in Figure 7. The depth
profiles for total particles and chlorophyll-containing particles were observed
in the Gulf of California. Total particle numbers are greater by almost a factor
of 10 in the upper 40 m and a factor of 100 greater at depths in excess of 50
m. Note that the shape of both profiles in the upper 40 m are similar. However,
below 40 m, the numbers of chlorophyll-containing particles decrease rapidly;
whereas the total number of particles is relatively constant with depth.

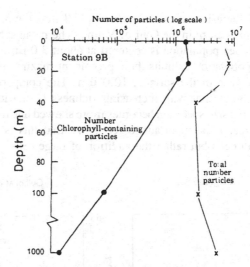

Figure 7. Vertical profile of chlorophyll-containing particles and total numbers of particles (>2.3–53 μm ESD) with depth. Data from the Gulf of California, Station 9B, Latitude 23°, 48.2′N, Longitude 108°, 20.2′W, Date 3/6/88, from USNS De-Steiguer.

There exists a consensus that a majority of particles in the oceans arise through some aspects of the process of primary production. This is demonstrated by particle observations in North Atlantic waters (Figure 8). These show that chlorophyll particle numbers and total particle numbers are strongly correlated with total chlorophyll biomass of these waters. Figure 9 presents two extremes in the size spectra of cells of chlorophyll-containing organisms. At one extreme are the populations observed in the oligotrophic Sargasso Sea. At the other extreme are populations observed in the waters of the continental shelf off New England.

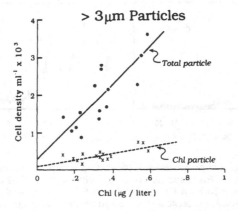

Figure 8. Particle (cell) density × 10^3.ml^{-1} vs. chlorophyll concentration (μg/liter) for particles >3 μm to 53 μm in diameter.

In both cases there is a subsurface chlorophyll maximum: at 58 m in the Sargasso Sea and at 24 m in the Gulf of Maine. The mean spherical diameter for the Sargasso Sea population is centered at about 3.0 μm. In contrast, the Gulf of Maine population exhibits three prominent maxima in cell size: 3.0, 5.0, and, in the case of the surface, 10.0 μm. The comparison of the two regions demonstrates that with increasing richness or nutrient availability, i.e., coastward, the size spectrum becomes more skewed due to the appearance of large cell sizes. It is important to recognize that this is not because of removal of small cells but rather the addition of large cells.

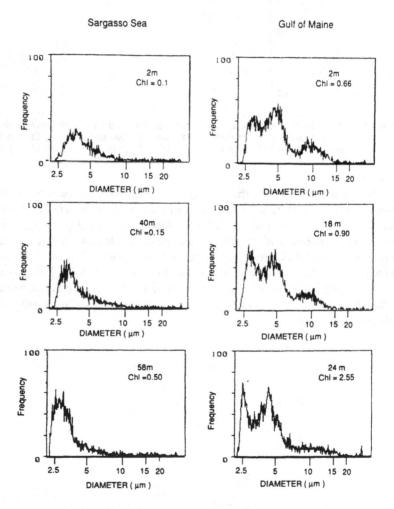

Figure 9. Size spectra of phytoplankton, (as distinguished by chlorophyll autofluorescence) for the Sargasso Sea (4 July, 1985) and the Gulf of Maine (20 July, 1985) obtained on a Becton-Dickinson FACS analyzer flow cytometer. The limit of detection is approximately 2.3 μm equivalent spherical diameters (ESD). From Yentsch, C.S. and Phinney, D.A., *Limnol. Oceanogr.* 34:1694–1705 (1989). With permission.)

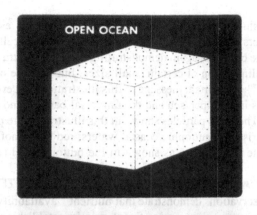

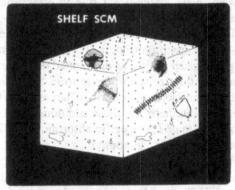

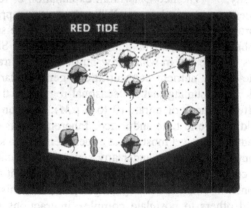

Figure 10. Relative particle sizes and numbers present in 10 μl of seawater. Open Ocean, Shelf Subsurface Chlorophyll Maximum (SCM), and red tide populations. (From Yentsch, C.M. and Spinrad, R.W., *Mar. Tech. Soc. J.* 21:58–68 (1987). With permission.)

In cartoon fashion Figure 10 depicts that the larger cells associated with rich coastal waters are present in reduced number, but their cell volume adds measurably to the biomass, again demonstrating that size spectra do not change because of the elimination of certain small sizes, but because of the addition of large sizes.[20] The important message of this illustration concerns the sparsity of large particles in the open sea and the universal background abundance of small particles. The large particles shown in this illustration are phytoplankton species characteristic of eutrophic coastal waters; red tide dinoflagellates and diatoms dominate the biomass as shown in the bottom panel (Figure 10).

5. BIOCHEMICAL FACTORS REGULATING CELL SIZE

The above observations demonstrate that nutrient "availability" or richness of natural ocean water plays a significant role in establishing the size and number of chlorophyll-containing cells.

The cell size distributions generated by the flow cytometer suggest the existence of a log-normal spectrum (Figure 9), which is common to all size spectra. It is most pronounced when data are plotted on a linear vs. log scale. The major differences in the size spectra arise from "nutrient availability," which is manifested by an increase of cells of larger sizes C_1 and C_2 over smaller cells (Figure 11). Furthermore, these additions to the basic size spectrum cause the variability in $a*$ (the specific absorption coefficient) observed with increasing concentrations of chlorophyll.

Thus, optical variability in particles results from species diversity, and changes in both time and space are lodged in the ecological theory of phytoplankton diversity.[21,22] The neo-Darwinian explanation of removal by natural selection does not seem adequate. What does seem appropriate is the concept of opportunistic growth. The important point is the surface-area-to-volume relationship and its impact on nutrient uptake.[22,23,31] Simple resource limitation models (Figure 11) in which low nutrient concentrations generate small cells while higher nutrient concentrations generate larger cells have been formulated.[23,24] Assuming that the seasonal regime of the upper layers of the oceans offers a range of nitrate (0 to 5 μM), one can visualize how size spectra could change in time and space. Inherent to this concept is that being large has advantages, yet most of the experimental data show that small cell sizes out-compete large cells under any simulated oceanic conditions.[25] Such data are in contrast to field observations, where nutrient-rich and highly productive regions are characterized by large species.[26] This apparent contradiction has led others to postulate complex interactions of growth and grazing by herbivores.[27,28]

Over time, some algal species have evolved into large, more complex cells, which provided cellular systems with more flexibility or adaptability, such as storage products, buoyancy regulation, resting stages, etc. All these factors, as well as others, become necessary for the survival of the species, especially in coastal and shelf ecosystems where rapid environmental change is often the case.[29] The relationship between turbulence and cell size appears to be one where most agree: turbulence convects nutrients which counteract the uptake problems associated with large cells, with low surface-area-to-volume.[14,30,31]

To say that information on evolutionary theory is required to explain the workings of particle optics may sound a bit extreme. However, those familiar with the difficulty of determining the causes of biogeographic boundaries recognize these formidable problems. Problems in identifying the cause of temporal and spatial change in particle optics (i.e., ocean optical geography) are just as severe. In the case of the biogeographer, the major tools are taxonomic, whereas the optical oceanographer uses ataxonomic tools.[32] It

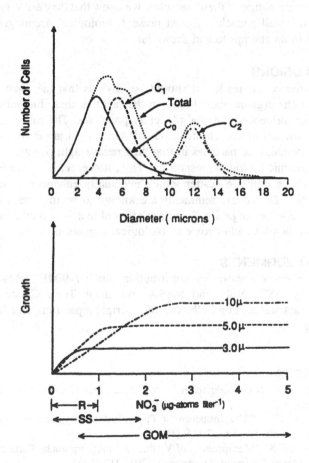

Figure 11. Concept of "bio-particle" size and nutrient relationships.
R = Recycling/regeneration of nutrients
SS = Sargasso Sea range
GOM = Gulf of Maine range
(From Yentsch, C.S. and Phinney, D.A., *Limnol. Oceanogr.*
34:1694–1705 (1989). With permission.)

remains to be seen whether these two disciplines will progress in parallel, converge, or diverge.

6. WHAT ARE THE MAJORITY OF PARTICLES?

As mentioned above, the numbers of fluorescing phytoplankton particles seldom exceed 50% of the total. The remainder of the particles are virtually uncharacterized. It is common practice to lump these under the term detritus.

There exists no shortage of candidates for what makes up "detritus". First, organic remnants of decomposition and grazing must constitute a major fraction. Bacteria and small heterotrophs are also likely candidates. Although we are unsure of the nature of these particles, we know that they are very abundant and often of small particle size. At present, biological oceanographers are challenged in an attempt to sort these out.

7. CONCLUSIONS

Even to the casual reader, it should be obvious that the most important particles are the organic ones, based on the fact that these biogenic particles have a major influence on global planetary processes. The role of submicron particles in the sea awaits clarification.[33,34] One might argue that suspended sediments are nuisance particles in that they reduce light penetration without inducing dynamic biological responses. This, however, is too severe in that sediments cause multiple scatter which eventually enhances absorption by phytoplankton. Moreover, sediments are known to be important in the adsorption of dissolved organics,[35] and chelation of metals and other substances known to be helpful and/or toxic to biological organisms.

ACKNOWLEDGMENTS

This is Bigelow Laboratory contribution number 90017. Research was supported by ONR, NSF, and NASA. We thank Terry Cucci and Dave Phinney for assistance, Peg Colby for manuscript preparation, and Jim Rollins for drafting.

REFERENCES

1. Van de Hulst, N.C. *Light Scattering by Small Particles* (New York: John Wiley & Sons, 1957).
2. Yentsch, C.S. "The Influence of Phytoplankton Pigments on the Color of Water," *Deep Sea Res.* 7:1–9 (1960).
3. Yentsch, C.S. "Measurement of Visible Light Absorption by Particulate Matter in the Ocean," *Limnol. Oceanogr.* 7:207–217 (1962).
4. Smith, R.C. and K.S. Baker. "The Bio-Optical State of Ocean Waters and Remote Sensing," *Limnol. Oceanogr.* 23:247–259 (1978).
5. Shibata, K. "Spectrophotometry of Intact Biological Materials. Absolute and Relative Measurements of their Transmission, Reflection and Absorption Spectra," *J. Biochem.* 45:599–623 (1958).
6. Yentsch, C.S. and D.W. Menzel. "A Method for the Determination of Phytoplankton Chlorophll and Phaeophytin by Fluorescence," *Deep Sea Res.* 10:221–231 (1963).

7. Lieberman, A. "Characterization of Particles in Liquid Suspension." *Tappi J.* June, 105–113 (1988).

8. Ackleson, S.G., W.M. Balch, and P.M. Holligan. "White Waters of the Gulf of Maine," *Oceanography* 1:18–22 (1988).

9. Ackleson, S.G., W.M. Balch, and P.M. Holligan. "AVHRR Observations of a Gulf of Maine Coccolithophore Bloom," *Photogramm. Eng. Remote Sensing* 55:473–474 (1989).

10. Sheldon, R.W. and T.R. Parsons. "A Continuous Size Spectrum for Particulate Matter in the Sea," *J. Fish. Res. Board Can.* 24:909–915 (1967).

11. Sheldon, R.W., A. Prakash, and W.H. Sutcliff, Jr. "The Size Distribution of Particles in the Ocean," *Limnol. Oceanogr.* 17:327–340 (1972).

12. Silvert, W. and T. Platt. "Energy Flux in the Pelagic Ecosystem: A Time-Dependent Equation," *Limnol. Oceanogr.* 23:813–816 (1978).

13. Silvert, W. and T. Platt. "Dynamic Energy-Flow Model of the Particle Size Distribution in Pelagic Ecosystems," *Am. Soc. Limnol. Oceanogr. Spec. Symp.* 3:754–763 (1980).

14. Legendre, L. and J. LeFevre. "Hydrodynamical Singularity as Controls of Recycled versus Export Production in the Oceans," in *Productivity of the Oceans, Present and Past*, Symp. Proc. (1989).

15. Legendre, L. and C.M. Yentsch. "Flow Cytometry and Image Analysis in Biological Oceanography and Limnology: an Overview," *Aquat. Sci. Issue* 10:501–510 (1989).

16. Yentsch, C.M., P.K. Horan, K. Muirhead, Q. Dortsch, E. Haugen, L. Legendre, L.S. Murphy, M.J. Perry, D.A. Phinney, S.A. Pomponi, R.W. Spinrad, M. Wood, C.S. Yentsch, and B.J. Zahuranec. "Flow Cytometry and Cell Sorting; a Powerful Technique for Analysis and Sorting of Aquatic Particles," *Limnol. Oceanogr.* 28:1275–1280 (1983).

17. Yentsch, C.M. and C.S. Yentsch. "Emergence of Optical Instrumentation for Measuring Biological Properties," *Oceanogr. Mar. Biol. Annu. Rev.* 22:55–98 (1984).

18. Phinney, D.A. and T.L. Cucci. "Perspectives in Aquatic Flow Cytometry," *Cytometry* 10:511–521 (1989).

19. Ward, B.B. "A Review of Immunological Methods for Oceanography," *Oceanography* 3:30–35 (1990).

20. Yentsch, C.M. and R.W. Spinrad. "Particles in Flow," *Mar. Tech. Soc. J.* 21:58–68 (1987).

21. Hutchinson, G.E. "The Paradox of the Plankton," *Am Nat.* 95:137–145 (1967).

22. Kilham, P. and S.S. Kilham. "The Evolutionary Ecology of Phytoplankton," in *The Physiological Ecology of Phytoplankton*, I. Morris, Ed. (New York: Blackwell Scientific, 1980), pp. 571–597.

23. Yentsch, C.S. "Some Aspects of the Environmental Physiology of Marine Phytoplankton: A Second Look," *Oceanogr. Mar. Biol. Annu. Rev.* 12:41–75 (1974).

24. Tilman, D. "Resource Competition Planktonic Algae: Experimental and Theoretical Approach," *Ecology* 58:338–348 (1977).

25. Geider, R.J., T. Platt, and J.A. Raven. "Size Dependence of Growth and Photosynthesis in Diatoms: A Synthesis," *Mar. Ecol. Prog. Ser.* 39:93–104 (1986).

26. Davis, C.O. "The Importance of Understanding Phytoplankton Life Strategies in the Design of Enclosure Experiments," in *Marine Mesocosms*, G.D. Grice and M.R. Reeves, Eds. (Berlin: Springer-Verlag, 1982), pp. 323–332.

27. Walsh, J.J. "Herbivory as a Factor in Patterns of Nutrient Utilization in the Sea," *Limnol. Oceanogr.* 21:1–13 (1976).

28. Sommer, U. "Phytoplankton Succession in Microcosm Experiments under Simultaneous Grazing Pressure and Resource Limitation," *Limnol. Oceanogr.* 33:1037–1054 (1988).

29. Smayda, T.J. "Phytoplankton Species Succession," in *The Physiological Ecology of Phytoplankton*, I. Morris, Ed. (New York: Blackwell Scientific, 1980), pp. 493–570.

30. Margalef, R. "Life Forms of Phytoplankton as Survival Alternatives in an Unstable Environment," *Oceanol. Acta* 1:493–509 (1978).

31. Malone, T.C. "Algal Size," in *The Physiological Ecology of Phytoplankton*, I. Morris, Ed. (New York: Blackwell Scientific, 1980), pp. 433–463.

32. Yentsch, C.S. and D.A. Phinney. "A Bridge between Ocean Optics and Microbial Ecology," *Limnol. Oceanogr.* 34:1694–1705 (1989).

33. Morel, A. and A. Bricaud. "Inherent Optical Properties of Algal Cells Including Picoplankton: Theoretical and Experimental Results," in *Photosynthetic Picoplankton*, T. Platt and W.K.W. Li, Eds. *Can. Bull. Fish. Aquat. Sci.* 214:521–559 (1988).

34. Koike, I., S. Hara, K. Terauchi, and K. Kogure. "Role of Sub-Micrometre Particles in the Ocean," *Nature* 345:242–244 (1990).

35. Baier, R.W., T.L. Cucci, and C.M. Yentsch. "Yellow substance absorption on clay," *Limnol. Oceanogr.* 34:949–957 (1989).

36. Yentsch, C.M. and S. Pomponi, "Automated Individual Cell Analysis in Aquatic Research," *Annu. Rev. Cytol.* 105:183–243 (1986).

PART III
Reaction and Transport of Particles in Aquatic Systems

CHAPTER 10

The Role of Particles and Colloids in the Transport of Radionuclides and Trace Metals in the Oceans

Bruce D. Honeyman

Environmental Science and Engineering, Colorado School of Mines, Golden, Colorado

and

Peter H. Santschi

Department of Marine Sciences, Texas A&M University, Galveston, Texas

TABLE OF CONTENTS

1.   Introduction 380
     1.1   The Emergence of Marine Surface Chemistry 381
     1.2   Isotopic Disequilibrium and Scavenging 382
     1.3   Laboratory Experiments of Trace Element
           Scavenging in the Oceans 385
     1.4   The Submicron Particle Pool 386

379

2.   Evidence for the Importance of Colloids in the Mobility
        of Radionuclides and Trace Metals 390

3.   Colloidal Pumping: The Coupling of Adsorption and
        Coagulation 391
        3.1   Coagulation of Homodisperse Particles 393
        3.2   Colloidal "Pumping" 395

4.   Application of the Colloidal Pumping Model to Oceanic
        Trace Element Scavenging 399
        4.1   Thorium Production as the Rate-Limiting Process 402
        4.2   Coagulation and Sedimentation 404
        4.3   Adsorption Coupled to Sedimentation 406
        4.4   Coagulation in High-Energy Dissipation
                Environments: Control by Shear Rates in the
                Water or at the Sediment-Water Interface 409
        4.5   The "Particle Concentration" Effect 410
        4.6   Control of Particle Mass Flux by Particle
                Coagulation 414

5.   Summary 415

Acknowledgments 416

Appendix: Glossary of Symbols 416

Note Added in Proof 418

References.. 418

1. INTRODUCTION

A central problem facing chemical oceanographers over the years has been to reconcile the relatively low measured trace element concentrations with the higher concentrations predicted from mineral solubility calculations. Research on the solubilities of carbonates and oxides, believed to be the controlling solid phases for most heavy metals, showed that seawater was undersaturated with respect to every solid phase considered. In the face of failure to find a solubility control for many trace elements, a new role for particles gradually began to emerge: their sorptive control of metal concentrations. It is this function that Turekian[1] described as "the Great Particle Conspiracy".

In 1954, Goldberg[2] coined the term "scavenging" to describe the removal of trace elements from the oceanic water column by their association with sinking particles and 2 years later Krauskopf[3] demonstrated, through a series of laboratory experiments, that sorption is a viable trace-element controlling process. One of the profound conclusions of Sillen's classic 1961 paper[4] is that the oceans' composition is related primarily to the relative efficiency of removal of the elements from the water column and not to the rate of elemental input by rivers. Thus, the major ions of ocean waters dominate seawater composition because they are not removed as efficiently as are many of the elements present at trace levels. While Sillen's article provided a framework for interpreting oceanic scavenging in terms of chemical processes, the idea did not reach maturation until a decade later with the advent of surface complexation models (SCMs).

1.1 The Emergence of Marine Surface Chemistry

Until the early 1970s most of the work on trace element/particle interactions focused on the interactions of ions with charged surfaces.[5] The development of surface reactivity models was in essence a series of attempts to explain deviations of observations from ideal ion-exchange behavior. Such models are physical models in the sense that the primary driving force for trace element/particle associations is considered to be separation of charge, e.g., a positively charged ion interacting with a negatively charged clay surface.

In the early 1970s, however, coordination chemists began to study the interaction of metal ions with metal-oxide surfaces such as γ-alumina and hydrous ferric oxide. The observation that the tendency of a metal ion to associate with a metal oxide surface correlates with the intensity of solution-phase hydrolysis product formation led to the concept of surface-complex formation.[6,7] Surface complexation models (SCMs) are still "physical" models in the sense that the charge of the sorbing ion and the surface of the sorbent are important factors. However, the possibility of chemical bond formation between surface-active solutes and binding "sites" on mineral surfaces was a new aspect, a key element in SCMs and one not included in ion-exchange models. The main difference between the various currently available surface complexation models is the way in which the free energy of adsorption is divided into electrostatic and chemical components.

Just as Sillen's model signaled the beginning of modern chemical oceanography, Schindler[8] firmly established the link between surface chemistry and the chemical composition of the ocean with his "zero-order" model for the removal of trace metals from the ocean waters. In his model, the relative rate of removal of trace elements from seawater is related to the intensity of interaction between trace elements and particle surface sites, given a constant particle flux. The model correctly predicts the order (but not the absolute value) of residence times of metals in the ocean, assuming SiO_2 as the model surface. Schindler's model was crucial for two reasons: (1) it provided a formal connection between solution-phase reactions, adsorption and trace

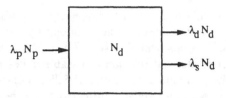

Figure 1. Schematic illustration of radionuclide and trace element scavenging. Subscripts p and d denote parent and daughter radionuclide pairs. The parent nuclide decays at a rate of $\lambda_p \cdot N_p$ producing N_d daughter atoms in the dissolved phase. N_d is removed from the dissolved phase through radioactive decay, $\lambda_d \cdot N_d$, and scavenging by suspended and settling particles, $\lambda_s \cdot N_d$. λ_p and λ_d are the parent and daughter radioactive decay constants (d^{-1}) and λ_s is a scavenging rate constant (d^{-1}). λ_s can be determined by the deficiency of daughter radionuclide activity, A_d, relative to that of the parent, A_p (Equation 1).

$$\lambda_p N_p = \lambda_d N_d + \lambda_s N_d \qquad (F1.1)$$

$$\lambda_d A_p = \lambda_d A_d + \lambda_s A_d \qquad (F1.2)$$

metal scavenging; and (2) he established the notion that marine particles are chemical species which compete against dissolved ligands for complexation with trace elements. Following Schindler, Balistrieri et al.[9] provided the next important step by investigating the extent to which variations in particle surface chemistry would affect trace element scavenging residence times.

1.2 Isotopic Disequilibrium and Scavenging

Concurrent with the development of scavenging models emphasizing chemistry was the application of uranium- and thorium-series radionuclide disequilibrium to problems of oceanic mixing and the estimation of particle and trace element residence times. The general interactions involved are illustrated in Figure 1. ^{238}U, a primordial radionuclide and parent of ^{234}Th, is conservative in ocean waters primarily because of the formation of uranyl-carbonate complexes which maintain the uranium in solution. Thorium, in contrast, is highly surface active and ^{234}Th is efficiently removed from the water column through sorption onto sinking particles. The rate constant for ^{234}Th scavenging, $\lambda_s(\lambda_s = \tau_s^{-1}; \tau_s =$ scavenging residence time), can be estimated from the ^{238}U:^{234}Th activity ratio, A_{238}/A_{234}

$$\lambda_s = \lambda_{234}\left(\frac{A_{238}}{A_{234}} - 1\right) \qquad (1)$$

where λ_{234} is the decay constant for ^{234}Th (i.e., 0.029 d^{-1}).

While there have been a number of variations on this simple box model, including the addition of the influence of mass transport,[10,11] interpretation of the results of such models depends on fundamental assumptions about the nature of the interactions between the scavenged element and particles, be they the scavenging process invoked or the assumed rate at which such processes occur.

The power of using radioactive disequilibrium between parent-daughter nuclides of the U-Th decay series as a tool for studying scavenging is that it adds an inherent kinetic element to scavenging models through the use of *in situ* radiochemical clocks. While the scavenging models of Schindler[8] and Balistrieri et al.[9] also yield a scavenging rate for trace elements, the rate aspect came from the residence time of particles as they sweep through the water column; a central aspect of those models is that the system is assumed to be in sorptive equilibrium. In contrast, radiochemists have interpreted radioactive disequilibrium in terms of the rate of sorptive reactions and, until Bacon and Anderson,[12] this was usually in terms of an irreversible binding of scavenged elements to particle surfaces. Their work marked an important point in scavenging studies based on radioactive disequilibrium between several dissolved mother/''particle-reactive'' daughter radionuclide pairs.

Bacon and Anderson applied three models for oceanic scavenging to observe distributions of ^{230}Th and ^{234}Th, with k_1, k_{-1} as rate constants (t^{-1}), and $K_d = Th_{part.}/Th_{diss.} = k_1/k_{-1}$. $Th_{diss.}$ represents thorium activity passing the filter employed to

Model I: Irreversible uptake

$$Th_{diss.} \xrightarrow{\lambda_s} Th_{part.}$$

$$\downarrow \text{Sedimentation } \lambda_p$$

(Str. A)

Model II: Irreversible uptake with fast particle removal

$$Th_{diss.} \xrightarrow{\lambda_s} Th_{fine\ particles} \xrightarrow{k_2} Th_{large\ particles}$$

$$\downarrow \text{Sedimentation } \lambda_p$$

(Str. B)

Model III: Reversible exchange

$$Th_{diss.} \underset{k_{-1}}{\overset{k_1}{\rightleftarrows}} Th_{part.}$$

$$\downarrow \text{Sedimentation } \lambda_p$$

(Str. C)

separate large particles from bulk solution and C_f the concentration of particles retained on the filter of thorium. Through their analysis they demonstrated that ocean-water column profiles, can only be explained in terms of reversible association of Th with particles and a sorption rate which is faster than the rate of removal of particulate matter (the latter on time-scales of months to years). Although one could argue that thorium is an extreme choice as an analog for other trace metals, Bacon and Anderson provided the needed step to reconcile kinetic scavenging models pioneered by radiochemists with the equilibrium scavenging models emphasizing the details of chemical interactions.

Figure 2 (Bacon and Anderson, 1982) shows values for the apparent adsorption rate constant, k_1, as a function of suspended matter mass concentration, C_f. Note (1) the broad correlation between k_1 and C_f; and (2), the extremely small values of the scavenging rate constants of 0.25 to 1.25 y^{-1}. The central question confronting chemical oceanographers over the last decade is whether data sets such as this reflect chemisorption reactions or physical processes (e.g., mass transfer or particle aggregation).

Bacon and Anderson also proposed a model in which Th can be associated with two different particle reservoirs, one which is slowly sinking and the other which is removed faster than the rate of association with sinking particles. While not specifying the chemical composition of the two particle pools, Bacon and Anderson nevertheless introduced an important concept: the existence of solid-phase ligands which have markedly different physical behavior among themselves. The general model of two marine particle size

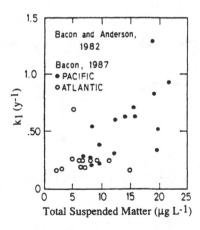

Figure 2. Adsorption rate coefficient, k_1, vs. total suspended matter, C_f, for the following reaction[12] (qq.v., Section 1.2):

$$Th_{diss.} \xrightarrow{\lambda_s} Th_{part.}$$

$$\downarrow \text{ Sedimentation}$$

$$\lambda_p$$

classes has been further developed[13,14] to include particle disaggregation as well as regeneration of thorium from particles.

1.3 Laboratory Experiments of Trace Element Scavenging in the Oceans

Oceanography has traditionally been a science rooted in field observations; marine chemistry has had a very strong emphasis on analytical, rather than experimental, chemistry. However, oceanic trace metal scavenging, perhaps more than any other sub-discipline in oceanography, has seen questions addressed through controlled laboratory experiments, in addition to field studies. In spite of their successes in producing reasonable phenomenological models of trace metal scavenging, marine surface chemistry and radiochemistry have led to two strongly contrasting views of the same process: surface chemistry describes the *equilibrium* relationships among chemical components and isotopic disequilibrium has produced *kinetic* models. Laboratory studies have provided a means to reconcile these views into a cohesive framework.

Balistrieri and Murray[15] were among the first to extend the techniques and chemical formalisms of surface chemists to the study of interactions at the seawater interface of natural particles through controlled laboratory experiments. While it is arguable whether one is able to 'mimic' real systems with laboratory studies, the work by Balistrieri, Murray, and others has demonstrated that such an approach can serve as an indispensable aid to understanding chemical processes controlling trace element/particle interactions.

Laboratory studies of the kinetic aspects of oceanic scavenging were initiated by Nyffeler et al.[16] and Li et al.[17] Figure 3 is an example of sorption data for a series of metals in the presence of a marine clay. The metal ions exhibit a wide variation in their equilibrium particle/solution distribution coefficients, K_d, and rate of approach to equilibrium. Figure 4 shows sorption rate constants derived by Nyffeler et al.[16] plotted as a function of equilibrium K_d by Jannasch et al.[18] The distribution coefficient, K_d, is, in this case, based on an operational definition of "dissolved", i.e., that fraction of the bulk sample which passes a 0.45 μm pore-sized filter. This definition ignores adsorption onto microparticulates passing the filter, as well as any filtration problems (e.g., Chapter 5, this volume) which might occur.

The relationship between the values of the rate constant and K_d, as well as the magnitude of the rate constants, begs two questions: (1) what is the meaning of this apparent extra-thermodynamic relationship? and (2) what processes are responsible for the observed rate constants? While laboratory experiments with natural particles have produced data which are consistent with field observations such as the particle concentration dependencies of partition coefficients and rate constants, the problem remains to explain laboratory and field data from theory.

An additional complication to unraveling the processes contributing to trace element scavenging is illustrated in Figure 5. Kinetic data for scandium sorption by natural marine particles[18] is shown resolved into a series of time

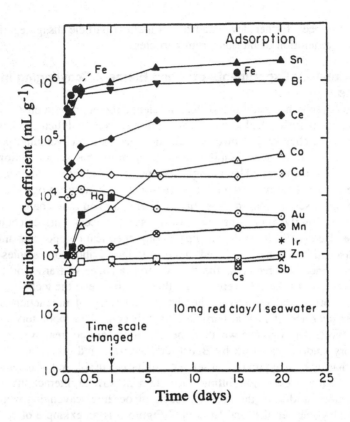

Figure 3. Log K_d vs. time.[17] "Solution"/particle partitioning of the radiotracer-tagged trace metals was determined by filtration with 0.4 μm filters. Reprinted with permission.

frames in which sorption, while always exhibiting first-order behavior, is described by a series of rate constants of progressively smaller values. The analysis by Jannasch et al.[18] suggests that either a finite number or a continuum of serial or parallel processes may control the scavenging rate. Interpretation of such kinetic data in terms of the underlying processes is also complicated by their mathematical equivalency of many process descriptions, e.g., first-order chemical reaction and diffusion.[19]

1.4 The Submicron Particle Pool

Until several years ago, trace element scavenging emphasized the interaction of surface-active species with particulate material of a few tenths of a micron in size or larger. Phase separation by filtration with ca. 0.45 μm filters stemmed from the need, in many oceanographic studies: (1) to know the amount of an element associated with potentially sinkable particles, and (2) to sample large (one to hundreds of liters) volumes of water without clogging filter membranes. Such protocols were bolstered by the generally held belief that the submicron particle pool is relatively small. Furthermore, the chemical nature of the colloid pool was largely a mystery, although it was generally assumed that colloids were just smaller units of the macroparticulate material.

Determination of the submicron particle-size spectrum and composition remains a difficult analytical problem. While new techniques such as laser Doppler velocimetry or laser correlation spectrometry are emerging as useful research tools for well-defined systems,[20] their application to natural systems, with generally low colloid mass concentrations and irregular particle shapes, is still in development.[21] Furthermore, a significant fraction of the colloid population likely is composed of organic macromolecules (e.g., acidic humic and fulvic acids and their derivatives, as well as neutral proteinaceous compounds) whose detection may not be amenable to laser techniques.

The existence and magnitude of this colloid pool is of significance to marine chemists for two reasons. First, the colloidal pool represents a large number of functional groups potentially available to complex with trace elements. The few measurements of colloid mass which are available indicate that it may be as large, if not larger, than the sum of particles of micron size or

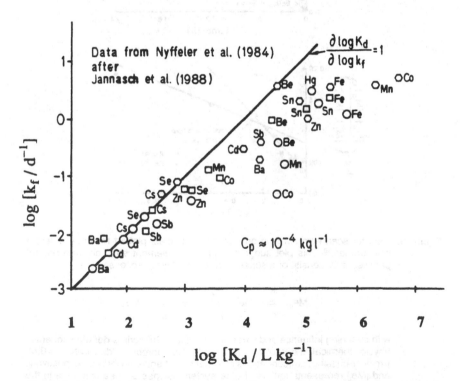

Figure 4. The correlation of particle-concentration normalized sorption rate constant, k_f (t^{-1}), derived from Nyffeler et al.,[16] with the distribution coefficient, K_d, after ca. 100 h, on a logarithmic scale.[18] Reprinted with permission.

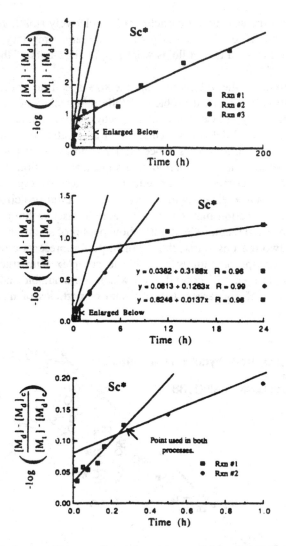

Figure 5. Data for scandium sorption onto marine particles, plotted for three different time frames.[18] This plot suggests that trace element sorption onto natural particles may consist of a series of rate controlling processes

$$Me_d \underset{k_{-1}}{\overset{k_1}{\rightleftarrows}} Me_1 \underset{k_{-2}}{\overset{k_2}{\rightleftarrows}} Me_2 \underset{k_{-3}}{\overset{k_3}{\rightleftarrows}} Me_3 \dots$$

with changing influence at different time scales. Subscripts denote sequential physicochemical states of Me. $[M_d]$ and $[M_d]_e$ represent "dissolved" (<0.45 μm filter-passing) concentrations of Me at time, t and equilibrium, respectively, and $[Me_t]$ represents total Me in the system. Slopes of linearized data in the regression calculations show that sorption, while exhibiting first-order behavior in consecutive time periods, is characterized by rate constants of progressively smaller value. Reprinted with permission.

greater.[22,23] Second, this pool of ill-defined material likely has a fate quite different than material in the large particle pool (e.g., their residence times are estimated to be orders of magnitude different); similarly, colloid-associated trace elements will behave in a manner quite different from species of that element which are truly dissolved or associated with large settleable particles.

Phase separation with standard filter sizes (≥ 400 nm) mostly removes carbonates, amorphous silica, plankton, and clays and only partially removes fragments of clays, crystalline and amorphous metal oxides, and high molecular weight organic material, such as proteins, lipids, carbohydrates, and humic acids. While the definition of 'dissolved' lies in the provenance of thermodynamics, the operational molecular weight (MW) cut-off limit, based upon utilization of membrane filters, is about 10^2 to 10^4, although such determinations are beset by subtle analytical problems.[24,25] Mass concentrations of suspended material range from 10^{-2} mg L^{-1} in the deep ocean to greater than a gram per liter in some rivers and estuaries.

Until quite recently, there were no estimates of the mass of the colloid pool, a pool for the most part consisting of large organic complexes, mineral fragments, and condensation nuclei. Morel and Gschwend[26] estimated, for river and lake waters, that the non-settleable fraction of suspended sediments comprised about 10% of the sediment mass. In contrast, Moran and Moore,[23] using cross-flow ultrafiltration, determined the size of the colloid pool in the North Atlantic surface ocean waters and found that it constituted twice the mass of settleable particles. Furthermore, Sugimura and Suzuki[22] showed, in a revolutionary paper, that using a high-temperature catalytic combustion technique, colloidal organic carbon with MW $\geq 4 \times 10^4$ was missed by older techniques and comprises up to 50% of the dissolved organic carbon (DOC). This discrepancy between dry, wet, and high temperature combustion techniques for DOC analysis was, however, noted before.[27]

Regardless, knowing mass concentration of the colloid and particle pools provides only a rough estimate of their relative influence on trace element behavior. Of greater importance are the types and quantity of trace-element complexing sites present in each pool and the intensity with which trace elements react with each site type. Colloidal and particulate trace-metal complexing material are markedly varied in their specific complexation capacities (i.e., moles of sites per gram material). α-SiO$_2$, for example, has about 1.5×10^{-5} mol g^{-1} sites,[28] a consequence of its relatively low specific surface area (3 to 4 m^2 g^{-1}). Amorphous silicas have specific surface areas ranging from about 100 to several hundred m^2 g^{-1}, depending on the formation process,[29] leading to specific site concentrations on the order of a millimole per gram. Marine organic matter (e.g., Balistrieri et al.[9]) has from 10^{-4} to 10^{-2} mol.g^{-1} of trace-metal complexing sites.

Figure 6 illustrates that the (1) type of surface site, (2) density of surface sites, and (3) interaction intensity of the metal are all critical parameters in controlling the distribution of trace elements between solution and particulate or colloid phases. This is exactly analogous to solution-phase complexation chemistry describing the distribution of a metal-ion among ligands.

In very broad terms, the oceanic water column contains three pools of trace-metal binding ligands: (1) macroparticles (that material a few tenths of a micron in size or larger, including marine 'snow' and fecal matter); (2) colloids; and (3) those ligands which are truly dissolved. Not only does the nature of these ligands change spatially and with time, but there is an exchange of material between the ligand pools, as well.

2. EVIDENCE FOR THE IMPORTANCE OF COLLOIDS IN THE MOBILITY OF RADIONUCLIDES AND TRACE METALS

Our understanding of the role that colloids play in the geochemical life of radionuclides is largely based on circumstantial evidence, although the role of submicron-sized particles in affecting the behavior of particle-reactive

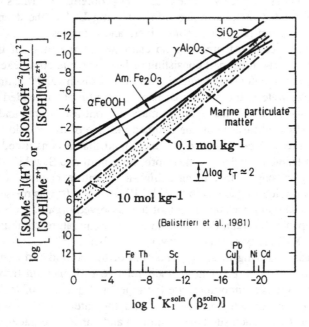

Figure 6. The tendency to form surface complexes, represented by the mass-action expressions for the formation of the surface complexes $SOMe^{z-1}$ and $SOMeOH^{z-2}$, is related to the intensity of solution-phase (hydroxo) complex formation. This observation led to the establishment of surface complexation models by Schindler and co-workers.[6,7] The residence time of radionuclides and trace elements in the water column, τ_T, is related to the fraction of Me associated with sinking particles, f_p, and the particle residence time, τ_p. An order of magnitude change in the fraction of sorbed Me corresponds to a decade change in residence time of the trace element, τ_T

$$\tau_T = \frac{\tau_p}{f_p} \tag{F6.1}$$

Sorption intensity changes with the nature of the trace element and the type and quantity of Me-sorbing particles. (Modified from Balistrieri, L.S. et al.[9])

solutes has long been recognized. Colloid transport through porous media has been reviewed by McDowell-Boyer et al.,[30] in surface waters by Buffle,[31] and the literature on the formation and properties of radiocolloidal dispersions summarized by Kepak.[32]

Buddemeier and Hunt[33] studied radionuclide transport in groundwaters pumped from a nuclear bomb test cavity and found that the transition and lanthanide radionuclides moved considerably farther than expected from the rate of transport of conservative species such as ^3H, scaled by the measured retardation factor. These nuclides were essentially all associated with the colloidal (<0.2 μm) fraction. Sheppard et al.[34] concluded that much of the mobility of radionuclides in soil waters is probably due to soil constituents in the 2 to 5 nm range. Recently, Penrose and co-workers[35] convincingly showed that colloidal material, most likely in the form of DOC, provided for the sub-surface transport of plutonium. Niven and Moore[36] examined the partitioning of ^{234}Th, added to seawater in the laboratory, among dissolved, colloidal, and particulate fractions and found that up to 64% of the total ^{234}Th was associated with the colloidal (1 nm to 0.2 μm) size fraction. According to Baskaran et al.,[37] up to 80% of natural ^{234}Th in Gulf of Mexico water is found in the colloidal fraction. Orlandini et al.[38] found in their study of an oligotrophic lake that submicron colloidal material can dominate the behavior of actinides. Nash and Choppin[39] estimated that in natural waters at 0.1 mg L^{-1} DOC (corresponding to humate site concentration = [Hu] = 4×10^{-7} mol L^{-1}) the ratio of [Th-Hu]:[Th$_{diss}$] would be about 4×10^{10}. Sequential extraction and ultrafiltration experiments by Santschi et al.[40,41] have shown that particle-reactive elements are involved in dynamic cycles of coagulation, aggregation, and disaggregation of particles and colloids in the water column of coastal marine ecosystems. This was shown by observing the partitioning of radiotracers of elements such as Sn, Hg, Cr, Fe, and Pa into colloidal fractions of a coastal marine model ecosystem. Last, but not least, Buchholtz et al.[42] demonstrated the aggregation effect of ionic and colloidal radionuclides during laboratory uptake experiments with marine sediments.

3. COLLOIDAL PUMPING: THE COUPLING OF ADSORPTION AND COAGULATION

The distribution of particle sizes in natural systems is the result of a number of processes which bring particles together (the coagulation mechanisms) or break aggregates apart. As particles are transferred throughout the particle-size spectrum, trace elements and radionuclides associated with those particles, either as sorbed species or intrinsic components of the particle matrix, are also moved between particle pools. The rate of trace element transfer, in such a "piggy-back" fashion, depends on the rate at which the particles, themselves, are moved throughout the particle size spectrum and the amount of trace elements associated with those particles.

Figure 7 is a schematic depiction of the component processes in colloidal pumping.[43] (a) Trace elements sorb onto colloidal particles; the particle/so-

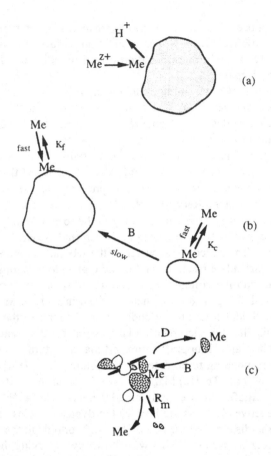

Figure 7. Schematic of colloidal pumping model: coagulation coupled to adsorption.[43] (a) Rapid sorption of trace elements onto particle surfaces. Formation of surface complexes is rapid compared to rates of particle-particle interactions. (b) Slower aggregation (indicated as B) of colloids with larger particles. Colloidal material includes organic macromolecules, mineral fragments and condensation nuclei. Large particles are distinguished from colloids by the comparative ease at which they can be separated from bulk solution, for example, with relatively large pore sized filters (0.45 μm). (c) Natural particle clusters most likely contain materials of varied size, history and composition. The transport of scavenged trace elements to their final depository on the ocean floor is likely circuitous as the material making up particles periodically cycle through various pools as the consequence of aggregation (B), disaggregation (D), and remineralization processes (R_m).

lution distribution of the trace elements is dependent on the general solution conditions (e.g., pH, ionic strength, temperature, etc.), the nature of trace-element complexing ligands, and the chemistry of the particle surfaces (refer to Section 1). (b) Colloidal particles are transferred to larger aggregates by some coagulation mechanism (e.g., Brownian motion, shear, differential settling, polymer bridging). If sorptive equilibrium is achieved on time-scales much smaller than the characteristic coagulation time, then the rate of appearance of a trace element in the large aggregate pool will be controlled by the coagulation rate.

3.1 Coagulation of Homodisperse Particles

It is important to recognize that coagulation is a kinetic process. Suspensions of colloids are never stable in a thermodynamic sense, primarily because of their large interfacial area. A dispersion is *kinetically* stable, when particles are permanently free.[44] Smoluchowski[45,46] laid the foundation of current co-agulation theory in his study of Brownian diffusion as the primary coagulation mechanism. He assumed that there is no interaction between particles except for a short-ranged force resulting in a permanent contact on the first encounter. If an energy barrier exists, coagulation will be slowed down. Smoluchowski related slow coagulation to rapid coagulation (diffusion-controlled coagula-tion, in the absence of an energy barrier) *via* the α-factor, the fraction of total collisions leading to permanent contact. Fuchs[47] related the energy of inter-action of two particles, V_c, to α by considering the particle-particle interaction as diffusion in a field of force. If V_c is greater than a few $k_B T$ (Boltzmann constant $\times$ absolute temperature), relatively stable dispersions will be found. At a V_c of approximately 15 $k_B T$, only 1 of 10^6 particle collisions will be successful in forming a stable aggregate.[48]

The degree of colloid stability is also often defined in terms of the stability ratio, W, the factor by which agglomeration is slower than in the absence of an energy barrier. The magnitude of W is related to the height of the energy barrier, V_c, and is equal to the reciprocal of α. When dealing with a single destabilization process, that is, homocoagulation (characterized by all particles having an energy barrier of the same magnitude), it is common to represent W on a relative scale. The least stable suspension is assigned a value of W = 1.00. The more stable the suspension, the higher the stability ratio.

Assume that at t_0 an experimental system contains particles which are initially completely dispersed (i.e., each particle existed as a discrete entity). With time, single particles (singlets) coagulate to form doublets, then triplets, quadruplets, and so on. Since the dominant coagulation mechanism for homodispersed particles smaller than 1 μm is collisions due to Brownian diffusion;[48,49] the change in the total number of particles per unit volume of fluid, N∞, may be written as

$$\frac{\partial N_\infty}{\partial t} = k_b N_\infty^2 \tag{2}$$

provided all interactions are efficient (W = 1), where k_b is the Brownian coagulation kernal ($= 4k_B T/3\mu = 5.5 \times 10^{-12}$ cm^3 s^{-1} at 25°C), and μ is the viscosity of the solution. This follows from Smoluchowski.[45] The solution to Equation 2 is

$$N_\infty = \frac{N_\infty^o}{1 + k_b N_\infty^o t} \tag{3}$$

For the general case where $W \neq 1$, the concentration of aggregates composed of k numbers of particles is given by

$$n_k = \frac{N_\infty^2 \left(\dfrac{t}{\tau_a}\right)^{(k-1)}}{\left(1 + \dfrac{t}{\tau_a}\right)^{(k+1)}} \tag{4}$$

where $\tau_a = W/(k_b N_\infty^0)$. τ_a represents the time needed for the initial number of particles to be reduced by half and W is the stability ratio, i.e., the factor by which actual coagulation is slower than by collisions due to Brownian diffusion. N_∞^0 is the initial particle number concentration. n_k is the number of particles in each class where $k = 1, 2, 3 \ldots$ indicates singlets, doublets, etc. Figure 8 shows the relative numbers of particles in each size class as a function of time. Note that (1) the group of singlets is always the largest class in terms of particle numbers; (2) dn_k/dt changes with time; and (3) although mass is conserved, the total number of particles, $N\infty$, decreases.

The use of this equation is subject to several significant assumptions which arguably make them inappropriate for use in most environmentally relevant

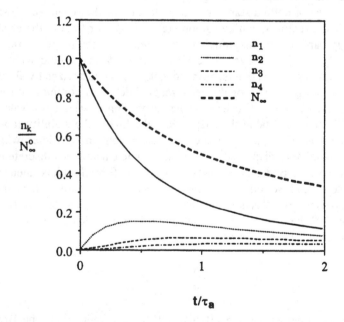

$$t/\tau_a$$

Figure 8. Time evolution of k-sized particle clusters. Calculations using Equation 4 where k represents the number of primary particles in each aggregate (k = 1: singlets; k = 2: doublets; etc.). The mass of each size group is $k \cdot n_k \cdot$ (mass per primary particle times the number of primary particles per aggregate). τ_a is the time needed for the number of particles to be reduced by a factor of two; i.e., it is the coagulation "half time." An assumption in these calculations is that the collision frequency function, k_b, is the same for all aggregate sizes.

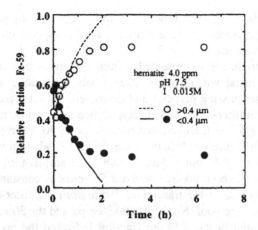

Figure 9. Relative amount of 0.25 μm hematite particles passing or retained by 0.4 μm Nucleopore® filters, as a function of time.[50] Hematite particles were labeled with ^{59}Fe as a tracer. The separation of particles at t = 0 represents coagulation of particles with filter surfaces (a constant artifact in these experiments). Model calculations were made with Equation 4 and: W = 7; $N_\infty^\bullet$ = 9.7 × 10⁷ particles mL⁻¹; k = 2. The symmetry of the relative mass vs. time curves indicates that hematite particles were transferred from the filter-passing (colloid) to the filter-retained, ("particulate") pools.

cases. These assumptions are (1) an initially monodisperse distribution; (2) that k_b is constant, which implies that collisions occur between like-sized particles (and aggregates). An immediate conclusion from this assumption is that only k = 1, 2, 4, 8, 16 . . . are allowed, contrary to the continuous function in k which is implied; and (3) that Brownian motion is the only destabilizing influence.

Figure 9 contains experimental results[50] for coagulation of spherical, 0.25 μm hematite particles. The figure shows the fraction of the total hematite mass either passing or retained by a 0.4 μm Nucleopore® filter. The distribution of hematite with time was followed through ^{59}Fe introduced to the hematite *lattice* as part of the hematite synthesis process. The curves are symmetrical: hematite particles lost from the <0.4 μm filter-passing fraction are transferred directly to the filter-retained fraction. The distribution of hematite particles with time was calculated using Equation 4, with k = 1 and 2 for the filter-passing and filter-retained fractions, respectively. Thus, singlets passed the filter while aggregates composed of two or more original particles were retained by the filter.

3.2 Colloidal "Pumping"

Just as the lattice-bound ^{59}Fe was transferred from the colloid (filter-passing) pool to the population of filter-retained aggregates, so would be sorbed trace elements. The rate of transfer of a sorbate (e.g., mol L⁻¹ d⁻¹) from the colloid to the 'particulate' pool will depend on the amount of a trace element associated with the colloids (e.g., mol g⁻¹ colloids) and the rate of colloid transfer (e.g., g $L^{-1} d^{-1}$). The overall process we call "colloidal pumping", as small colloids and their associated trace elements are "pumped" up the size spectrum through coagulation.

Consider again the results presented in Figure 9 but include a sorbing metal. Species of the metal will reside in three pools: associated with the filter-retained and filter-passing particles and dissolved. As filter-passing particles coagulate to form filter-retained particles, sorbed species are transferred from the filter-passing to the filter-retained particle pool. An example of this scenario is shown in Figure 10a.[50] In this case, ^{113}Sn(IV) adsorption is essentially instantaneous (i.e., <5 min.). Since ^{113}Sn is in sorptive equilibrium with hematite particles in both pools, dissolved Sn remains constant and the only change in the system is the transfer of ^{113}Sn from the filter-passing to the filter-retained particle pool. No detectable ^{59}Fe passed the 30 nm filter so any ^{113}Sn activity found in the <30 nm fraction indicated the presence of truly dissolved species of ^{113}Sn. This coagulation-controlled transfer of ^{113}Sn from the filter-passing to the filter-retained particle pool is an illustration of colloidal pumping.

Figure 10b[50] presents distribution coefficients for ^{113}Sn sorption on hematite as a function of time. The closed symbols, K_d, show sorption kinetic behavior typical of what is observed for many types of systems[16-18] (refer to Figure 3). As is, these results are at odds with both physical/chemical adsorption theory and metal sorption mechanism studies[51] in which metal ion adsorption is completed rapidly (see Chapter 13, this volume[52]) with characteristic times on the order of milliseconds to seconds. A number of hypotheses have been advanced to explain such slow sorption kinetics including mass transport control, slow ligand exchange and particle aggregation. The open circles of Figure 10b show the calculated partitioning coefficient, K_p, when an unambiguous distinction has been made between sorbed and dissolved species of Sn. It is clear that the kinetic effect in this case is due solely to the coagulation of hematite particles.

Suppose that the system described above contains a suite of radioactive trace metals rather than just ^{113}Sn. The order of the rate of appearance of the trace metals in the filter-retained particle pool will depend on the relative amount of each trace metal which is associated with colloids. This amount, for each trace metal, is equal to the fraction of the trace metal associated with colloids, f_c, multiplied by the total amount of trace metal in the system, Me_T. Our simple system is composed of colloid and particle phases and metals associated with these phases, as well as the solution phase. The fraction of metal associated with colloids, f_c, is a function of the conditional partitioning constants for trace metal sorption on colloids and particles, K_c and K_f, respectively, and the number of trace-metal binding sites in the colloid and particle pools.

Figure 11 shows the results[50] of colloidal pumping experiments for a series of metal ions with log K_c values (L kg^{-1}) ranging from 5.5 to 4.3 (K_c = experimentally determined partition coefficient of species between colloidal particles and solution). The line labeled [59]Fe (Figure 11a) indicates the fractional appearance of colloidal hematite in the filter-retained particle pool. As K_c decreases so does the rate constant for the disappearance of Me from the <0.4 μm pool. Figure 11b contains a plot of the colloidal pumping rate constant, k_{cp}, the rate constant for the loss of radionuclide activity from the

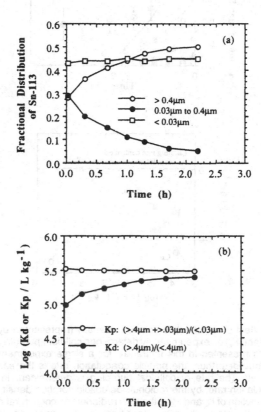

Figure 10. Example of colloidal pumping: [113]Sn(IV) distribution in a coagulating hematite suspension.[50] This figure shows the results of experiments on the distribution of chemical species of Sn(IV) between dissolved, colloid and particulate pools. Time t = 0 represents the point of addition of 0.25 μm ^{59}Fe-labeled hematite particles to a particle-free [113]Sn solution. Slurry samples were sequentially filtered through 0.4 and 0.03 μm Nucleopore filters. No ^{59}Fe label passed the 0.03 μm filter so [113]Sn activity in the <0.03 μm fraction was considered truly dissolved. (a) The distribution of [113]Sn between dissolved, colloid, and particle pools. Note that sorptive equilibrium is rapidly attained (i.e., the dissolved [113]Sn activity is time-independent). However, with only the 0.4 μm filter-retained fraction as a reference frame, "sorption" appears to be slow, reaching equilibrium only after several hours. (b) Comparison of distribution coefficients. Closed symbols: an operationally-defined "dissolved" fraction. Open symbols: ratio of sorbed (colloidal + filter-retained particles) to truly dissolved Sn. Thus, particle aggregation combined with an operational definition of "dissolved" gives the appearance that sorption is slow whereas, in this case, it is, in fact, quite rapid.

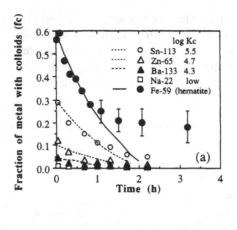

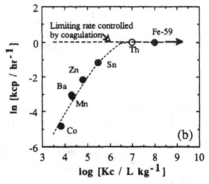

Figure 11. The effect of radionuclide sorption intensity (represented by the conditional constant, K_c) on experimentally determined colloidal-pumping rates, k_{cp}[50]. The results presented in this figure are for a single experiment with a suite of radionuclides; thus, the particle coagulation rate is the same for all radionuclides. The rate of colloidal pumping is determined, in addition to the coagulation rate, by the radionuclide/colloid sorption density, Γ_c. Γ_c, in turn, is a function of K_c and Me_T, the total radionuclide concentration. (a) The lower the K_c, the slower the rate of loss from the 0.4 μm filter-passing fraction. (b) ln k_{cp} vs. log K_c where k_{cp} is the colloidal pumping rate determined from the initial rate (at time $t \leq 1$ h) of transfer. Data included are those from Figure 11a as well as data for ^{109}Cd, ^{137}Cs, ^{54}Mn, and ^{60}Co. The increase in values for k_{cp}, with those for K_c, are consistent with other such correlations reported in the literature.[18] The open symbol is the calculated colloidal-pumping rate for thorium. Because thorium is so highly particle reactive, it is possible that thorium isotopes may find use as *in situ* coagulometers. All model lines were calculated using values for (1) hematite coagulation rate and (2) equilibrium sorption constants.

0.4 μm filter-passing fraction, for each radionuclide studied, as a function of K_c. The increase in ''sorption'' rate constant with K_c is similar to the correlations between K_d and pseudo-first-order sorption rate coefficients derived by Jannasch et al.[18] from the data of Nyffeler et al.[16] (Figure 4). The range in rate coefficients found in Figure 11b is consistent with those reported for other laboratory systems but are much smaller than rate coefficients reported for studies of actual sorption mechanisms.[51-53]

4. APPLICATION OF THE COLLOIDAL PUMPING MODEL TO OCEANIC TRACE ELEMENT SCAVENGING

The idea that coagulation plays an important role in regulating the composition of aquatic systems has been in the geochemical literature for some time, although the nature of the contribution has undergone significant evolution. Hahn and Stumm[54] suggested that coagulation controls the distribution of particles, and, therefore, the availability of reactive mineral surfaces, in many systems. The effect of particle-particle interactions on the rate of scavenging of metals from the solution phase by particles was first suggested by Tsunogai and Minagawa[55] as an *ad hoc* rationale to explain data for U/Th disequilibrium and has been a recurrent theme over the last decade.[43,56]

Figure 12 depicts the current view of particle dynamics in the ocean. Such models were developed through the use of dissolved U/Th-series isotopes and their particle-reactive daughter products which serve as particle tracers. Bacon et al.[13] and Nozaki and co-workers[14] have developed scavenging models incorporating two particle size classes: one composed of particles large enough to settle and a second of non-settleable particles (Figure 12a). Filter-passing trace elements, Me_{d+c}, undergo exchange reactions with the small particle pool (the small particles in this case are still relatively large, $0.45-10$ μm). It is important to note that what is often called filter passing "dissolved" Me includes Me associated with colloids. The distribution of Me among sorbed and "dissolved" fractions is controlled by aggregation and disaggregation reactions. Because the small particles do not settle, the vertical flux of Me is controlled by small/large particle interactions. Clegg and Whitfield[57] (Figure 12b) have included in the model a mineralization rate constant, r_m, for all reactions originating from particle pools.

While the models depicted in Figures 12a and b provide a framework for evaluating trace metal scavenging data, and in particular that from uranium and thorium series isotopic disequilibrium, the correspondence between field measurements of rate constants for mass transfer, e.g., k_1 and k_2, and the postulated processes is not yet resolved. The model depicted in Figure 12c includes a pool of colloidal particles. This model[43] is able to explain all anomalous features of U/Th disequilibrium data.

Consider Figure 13.[58] R'_f (the ordinate) is a pseudo-first-order "sorption" rate constant derived from U/Th disequilibrium and Th partitioning between filter-retained particles and an operationally defined "dissolved" phase (i.e., the filter-passing portion of the bulk solution). R'_f is presumed to represent the process described by k_1 in Figure 12. Note that R'_f, from a variety of sources and oceanic environments, ranges from 10^{-3} d^{-1} in the deep ocean to greater than 1 d^{-1} in coastal environments.

At least two questions arise from these data: (1) Are the sorption times consistent with physicochemical theory and laboratory observations? (2) Is

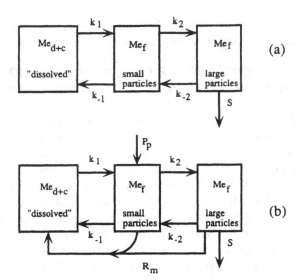

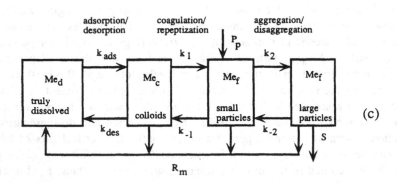

Figure 12. Schematic illustration of the current view of particle dynamics in the ocean. (a) Bacon and co-workers[13,14] have determined the rate constants for the interactions shown. k_1, for example, represents the rate constant for the appearance of Th radionuclides in the non-settleable particle pool. While the transfer process is assumed to be sorption, the speciation of Me is based on an operational definition of "dissolved". The notation, Me_{d+c}, is to differentiate an operational definition of "dissolved" from truly dissolved species of Me(Me_d). (b) Clegg and Whitfield[57] and others have added a term for a remineralization reaction, R_m, which can originate from all particle pools and a source term for small particles due to primary productivity, P_p. In these models the particles considered are relatively big, micron-sized or larger. (c) Honeyman and Santschi[43] and others are proposing a colloidal phase which can take up trace elements and radionuclides rapidly through sorption reactions but which coagulates relatively slowly with the small particle pool. Coagulation of colloids can thus become a rate controlling step in radionuclide scavenging by small particles. Thus, the "scavenging" step of models (a) and (b) is composed of a rapid sorption step, here designated k_{ads}, and a slower aggregation step, k_1.

the observed relationship between the sorption rate constant, R_f, and filter-retained particle concentration, C_f, expected if scavenging is controlled by sorption processes?

Laboratory studies of metal-ion adsorption onto well-characterized metal-oxide surfaces[51,53] have shown that sorption is rapid. Equilibrium is achieved on the order of milliseconds to seconds at milli- to micromolar activities of surface sites and metal-ions, respectively, and sorption rate shows a first-order dependence on surface site concentration. It is possible, but unlikely, that the measured sorption rate constants in the ocean are small simply because of the low reactant concentrations (in deep ocean water, the total Th concentration, Th_T is about 10^{-14} mol L^{-1} and marine particulate matter may provide between site concentrations of 10^{-9} to 10^{-7} mol.L^{-1}). An additional

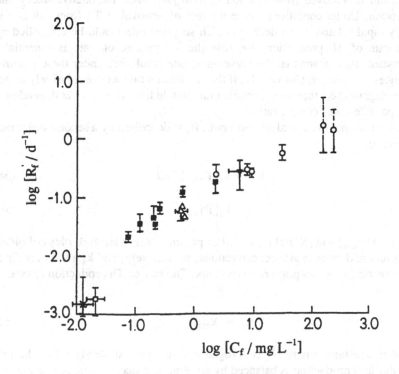

Figure 13. A plot of the "scavenging" rate constant, log R_f' (which is here identical to k_1 of Figure 12) vs. the filter-retained particle concentration, log C_f, using oceanographic data from the literature[12,14,59-63] Scavenging, for this data set, represents the appearance of Th in the small particle pool.

$$R_f = (k_{-1} + \lambda_{234} + \lambda_p) \left\{ \frac{{}^A Th_{part}}{{}^A Th_{diss}} \right\} \qquad (F13.1)$$

where k_{-1} is a release or desorption rate constant (assuming 0.007 d^{-1}) and λ_{234} and λ_p are the ^{234}Th decay and particle removal rate constants, respectively, and A represents thorium activity (From Honeyman et al.[58]. Reprinted with permission.

complication is the possibility of slow solution-phase ligand-exchange processes. For example, Hering and Morel[64] showed that a system containing a suite of Me-complexing ligands can slowly approach equilibrium under certain circumstances even if the rates of individual complexation reactions are relatively fast (Figure 14). Their data confirm previous observations of the same phenomenon by Raspor et al.[65]

4.1 Thorium Production as the Rate-Limiting Process

One of the great advantages of using disequilibrium between U/Th decay series isotopes as a tool for studying scavenging rates is that the thorium source term is precisely known.

While uranium in ocean water is present in constant proportion to salinity, thorium is removed from solution by two processes: radioactive decay and sorption. Under conditions where the rate of removal of Th from solution is very rapid relative to its decay rate, Th sorption rate would be controlled by the rate of Th production. Because the Th production rate is essentially constant, this means that the scavenging rate would be, under these circumstances, zero order. Conversely, if the production rate were excessively large, one might expect that the scavenging rate would have a first-order dependence on particle (site) concentration.

Assume that the actual sorption rate, R_s is described by a second-order rate expression

$$R_S = k_S[Th_{diss}][XOH] \tag{5a}$$

$$= k_1[Th_{diss}] \tag{5b}$$

where $[Th_{diss}]$ and $[XOH]$ are the filter-passing (truly dissolved plus colloidal) thorium and particle site concentrations, respectively, and k_1 and k_s are first- and second-order sorption rate constants. The rate of Th production (mol L^{-1} s^{-1}) is

$$R_p = \lambda_{234}A_{238}/N_A \tag{6}$$

and is constant, where N_A is Avogadro's number. At steady state, the rate of thorium production is balanced by sorption and decay rates. A rate expression for ^{234}Th concentration in the solution phase is

$$\partial[Th_{diss}]/\partial t = \lambda_{234}A_{238}/N_A - k_S[Th_{diss}][XOH] - \lambda_{234}[Th_{diss}]$$

$$= \text{production} - \text{sorption} - \text{decay}$$

$$= 0, \text{ at steady state} \tag{7}$$

The three terms in Equation 7 are plotted as a function of site concentration, $[XOH]_T$, in Figure 15, for two values of the second-order sorption rate con-

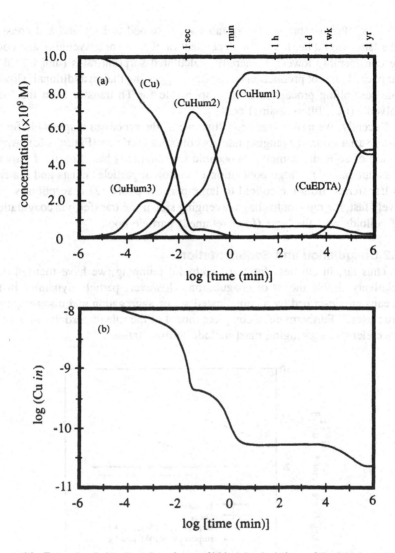

Figure 14. Example of slow ligand exchange.[64] Model calculations of the initial reaction of Cu with humic acid (represented as a discrete ligand) and CaEDTA followed by a pseudo-equilibrium exchange between Cu-humate and CaEDTA. The formation rates of all Cu-humate complexes are taken as equal and the dissociation rate constants are inversely proportional to their equilibrium stability constants. The rate of formation of Cu-humate species is governed by the concentration of free humate ligands: the fastest reaction occurs at the weakest site. The system slowly reaches equilibrium because of the low availability of inorganic Cu for complexation with CaEDTA. Cu availability is low because strong humate binding sites are in excess of Cu. Inorganic copper activity was determined by amperometric measurements and includes Cu hydroxy, carbonato, and chloride complexes. (a) Distribution of Cu complexes; (b) inorganic Cu activity as a function of log (t). Reprinted with permission.

stant, k_s. Production and decay rates are assumed to be given and constant (the latter would be, in reality, dependent on the rate of scavenging and could be considerably lower). Generally, calculated sorption rates (mol L^{-1} d^{-1}) far exceed rates of production and decay, suggesting that an additional, slower, rate-controlling process must be responsible for Th transfer from the "dissolved" (i.e., filter-passing) pool.

Recently, we have suggested[43] that one of the processes responsible for the large variation in scavenging rates is colloidal pumping (Figure 12c). Application of colloidal pumping to oceanic scavenging is based on the following assumptions: (1) a large population of colloidal particles exists and material is transferred from the colloid to large-particle pools; (2) if sorption is relatively fast, the rate-controlling scavenging step is the transfer, via coagulation, of colloids with the large (filter-retained) particle pool.

4.2 Coagulation and Sedimentation

Thus far, in our description of colloidal pumping, we have focused on a relatively simple model of coagulation. However, particle dynamics in the oceans are described by a complicated set of aggregation and disaggregation processes.[49] Furthermore, a complete model of the role of particles in oceanic trace element scavenging must include sedimentation.

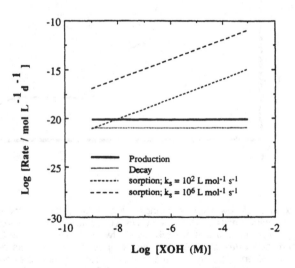

Figure 15. Comparison of thorium sorption and production rates as a function of particle site concentration, [XOH], and second-order sorption rate constant, k_s. Production and decay rates are essentially constant but sorption rate depends on the concentration of reactive particle sites, dissolved ^{234}Th concentration (ca. 10^{-19} mol L^{-1}) and k_s. Reported values for k_s, for a variety of metal ions, range from 10^2 to 10^7 L mol^{-1}·s^{-1} (Hayes and Leckie[53] and references therein). The range in particle concentrations for the computation is 10^{-8} to 10^{-2} kg L^{-1} and we have assumed that ca. 0.1 mol of sites per kg particles are available to form complexes with ^{234}Th species. It is clear that, for most of the simulation conditions, steady-state (Equation 7 = 0) is not possible without invoking a slower, rate-determining process controlling the transfer of "dissolved" Th from the filter-passing to filter-retained particle pools.

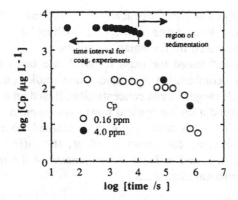

Figure 16. Log C_p vs. log (time) for hematite coagulation[50] for two different values of total particle concentration, C_p. Particles initially homodisperse and non-settleable coagulate (qq.v., Figure 9) until aggregates become large enough to settle. At this point, mass removal from the system occurs: the lower the initial particle concentration, the longer the time period required for the onset of sedimentation. Farley and Morel[69] have derived a semi-empirical model to describe the effect of coagulation on sedimentation.

Figure 16 is a plot of suspended hematite mass as a function of time for a system initially containing 4 ppm of spherical, 0.25 μm hematite particles. Suspended mass is conserved until about $10^{3.5}$ s (0.9 h). During this period, particles are coagulating but the only particle transfer is between particle aggregate groups (qq.v., Figure 9). This is the time period covered by the coagulation experiments described earlier. Eventually, however, aggregates become too large to remain in suspension and settle out of the system. It is this process, sedimentation driven by coagulation, that is of interest for oceanic scavenging.

The general dynamic equation for the transport of particles in flow systems is described by Friedlander[66] and Jeffrey[67] and consists of terms for advection (i.e., flow as current), Brownian and turbulent diffusion, settling of particles, and the increase and decrease of particle numbers of certain size classes as the consequence of coagulation and particle breakup.

No general analytical solution exists for the dynamic equation coupling coagulation with sedimentation: it must be solved using either simplified analytical solutions[68] or numerically. A critical aspect is the solution to the coagulation terms. One approach is to use similarity transforms that reduce the coagulation equation to a function of one independent but observable variable, e.g., filter retained particle concentration, C_f. A characteristic of this approach is the assumption that the particle-size distribution is "self-preserving". Such a distribution is one in which a coagulating suspension, after a long time of evolution, develops a particle size distribution whose shape is related to the mechanism governing coagulation. An assumption inherent in this approach is that the particle-size spectrum consists of subranges dominated by a single coagulation mechanism. For example, particle coagulation in systems of low particle concentration (≤ 1 mg L^{-1}), small particle size (≤ 1 μm) and low shear rate (G ≤ 1 s^{-1}), such as large parts of the deep ocean, will be dominated by Brownian coagulation.

Farley and Morel[64] tested the mass removal rate law and the similarity approach through a combination of sedimentation studies and numerical simulations over a wide range in mass concentration. Based on their experimental results, they proposed a semi-empirical power law expression which has the form of the overall mass removal expression obtained from similarity arguments.[70] It contains terms for Brownian, shear, and differential settling coagulation mechanisms, and includes the exponents of the three power laws obtained from numerical simulation

$$\frac{\partial C_f}{\partial t} = -B_{ds}C_p^{2.3} - B_{sh}C_p^{1.9} - B_bC_p^{1.3} \tag{8}$$

where C_f is the measurable, i.e., filter-retained, particle mass concentration, C_p the total mass concentration of particles, and B_b, B_{sh}, and B_b are coagulation coefficients for the Brownian, shear, and differential settling mechanisms, respectively. Broken exponents of C_f result from the parametrization of coagulation with respect to particle concentration instead of particle numbers and their distribution in the different size classes. It is important to recognize that while particles are removed through gravitational sedimentation because they exist as clusters too large to remain in suspension, the rate at which such clusters are formed is governed by the rate of coagulation.

4.3 Adsorption Coupled to Sedimentation

When particles are removed from a system through sedimentation, so is their sorbed trace-element load. The sedimentation transport of trace elements is equal to the sedimentation rate (e.g., g.L^{-1}.d^{-1}) times the trace element adsorption density (e.g., mole sorbed per gram particles).

Figure 17[43] contains colloidal pumping model results for the data compiled by Honeyman et al.[58] The modeling is based on the assumption that the mass loss of particles due to sedimentation is balanced by the supply of colloidal particles to the settleable-particle pool. Considering, for illustrative purposes, only the term for Brownian coagulation in Equation 8, this assumption is expressed as follows

$$-\frac{\partial C_c}{\partial t} = \frac{\partial C_f}{\partial t} = -B_bC_p^{1.3} \tag{9a}$$

$$= -B_b'C_p \text{ with } B_b' = B_bC_p^{0.3} \tag{9b}$$

where C_c represents the mass concentration of filter-passing colloids. It should be noted that this is a caveat in the application of Farley and Morel's model, which was developed for non-steady-state batch reactors, to a natural system

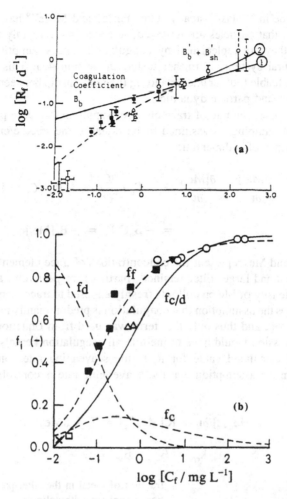

Figure 17. Application of Farley and Morel's sedimentation model to oceanic Th scavenging.[43] Particle coagulation and sedimentation is coupled to Th sorption onto colloidal particles. (a) Data are the same as shown in Figure 13. Critical parameter values include: Brownian collision efficiency factor, $\alpha = 0.5$ ($-$), particle density, $\rho_p = 2.5$ g cm^{-3} and mean depth, h = 6 m (the average depth of a water column that sorbed ^{234}Th atom will see before decay). Solid lines represent the scavenging rate constant when a trace element or radionuclide in the filter-passing fraction is completely associated with colloidal particles: (1) Brownian coagulation; (2) Brownian + shear coagulation. In other words, only particle-particle interactions are considered. The dashed line includes the effect of Th solution/colloid partitioning on scavenging rate. (b) Calculated distribution of Th between dissolved, f_d, colloidal, f_c, and filter-retained, f_f, fractions. $f_{c/d}$ represents the portion of Th passing a filter but which is associated with colloids. Note that as $f_{c/d}$ approaches 1, that is, when little truly dissolved Th exists, that the scavenging rate constant approaches the maximum rate of coagulation + sedimentation. Actual values of f_f correspond to the scavenging rates and sources listed in Figure 13. Reprinted with permission.

assumed to be in "quasi" steady state. Farley and Morel[69] have suggested, for example, that particles are removed, at times, more rapidly by sedimentation than they are replenished by coagulation. The assumption of steady state in natural systems is further tested in Section 4.6. This model thus provides a valuable tool for understanding the relationship between scavenging rate constants and particle dynamics.

At steady state, the rate of trace element supply to the large particle pool, via colloidal pumping, is assumed to be equal to the trace element loss as the consequence of sedimentation

$$-\frac{\partial[\text{Me}_c]}{\partial t} = \frac{\partial[\text{Me}_f]}{\partial t} = -\Gamma_c \cdot \frac{\partial C_c}{\partial t} = -B_b \cdot C_p^{1.3} \cdot \Gamma_c \qquad (10a)$$

$$= -B_b' C_p \Gamma_c = -B_b'[\text{Me}_f] \qquad (10b)$$

where Me_c and Me_f represent the concentrations of trace elements associated with colloidal and large, filter-retained, particles, respectively, and Γ_c is the adsorption density of Me on colloids freshly exposed to trace elements. Equation 10 carries the assumption that coagulation is predominantly due to Brownian coagulation, and thus only that term was used from Equation 8. A more general expression would have to include all coagulation kernels from Equation 8. The calculated value for R_f is the scavenging rate constant which follows from the assumption that the scavenging rate is controlled by coagulation, i.e.,

$$\partial[\text{Me}_{d+c}]/\partial t = R_f[\text{Me}_{d+c}] = f_{c/d} k_{cp}[\text{Me}_{d+c}] \qquad (11a)$$

$$k_{cp} = B_b' + B_{sh}' + \dots \qquad (11b)$$

where $f_{c/d}$ = fraction of metal in the filter-passing solution associated with colloids
       k_{cp} = colloid pumping rate constant (time^{-1})
       B_b', B_{sh}' = pseudo-first order rate constants for Brownian shear coagulation

and the subscript "d+c" denotes the filter-passing (operationally defined "dissolved") portion of bulk solution.

Lines 1 and 2 in Figure 17 represent upper limit calculations for different coagulation mechanisms: (1) Brownian; (2) Brownian + shear.

Figure 17b shows the calculated distribution of Th among the dissolved, colloid, and large particle pools. These calculations are based on the assumption that the intensity of Th/marine particle interactions does not change with particle concentration; however, the fraction of total Th which is sorbed *is* a function of C_f and C_c. (Note: Section 4.5 describes the way in which the relationship between C_f and C_c used in these calculations was derived.)

At low values of C_f, most of the filter-passing Th is truly dissolved. As C_f and C_c increase, so does the fraction of Th which is associated with colloids. Eventually, at a C_f of approximately 10 mg L^{-1}, truly dissolved Th is an insignificant proportion of the total Th mass. Critical, though, is the observation that a significant amount of filter-passing Th is associated with colloids. Recently, model predictions for Th isotopes and Al have been verified (for aluminum in the ocean: B. Moran, personal communication; Th in Gulf of Mexico waters: M. Baskaran, personal communication).

4.4 Coagulation in High-Energy Dissipation Environments: Control by Shear Rates in the Water or at the Sediment-Water Interface

Coastal regions, with their high energy dissipation rates and their relatively high particle concentrations and fluxes, are areas where turbulent mixing energies could significantly affect trace metal scavenging rates by colloidal pumping onto particles.

The rate constant for scavenging, λ_s, obtained from Equation 1, can be compared to the rate constants for coagulation/sedimentation, calculated by Equations 8 through 10. At lower particle concentrations and sizes,[69] only the Brownian and shear coagulation terms are important. Shear coagulation rates (with kernel B_{sh}) which depend on the rate of energy dissipation[71,72] can be estimated from first principles as shown below. The shear coagulation kernel, B_{sh}, contains the current shear factor, G (in units of s^{-1}), which is a function of the energy dissipation rate, ϵ

$$G = \sqrt{0.3 \frac{\epsilon}{v_f}} \qquad (12)$$

where v_f is the kinematic viscosity.[72]

For the purpose of calculating shear coagulation rates, it is crucial how ϵ is calculated: (a) average value over the water column, calculated from energy loss through small scale turbulence in the water; (b) value in the viscous boundary layer, where most energy is dissipated by friction with the bottom. There are orders of magnitude difference between the two estimates:

(a) Water column averaged values of ϵ can be calculated using equations given by Garrett et al.[73]

$$\epsilon = C_d \frac{U^3}{h} \qquad (13)$$

where C_d is the drag coefficient (dimensionless, with a value of $\approx 2.5 \times 10^{-3}$), U is the average current velocity (m s^{-1}) and h is the depth of the homogeneous water column (m).

(b) The energy dissipation rate, ϵ, in the viscous sublayer can be estimated from Reference 71

$$\epsilon = \frac{u_*^2}{\upsilon_f} \tag{14}$$

where u_* is the friction velocity (m s^{-1}), which can be visualized as the current velocity at an interface (with $u_* = U_1(C_d)^{1/2}$; U_1 is the current velocity at 1.0 m depth).

Kinetic energy in shallow coastal environments is mostly dissipated near the sediment-water interface, thereby imparting momentum to particles causing impaction, coagulation, and resuspension. Energy dissipation, and therefore, shear coagulation, is consequently higher near the interface than in the water column. There is some recent evidence that interfacial shear coagulation is increasing coagulation rates in coastal environments and thus drastically changing the particle size distribution.[72] Water column averaged values for ϵ provide, however, lower limits for the calculation of shear coagulation rates. ϵ values, taken from the literature, range from 6×10^{-4} m^2 s^{-3} for the interfacial region of Salem Sound, MA,[71] 3×10^{-4} m^2 s^{-3} for the water column of the Great Bay Estuary in New Hampshire[74] to 2.5×10^{-5} [75] and $2 \times 10^{-7} - 1 \times 10^{-4}$ m^2 s^{-3} [76] for the waters of Narragansett Bay, R.I. For Galveston Bay waters, we estimate values for ϵ of 1 to 2×10^{-4} m^2 s^{-3}, depending on wind and tidal conditions.

Data sets for ^{234}Th uptake rates by particles in coastal waters with different energy dissipation rates (10^{-5} to 10^{-3} m^2 s^{-3}) but with similar particle concentrations (i.e., 10^1 mg L^{-1}), were used to further test the predictions of the scavenging model of Honeyman and Santschi.[43] Included are data from Galveston Bay, TX,[37] the Irish Sea,[77,78] and Narragansett Bay, R.I.[63,79] The high energy dissipation rates, ϵ, in the first two areas cause particle coagulation rates to be dominated by turbulent shear above the sediment-water interface, while in the third area, where energy dissipation rates are lower, coagulation rates are mainly given by the Brownian mechanism.

Assuming reasonable (instead of maximum) estimates of the coagulation parameters, i.e., a collision efficiency factor, α, of 0.1,[74] and an effective particle density in the ocean of 1.3 g cm^{-3}, the scavenging rate constants reported from the measurement of the disequilibrium between ^{234}Th and ^{238}U in the water column of the Irish Sea, Galveston Bay, and Narragansett Bay can be reasonably well predicted using our model (Figure 18). The relatively high value of ϵ of 2.0×10^{-3} m^2 s^{-3} needed for the model fit is appropriate for the viscous sublayer of a coastal system where 5 to 15 m tides cause exceptionally high rates of energy dissipation[73,80] and thus shear coagulation.[72]

4.5 The "Particle Concentration" Effect

Another way in which colloids have been implicated in affecting descriptions of sorption is as an artifact in experimental determinations of sorption binding constants, or conditional parameters such as K_d. The "particle concentration effect" is the observed decrease in the value of apparent partitioning "constants" with increasing particle concentration, C_f. Even though this effect is an artifact in experimental systems in the laboratory, it also reflects real changes in the distribution of sorbed species in natural systems with differing particle concentrations.[81] Therefore, this effect is mentioned here in some greater detail.

Two schools of thought have emerged as explanations for the anomalous behavior. One line of thought[82] holds that particle-particle interactions produce an interfacial environment which is relatively unfavorable to the formation of surface complexes. The higher the particle concentration, the greater the frequency of particle-particle interactions and the more the equilibrium is shifted toward solution-phase species. A second school of thought[26] is that an operationally defined 'dissolved' phase may contain trace-elements associated with colloids as well as those which are truly dissolved. The presence of colloids in the "dissolved" phase reduces the partitioning constant from its "true" value; the magnitude of the reduction depends on the intensity of trace-element binding with colloids and the amount of colloids present.

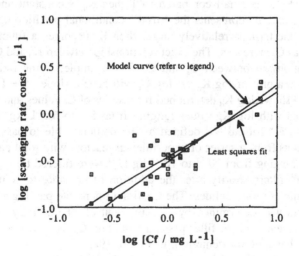

Figure 18. Log of the observed and modeled scavenging rate constant, λ_s, as a function of log filter retained particle concentration, C_f, for the Irish Sea. Data are from Kershaw et al.[77,78] For Narragansett Bay, with $C_f = 1-10$ mg L^{-1}, the measured [63,79] and predicted values of λ_s are 0.1–1 and 0.1–0.6 d^{-1}, and for Galveston Bay (with $C_f = 30$ mg L^{-1}), 1.5 and 1.4 d^{-1}, respectively.[37] The model curve represents a shear rate of 25 s^{-1}, which is equivalent to an energy dissipation rate of 2.3×10^{-3} m^2 s^{-3}.

Consider, again, three pools in which metal ions, Me, may reside: associated with colloid, bound to large particles and truly dissolved (Me_c, Me_f, and Me_d, respectively). Conditional association constants can be written to express the relative affinity of Me for the two particle pools:

$$K_c = \frac{[Me_c]}{[Me_d] \cdot C_c} \tag{15a}$$

$$K_f = \frac{[Me_f]}{[Me_d] \cdot C_f} \tag{15b}$$

K_c and K_f are conditional because they depend on the general solution composition (pH, ionic strength, etc.) and the presence of Me-complexing ligands. A K_d can be expressed as given in Equation 15a, b[26,43,83]

$$K_d = \frac{K_f}{1 + K_c \cdot C_c} = \frac{[Me_f]}{([Me_d] + [Me_c]) \cdot C_f} \tag{16}$$

If the term $K_c \cdot C_c$ is small compared to 1 then K_d is constant and equal in value to K_f, i.e., it represents the "true" conditional partitioning constant. If, instead, the term is relatively large, then K_d becomes a function of C_f, decreasing as C_f increases. The exact relationship between K_d and C_f depends on the relationship between C_c and C_f. For example, if the ratio $C_c : C_f$ is constant then a plot of log K_d vs. log C_f will have a slope of -1.

Figure 19 shows log K_d determined for a series of C_f values and for several radionuclides with log K_c values ranging from 5.4 to 6.3 L kg^{-1}. In these experiments, a "colloid" is defined as the particles able to pass a 0.4 μm filter. Aliquots from a series of experimental reactors with total particle concentrations ranging from 10^{-7} to 10^{-3} kg L^{-1} were filtered through 0.4 μm Nucleopore® filters shortly after the addition of hematite to solutions containing a suite of radionuclides. The fraction of hematite passing and retained by the filters was determined by measurement of ^{59}Fe activity. The log of the mass concentration of filter-passing hematite, C_c, is shown as a function of log total hematite concentration in Figure 19a.

The effect of colloidal hematite on calculated distribution coefficients is demonstrated in Figure 19b, using ^{113}Sn, ^{109}Cd, and ^{133}Ba as examples. K_d is defined by Equation 16 and is the ratio of filter-retained to filter-passing radionuclide activities, divided by the filter-retained particle mass concentration, C_f. K_c (Equation 15) is the conditional partitioning constant for ^{113}Sn, ^{109}Cd, and ^{133}Ba sorbing on the hematite particles used in these experiments. The higher value of K_c for these experiments is a consequence of the higher system pH (8.0 compared to 7.5). Although the value of K_c is constant, log K_d exhibits a wide range in its value. At low C_p, K_d approaches the value of

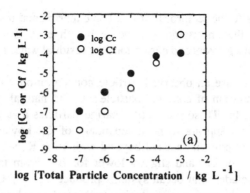

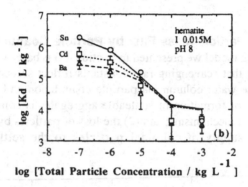

Figure 19. Log K_d vs. log C_p for a suite of radionuclides[50] as an illustration of the particle concentration effect. The data come from laboratory studies using 0.25 μm spherical hematite particles and a continuously varying ratio between filter-retained (C_f) and filter-passing (C_c) hematite particles. The bulk solution was filtered with 0.4 μm Nucleopore filters (a) Variation of the mass of filter-passing (colloidal) hematite with total particle mass, C_p. (b) Log of distribution coefficients for ^{113}Sn (IV), ^{109}Cd (II), and ^{133}Ba (II) sorbing on hematite as a function of log total particle mass, log C_p. K_d values are based on the ratio of 0.4 μm filter-retained and filter-passing radionuclide activities times the reciprocal of the filter-retained mass, C_f of hematite. Model calculations are found on application of Equation 15 with K_c values of $10^{6.3}$ $10^{5.7}$ and $10^{5.4}$ for Sn, Cd, and Ba, respectively, and the measured colloid mass concentration.

the "true" conditional partitioning constant, K_c, and decreases in value with increasing C_p. This change in value reflects the increasing importance of the colloid-associated radionuclide pool as C_p increases (the term K_cC_c is also increasing in value as shown in Figure 19a). At low C_p, truly dissolved Sn(IV) is the dominant fraction of Sn(IV) ions passing the 0.4 μm filter. Thus, log K_d approaches the "limiting" value of log K_c because the artifact of filter-passing (colloidal) radionuclide species is negligibly small. The model simulations for log K_d as a function of log C_p were completed assuming that K_c is equal to K_f, and that K_c is constant in value over the range in C_p studied. Model results indicate that at increasing values of C_p, the term $C_c·K_c$ becomes large relative to 1, and the log K_d values for the three metal ions will then be close in value to the reciprocal of C_c. The experimental results follow this trend, although the uncertainties associated with each data point, due to cumulative counting errors and corrections for residual water retained by filter surfaces, are large.

It is possible to use an observed particle concentration effect to estimate the mass concentration of colloidal particles, C_c, in natural systems. Figure 20 shows log K_d for Th sorption onto marine particles as a function of log C_f. At low C_f, K_d appears to be independent of C_f, however, as log C_f increases a critical point is reached beyond which log K_d vs. log C_f falls on a slope of -0.7 for Th, and slightly lower for Be. From this slope it can therefore be concluded that ocean systems produce $C_c = f(C_f^{0.7})$. The exact physical or chemical processes producing such a relationship are, as yet, not known.

4.6 Control of Particle Mass Flux by Particle Coagulation

The scavenging model we presented in Section 4.3 is based on the following assumptions: (1) that scavenging rate constants reflect sorbed trace element removal from the water column via particle coagulation and sedimentation, whereby the rate of formation of settleable aggregates is controlled by individual coagulation mechanisms; and (2) the loss of particles by sedimentation is balanced by supply of colloidal particles to the settleable particle

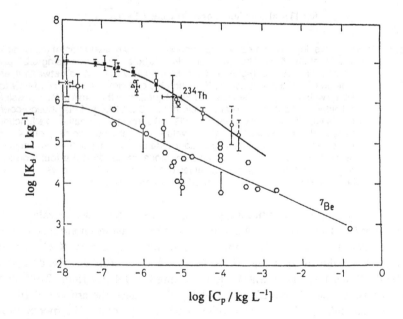

Figure 20. Log K_d vs. log C_f for [234]Th and [7]Be. Data sources for Th are given in Figure 13. [7]Be:[16,17,84-87] These data indicate that the "particle concentration effect" may extend over many orders of magnitude. (From Honeyman et al.[58]. Reprinted with permission.)

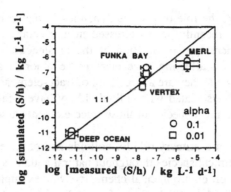

Figure 21. Log of simulated mass removal rate, S/h (kg L^{-1} d^{-1}) vs. log of measured mass removal rate, S/h (kg L^{-1} d^{-1}). Predicted rate was calculated using the sedimentation model of Farley and Morel.[69] Deep Sea/Sargasso Sea: h = 3200 m, C_f = 10^{-8} kg L^{-1}; Funka Bay: h = 30–45 m, C_f = 6 × 10^{-6} kg L^{-1}; MERL/Narraganset Bay: h = 5 m, C_f = 3–10 × 10^{-6} kg L^{-1}. Error bars on predicted sedimentation rates reflect the range in values for h and C_f.

pool. Another way in which to verify such a model is a comparison of measured and calculated, depth-normalized sedimentation rates, S/h (with S = sediment accumulation rate, h = mean depth).

Figure 21 contains such a comparison. We examined and simulated the mass loss rate, S/h (kg L^{-1} d^{-1}), for three environments. These are (1) the deep ocean and Sargasso Sea[88] (S/h ca. $10^{-11.2}$ kg L^{-1} d^{-1}); (2) Funka Bay[62] (S/h ca. $10^{-7.6}$ kg L^{-1} d^{-1}); and (3) MERL experiments to simulate Narragansett Bay, R.I.[79] (S/h ca. $10^{-5.5}$ kg L^{-1} d^{-1}). Mass removal rates were calculated using the sedimentation model of Farley and Morel.[69] Calculations were made with α equal to 0.1 or 0.01 and an effective particle density, ρ_p — ρ_{sw}, of 1 g cm^{-3}, with ρ_p = *in situ* particle density and ρ_{sw} = seawater density. All other model parameters, except for h (water column depth), and C_f (particle concentration) are as given in Figure 17.

The agreement between simulated and measured particle fluxes in these very different hydrodynamic and particle dynamic regimes is encouraging. It needs to be further verified as more data becomes available.

5. SUMMARY

The last 40 years has produced a considerable evolution in our understanding of the processes controlling the behavior of many trace elements and radionuclides in the oceans. The emergence of aquatic surface chemistry 20 years ago provided a powerful framework for understanding many observations of trace metal behavior in natural aquatic systems.

Until recently, the environmental behavior of trace elements could be couched almost entirely in chemical terms. While it has long been recognized that the fate of trace elements in natural systems is, to a large extent, controlled by sorption processes, the role of particle dynamics in regulating trace element behavior has only recently been addressed in any detail. Of particular importance is the increasing recognition of the existence and importance of colloidal particles in affecting trace element behavior. The importance of colloids is twofold: (1) they are a large pool of trace-element binding ligands; (2) trace elements associated with colloids likely have a fate different from truly dissolved chemical species or those trace elements associated with macroparticles.

The emergence of a primary role for colloids and particle dynamics in the distribution of chemical species throughout natural aquatic systems may have a profound effect on environmental chemistry. For example, the likelihood of transport of colloidal particles in surface and ground waters will force the casting aside of many closely held shibboleths regarding the surface and subsurface transport of chemical contaminants. The need to characterize the natural colloid pool will also result in stringent sampling protocols, higher standards for analytical environmental chemistry, and the development of new analytical techniques.

ACKNOWLEDGMENTS

We are grateful for the careful reviews of this manuscript by J. Buffle, H.P. van Leeuwen, and A. Tessier. The material reported in this paper is based in part upon work supported by the Texas Advanced Research Program (Grant #4697), the Texas Institute of Oceanography (Grant #18183), and the U.S. National Science Foundation (Grant #OCE-9012103).

APPENDIX: GLOSSARY OF SYMBOLS

A Activity (decays $\text{min}^{-1} \text{ L}^{-1}$)

B_b Brownian coagulation coefficient ($\text{L}^{0.3} \text{ mg}^{-0.3} \text{ g}^{-1}$)

B_{sh} Shear coagulation coefficient ($\text{L}^{0.9} \text{ mg}^{-0.9} \text{ s}^{-1}$)

B_{ds} Differential settling coefficient ($\text{L}^{1.3} \text{ mg}^{-1.3} \text{ s}^{-1}$)

B'_b Pseudo-first-order Brownian coagulation coefficient (d^{-1})

B'_{sh} Pseudo-first-order shear coagulation coefficient (d^{-1})

C_f Mass concentration of particles retained by a filter (kg L^{-1})

C_c Mass concentrations of particles passing a filter (i.e., "colloids") (kg L^{-1})

C_d Drag coefficient ($-$)

C_p Total particle mass concentration $= C_c + C_f$

DOC   Dissolved organic carbon (mg L^{-1})

f_c Fraction of metal association with colloidal material

f_d Fraction of metal in dissolved species

$f_{c/d}$ Fraction of metal which is operationally defined as part of the "dissolved" phase (e.g., filter passing) but which is associated with colloids

f_f Fraction of metal associated with large (i.e., settleable) particles

G Shear rate (s^{-1})

h   Height of homogeneous water column (m)

k   Number of particles in a cluster $(1,2, \dots)$

k_b Brownian coagulation kernel $(m^2 \, s^{-1})$

k_B Boltzmann constant $(J^\circ \, K^{-1})$

k_1 First-order uptake or sorption rate constant (d^{-1})

k_{-1} First-order release or desorption rate constant (d^{-1})

k_2 First-order aggregation rate constant (d^{-1})

k_{-2} First-order disaggregation rate constant (d^{-1})

k_{cp} Pseudo-first-order colloidal pumping rate constant (d^{-1})

k_s Second-order sorption rate constant $(L \, mol^{-1} \, s^{-1})$

K_c Colloid/solution partition coefficient $(L \, kg^{-1})$

K_d Empirical particle/solution partition coefficient, obtained from filtration, with colloids in filter-passing solution $(L \, kg^{-1})$

K_f Particle/solution partition coefficient $(L \, kg^{-1})$

K_p Overall partitioning coefficient $(L \, kg^{-1})$

Me_c Sum of all species of Me associated with colloids (M)

Me_d Sum of all truly dissolved species of Me (M)

Me_f Sum of all species of Me associated with filter retained particles (M)

Me_{d+c} Concentration of the species of element Me passing a filter or otherwise operationally considered to be "dissolved"; the sum of colloidal and truly dissolved species of Me $(= Me_d)$ (M)

n_k Number of aggregates composed of k of primary particles (number mL^{-1})

N_∞^0 Number concentration of monodisperse particles initially in system (number mL^{-1})

N_∞ Total number of particles at some time, t (number mL^{-1})

R_f' First-order ^{234}Th uptake or sorption rate constant (d^{-1})

r_m First-order mineralization rate constant (d^{-1})

R_p Thorium production rate $(mol \, L^{-1} \, d^{-1})$

R_s Sorption rate $(mol \, L^{-1} \, d^{-1})$

S/h   Sedimentation flux $(kg \, L^{-1} \, d^{-1})$

Th_{d+c} Concentration of species of Th isotopes passing a filter or otherwise operationally considered to be "dissolved"; the sum of colloidal and truly dissolved species of Th

$u_.$ Friction velocity $(m \, s^{-1})$

U   Current velocity $(m \, s^{-1})$

U_1 Current velocity at 1 m depth $(m \, s^{-1})$

V_c Energy of interaction of particles (J)

W   Stability ratio $(= \alpha^{-1})$ $(-)$

XOH   Concentration of particle surface sites (M)

α Collision efficiency factor $(= W^{-1})$ $(-)$

ϵ Energy dissipation rate $(m^2 \, s^{-3})$

Γ_c Sorption density of Me on colloids $(mol \, kg^{-1})$

λ_s Irreversible, first-order scavenging rate constant (d^{-1})

λ_p Particle removal rate constant (d^{-1})

λ_d Radioactive decay constant (d^{-1})
τ_a Coagulation; 'half' time (h)
υ_f Kinematic viscosity $(m^2 \ s^{-1})$

NOTE ADDED IN PROOF

Several very recent articles have appeared which should be cited here. Whitehouse et al.[89,90] have demonstrated the association of selected trace metals with marine colloidal water. Hurd and Spencer,[91] in their AGU Monograph, include descriptions of state-of-the-art ultrafiltration techniques for the collection of colloidal material in marine systems.

REFERENCES

1. Turekian, K. "The Fate of Metals in the Oceans," *Geochim. Cosmochim. Acta* 41:1139–1144 (1977).
2. Goldberg, E.D. "Marine Geochemistry. I. Scavengers of the Sea," *J. Geol.* 62:249–265 (1954).
3. Krauskopf, K. "Factors Controlling the Concentrations of Thirteen Trace Metals in Seawater," *Geochim. Cosmochim. Acta* 12:331–344 (1956).
4. Sillen, H.G. *Oceanography*, Sears, M., Ed. (Washington, D.C.: American Association for the Advancement of Science, 1961) p. 549.
5. Thomas, G.W. "Historical Developments in Soil Chemistry: Ion Exchange," *Soil Sci. Soc. Am. J.* 41:230–238 (1977).
6. Schindler, P.W., B. Furst, R. Dick, and P.U. Wolf. "Ligand Properties of Surface Silanol Groups. I. Surface Complex Formation with Fe^{3+}, Cu^{2+} and Pb^{2+}," *J. Colloid Interface Sci.* 55:469–475 (1976).
7. Stumm, W., H. Hohl, and F. Dalang. "Interaction of Metal Ions with Hydrous Oxide Surfaces," *Croat. Chem. Acta* 48:491–498 (1976).
8. Schindler, P.W. "Removal of Trace Metals from the Oceans: A Zero Order Model," *Thalassia Jugosl.* 11:101–111 (1975).
9. Balistrieri, L.S., P.G. Brewer, and J.W. Murray, "Scavenging Residence Times of Trace Metals and Surface Chemistry of Sinking Particles in the Deep Ocean," *Deep Sea Res.* 28:101–121 (1981).
10. Bacon, M.P., D.W. Spencer, and P.G. Brewer. "^{210}Pb, ^{226}Ra and $^{210}Po/^{210}Pb$ Disequilibria in Seawater and Suspended Particulate Matter, *Earth Planet. Sci. Lett.* 32:277–296 (1981).
11. Brewer, P.G., Y. Nozaki, D.W. Spencer, and A.P. Fleer. "Sediment Trap Experiments in the Deep North Atlantic: Isotopic and Elemental Fluxes," *J. Mar. Res.* 38:703–728 (1980).
12. Bacon, M.P. and R.F. Anderson. "Distribution of Thorium Isotopes between Dissolved and Particulate Forms in the Deep Sea," *J. Geophys. Res.* 87:2045–2056 (1982).
13. Bacon, M.P., C.-H. Huh, A. Fleer, and W.G. Deuser. "Seasonality in the Flux of Natural Radionuclides and Plutonium in the Sargasso Sea," *Deep Sea Res.* 32:273–286 (1985).
14. Nozaki, Y., H.-S. Yang, and M. Yamada. "Scavenging of Thorium in the Ocean," *J. Geophys. Res.* 87:2045–2056 (1987).

15. Balistrieri, L.S. and J.W. Murray. "The Surface Chemistry of Goethite (α-Fe$_2$O$_3$) in Major Ion Seawater," *Am. J. Sci.* 281:788–806 (1981).

16. Nyffeler, U.P., Y.-H. Li, and P.H. Santschi. "A Kinetic Approach to Describe Trace-Element Distribution between Particles and Solution in Natural Systems," *Geochim. Cosmochim. Acta* 48:1513–1522 (1984).

17. Li, Y.-H., L. Burkhardt, M. Buchholtz, P. O'Hara, and P.H. Santschi. "Partition of Radiotracers between Suspended Particles and Seawater," *Geochim. Cosmochim. Acta* 48:2011–2019 (1984).

18. Jannasch, H.W., B.D. Honeyman, L.S. Balistrieri, and J.W. Murray. "Kinetics of Trace Element Uptake by Marine Particles," *Geochim. Cosmochim. Acta* 52:567–577 (1988).

19. Nkedi-Kizza, P., W. Biggar, H.M. Selim, M.T. van Genuchten, P.J. Wierenga, J.M. Davidson, and D.R. Nelson. "On the Equivalence of Two Conceptual Models for Describing Ion Exchange during Transport through an Aggregated Oxisol," *Water Resour. Res.* 20:1123–2056 (1984).

20. Amis, E.J., C.C. Han, and Y. Matsushita. *Polymer* 25:650–658 (1984).

21. Honeyman, B.D., A.W. Adamson, and J.W. Murray. "The Nature of Reactions on Marine Particle Surfaces," *Appl. Geochem.* 3:19–26 (1988).

22. Sugimura, Y. and Y. Suzuki. "A High Temperature Catalytic Oxidation Method for the Determination of Non-Volatile Dissolved Organic Carbon in Seawater by Direct Injection of a Volatile Sample," *Mar. Chem.* 24:105–131 (1988).

23. Moran, S.B. and R.M. Moore. "The Distribution of Colloidal Aluminum and Organic Carbon in Coastal and Open Ocean Waters off Nova Scotia," *Geochim. Cosmochim. Acta* 53:2513–2524 (1989).

24. Buffle, J. in *Characterization of Environmental Particles*, J. Buffle and H.P. van Leeuwen, Eds., IUPAC Environmental Analytical Chemistry Series, Vol. I. (Chelsea, MI: Lewis Publishers, in press), Chap. 5.

25. DeGueldre, C.B. Baeyens, W. Goerlich, J. Riga, J. Verbist, and P. Stadelmann. "Colloids in Water from a Subsurface Fracture in Granite Rock, Grimsel Test Site, Switzerland," *Geochim. Cosmochim. Acta* 53:603–610 (1989).

26. Morel, F.M.M. and P.M. Gschwend. "The Role of Colloids in the Partitioning of Solutes in Natural Waters," in *Aquatic Surface Chemistry: Chemical Processes at the Particle-Water Interface,* W. Stumm, Ed. (John Wiley & Sons: New York, 1987), chap. 15.

27. Williams, P.J. Le B. "Biological and Chemical Aspects of Dissolved Organic Matter in Seawater." in *Chemical Oceanography,* Vol. 2, J.P. Riley and R. Chester, Eds. (New York: Academic Press, 1975) pp. 301–363.

28. Benjamin, M.M. "Effects of Competing Metals and Complexing Ligands on Trace Metal Adsorption at the Oxide/Solution Interface," Ph.D. thesis, Stanford University, Stanford, CA (1978).

29. Kent, D.B. and M. Kastner. "Mg^{2+} Removal in the System Mg^{2+}-Amorphous SiO$_2$-H$_2$O by Adsorption and Mg^{2+}-Hydroxysilicate Precipitation," *Geochim. Cosmochim. Acta* 49:1123–1136 (1985).

30. McDowell-Boyer, L.M., J.R. Hunt, and N. Sitar. "Particle Transport through Porous Media," *Water Resour. Res.* 22:1901–1921 (1986).

31. Buffle, J. *Complexation Reactions in Aquatic Systems. An Analytical Approach* (Chichester: Ellis Horwood, 1987).

32. Kepak, F. "Behavior of Carrier-Free Radionuclides," in *Radionuclide Techniques and Applications,* A. Evans and M. Muramatsu, Eds. (New York: Marcel Dekker, 1977), chap. 8.

33. Buddemeier, R.W. and R.J. Hunt. "Transport of Colloidal Contaminants in Groundwater: Radionuclide Migration at the Nevada Test Site," *Appl. Geochim.* 3:535–548 (1988).

34. Sheppard, J.C., M.J. Campbell, T. Cheng, and J.A. Kittric. "Retention of Radionuclides by Mobile Humic Compounds and Soil Particles," *Environ. Sci. Technol.* 11:1349–1353 (1980).

35. Penrose, W.R., W.L. Polzer, E.H. Essington, D.M. Nelson, and K.A. Orlandini. "Mobility of Plutonium and Americium through a Shallow Aquifer in a Semiarid Region," *Environ. Sci. Technol.* 24:228–233 (1990).

36. Niven, S.E.H. and R.M. Moore. "Effect of Natural Colloidal Matter on the Equilibrium Adsorption of Thorium in Seawater," in *Radionuclides: A Tool for Oceanography*, J. Guary, P. Guegueniat and R.J. Pentreath, Eds. (New York: Elsevier Applied Science, 1987), p. 111–120.

37. Baskaran, M., P. H. Santschi, G. Benoit, and B. D. Honeyman. "Colloidal Thorium Isotopes and Marine Scavenging in the Gulf of Mexico," *Geochim. Cosmochim. Acta,* accepted for publication (1991).

38. Orlandini, K.A., W.R. Penrose, B.R. Harvey, M.B. Lovett, and M.W. Findlay. "Colloidal Behavior of Actinides in an Oligotrophic Lake," *Environ. Sci. Technol.* 24:706–712 (1990).

39. Nash, K.L. and G.R. Choppin. "Interaction of Humic and Fulvic Acids with Th(IV)," *J. Inorg. Nucl. Chem.* 42:1045–1050 (1980).

40. Santschi, P.H., M. Amdurer, D. Adler, P. O'Hara, Y.-H. Li, and P. Doering. "Relative Mobility of Radioactive Trace Elements across the Sediment-Water Interface in MERL Model Ecosystems of Narragansett Bay," *J. Mar. Res.* 45:1007–1048 (1987).

41. Santschi, P.H., Y.-H. Li, D.M. Adler, M. Amdurer, J. Bell, and U.P. Nyffeler. "The Relative Mobility of Natural (Th, Pb and Po) and Fallout (Pu, Am and Cs) Radionuclides in the Coastal Marine Environment: Results from Model Ecosystems and Narragansette Bay," *Geochim. Cosmochim. Acta* 47:201–210 (1983).

42. Buchholtz, M., P.H. Santschi, and W.S. Broecker. "Comparison of Radiotracer K_d Values from Batch Adsorption Experiments with *in situ* Determination in the Deep Sea Using the Manop Lander: The Importance of Geochemical Mechanisms in Controlling Ion Uptake and Migration," in *Application of Distribution Coefficients to Radiological Assessment Models*, T.H. Selby and C. Myttenaere, Eds. (London, Elsevier Applied Science Publishers, 1986). p. 192–206.

43. Honeyman, B.D. and P.H. Santschi. "A Brownian-Pumping Model for Oceanic Trace Metal Scavenging: Evidence from Th Isotopes," *Deep Sea Res.* 47:951–992 (1989).

44. Overbeek, J.Th.G. "Recent Developments in the Understanding of Colloid Stability," *Colloid. Interface Sci.* 1, 431–445: *Proc. 50th Int. Conf. Colloid Surf. Sci. Symp.*, M. Kerker, Ed. (1976).

45. Smoluchowski, M. "Drei Vortraege ueber Diffusion, Brownsche Molekularbewegung und Koagulation von Kolloidteilchen," *Phys. Z.* 17:557–571 (1916).

46. Smoluchowski, M. "Versuch einer mathematischen Theorie der Koagulationskinetik kolloider Loesungen," *Z. Phys. Chem.* 92:129–169 (1917).

47. Fuchs, N. "Ueber die Stabilitaet und Aufladung der Aerosole," *Z. Phys.* 89:736–743 (1934).

48. Stumm, W. and J.J. Morgan. *Aquatic Chemistry: An Introduction Emphasizing Chemical Equilibrium in Natural Waters* (New York: John Wiley & Sons, 1981), p. 780.

49. McCave, I.N. "Size Spectra and Aggregation of Suspended Particles in the Deep Ocean," *Deep Sea Res.* 31:329–352 (1984).

50. Honeyman, B.D. and P.H. Santschi. "Coupling Adsorption and Particle Aggregation: Laboratory Studies of "Colloidal Pumping" using ^{59}Fe-Labeled Hematite," *Environ. Sci. Technol.* 25:1739–1747 (1991).

51. Yasunaga, T. and T. Ikeda. "Adsorption-Desorption Kinetics at the Metal-Oxide-Solution Interface Studied by Relaxation Methods," in *Geochemical Processes at Mineral Surfaces*. J.A. Davis and K.F. Hayes, Eds. (Washington, D.C.: American Chemical Society Symp. Ser. No. 323, 1986), chap. 12.

52. van Leeuwen, H.P. in *Characterization of Environmental Particles*, J. Buffle and H.P. van Leeuwen, Eds., IUPAC Environmental Analytical Chemistry Series, Vol. I. (Chelsea, MI: Lewis Publishers, in press), chap. 13.

53. Hayes, K.F. and J.O. Leckie. "Mechanism of Lead-Ion Adsorption at the Goethite-Water Interface," in *Geochemical Processes at Mineral Surfaces*, J.A. Davis and K.F. Hayes, Eds. (American Chemical Society Symp. Ser. No. 323, 1986), chap. 7.

54. Hahn, H. and W. Stumm. "The Role of Coagulation in Natural Waters," *Am. J. Sci.* 268:354–368 (1970).

55. Tsunogai, S. and M. Minagawa. "Settling Model for the Removal of Insoluble Chemical Elements from Water," *Geochim. J.* 12:483–490 (1978).

56. Santschi, P.H., Y.-H. Li, U.P. Nyffeler, and P. O'Hara. "Radionuclide Cycling in Natural Waters: Relevance of Scavenging Kinetics," in *Sediments and Water Interactions*, P.G. Sly, Ed. (New York: Springer-Verlag, 1986), chap. 17.

57. Clegg, S.L. and Whitfield, M. "A Generalized Model for the Scavenging of Trace Metals in the Open Ocean. I. Particle Cycling," *Deep Sea Res.* 37:809–832 (1990).

58. Honeyman, B.D., L.S. Balistrieri, and J.W. Murray. "Oceanic Trace Metal Scavenging: The Importance of Particle Concentration," *Deep Sea Res.* 35:227–246 (1988).

59. Coale, K.H. and K.W. Bruland. "^{234}Th:^{238}U Disequilibria within the California Current," *Limnol. Oceanogr.* 30:189–200 (1985).

60. McKee, B.A., D.J. DeMaster, and C.A. Nittrouer. "The Use of ^{234}Th/^{238}U Disequilibria to Examine the Fate of Particle Reactive Species on the Yangtze Continental Shelf," *Earth Planet. Sci. Lett.* 68:431–442 (1984).

61. McKee, B.A., D.J. DeMaster, and C.A. Nittrouer. "Temporal Variability in the Partitioning of Thorium between Dissolved and Particulate Phases on the Amazon Shelf: Implications for the Scavenging of Particle-Reactive Species," *Cont. Shelf Res.* 6:87–106 (1986).

62. Minagawa, M. and S. Tsunogai. "Removal of ^{234}Th from a Coastal Sea: Funka Bay, Japan," *Earth Planet. Sci. Lett.* 47:51–64 (1980).

63. Santschi, P.H., Y.-H. Li, and J. Bell. "Natural Radionuclides in Narragansett Bay," *Earth Planet. Sci. Lett.* 45:201–213 (1979).

64. Hering, J.G. and F.M.M. Morel. "Slow Coordination Reactions in Seawater," *Geochim. Cosmochim. Acta* 53:611–618 (1989).

65. Raspor, B., H.W. Nünberg, and P. Valenta. "Voltammetric Studies on the Stability of the Zn(II)-Chelates with NTA and EDTA and the Kinetics of their Formation in Lake Ontario Water," *Limnol. Oceanogr.* 26:54–66 (1981).

66. Friedlander, S.K. *Smoke, Dust and Haze.* (New York: John Wiley & Sons, 1977) p. 317.

67. Jeffery, D.J. "Aggregation and Breakup of Clay Flocs in Turbulent Flow," *Adv. Colloid Interface Sci.* 17:213–218 (1982).

68. Hunt, J.R. "Particle Dynamics in Seawater: Implications for Predicting the Fate of Discharged Particles, *Environ. Sci. Technol.* 16:303–309 (1982).

69. Farley, K.J. and F.M.M. Morel. "Role of Coagulation in Sedimentation Kinetics," *Environ. Sci. Technol.* 20:187–195 (1986).

70. Farley, K.J. "Sorption and Sedimentation Mechanisms of Trace Metal Removal," Ph.D. thesis, Massachusetts Institute of Technology, Cambridge, MA (1984).

71. Fisher, H.B., J.E. List, R.C. Koh, R.C. Imberger, and N.H. Brooks. *Mixing in Inland and Coastal Waters* (New York: Academic Press, 1979).

72. Newman, K., S.L. Frankel, and K.D. Stolzenbach. "Flow Cytometric Detection and Sizing of Fluorescent Particles Deposited at a Sewage Outfall Site," *Environ. Sci. Technol.* 24:513–518 (1990).

73. Garrett, C.J.R., J.R. Keeley, and D.A. Greenberg. *Atmos. Ocean,* 16:403–423 (1978).

74. Edzwald, J.K., J.C. Upchurch, and C.R. O'Melia. "Coagulation in Estuaries," *Environ. Sci. Technol.* 8:58–63 (1974).

75. Levine, E.R. and K.E. Kenyon. "The Tital Energetics of Narragansett Bay," *J. Geophys. Res.* 80:1683–1888 (1975).

76. Nixon, S.W., D. Alanso, M.E.Q. Pilson, and B.A. Buckley. in Microcosms in Ecological Research. DOE Symp. Ser. No. 52, USADE TIC.CONF-781101. G.P. Giesy, Ed. (Washington, D.C.: U.S. Department of Energy, 1980) pp. 818–849.

77. Kershaw, P. and A.J. Young. *Environ. Radioactivity,* 6:1–23 (1988).

78. Kershaw, P., P.A. Gurbutt, A.K. Young, and D.J. Allington. In *Radionuclides: A Tool for Oceanography.* J. Guary, P. Guegueniat, and R.J. Pentreath, Eds. (New York: Elsevier Applied Science, 1989), pp. 131–142.

79. Santschi, P.H., D. Adler, M. Amdurer, Y.-H. Li, and J. Bell. "Thorium Isotopes as Analogues for "Particle-Reactive" Pollutants in Coastal Marine Environments," *Earth Planet. Sci. Lett.* 47:327–335 (1980).

80. Taylor, G.I. "Tidal Friction in the Irish Sea," *Phil. Trans. R. Soc. London A,* 220:1–93 (1919).

81. Honeyman, B.D. and P.S. Santschi. "Critical Review: Metals in Aquatic Systems," *Environ. Sci. Technol.* 22:862–871 (1988).

82. Di Toro, D.M., J.D. Mahony, P.R. Kirchgraber, A.L. O'Byme, L.R. Pasquale, and D.C. Piccirilli. "Effects of Nonreversibility, Particle Concentration and Ionic Strength on Heavy Metal Adsorption," *Environ. Sci. Technol.* 20:55–61 (1986).

83. Higgo, J.J.W. and L.V.C. Rees. "Adsorption of Actinides by Marine Sediments: Effect of the Sediment/Water Ratio on the Measured Distribution Coefficient," *Environ. Sci. Technol.* 20:483–490 (1986).

84. Hawley, N., J.A. Robbins, and B.J. Eadie. "The Partitioning of ^{7}Be in Fresh Water," *Geochim. Cosmochim. Acta* 50:1127–1132 (1986).

85. Buchholtz, M. "Radionuclide Mobility across the Sediment-Water Interface in the Deep Sea," Ph.D. thesis, Columbia University, New York (1987).

86. Bloom, N. and E.A. "Solubility Behavior of Atmospheric ^{7}Be in the Marine Environment," *Mar. Chem.* 12:323–331 (1983).

87. Olsen, C.R., I.L. Larsen, R.H. Brenster, N.H. Cutshall, R.F. Bopp, and H.J. Simpson, "A Geochemical Assessment of Sedimentation and Contaminant Distributions in the Hudson-Raritan Estuary," (Springfield, VA:NOAA Tech. Rep. NOS OMS 2, N.T.I.S. (1984).

88. Anderson, R.F., M.P. Bacon, and P. Brewer. "Removal of ^{230}Th and ^{231}Pa from the Open Ocean," *Earth Planet. Sci. Lett.* 62:7–23 (1983).

89. Whitehouse, B. G., G. Petrick, and M. Ehrhardt. "Cross-flow filtration of colloids from Baltic seawater," *Water Res.*, 20:1599–1601 (1986).

90. Whitehouse, B. G., P. A. Yeats, and P. M. Strain. "Cross-flow filtration of colloids from aquatic environments," *Limnol. Oceanogr.*, 35(6):1368–1375 (1990).

91. Hurd, D. C., and O. W. Spencer. "Marine Particles: Analysis and Characterization". Geophysical Monograph 63, American Geophysical Union, Washington, D.C. (1991).

CHAPTER 11

Sorption of Trace Elements on Natural Particles in Oxic Environments

André Tessier

INRS-Eau, University of Quebec, Sainte-Foy, Quebec, Canada

TABLE OF CONTENTS

1. Introduction 426

2. Metal Binding Properties of Sediments — Similarities
   with Adsorption 428

3. Distribution Coefficient 429

4. Sedimentary Phases Responsible for Trace Element
   Sorption...432

5. Competitive Sorption Models 434

6. Estimation of Sorption Constants of Natural Particulate
   Material...440

7. Concluding Remarks 446

Acknowledgments 447

References 447

1. INTRODUCTION

A large body of measurements indicates that an important fraction of trace elements (e.g., As, Cu, Ni, Pb, Zn) introduced into the aquatic environment is found associated with suspended or bottom sediments;[1] this association can change their chemical reactivity and biological availability.[2] The processes that govern the scavenging of trace elements by particulate matter and their release to the ambient water when the environmental conditions are changed must be understood if the impacts of trace elements on the environment are to be predicted.

There are indications that precipitation of sulfides controls trace metal concentrations in anoxic waters.[3-5] In contrast, oxic surface waters generally do not appear to be saturated with respect to trace metal solid phases. Indeed, solubility calculations based on the known phases of these solids generally show undersaturation;[6-8] examples are shown in Figure 1 for zinc and nickel in lakes of various pH. Notable exceptions to this general pattern would be Fe(III) and Mn(IV), with respect to their respective oxyhydroxide phases. In oxic environments, reactions other than precipitation, including adsorption, absorption, surface precipitation, and coprecipitation, must be considered. The general term "sorption" will be used in this chapter for these reactions which can hardly be distinguished in natural waters.[10] The term "adsorption" will be restricted to the reaction of a solute with functional groups present at the surface of a solid and inside porous structures in well defined systems.[97] Note that often *adsorption* constants, measured in well-controlled laboratory conditions, are used to model *sorption* processes occurring in natural systems. This may lead to ambiguous interpretations which are discussed in this paper.

Suspended particulates and bottom sediments are complex mixtures of various components including: (i) residues of weathering and erosion such as clays and other aluminosilicates, iron and aluminum oxyhydroxides; (ii) substances produced by biological activity, both organic (living organisms and biological detritus, tubes of animals, humic substances) or inorganic (car-

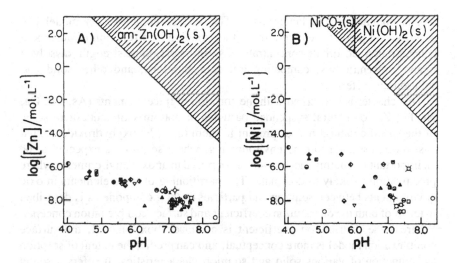

Figure 1. Solubility diagrams for zinc and nickel. Carbonate and oxide solid phases were considered in these graphical representations. The necessary solubility products and formation constants of the inorganic complexes were obtained from Reference 9; complexation by natural organic ligands was not taken into account because of the lack of relevant equilibrium constants. Total dissolved zinc and nickel concentrations, obtained at 41 sites (oxic) in 26 lakes of various pH values, are shown as points with various symbols. All the lake waters show a large undersaturation with respect to the carbonate and oxide phases.

bonates, phosphates and silica); and (iii) diagenetic products including iron and manganese oxyhydroxides, in the upper layer of the sediments and sulfides, in deeper strata. Simple model compounds chosen to mimic components of natural sediments (e.g., Fe, Mn, Si, Al oxyhydroxides; clays; humic substances) have been shown, in laboratory experiments, to adsorb trace elements.[11-14] Among these, the hydrous metal oxides have been particularly studied for their adsorption properties because of their ubiquiteness in natural environments and of their expected important role in influencing dissolved trace element concentrations in natural waters.

Several models have been developed for interpreting the adsorption of trace elements at oxide surfaces. The most widely accepted one, the surface complexation model, is based on an extension of solution coordination chemistry to the description of interactions of ions with reactive surface sites present at the solid-solution interface.[15-19] For the thermodynamic description of surface complexation, change in the Gibbs free energy of adsorption (ΔG_{ads}) is split into two terms: one accounts for the formation of a chemical bond between the adsorbate and the reactive surface sites ($\Delta G_{intrinsic}$), and the other accounts for the electrostatic energy of interactions due to the charges that have developed at the surface of the oxide ($\Delta G_{coulombic}$). The various versions of the surface complexation model that have been proposed differ in the relative weight of the chemical and electrostatic terms in the free energy of adsorption; yet, it has been shown that they all explain equally well the experimental results obtained in the laboratory.[20] The surface complexation model can describe and predict the extent of surface binding as a function of adsorbent concentration, adsorbate concentration, solution pH, ionic strength, dissolved ligand concentrations, competing ion concentrations, and other solid and solution characteristics.

This chapter is concerned with the sorption of trace elements (As, Cd, Cu, Ni, Pb, Zn) on natural suspended particles or bottom sediments or to some of the specific natural phases present in them (e.g., Fe oxyhydroxides). Discussion is restricted to oxic environments, where sorption is expected to be an important regulating mechanism, as opposed to anoxic environments where precipitation is likely to dominate. The partitioning of trace elements in oxic environments between solution and particles or their components is described in terms of both the distribution coefficient and surface complexation concepts. Whereas the distribution coefficient is a "black box" model, the surface complexation model is more conceptual, and can predict the extent of sorption as a function of various solid and solution characteristics. It offers a sound framework for interpreting field observations.

2. METAL BINDING PROPERTIES OF SEDIMENTS — SIMILARITIES WITH ADSORPTION

Similarities between binding of trace elements to sediments and adsorption of these elements to hydrous metal oxides have been noted in several studies. Lion et al.,[21] in a well thought out laboratory experiment, have studied the binding of Cd and Pb to oxidized surficial estuarine sediments from San Francisco Bay and have compared it with the adsorption of the same metals on synthetic amorphous iron (am-Fe) oxyhydroxides (Figure 2). Typically, adsorption of trace metals onto iron oxyhydroxides increases from near 0% to near 100% as the pH increases through a narrow critical range of 1 to 2 pH units (the so-called adsorption edge; e.g., Figure 2b). Figure 2 reveals that the curves of percent Cd bound to the natural sediments as a function of pH are qualitatively similar to those of Cd adsorbed on the synthetic iron oxyhydroxides (compare Figure 2a and 2c with 2b and 2d, respectively); differences in steepness among the curves were attributed by the authors to differences in binding intensities, site densities, and reaction stoichiometry between the two solids.[21] Similarities between binding to sediments and adsorption to am-Fe oxyhydroxides were also observed for Pb in the same study. Similar sigmoidal curves have been obtained in laboratory experiments for the binding of Cd, Cu, and Zn to bottom sediments from the Meuse and the Rhone rivers or to suspended matter from the Gironde estuary,[22-24] and for the binding of Cu and Zn to interfacial sediments from a marine site.[25,26]

Figure 2 also shows that the sorption edges for both natural sediments and synthetic am-Fe oxyhydroxides shift in the same direction when total metal (Figure 2c, d) or solid concentrations (Figure 2a, b) are changed in the system. A shift to lower pH with decreasing total metal concentration at a given solid concentration (or with increasing solid concentration at a given metal concentration) has been attributed by the authors to the presence of a range of sites of varying binding intensities. Collectively, these observations suggest that adsorption occurs in natural sediments.

3. DISTRIBUTION COEFFICIENT

Partitioning of trace metals between solution and particulate matter has most often been expressed using distribution (or partition) coefficients, $K_D(L.g^{-1})$:

$$K_D = \frac{\{M\}}{[M]} \qquad (1)$$

where $\{M\}$ (mol.g^{-1}) and $[M]$ (mol.L^{-1}) represent the total trace metal concentrations in the particulate and dissolved phases respectively at equilibrium. It should be noted that K_D is an operational coefficient, and not an equilibrium constant. Values of K_D have been obtained in the following ways:

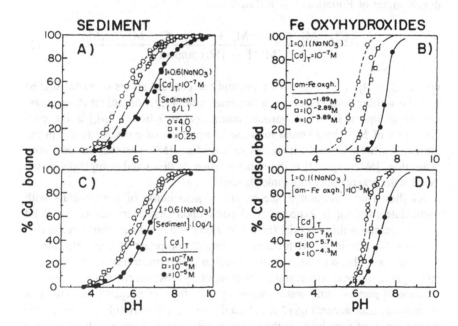

Figure 2. Sorption of Cd on an estuarine sediment (A and C) or on synthetic amorphous iron oxyhydroxides (B and D) at various solid or cadmium concentrations. In these experiments, known concentrations of sediments from San Francisco Bay or of am-Fe oxyhydroxides were suspended in water and spiked with Cd. After a given equilibration period, solids were removed from solution by centrifugation and the remaining dissolved Cd concentration and equilibrium pH were measured. (Modified from Lion, L. W. et al. *Environ. Sci. Technol.* 16:660–666 (1982)

(i)   From direct measurements of dissolved and particulate trace metal concentrations in the water column — {M} was measured on material collected in sediment traps,[27-32] or obtained by difference between metal concentrations measured in unfiltered and filtered[33] or in uncentrifuged and centrifuged[34] water samples; [M] was measured after centrifugation or filtration

(ii) From direct measurements of {M} in the surficial bottom sediments and [M] in either the overlying water[8,35,36] or the porewater[33]

(iii) In laboratory batch experiments where metal aliquots (usually radiotracers) were added to a suspension of natural solid material, which was either surficial bottom sediments[26,28,29,37-40] or suspended material collected in sediment traps[37] or by centrifugation[41] — [M] and {M} were measured after an incubation time, followed by a separation of the dissolved and particulate phases by centrifugation or filtration

Values of K_D reported for various environments are shown in Figure 3 for Cd, Cu, Ni, Pb, and Zn. A striking feature in Figure 3 is the great variability in K_D for a given metal which can be better understood by a more explicit development of Equation 1 as follows:

$$K_D = \frac{\{C_1 - M\} + \{C_2 - M\} + \cdots + \{C_n - M\} + \{M_r\}}{[M^{z+}] + [M(complex)]} \qquad (2)$$

where $\{C_1 - M\} \ldots \{C_n - M\}$ represent the concentrations of metal sorbed on the n sorbing components of particulate matter (e.g., organic matter; clays, oxyhydroxides of iron, manganese, aluminum, or silicon); $\{M_r\}$ is the concentration of M tightly bound to some components of particles (e.g., in the lattice of silicates, resistant oxides or sulfides); $[M^{z+}]$ is the free ion concentration; $[M(complex)]$ is the concentration of dissolved complexes of the metal with organic and inorganic ligands.

As shown by Equation 2, the extent of association of trace metals with particulate material is expected to depend upon characteristics of the solid phases present in this material (concentrations of the various components that can sorb M; affinity of M for these sediment components; concentration of M_r) and upon characteristics of the solution (pH; concentration of inorganic and organic ligands; concentration of competing ions; temperature). The composition of particles and water chemistry vary from one aquatic environment to another, and accordingly, K_D should show great variability. One of the main sources of variability is the solution pH; indeed, Figure 3 shows a clear tendency of K_D to increase with pH. This general behavior is in agreement with laboratory experiments that show increase with pH in the adsorption of trace metals on various solids resembling sediment components: humic material;[13] clays;[11] oxyhydroxides of iron,[12] manganese,[14] silicon,[42] and aluminum.[43] Water chemistry and particle composition vary seasonally, notably due to variations in primary production and decomposition of organic matter. These variations can explain the variability observed for K_D at a given site.[30]

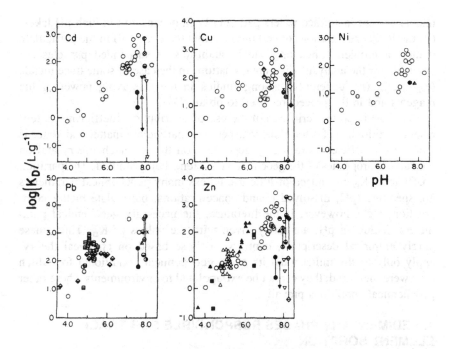

Figure 3. Distribution coefficients of Cd, Cu, Ni, Pb, and Zn reported in the literature for various aquatic environments. Data are from References 26 (△); 39 (●); 37 (▽); 29 (□ *in situ,* ■ laboratory); 30 (x); 31 (◒); 32 (⊗); 8, 35, and 36 (○); 27 (⊟ *in situ,* ◈ laboratory); 34 (◑); 38 (◓); 41 (◆); 28 (△ *in situ,* ▲ laboratory); 33 (◆ suspended sediments, ; ◈ bottom sediments, ◇ laboratory); 40 (▼). Vertical lines indicate ranges of values reported by the authors.

Contrary to what is expected from Figure 2a, b, the partition coefficients have also been reported to decrease, at a given pH value, with increasing particle concentration.[40,44-47] The reasons for this particle concentration effect are still obscure and various interpretations are discussed in Chapter 10, this volume. In addition, at a given pH value, the partition coefficient values obtained from laboratory experiments with radiotracers are usually lower than those obtained from field measurements. Many factors can explain this observation.[45] High suspended particle concentrations are normally used in laboratory experiments compared to suspended particle loading in lakes and oceans, and, as mentioned above, particle concentration may affect the observed K_D value. The equilibration times used in radiotracer experiments (a few hours to a few days) are shorter than the residence times of particles in natural waters and therefore, there may be insufficient time to incorporate radiotracers in lattice positions (e.g., $\{M_r\}$ in Equation 2); in contrast, in the determination of field-derived values of K_D, these tightly bound trace metals are at least partly dissolved when strong acids are used to attack the particulate material. Finally, some of the scatter observed in Figure 3 might be due to experimental difficulties such as (i) the low concentration of dissolved trace metals present in surface waters particularly in oceans and in high pH lakes; (ii) the low concentrations of certain trace metals (e.g., Cd) in the particulate matter, a problem when only small quantities of suspended particles are available for the analysis; (iii) contamination, in the case of some trace metals (e.g., Zn); (iv) differences among studies in the dissolving power of the reagents and in the procedures used to obtain {M}.

The most important criticism of the use of distribution coefficients to represent partitioning of a trace element between particulate matter and water is that these coefficients are highly dependent on the water chemistry and on the surface properties of the geochemical systems being studied. The partition coefficients K_D are indeed of little use unless many geochemical parameters are specified (pH, dissolved ligand concentrations, particulate matter composition, etc.); however, in the literature, the necessary geochemical parameters, including pH, are rarely given with the values of K_D. Thus, these purely empirical descriptors, contrary to those based on chemical theory, apply only to the milieu (and its exact geochemical conditions) for which they were measured; they cannot be extrapolated to environments where other geochemical conditions prevail.

4. SEDIMENTARY PHASES RESPONSIBLE FOR TRACE ELEMENT SORPTION

Oxidized sediments contain many components that can, in principle, sorb trace elements. These include clays, hydrous oxides of aluminum, iron, manganese and silicon, carbonates, and various types of organic matter. Although the identity of the sediment components primarily responsible for sorption has not been definitely established, recent studies using various approaches suggest that some sedimentary substrates might be more important than others in the sorption of trace elements.

Luoma and Bryan[48] have determined statistical relationships between concentrations of extractable trace metals and those of extractable sediment components (i.e., extractable iron, manganese, organic matter, calcium) of 50 oxidized surficial estuarine sediments presenting a wide range of sedimentary trace metal and sediment component concentrations (variations of 1 to 3 orders of magnitude). Their major findings were that the various sediment components competed for a given trace metal; extractable iron (0.2 M ammonium oxalate in 0.2 M oxalic acid or 1 M HCl), presumably iron(III) oxyhydroxides, was more important than total Fe in binding Ag, Cd, Cu, Pb, and Zn; organic matter extracted with 0.1 M NaOH, presumably humic acids, was highly important in binding Ag and Cu; Mn(IV) oxides were also found to be binding substrates, although to a lesser extent than the other two components, for Cd, Co, Cu, Pb, and Zn; in contrast, $CaCO_3$ appeared to be a diluter, since it correlated negatively with all metals except Co.

Using a different approach, Lion et al.[21] have examined trace metal adsorption on oxidized surficial estuarine sediments before and after removing, by chemical attack, various components from the sediments. Decreases in adsorption were observed after the stripping of certain sediment components. The results (Figure 4) suggest that iron and manganese oxides and organic matter are important phases for the binding of lead and cadmium in the sediments.

Correlations were also observed between the concentrations of particulate trace metals and the abundance of certain components in the settling particles in lake water column;[30-32] they agreed well with the conclusions obtained by Luoma and Bryan for bottom estuarine sediments. Indeed, trace metals were found to be associated, in the lake water column, mainly with organic matter (phytoplankton in this case) and, to a lesser extent, with Fe and Mn oxides. Trace metal concentrations were negatively correlated with those of Ca, and extrapolation of the linear regressions to pure calcite led to trace metal concentrations of approximately 0. The negligible role of calcium carbonate in the sorption of trace metals was also confirmed by calculating the specific surface area of each of the components present in typical suspended matter from Swiss Alp lakes;[32] despite its great abundance in the settling seston, $CaCO_3$ represented only a negligible percentage of its total surface area.

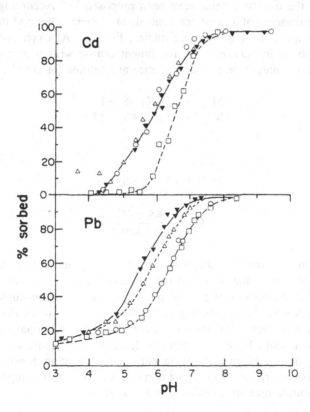

Figure 4. Sorption of cadmium and lead on unaltered estuarine sediment (▼), and after treating the sediment with $MgCl_2$(△), $NH_2OH.HCl$ to remove Fe oxyhydroxides (○) and H_2O_2 to remove organic matter (□). (Modified from Lion, L. W. et al. *Environ. Sci. Technol.* 16:660–666 (1982).

At a given pH, trace metal binding intensities determined in the laboratory are of similar magnitude for both Fe and Al oxyhydroxides, but they exceed the values for binding on Si oxides by several orders of magnitude.[32,49] These observations suggest that silicon oxides are probably not important sorbing components in sediments. It is generally thought that clays act only as a support for oxide or organic matter deposition, i.e., they are not involved directly in sorption processes.[50,51]

To summarize, available information suggests that (i) organic matter and hydrous oxides of Fe and Mn (and perhaps Al) are the main sediment components responsible for sorption, and (ii) there is competition between these substrates for sorbing trace metals. This latter point is supported by empirical studies on the extraction of oxidized sediments using various chemical reagents that have consistently shown that trace elements bind to more than one component of particulate matter.[1]

5. COMPETITIVE SORPTION MODELS

Models based on the competitive sorption of trace metals by various components of the oxic sediments have been proposed.[52,53] According to these models, partitioning of a given trace metal, M, among water and the various sorbing components (e.g., organic matter, Fe, Mn, Al oxyhydroxides, as discussed above in Section 4) of a sediment can be written in a simplified manner where charges on the solid species are omitted for simplicity:

$$M^{z+} + \equiv S_1 \underline{K_1} \equiv S_1 - M$$
$$M^{z+} + \equiv S_2 \underline{\overline{K_2}} \equiv S_2 - M$$

$$\cdot \qquad \cdot \qquad \cdot$$
$$\cdot \qquad \cdot \qquad \cdot \qquad (3)$$
$$\cdot \qquad \cdot \qquad \cdot$$

$$M^{z+} + \equiv S_n \underline{K_n} \equiv S_n - M$$

$$K_i = \frac{\{\equiv S_i - M\}}{\{\equiv S_i\}[M^{z+}]} \qquad (4)$$

The notation "$\equiv$" refers to sorption *sites*, whereas {} and [] refer to concentrations of the solid and dissolved species respectively; $\{\equiv S_i\}$ and $\{\equiv S_i - M\}$ are the concentrations $(mol.g^{-1})$ of free sites and of sites occupied by the metal respectively for the sorbing component i (i = 1 to n); $[M^{z+}]$ is the concentration of free metal ion in solution; K_i is a conditional equilibrium constant (which should be a function of pH, temperature, ionic strength, etc.) for the sorption of M on that component. At low sorption densities, that is when the concentration of occupied sorption sites is small compared to the total concentration of sites available, the condition:

$$\{\equiv S_i\} \approx \{\equiv S_i\}_t \qquad (5)$$

should be fulfilled, where $\{\equiv S_i\}_t$ is the total concentration of site of the sorbing component i, which can be related to the concentration of component i by the following equation:

$$\{\equiv S_i\}_t = N_i \cdot \{C_i\} \tag{6}$$

where N_i is the number of moles of sites per mole of component i and $\{C_i\}$ is the concentration (mol.g^{-1}) of component i in the sediments. Since M is assumed to be associated with the sorption sites in the ratio 1:1, one can also write for the sake of consistency:

$$\{\equiv S_i - M\} = \{C_i - M\} \tag{7}$$

where $\{C_i - M\}$ is the concentration (mol.g^{-1}) of M associated with component i. Substitution of relations 5 to 7 into Equation 4 leads to:

$$K_i = \frac{\{C_i - M\}}{N_i\{C_i\}[M^{z+}]} \tag{8}$$

A mass law equation like Equation 8 can be written for each sorbing component of the particulate matter. Together with the appropriate mass law equations for dissolved species (acid-base; complexation) and the mass balance equations, this constitutes a competitive sorption model. This kind of model allows the calculation of the distribution of trace metals among water and the various sorbing components of sediments. Mathematically, these models are similar to those used to calculate the distribution of a dissolved metal among a mixture of dissolved ligands. Necessary input data for such models include (i) water composition (pH, concentrations of cations and anions); (ii) the abundance of each of the sorbing components of the natural sediments ($\{C_i\}$); (iii) their binding capacities or densities of sites (N_i); and (iv) their binding intensities (K_i). An important assumption behind this kind of model is the additivity rule, i.e., there is no component-component interaction in a mixture of various sorbing components and the total concentration of metal sorbed can be predicted from experiments performed with pure (single) sorbing phases.

Several researchers have attempted to predict trace metal partitioning with competitive sorption models. Wiley and Nelson[54] have predicted the partitioning of Cd among sorbents present in freshwater lake sediments using estimated phase abundances (clays, organic matter, Fe and Mn oxyhydroxides) together with conditional constants determined in the laboratory with synthetic phases and lake waters. Their conditional constants, $^*K_i$ can be written:

$$^*K_i = \frac{\{C_i - M\}}{[M]_t\{C_i\}} \tag{9}$$

where $[M]_t$ is the total concentration of dissolved M. Comparison of Equations 8 and 9 shows that in this approach, the values of K_i and N_i are lumped together in the conditional constants $*K_i$ which takes also into account water chemistry, in particular the degree of complexation of M by dissolved complexants. The competitive sorption model predicted that sedimentary Fe oxyhydroxides should dominate the sorption of cadmium in freshwater sediments. Using a similar approach, Davies-Colley et al.[55] have also predicted the partitioning of Cd and Cu among sorbing components of estuarine sediments; the conditional constants ($*K_i$) for the model phases were, in this case, measured in seawater. The competitive sorption model predicted that in estuarine sediments the sorption of cadmium should be mostly effected by Fe oxyhydroxides, whereas that of copper should be dominated by humic acids. In an attempt to verify model prediction, they have measured, in laboratory experiments, the total metal uptake of Cd and Cu by estuarine sediments and have compared it with its model prediction. The results where expressed in terms of a global conditional constant for the whole sediment, K_{total}, which can be written:

$$
K_{total} = \frac{\{Sed-M\}}{[M]_t \sum_{i=1}^{m} \{C_i\}} = \frac{\sum_{i=1}^{m} {}^*K_i\{C_i\}}{\sum_{i=1}^{m} \{C_i\}} \tag{10}
$$

where the second equality implies that $[M]_t = [M^{z+}]$ (no complexation by dissolved complexants), and where $\{Sed-M\}$ is the concentration of total M sorbed to the sediment, i.e., the sum of the concentrations of metal sorbed to the n individual phases of the sediment. The results show a semi-quantitative agreement between predicted and measured K_{total} (Figure 5). This does not, however, constitute a satisfactory verification of the model performance in predicting partitioning of a trace element among phases.

Luoma[56] has attempted a more adequate verification of the prediction of the competitive sorption model. He has calculated the partitioning of Cu between organic matter and Fe and Mn oxyhydroxides for 24 estuarine sediments of various physicochemical characteristics. The equilibrium constants and site density values for sediment components were taken from adsorption experiments performed in the laboratory with model substrates,[53,55] whereas the concentrations of the sorbing sediment components (organic matter, Fe and Mn oxyhydroxides) were obtained by chemical extractions of the sediments. Copper partitioning calculated with the competitive sorption model showed some agreement with the distribution of this metal obtained empirically from chemical extractions of the sediments with reagents of varying strength. The agreement between predictions and measurements were best when inorganic components were predominant, and the poorest when both component types (organic and inorganic) were present in similar concentrations (Figure 6). Both methods indicated that Cu binds to both organic and

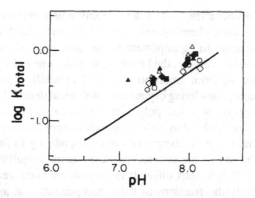

Figure 5. Comparison of Cd sorption by estuarine sediments with competitive sorption model predictions. The points represent the global K values for an experiment performed in the laboratory with an estuarine sediment and seawater at 32% salinity at pH8. The line represents the prediction of the global K based on determination of phase abundances and individual conditional constants. Units of K_{total} in $L \cdot g^{-1}$. (Modified from Davies-Colley, R. J. et al. *Environ. Sci. Technol.* 18:491–499 (1984).

inorganic sediment components in proportions depending on the relative concentrations of the sorbing components. However, as pointed out by Luoma,[56] assessment of the validity of the model predictions is difficult since the trace element partitioning obtained with extraction methods is uncertain due to problems inherent to these methods (non selectivity of extractants; readsorption).

The modeling of the behavior of sediment-bound metals is far less advanced than is that of dissolved species. Problems associated with the use of competitive sorption models are discussed below.

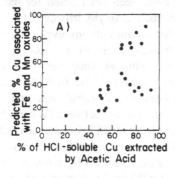

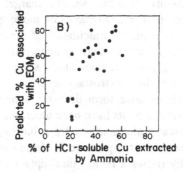

Figure 6. Comparison of the percentage of total sedimentary Cu extracted with acetic acid (A) or with ammonia (B) with the percentages predicted by the competitive sorption model to be associated with iron and manganese oxyhydroxides or extractable organic matter (EOM), respectively. Concentrations of Cu extracted with HCl, acetic acid, and ammonia are presumed to approximate those of total metal, metal associated with Fe and Mn oxyhydroxides and EOM, respectively. (Modified from Luoma, S. N. *Mar. Chem.* 20:45–59 (1986).

Component abundance — The determination of sorbing component concentrations relies upon chemical extraction techniques. Each of the sediment components that seem to be important for sorbing trace elements occur in a variety of forms. For example, the Fe oxyhydroxides present a whole spectrum of crystalline forms, from amorphous to well crystallized, whereas natural organic matter comprises living organisms, biological debris, humic and fulvic compounds, proteins, peptides, polysaccharides, and other compounds;[57] the extent of trace element binding to a component depends strongly on its exact form.[53] Furthermore, these compounds have a tendency to form coatings on particles, thus armoring the covered particles against equilibration with the aqueous phase.[58-60] It is thus difficult to choose chemical reagents that will dissolve selectively the fractions of these components that are reactive with respect to the sorption of trace elements.

Binding intensity — Conditional equilibrium constants such as those defined by Equation 9 can be measured and used as inputs to competitive sorption models.[54,55] These conditional constants need, however, to be determined in a solution of similar composition (pH, ligand nature and concentrations, ionic strength, temperature) as the system being modeled, which is especially problematic for freshwaters since their chemistry varies from one site to another. More importantly, the conditional constants should be measured for the exact sorbing solids present in natural sediments at the right metal:sorbent ratio. Such measurements are scarce (see Section 6 following); the binding constants are usually determined for laboratory prepared sorbing phases (model phases) chosen to mimic sediment components.[52,54,55] The exact nature of sedimentary sorbing components is, however, largely unknown, and the phases used in the laboratory may prove to poorly resemble them.

An alternative approach to the measurement of conditional constants involves their calculation from intrinsic equilibrium constants which have been determined in laboratory experiments.[61,62] Intrinsic constants are hypothetical constants for a null surface charge; they have been determined for the adsorption of trace elements on model phases such as oxyhydroxides of Al, Si, Fe, and Mn.[16,63] Calculation of the conditional constants involves correcting the intrinsic constants by appropriate electrostatic terms to take into account the charges developed at the surfaces of the sediment components.[16] Calculation of this electrostatic term for natural waters poses difficulties. Additional problems arise form differing conditions between laboratory and field. Intrinsic constants have been determined at temperatures between 20 and 25°C, whereas temperatures in the natural systems can be much lower and adsorption depends on temperature.[64] Monitoring adsorption as a function of time typically shows a rapid initial uptake ($\approx$ minutes) followed by a much slower reaction which may last several days. "Equilibration" times from a few minutes[65,66] to a few hours[12,67] have been chosen for the determination of intrinsic constants; these reaction times are much shorter than those prevailing in natural systems (e.g., residence time in the water column; burial or mixing time in sediments). Longer reaction times would favor an increase in the values of the adsorption constants. Dynamic aspects of adsorption are discussed in Chapter 13 of this book. Studies over wide ranges of metal concentrations strongly suggest that oxyhydroxides,[25,68] and humic or fulvic material[69,70] comprise various binding sites, the binding energies of which may vary by many orders of magnitudes; the influence of heterogeneity in binding sites on the binding properties of particles is discussed in Chapter 12 of this book. This heterogeneity in binding sites can result in large variations in the observed intrinsic constants with adsorption density. Typically, at a given pH, a plot of the logarithm of adsorption constant (K) as a function of the logarithm of adsorption density (Γ) shows two regions: a first one, at low density of adsorption, where K does not vary with increasing Γ (because high energy sites are in excess), and a second one, at higher density of adsorption, where K decreases with Γ (because low energy sites are being occupied). Laboratory experiments tend to use higher trace metal concentrations than those present in natural waters, and therefore laboratory-derived adsorption constants tend to be too small when they are applied to natural systems.

Binding capacity — The density of surface sites of a given phase can in principle be determined[53] by various methods including fast tritium exchange,[71] acid-base titration,[23,72] metal adsorption isotherms at constant pH,[14] or calculations based on the specific surface area of the solid and that occupied by the adsorbed metal.[73] Site densities for single synthetic phases (e.g., amorphous Fe oxyhydroxides; goethite) determined by various methods are not always consistent. This probably reflects both the complexity of the single substrates, and the limitations of the methods available.[56] More importantly, the techniques developed for single-component systems cannot be used directly to determine the site density of a particular component within a mixture of components such as occurs in a natural sediment. To determine the site density value of a given component of a natural sediment, it is necessary to isolate that component from the sediment matrix while preserving its integrity. Recent experiments where inert surfaces were inserted vertically in lake sediments have shown that it is possible to collect separately Fe and Mn oxyhydroxides formed diagenetically in the sediments;[74] this method of isolation with minimal perturbation could allow measurement of site densities of specific components of natural sediments. For the moment, any calculation with the competitive sorption models must rely upon estimates of site densities obtained in the laboratory on model sediment components; however, there is no guarantee that they will apply to natural systems.

Additional problems — A main assumption behind the competitive sorption model is that equilibrium conditions prevail for the partitioning of trace metals between the water and the various components that can bind it; kinetic information to support such a hypothesis is presently lacking. The model assumes also that all the components behave independently; it is well known, however, that sedimentary particles are not pure substrates, but rather aggregates of many substrates (e.g., clays coated with organic matter and/or iron oxyhydroxides).[50,51] To what extent aggregation influences binding of a trace metal on given components is not yet well known; some of the effects of aggregation are however discussed in Chapter 10 of this book. Finally, for the determination of equilibrium constants such as in Equations 4 or 9, separation between dissolved and particulate metal is generally performed. This is often done by filtration which may be affected by a number of artefacts (Chapter 5 of this book).

6. ESTIMATION OF SORPTION CONSTANTS OF NATURAL PARTICULATE MATERIAL

Due to the uncertainties associated with the use of laboratory-derived adsorption constants to make predictions for natural waters, several studies have attempted to estimate sorption constants from *in situ* measurements. In these studies, surface complexation concepts, which are useful for describing adsorption of trace elements at the surface of hydrous metal oxides,[16,17,42,43] have been used to relate concentrations of trace elements associated with iron oxyhydroxides present in natural particulate material to the dissolved concentrations of these elements.[8,35,75-77] Although it has not been demonstrated that adsorption is the main process involved in sorption on natural particulate matter, it is plausible that surface complexation concepts apply, since components of natural particles contain functional groups that can be considered as ligands. This theory provides a sound framework for examining trace element sorption on natural particulate matter. The basic equations that were used in these field studies, and the definitions of the various parameters, are given in Table 1.

Combining Equations 12 and 14 through 18 from Table 1 leads to:

$$K_M = \frac{N_{Fe} \, ^*K_M}{[H^+]^{m+1}} = \frac{\{Fe-M\}}{\{Fe-ox\}[M^{z+}]} \tag{19}$$

and

$$K_A = N_{Fe} \, ^*K_A[H^+]^m = \frac{\{Fe-A\}}{\{Fe-ox\}[A^{n-}]} \tag{20}$$

for cations (M^{z+}) and anions (A^{n-}), respectively.

Field measurement of sorption constants for natural iron oxyhydroxides — Field-derived values for the conditional constants (K_M and K_A) were obtained from *in situ* measurements of the variables appearing on the right-hand side of Equations 19 and 20 for the sorption of As, Cd, Cu, Ni, Pb, and Zn on Fe oxyhydroxides present in oxic bottom sediments of Canadian lakes.[8,35,76,77] The values of {Fe-M}, {Fe-A}, and {Fe-ox} were obtained by extracting the oxic lake bottom sediments with a reducing reagent ($NH_2OH.HCl$) designed to dissolve amorphous iron oxyhydroxides together with associated trace elements. The concentrations of M^{z+} and A^{n-} were calculated from total dissolved concentrations of the trace elements in the water overlying the

Table 1. Basic Equations for Describing Surface Complexation of Cations and Anions on Iron Oxyhydroxides

For a cation (M^{z+}):[78]

$$\equiv Fe\text{–}OH + M^{z+} + mH_2O \xrightarrow{*K_M} \equiv Fe\text{–}OM(OH)_m^{z-m-1} + (m + 1) H^+ \qquad (11)$$

$$*K_M = \frac{\{\equiv Fe\text{–}OM(OH)_m^{z-m-1}\}[H^+]^{m+1}}{\{\equiv Fe\text{–}OH\}[M^{z+}]} \qquad (12)$$

For an anion (A^{n-}):[78]

$$\equiv FeOH + A^{n-} + mH^+ \xrightarrow{*K_A} \equiv FeOH_{m+1}\text{–}A^{m-n} \qquad (13)$$

$$*K_A = \frac{\{\equiv FeOH_{m+1} - A^{m-n}\}}{\{\equiv Fe\text{–}OH\}[A^{n-}][H^+]^m} \qquad (14)$$

where $*K_M$ and $*K_A$ are apparent overall equilibrium constants for the adsorption of the cationic and anionic adsorbates, respectively; {} and [] represent concentrations in the solid and solution phases, respectively, whereas "$\equiv$" refers to adsorption sites, either free ($\{\equiv Fe\text{–}OH\}$) or occupied by M or A. At low density of adsorption:

$$\{\equiv Fe\text{–}OH\} \approx \{\equiv Fe\text{–}O\text{–}\}_t \qquad (15)$$

where $\{\equiv Fe\text{–}O\text{–}\}_t$ is the total concentration of sites which in turn can be expressed as:

$$\{\equiv Fe\text{–}O\text{–}\}_t = N_{Fe} \cdot \{Fe\text{–}ox\} \qquad (16)$$

where N_{Fe} is the number of mole of sites per mole of Fe oxyhydroxides, and $\{Fe\text{-}ox\}$ is the concentration of iron oxyhydroxides. The concentration of occupied sites is related to the concentration of M, $\{Fe\text{-}M\}$, and A, $\{Fe\text{-}A\}$, associated with the Fe oxyhydroxides:

$$\{\equiv Fe\text{–}OM(OH)_m^{z-m-1}\} = \{Fe\text{-}M\} \qquad (17)$$

$$\{\equiv Fe\text{–}OH_{m+1} - A^{m-n}\} = \{Fe\text{-}A\} \qquad (18)$$

sediments (measured by *in situ* dialysis; 0.2 μm porosity)[79] and the appropriate equilibrium constants for inorganic complexation and for the dissociation of arsenic acid. Possible complexation by natural organic ligands (e.g., fulvic or humic acids) was not considered because of the lack of relevant equilibrium constants. These extensive studies covered 40 sites in lakes chosen to represent a variety of geological settings, lake pH values, and trace element concentrations in the sediments and in the overlying waters. Equation 19 predicts that a plot of log K_M vs. pH should yield a straight line of slope m + 1 and an intercept of log $(N_{Fe} \cdot *K_M)$; similarly, a plot of log K_A vs. pH should yield, according to Equation 20, a slope of $-m$ and an intercept of log $(N_{Fe} \cdot *K_A)$.

Figure 7 shows that the relationship between log K_M and pH is linear for each metal studied, in agreement with the model depicted by Equation 19; the linear regression equations obtained are given in Table 2. The slopes of the regression lines are very close to 1 for Cd and Ni but somewhat higher for Zn (1.2) and lower for Pb (0.8) and Cu (0.7). The lower slopes for Cu and Pb are attributed partly to the unaccounted influence of organic complexes of these metals in the estimation of $[M^{z+}]$, which leads to an underestimation of K_M, according to Equation 19;[8] this effect should be greater at the higher pH values where organic complexation is favored, thus leading to a decrease in the slope of log K_M vs. pH. The proton stoichiometry found from field measurements (Table 2) is in general agreement with the surface complexation model for adsorption of trace metals on iron oxyhydroxide surfaces, which suggests that a combination of surface complexation reactions involving the release of one or two protons per M adsorbed should occur.[16] Another point of agreement between the field-derived sorption constants and the laboratory-

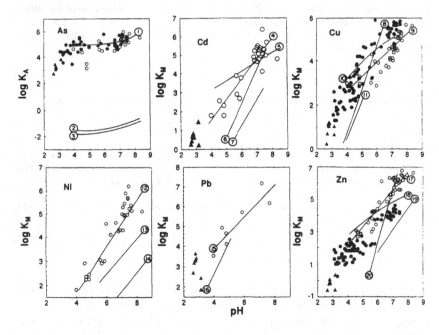

Figure 7. Comparison of K_M and K_A values from field and laboratory studies for As, Cd, Cu, Ni, Pb, and Zn. The points were obtained from field measurements in the Carnon River, England (●; References 75 and 84) in various Canadian Lakes (○; References 8, 35 and 36) and in streams affected by acid mine drainage (▲; Reference 85). Numbers on the curves correspond to the following studies: 4, 9, 12, 15, and 17 are linear regression lines for the Canadian lakes (see Table 2 for the regression equations); 1 (Reference 87); 3 (Reference 88); 4 (Reference 89); 6, 8 and 20 (Reference 68); 7 and 11 (Reference 90); 5 (Reference 66); 10 and 18 (Reference 65); 13 (Reference 91); 14 (Reference 92); 16 (Reference 93); 19 (Reference 14). All laboratory data were obtained with am-Fe oxyhydroxides, except curve 14 which was obtained with goethite. Units of K_M and K_A in L.mol^{-1}.

Table 2. Linear Regression Equations Describing the Field-Derived Sorption Constants Obtained for Canadian Lakes

General Equation

$$\log K_M = (m+1)\, pH + \log N_{Fe} \cdot {}^*K_M$$

Cadmium

$$\log K_M = 1.03\, pH - 2.44 \quad (r^2 = 0.80;\ n = 26)$$

Copper

$$\log K_M = 0.64\, pH + 0.10 \quad (r^2 = 0.75;\ n = 39)$$

Nickel

$$\log K_M = 1.04\, pH - 2.29 \quad (r^2 = 0.87;\ n = 29)$$

Lead

$$\log K_M = 0.81\, pH + 0.67 \quad (r^2 = 0.81;\ n = 7)$$

Zinc

$$\log K_M = 1.21\, pH - 2.83 \quad (r^2 = 0.89;\ n = 41)$$

derived adsorption constants is the sequence of increasing values of the constants; at low pH values, where the effects of complexation by organic matter are minimum, the sequence of the field-derived K_M values is Pb > Cu > Zn > Ni ≈ Cd, i.e., similar to that reported for adsorption of these metals on synthetic iron oxyhydroxides in well-defined media.[80] In addition, the field sorption constants are correlated with the first hydrolysis constants of the metals (Figure 8). Such correlations are expected, given the similarity between the dissolved ligand H–OH and the surface ligand ≡Fe–OH, and have been reported for laboratory-derived adsorption constants.[81,82]

In contrast to trace metals, a plot of $\log K_A$ vs. pH yields a slope close to zero for arsenic in the pH range of the lakes studied (Figure 7). Assuming, from thermodynamic calculations and experimental evidence,[83] that arsenic

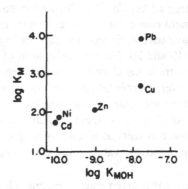

Figure 8. Relationship between field-derived constants for the sorption of trace metals on natural Fe oxyhydroxides (K_M) and the first hydrolysis constants of these metals (K_{MOH}). The K_M values were calculated with the linear regression equations given in Table 2. A low pH of 4 was chosen to minimize the effects of complexation by organic matter which was not taken into account in the calculation of K_M. Units of K_M and K_{MOH} in L.mol⁻¹.

associated with natural iron oxyhydroxides as well as dissolved As in the overlying waters are essentially As(V), Equation 13 from Table 1 can be written:

$$\equiv FeOH + H_2AsO_4^- + mH^+ = \equiv FeOH_{m+1} - AsO_3^{m-1} + H_2O \quad (21)$$

Thus, if the assumption concerning the redox state of As holds, the almost nil slope (i.e., $m \approx 0$) found in the pH range 4 to 8.4 for the Canadian lakes (Figure 7) suggests that $\equiv FeOAsO_3H^-$ is the dominant species of As associated with the Fe oxyhydroxides of the surficial sediments of these lakes.[77]

Johnson[75,84] has also determined field-derived constants using Equations 19 and 20 for the sorption of As, Cu, and Zn on Fe oxyhydroxides present in suspended particulate matter of the Carnon River system in England. The study took place in a part of the river influenced by mine water drainage, where suspended matter was highly enriched in amorphous iron oxyhydroxides and contained low levels of organic matter; trace element concentrations were high and pH varied spatially between 3.1 and 7.5. These conditions were highly favorable to the determination of constants for the sorption of trace elements on natural iron oxyhydroxides. River water samples were filtered, and $[M^{z+}]$ and $[A^{n-}]$ were calculated by taking into account inorganic ligand concentrations and dissociation of arsenic acid; {Fe-M}, {Fe-A}, and {Fe-ox} were estimated by attacking the particulate material retained on the filters with 1.0 M HCl. The values of K_M and K_A obtained by Johnson for As, Cu, and Zn are shown in Figure 7.

Chapman et al.[85] have sampled water and iron-rich concretions deposited in the bed sediments at various sites in streams affected by acid mine drainage (pH 2.6 to 3.1); for each site, they reported the pH values, the concentrations of dissolved trace elements (As, Cd, Cu, Pb, Zn) in the water samples, and the ratios of trace element concentrations over that of iron concentrations in the concretions. We have used their reported figures to calculate K_M and K_A values with Equations 19 and 20. For the calculation of K_M, it was assumed that the total dissolved trace metal concentrations reported equaled $[M^{z+}]$ since, at pH of 3 and lower, complexation in solution should be insignificant; their ratio of M:Fe in the concretions was equated to {Fe-M}/{Fe-ox}. For the estimation of K_A, $[H_2AsO_4^-]$ was calculated from their reported values of [As] and the dissociation constants of arsenic acid ($pK_1 = 2.2$, $pK_2 = 7.0$, $pK_3 = 11.6$).[86] The K_M and K_A values thus calculated are also plotted in Figure 7.

It is noteworthy that, for a given trace element, all K_M (and K_A) values compare reasonably well, despite differences in site (river, stream, or lakes), type of natural solid phase studied (Fe-rich suspended matter, iron concretions or oxic bottom sediments) and methodology (filtration, *in situ* dialysis; leaching with HCl or $NH_2OH \cdot HCl$). The greater variability observed for Cu than for the other trace metals can be mainly attributed to inter-site variation in the nature and the amount of dissolved organic matter; the organic ligands are expected to have more influence on the chemistry of Cu than on that of other trace metals.

Description of trace element binding to natural particulate matter by sorption constants based on surface complexation concepts represents an improvement over the use of purely empirical values of K_D. The chemical theory on which they are based allows one to generalize and to make predictions that can be tested. Comparison of Figures 3 and 7 suggests also that the use of K_M and K_A values, instead of K_D, reduces scatter. This is, however, not surprising since the conditional sorption constants (K_M or K_A) are determined for a given sorbing component rather than for whole sediments with undefined and variable composition. In addition, solution chemistry (pH, inorganic ligand concentrations) is taken into account in the calculation of K_M and K_A, which is usually not the case when K_D is used. The approach used to determine field values of K_M and K_A present, however, some weaknesses. As discussed above (see Section 5), it is difficult to find a reagent that can dissolve selectively, from a complex matrix like a natural sediment, the fraction of iron oxyhydroxides which is in sorption equilibrium with the ambient water. It is even more difficult to extract selectively from the sediments the trace elements that were associated with the iron oxyhydroxides. For example, the reagents generally used for dissolving these compounds also dissolve manganese oxyhydroxides (and thus the bound trace elements). Another problem associated with extractions is the possible readsorption of the extracted trace elements onto remaining solid phases of the sediment.[94,95] Other factors that can contribute to the scatter of K_M and K_A at a given pH include variations among stations in the values of N_{Fe}, in the density of adsorption and in the concentration of particles.

Comparison with laboratory-derived constants — For comparative purposes, values of conditional adsorption constants were calculated from literature data for the adsorption of trace elements on model iron oxyhydroxides as a function of solution pH. At a given pH, the concentrations of trace element adsorbed and that remaining in solution were estimated graphically from the published adsorption curves (examples of such sigmoid curves are given in Figure 2b and d) and the accompanying experimental data. These two values were used together with the total iron concentration added to the system to calculate conditional adsorption constants with Equations 19 or 20, i.e., on the same basis as the field-derived sorption constants. Figure 7 compares the field-derived constants with those obtained in the laboratory. Even if the pH dependence of K_M obtained from field measurements (Figure 7) and their sequence (Figure 8) resemble those obtained in laboratory experiments with iron oxyhydroxides, the field- and laboratory-derived values of K_M themselves differ at a given pH in some cases by several log units. A striking point in Figure 7 is that the K_M values obtained by various researchers for the adsorption of a given metal on the "same" substrate (e.g., am-Fe oxyhydroxides) vary by several orders of magnitude at a given pH; this situation makes the comparison between field and laboratory results difficult. These large differences among laboratory experiments are probably due in part to the use of various experimental conditions (time of equilibration; temperature; ionic strength; preparation of the sorbent and its aging; concentration of trace element), and in part to experimental artifacts such as inadequate phase separation, contamination, and losses.

Many factors can explain the differences between field- and laboratory-derived K_M values. The temperature in natural systems ($\approx 4-25°C$) can differ appreciably from that for the laboratory experiments (room temperature; $\approx 20-25°C$) and temperature affects adsorption.[64] Reactions in natural systems can proceed for much longer times than in laboratory experiments for which equilibrium is usually assumed to be attained within minutes or hours. The surface area and porosity of solid phases will influence their adsorption characteristics. For example, Crosby et al.[96] have identified significant differences in surface area and porosity between fresh and aged iron oxyhydroxides, and between those derived from Fe(III) and Fe(II) sources. Most laboratory studies involve the precipitation of high Fe(III) concentrations (much higher than in natural waters) while natural iron oxyhydroxides are more commonly derived from the oxidation of Fe(II).[96] Laboratory experiments involve pure iron oxyhydroxide forms, whereas in natural systems these pure compounds probably are not present; Al, Si, P, and organic matter can be associated with natural iron oxyhydroxides and influence their adsorption characteristics. Iron oxyhydroxides form coatings on particles in natural systems whereas they are present as discrete particles in most laboratory experiments; the formation of coatings on particles may enhance the surface area. Studies over wide ranges of metal concentrations in well-defined systems suggest that the surfaces of iron oxyhydroxides comprise various binding sites, the binding energy of which may vary by several orders of magnitude.[68] This can result in a variation of the binding constants with surface coverage. Competitive interaction of high concentrations of cations (or anion) can also influence the binding constant value observed for a given trace element. Dissolved ligands may compete with the solid phases for binding a metal ion and thus reduce its adsorption; alternatively, the complex formed may adsorb strongly onto some solid phases and thus enhance the metal adsorption.

7. CONCLUDING REMARKS

Concentrations of trace elements in oxic natural waters are usually much lower than those predicted by solubility calculations, at least those based on the known solid phases of these elements. This observation has led to suggestions that other "unknown" reactions, described herein under the general term sorption, are responsible for the low dissolved trace element concentrations found in oxic natural waters. Identification of the exact reactions involved is needed to enable the formulation of predictive models based on accepted chemical theory.

Distribution coefficients are often used to describe the solid-solution partitioning of trace elements. A compilation of data obtained for various marine and aquatic environments reveals great variability in the value of this coefficient for a given trace element. Thus, it is concluded that empirical coefficients like K_D are of little use in the quantitative description of natural waters or for predictive purposes. In this respect, recourse to sound chemical theories, such as those based on surface complexation concepts, seems to be a more promising approach. Since these models are based on fundamental chemical properties of trace elements, they should allow predictions to be made concerning trace element behavior under conditions differing from those in which the measurements were made. Some field experiments suggest that the surface complexation theory applies to the sorption of trace elements on natural iron oxyhydroxides since the proton stoichiometry found from field measurements is similar to expectations based on laboratory measurements made in well-defined systems; the sequence of the equilibrium constants for the trace metals are the same for field and laboratory experiments; both field- and laboratory-derived equilibrium constants are correlated with the hydrolysis constants of the metals.

If adsorption proves to be the main chemical process responsible for the low dissolved trace element concentrations observed in natural systems, then substantial efforts should be directed at reconciling field and laboratory adsorption measurements. The experimental conditions used in the laboratory experiments should be critically examined for their closeness to natural conditions (e.g., reaction time, preparation of the model phase, ionic strength, temperature) and for their absence of artifacts (e.g., adequate solid-liquid separation). The exact forms of natural phases responsible for adsorption should be identified, and their adsorption characteristics determined (site density, pH of zero point of charge, intrinsic acidity constants, trace element adsorption constants).

ACKNOWLEDGMENTS

Financial support from the Québec Fond pour la Formation de Chercheurs et l'Aide à la Recherche and the Natural Sciences and Engineering Research Council of Canada is acknowledged. The author thanks J. Buffle, G. Den Boef, R. Gächter, B. Griepink, L. Hare, L. K. Koopal, H. P. van Leeuwen, and G. H. Nancollas for their critical comments.

REFERENCES

1. Förstner, U. and G. T. W. Wittmann. *Metal Pollution in the Aquatic Environment*, 2nd ed. (Berlin: Springer-Verlag, 1981).
2. Luoma, S. N. "Bioavailability of Trace Metals to Aquatic Organisms — A Review," *Sci. Total Environ.* 28:1–22 (1983).
3. Carignan, R. and J. O. Nriagu. "Trace Metal Deposition and Mobility in the Sediments of Two Lakes Near Sudbury, Ontario," *Geochim. Cosmochim. Acta* 49:1753–1764 (1985).
4. Jacobs, L. and S. Emerson. "Trace Metal Solubility in an Anoxic Fjord," *Earth Planet. Sci. Lett.* 60:237–252 (1982).

5. Jacobs, L., S. Emerson, and J. Skei. "Partitioning and Transport of Metals Across the O_2/H_2S Interface in a Permanently Anoxic Basin: Framvaren Fjord, Norway," *Geochim. Cosmochim. Acta* 49:1433–1444 (1985).

6. Krauskopf, K. B. "Factors Controlling the Concentrations of Thirteen Rare Metals in Sea-Water," *Geochim. Cosmochim. Acta* 9:1–32B (1956).

7. Schindler, P. W. "Heterogeneous Equilibria Involving Oxides, Hydroxides, Carbonates, and Hydroxide Carbonates," in *Equilibrium Concepts in Natural Water Systems,* W. Stumm, Ed. (Washington, D.C.: American Chemical Society, 1967), Adv. Chem. Ser. 67:196–221.

8. Tessier, A., R. Carignan, and N. Belzile. "Reactions of Trace Elements Near the Sediment-Water Interface in Lakes," in *Transport and Transformation of Contaminants near the Sediment-Water Interface,* J. DePinto and W. Lick, Eds. (Berlin: Springer Verlag, submitted for publication).

9. Smith, R. M. and A. E. Martell. *Critical Stability Constants.* (New York: Plenum Press, 1977).

10. Honeyman, B. D. and P. H. Santschi. "Metals in Aquatic Systems," *Environ. Sci. Technol.* 22:862–871 (1988).

11. Farrah, H. and W. F. Pickering. "Influence of Clay-Solute Interactions on Aqueous Heavy Metals Ion Levels," *Water Air Soil Pollut.* 8:189–197 (1977).

12. Benjamin, M. M. and J. O. Leckie. "Competitive Adsorption of Cd, Cu, Zn, and Pb on Amorphous Iron Oxyhydroxide," *J. Colloid Interface Sci.* 83:410–419 (1981).

13. Beveridge, A. and W. F. Pickering. "Influence of Humate-Solute Interactions on Aqueous Heavy Metal Ion Levels," *Water Air Soil Pollut.* 14:171–185 (1980).

14. Dempsey, B. A. and P. C. Singer. "The Effects of Calcium on the Adsorption of Zinc by MnOx(s) and Fe(OH)3(am)," in *Contaminants and Sediments.* Vol. 2, R. A. Baker, Ed. (Ann Arbor, MI: Ann Arbor Science Publishers, 1980), pp. 333–352.

15. Stumm, W., C. P. Huang, and S. R. Jenkins. "Specific Chemical Interaction Affecting the Stability of Dispersed Systems," *Croat. Chim. Acta* 42:223–245 (1970).

16. Schindler, P. W. "Surface Complexes at Oxide-Water Interfaces," in *Adsorption of Inorganics at Solid-Liquid Interfaces,* M. A. Anderson and A. J. Rubin, Eds. (Ann Arbor, MI: Ann Arbor Sciences Publishers, 1981), pp. 1–49.

17. Davis, J. A. and J. O. Leckie. "Surface Ionization and Complexation at the Oxide/Water Interface. II. Surface Properties of Amorphous Iron Oxyhydroxide and Adsorption of Metal Ions," *J. Colloid Interface Sci.* 67:90–107 (1978).

18. Van Riemsdijk, W. H., J. C. M. de Wit, L. K. Koopal, and G. H. Bolt. "Metal Ion Adsorption on Heterogenous Surfaces: Adsorption Models," *J. Colloid Interface Sci.* 116(2):511–522 (1987).

19. Hiemstra, T., W. H. Van Riemsdijk, and G. H. Bolt. "Multisite Proton Adsorption Modeling at the Solid/Solution Interface of (hydr)oxides: A New Approach," *J. Colloid Interface Sci.* 133(1):91–104 (1989).

20. Westall, J. and H. Hohl. "A Comparison of Electrostatic Models for the Oxide/Solution Interface," *Adv. Colloid Interface Sci.* 12:265–294 (1980).

21. Lion, L. W., R. S. Altmann, and J. O. Leckie. "Trace-Metal Adsorption Characteristics of Estuarine Particulate Matter: Evaluation of Contributions of Fe/Mn Oxide and Organic Coatings," *Environ. Sci. Technol.* 16:660–666 (1982).

22. Mouvet, C. and A. C. M. Bourg. "Speciation (including adsorbed species) of Copper, Lead, Nickel and Zinc in the Meuse River. Observed Results Compared to Values Calculated with a Chemical Equilibrium Computer Program," *Water Res.* 17:641–649 (1983).

23. Bourg, A. C. M. "Role of Fresh Water/Sea Water Mixing on Trace Metal Adsorption Phenomena," in *Trace Metals in Sea Water*, C. S. Wong, E. Boyle, K. W. Bruland, J. D. Burton, and E. D. Goldberg, Eds. (New York: Plenum Press, 1983), pp. 195–208.

24. Bourg, A. C. M. and C. Mouvet. "A Heterogeneous Complexation Model of the Adsorption of Trace Metals on Natural Particulate Matter," in *Complexation of Trace Metals in Natural Waters*, C. J. M. Kramer, and J. C. Duinker, Eds. (The Hague, Netherlands: Martinus Nijhoff/Dr. W. Junk, 1984), pp. 267–278.

25. Balistrieri, L. S. and J. W. Murray. "Metal-Solid Interactions in the Marine Environment: Estimating Apparent Equilibrium Binding Constants," *Geochim. Cosmochim. Acta* 47:1091–1098 (1983).

26. Balistrieri, L. S. and J. W. Murray. "Marine Scavenging: Trace Metal Adsorption by Interfacial Sediment from MANOP Site H," *Geochim. Cosmochim. Acta* 48:921–929 (1984).

27. White, J. R. "The Particle-Solution Chemistry of Lead in Acidic Lake Systems," in *Chemical Quality of Water and the Hydrologic Cycle*, R. C. Averett and D. M. McKnight, Eds. (Chelsea, MI: Lewis Publishers, 1987), pp. 211–234.

28. White, J. R. and C. T. Driscoll. "Zinc Cycling in an Acidic Adirondack Lake," *Environ. Sci. Technol.* 21:211–216 (1987).

29. Santschi, P. H., U. P. Nyffeler, R. F. Anderson, S. L. Schiff, P. O'Hara, and R. H. Hesslein. "Response of Radioactive Trace Metals to Acid-Base Titrations in Controlled Experimental Ecosystems: Evaluation of Transport Parameters for Application to Whole-Lake Radiotracer Experiments," *Can. J. Fish. Aquat. Sci.* 43:60–77 (1986).

30. Sigg, L. "Metal Transfer Mechanism in Lakes; the Role of Settling Particles," in *Chemical Processes in Lakes*, W. Stumm, Ed. (New York: Wiley-Interscience, 1985), pp. 283–310.

31. Sigg, L., M. Sturm, J. Davis, and W. Stumm. "Metal Transfer Mechanisms in Lakes," *Thalassia Jugosl.* 18:293–311 (1982).

32. Sigg, L., M. Sturm, and D. Kistler. "Vertical Transport of Heavy Metals by Settling Particles in Lake Zurich," *Limnol. Oceanogr.* 32:112–130 (1987).

33. Young, T. C., J. V. DePinto, and T. W. Kipp. "Adsorption and Desorption of Zn, Cu, and Cr by Sediments from the Raisin River (Michigan)," *J. Great Lakes Res.* 353–366 (1987).

34. Müller, B. and L. Sigg. "Interaction of Trace Metals with Natural Particle Surfaces: Comparison between Adsorption Experiments and Field Measurements," *Aquat. Sci.* 52:75–92 (1990).

35. Tessier, A., R. Carignan, B. Dubreuil, and F. Rapin. "Partitioning of Zinc between the Water Column and the Oxic Sediments in Lakes," *Geochim. Cosmochim. Acta* 53:1511–1522 (1989).

36. Tessier, A. Unpublished results (1990).

37. Nyffeler, U. P., Y. H. Li, and P. H. Santschi. "A Kinetic Approach to describe Trace-Element Distribution between Particles and Solution in Natural Aquatic Systems," *Geochim. Cosmochim. Acta* 48:1513–1522 (1984).

38. Santschi, P. H., U. P. Nyffeler, P. O'Hara, M. Buchholtz, and W. S. Broecker. "Radiotracer Uptake on the Sea Floor: Results from the MANOP Chamber Deployments in the Eastern Pacific," *Deep Sea Res.* 31:451–468 (1984).

39. Balistrieri, L. S. and J. W. Murray. "The Surface Chemistry of Sediments from the Panama Basin: the Influence of Mn Oxides on Metal Adsorption," *Geochim. Cosmochim. Acta* 50:2235–2243 (1986).

40. Li, Y. H., L. Burkhardt, M. Buchholtz, P. O'Hara, and P. H. Santschi. "Partition of Radiotracers between Suspended Particles and Seawater," *Geochim. Cosmochim. Acta* 48:2011–2019 (1984).

41. McIlroy, L. M., J. V. DePinto, T. C. Young, and S. C. Martin. "Partitioning of Heavy Metals to Suspended Solids of the Flint River, Michigan," *Environ. Toxicol. Chem.* 5:609–623 (1986).

42. Schindler, P. W., B. Fürst, R. Dick, and P. U. Wolf. "Ligand Properties of Surface Silanol Groups. I. Surface Complex Formation with $Fe3+$, $Cu2+$, $Cd2+$, and $Pb2+$," *J. Colloid Interface Sci.* 55:469–475 (1976).

43. Stumm, W., H. Hohl, and F. Dalang. "Interaction of Metal Ions with Hydrous Oxide Surfaces, *Croat. Chim. Acta* 48:491–504 (1976).

44. Aston, S. R. and E. K. Duursma. "Concentration Effects on ^{137}Cs, ^{65}Zn, ^{60}Co and ^{106}Ru Sorption by Marine Sediments, with Geochemical Implications," *Neth. J. Sea Res.* 6:225–240 (1973).

45. Balls, P. W. "The Partitioning of Trace Metals between Dissolved and Particulate Phases in European Coastal Waters: a Compilation of Field Data and Comparison with Laboratory Studies," *Neth. J. Sea Res.* 23:7–14 (1989).

46. DiToro, D. M., J. D. Mahony, P. R. Kirchgraber, A. L. O'Byrne, L. R. Pasquale, and D. C. Piccirilli. "Effects of Nonreversibility, Particle Concentration, and Ionic Strength on Heavy Metal Sorption," *Environ. Sci. Technol.* 20:55–61 (1986).

47. Honeyman, B. D., L. S. Balistrieri, and J. W. Murray. "Oceanic Trace Metal Scavenging: The Importance of Particle Concentration," *Deep Sea Res.* 35:227–246 (1988).

48. Luoma, S. N. and G. W. Bryan. "A Statistical Assessment of the Form of Trace Metals in Oxidized Estuarine Sediments Employing Chemical Extractants," *Sci. Total Environ.* 17:165–196 (1981).

49. Jenne, E. A. and J. M. Zachara. "Factors Influencing the Sorption of Metals," in *Fate and Effects of Sediment-Bound Chemicals in Aquatic Systems*, K. L. Dickson, A. W. Maki, and W. A. Brungs, Eds. (New York: Pergamon Press, 1984), pp. 83–98.

50. Jenne, E. A. "Controls on Mn, Fe, Co, Ni, Cu, and Zn Concentrations in Soils and Water: The Significant Role of Hydrous Mn and Fe Oxides," in *Trace Inorganics in Water*, R. F. Gould, Ed. (Washington, D.C.: American Chemical Society, 1968), Adv. Chem. Ser. 73:337–387.

51. Jenne, E. A. "Trace Element Sorption by Sediments and Soil — Sites and Processes," in *Symposium on Molybdenum in the Environment*, Vol. 2, W. Chappell and K. Petersen, Eds., (New York: Marcel Dekker, 1977), pp. 425–553.

52. Oakley, S. M., P. O. Nelson, and K. J. Williamson. "Model of Trace-Metal Partitioning in Marine Sediments," *Environ. Sci. Technol.* 15:474–480 (1981).

53. Luoma, S. N. and J. A. Davis. "Requirements for Modeling Trace Metal Partitioning in Oxidized Estuarine Sediments," *Mar. Chem.* 12:159–181 (1983).

54. Wiley, J. O. and P. O. Nelson. "Cadmium Adsorption by Aerobic Lake Sediments," *J. Environ. Eng.* 110:226–243 (1984).

55. Davies-Colley, R. J., P. O. Nelson, and K. J. Williamson. "Copper and Cadmium Uptake by Estuarine Sedimentary Phases," *Environ. Sci. Technol.* 18:491–499 (1984).

56. Luoma, S. N. "A Comparison of Two Methods for Determining Copper Partitioning in Oxidized Sediments," *Mar. Chem.* 20:45–59 (1986).

57. Buffle, J. *Complexation Reactions in Aquatic Systems,* (Chichester: Ellis Horwood, 1988).

58. Hunter, K. A. and P. S. Liss. "The Surface Charge of Suspended Particles in Estuarine and Coastal Waters," *Nature* 282:823–825 (1979).

59. Hunter, K. A. "Microelectrophoretic Properties of Natural Surface-Active Organic Matter in Coastal Seawater," *Limnol. Oceanogr.* 25:807–822 (1980).

60. Neihof, R. A. and G. I. Loeb. "The Surface Charge of Particulate Matter in Seawater," *Limnol. Oceanogr.* 17:7–16 (1972).

61. Jenne, E. A., D. M. DiToro, H. E. Allen, and C. S. Zarba. "An Activity-Based Model for Developing Sediment Criteria for Metals: I. A New Approach," in *Proc. Int. Conf. on Chemicals in the Environ.*, J. N. Lester, R. Perry, and R. M. Sterritt, Eds. (London: Selper Ltd., 1986), pp. 560–568.

62. Shea, D. "Developing National Sediment Quality Criteria," *Environ. Sci. Technol.* 22(11):1256–1261.

63. Dzombak, D. A. and F. M. M. Morel. *Surface Complexation Modeling. Hydrous Ferric Oxide,* (New York: Wiley-Interscience, 1990).

64. Fokkink, L. G. J., A. De Keizer, and J. Lyklema. "Temperature Dependence of Cadmium Adsorption on Oxides. I. Experimental Observations and Model Analysis," *J. Colloid Interface Sci.* 135(1):118–131 (1990).

65. Millward, G. E. and R. M. Moore. "The Adsorption of Cu, Mn, and Zn by Iron Oxyhydroxide in Model Estuarine Solutions," *Water Res.* 16:981–985 (1982).

66. Millward, G. E., "The Adsorption of Cadmium by Iron(III) Precipitates in Model Estuarine Solutions," *Environ. Technol. Lett.* 1:394–399 (1980).

67. Balistrieri, L. S. and J. W. Murray. "The Adsorption of Cu, Pb, Zn, and Cd on Goethite from Major Ion Seawater," *Geochim. Cosmochim. Acta* 46:1253–1265 (1982).

68. Benjamin, M. M. and J. O. Leckie. "Multiple-Site Adsorption of Cd, Cu, Zn, and Pb on Amorphous Iron Oxyhydroxide," *J. Colloid Interface Sci.* 79:209–221 (1981).

69. Gamble, D. S., A. W. Underdown, and C. H. Langford. "Copper(II) Titration of Fulvic Acid Ligand Sites with Theoretical, Potentiometric, and Spectrophotometric Analysis," *Anal. Chem.* 52:1901–1908 (1980).

70. Buffle, J., R. S. Altmann, M. Filella, and A. Tessier. "Complexation by Natural Heterogeneous Compounds: Site Occupation Distribution Functions, a Normalized Description of Metal Complexation," *Geochim. Cosmochim. Acta* 54:1535–1553 (1990).

71. Yates, D. E. and T. W. Healy. "The Structure of the Silica/Electrolyte Interface," *J. Colloid Interface Sci.* 55:9–19 (1976).

72. Huang, C. P. and W. Stumm. "Specific Adsorption of Cations on Hydrous γ-Al_2O_3," *J. Colloid Interface Sci.* 43:409–420 (1973).

73. James, R. O. and T. W. Healy. "Adsorption of Hydrolysable Metal Ions at the Oxide-Water Interface. III. A Thermodynamic Model of Adsorption," *J. Colloid Interface Sci.* 40:65–81 (1972).

74. Belzile, N., R. DeVitre, and A. Tessier. "In Situ Collection of Diagenetic Iron and Manganese Oxyhydroxides from Natural Sediments," *Nature* 340:376–377 (1989).

75. Johnson, C. A. "The Regulation of Trace Element Concentrations in River and Estuarine Water Contaminated with Acid Mine Drainage: The Adsorption of Cu and Zn on Amorphous Fe Oxyhydroxides," *Geochim. Cosmochim. Acta* 50:2433–2438 (1986).

76. Tessier, A., F. Rapin, and R. Carignan. "Trace Metals in Oxic Lake Sediments: Possible Adsorption onto Iron Oxyhydroxides," *Geochim. Cosmochim. Acta* 49:183–194 (1985).

77. Belzile, N. and A. Tessier. "Interactions between Arsenic and Natural Sedimentary Iron Oxyhydroxides," *Geochim. Cosmochim. Acta* 54:103–109 (1990).

78. Benjamin. M. M., K. F. Hayes, and J. O. Leckie. "Removal of Toxic Metals from Power-Generation Wastes Streams by Adsorption and Coprecipitation," *J. Water Pollut. Control Fed.* 54(11):1472–1481 (1982).

79. Hesslein, R. H. "An In Situ Sampler for Close Interval Pore Water Studies," *Limnol. Oceanogr.* 21:912–914 (1976).

80. Leckie, J. O., D. T. Merrill, and W. Chow. "Trace Element Removal from Power Plant Wastestreams by Adsorption/Coprecipitation with Amorphous Iron Oxyhydroxide," in *AICHE Symp. Ser.* 81:28–42 (1983).

81. Balistrieri, L., P. G. Brewer, and J. W. Murray. "Scavenging Residence Times of Trace Metals and Surface Chemistry of Sinking Particles in the Deep Ocean," *Deep Sea Res.* 28A:101–121 (1991).

82. Schindler, P. W. "Removal of Trace Metals from the Oceans: A Zero Order Model," *Thalassia Jugosl.* 11(1/2):101–111 (1975).

83. De Vitre, R., N. Belzile, and A. Tessier. "Speciation and Adsorption of As on Diagenetic Iron Oxyhydroxides," *Limnol. Oceanogr.* (in press).

84. Johnson, C. A. "The Sources, Dispersion and Speciation of Trace Elements Derived from Acid Mine Drainage in the Carnon River and Restronguet Creek, Cornwall," Ph.D. dissertation, Imperial College, London University (1984).

85. Chapman, B. M., D. R. Jones, and R. F. Jung. "Processes Controlling Metal Ion Attenuation in Acid Mine Drainage Streams," *Geochim. Cosmochim. Acta* 47:1957–1973 (1983).

86. Ferguson, J. F. and J. Gavis. "A Review of the Arsenic Cycle in Natural Waters," *Water Res.* 6:1259–1274 (1972).

87. Pierce, M. L. and C. B. Moore. "Adsorption of Arsenite and Arsenate on Amorphous Iron Hydroxide," *Water Res.* 16:1247–1253 (1982).

88. Hingston, F. J. "Specific Adsorption of Anions on Goethite and Gibbsite," Ph.D. thesis, University of W. Australia (1970).

89. Hingston, F. J., A. M. Posner, and J. P. Quirk. "Competitive Adsorption of Negatively Charged Ligands on Oxide Surfaces," *Disc. Faraday Soc.* 52:334–342 (1971).

90. Kinniburgh, D. G. and M. L. Jackson. "Cation Adsorption by Hydrous Metal Oxides and Clays," in *Adsorption of Inorganics at Solid-Liquid Interfaces*, M. A. Anderson and A. J. Rubin, Eds., (Ann Arbor, MI: Ann Arbor Science Publishers, 1981), pp. 91–160.

91. Leckie, J. O., A. R. Appleton, N. B. Ball, K. F. Hayes, and B. D. Honeyman. "Adsorptive Removal of Trace Elements from Fly-Ash Pond Effluents onto Iron Oxyhydroxide," Electric Power Research Institute Rep. EPRI-RP-910-1 (1984), Palo Alto, CA.

92. Theis, T. L. and R. O. Richter. "Adsorption Reactions of Nickel Species at Oxide Surfaces," in *Particulates in Water: Characterization, Fate, Effects, and Removal*, M. C. Kavanaugh and J. O. Leckie, Eds. (Washington, D. C.: American Chemical Society, 1980), Adv. Chem. Ser., 189:73–96.

93. Benjamin, M. M. and J. O. Leckie. "Effects of Complexation by Cl, SO_4, and S_2O_3 on Adsorption Behavior of Cd on Oxide Surfaces," *Environ. Sci. Technol.* 16:162–170 (1982).

94. Rendell, P. S., G. E. Batley, and A. J. Cameron. "Adsorption as a Control of Metal Concentrations in Sediment Extracts," *Environ. Sci. Technol.* 14:314–318 (1980).

95. Belzile, N., P. Lecomte, and A. Tessier. "Testing Readsorption of Trace Elements during Partial Chemical Extractions of Bottom Sediments," *Environ. Sci. Technol.* 23:1015–1020 (1989).

96. Crosby, S. A., D. R. Glasson, A. H. Cuttler, I. Butler, D. R. Turner, M. Whitfield, and G. E. Millward. "Surface Areas and Porosities of Fe(III)- and Fe(II)-Derived Oxyhydroxides," *Environ. Sci. Technol.* 17:709–713 (1983).

97. Everett, D. H. "Definitions, Terminology and Symbols in Colloid and Surface Chemistry," *Pure Appl. Chem.* 31:578–638 (1972).

CHAPTER 12

ION BINDING BY NATURAL
HETEROGENEOUS COLLOIDS

W. H. van Riemsdijk

Department of Soil Science and Plant Nutrition, Wageningen Agricultural University, Wageningen, The Netherlands

and

L. K. Koopal

Department of Physical and Colloid Chemistry, Wageningen Agricultural University, Wageningen, The Netherlands

TABLE OF CONTENTS

1.   Introduction 457

2.   Mineral and Organic Colloids...................................460

3.   Ion Binding to Homogeneous Colloids...........................461
      3.1   Introduction 461
      3.2   Proton Binding 462

3.3   Ion Complexation 464
3.4   Ion Exchange.. 466
3.5   Specific Adsorption on Independent Sites 467

4.   Ion Binding to Heterogeneous Colloids 467
      4.1   General Aspects of Heterogeneity 467
      4.2   Proton Binding to Heterogeneous Particles 468
      4.3   Metal Ion Binding to Heterogeneous Particles........... 470

5.   Electrostatic Effects and Particle Characterization 471
      5.1   General Aspects 471
      5.2   Electrostatic Modeling and Particle
             Characteristics.. 473

6.   Heterogeneity Analysis 475
      6.1   Introduction 475
      6.2   "Monocomponent" Normalized Freundlich
             Equations... 476
      6.3   "Multicomponent" Normalized Freundlich
             Equations... 478
      6.4   Numerical Inversion of the Integral Binding
             Equation.. 480
      6.5   Analytical Inversion of the Integral Equation........... 480
             6.5.1 LIA Methods 481
             6.5.2 Affinity Spectrum.............................. 485
             6.5.3 DEF Method 486

7.   Concluding Remarks.. 488

Acknowledgments 488

References... 488

1. INTRODUCTION

Ion binding to colloids is an important process affecting the activity of many species in the environment. Components with a high specific surface area are of particular importance. Three groups of colloidal particles are of relevance in this respect: the layered aluminum silicates, phyllosilicates or clay minerals, the metal (hydr)oxides, especially those of iron, aluminum, and manganese, and the organic particles, such as the fulvic and humic acids. A characteristic common to all three groups is that the particle dimensions may be extremely small. The thickness of an individual montmorillonite clay platelet is about 1 nm, its diameter is on the order of 0.1 μm and its specific surface area is about 800 $m^2.g^{-1}$. [1] Crystalline iron oxides in soils, such as goethite and hematite, have diameters ranging from 10 to 80 nm,[2] whereas amorphous hydrous ferric oxide particles may be even smaller, i.e., a diameter of 3 nm, and exhibit a specific surface area of around 600 $m^2.g^{-1}$. [3] Fulvic acids are relatively small macromolecules with a mean diameter in the order of 1 nm [4], a molecular weight of 500 to 1200 $g.mol^{-1}$ [5] and an extremely large specific surface area in the order of a few thousand $m^2.g^{-1}$, [4] All these colloids are heterogeneous with respect to their binding properties towards ions due to the presence of a (large) variety of surface groups. In natural systems these colloids often occur as mixtures or aggregates leading to an even larger heterogeneity. In soils, metal (hydr)oxides may occur as coatings on clay particles or sand grains or they may be intimately mixed with soil organic matter. In aqueous systems heterogeneous flocs may occur in which several types of reactive particles may be present. Metal (hydr)oxides in aqueous systems are often coated with organic macromolecules leading to an overall negative zeta potential,[6] although the metal (hydr)oxide particle itself may exhibit a positive zeta potential.

With regard to ion binding to colloidal particles distinction can be made between primary charge determining ions, mostly protons, specifically adsorbing ions, and indifferent ions. The indifferent ions adsorb in the diffuse double layer, whereas specifically adsorbing ions are predominantly bound in a monolayer adjacent to the surface. For metal (hydr)oxides, clay edges, and humic and fulvic acids, binding of protons leads to a primary particle charge that is strongly dependent on pH. The charge leads to an electric field around the particles, this field in turn affects the proton adsorption and depends on the concentration of indifferent salt. Specific adsorption of ions other than protons is the net result of an ion specific chemical attraction and an electrostatic repulsion or attraction due to the electric field. Therefore, specific adsorption of ions is affected by the type of reactive groups being present on the colloid, and the electrostatic potential in the vicinity of the adsorption sites. Other important factors influencing ion adsorption are the interaction with other specifically adsorbing species, and the speciation of the specifically adsorbing ion in the bulk solution, e.g., complex formation between a metal ion and dissolved ligands.

With respect to the electrostatics, it should be noted that the change in overall particle charge is the net effect of the charge added by the specifically adsorbing ion and the concomitant net release or uptake of protons. The concomitant release or uptake of protons generally mitigates the charge added to the surface by specific adsorption. Specific cation adsorption (at constant pH) leads to a net release of protons from the particle, whereas specific anion adsorption results in a net adsorption of protons. Nevertheless, progressive specific adsorption leads to an overall change in particle charge which makes further adsorption less favorable.

Chemical heterogeneity also leads to an affinity decrease with progressive adsorption since the higher affinity sites are mainly responsible for ion binding at low coverage and the lower affinity sites dominate the binding process at higher coverages. The fact that electrostatic and chemical heterogeneity effects operate simultaneously complicates interpretation of ion binding data to heterogeneous particles considerably. The electrostatic effect on ion binding is interpreted as an (apparent) heterogeneity when the electrostatics are not explicitly taken into account.

The intrinsic chemical heterogeneity of a particle derives from the presence of reactive groups which differ in their intrinsic affinities for an adsorbing species. The differences in affinity may result from different types of reactive surface groups, for example, singly, doubly, or triply coordinated surface oxygen groups present on surfaces of metal (hydr)oxide particles, or from a variation in the chemical environment of a specific type of reactive group. An example of the effect of a variation in chemical environment on the proton affinity is presented in Table 1 for a series of organic molecules with one or more carboxylic groups. The protonation constants in Table 1 differ as much as 4.5 log K units. In general, electron donating groups enhance the protonation constants or the proton affinity; electron accepting groups decrease it. The proton affinity of a carboxylic acid group is also greatly enhanced if the molecule contains a second acid group that is dissociated. The latter increase in affinity is partly caused by the negative electric field that is present around the molecule due to the already dissociated group. If a particle surface would exhibit carboxylic acid groups with proton affinity constants covering the range of values given in Table 1, a considerable intrinsic plus apparent heterogeneity would be present as a result of the differences in intrinsic proton affinity and of the electrostatic effects. If the electrostatic effects would be taken into account separately a narrower distribution of intrinsic proton affinities would result.

Our knowledge about the intrinsic chemical heterogeneity of natural colloidal particles is at present very limited. However, it is, in principle, possible to derive the affinity distribution using adsorption measurements. Since the proton is the primary adsorbing species for most natural particles under environmental conditions, analysis of proton binding curves in terms of heterogeneity is a useful tool, as a first step, in the characterization of the heterogeneity of natural colloids. Moreover, proton adsorption is, compared to

Table 1. Protonation constants of Various Carboxylic Acid Groups

Organic Acid	log K_1	log K_2	log K_3
Formic	3.75		
Acetic	4.75		
n-butyric	4.81		
Octanoic	4.89		
Oxalic	4.19	1.23	
Malonic	5.69	2.83	
L-tartaric	4.34	2.98	
Malic	5.11	3.40	
Dihydroxy malic	1.92		
Citric	6.39	4.77	3.14
Picric	0.38		
α-Naphtoic	3.70		
Benzoic	4.19		
Dihydroxy benzoic (2,2)	2.94		
Dihydroxy benzoic (2,5)	2.97		
Dihydroxy benzoic (3,4)	4.48		
Dihydroxy benzoic (3,5)	4.04		

Note: Log K_1, log K_2, and log K_3 correspond to the first, second, or third protonation constant, respectively.

adsorption of other ions, a relatively simple process for which the overall adsorption equation can be derived relatively easy.

The heterogeneity analysis of ion binding data started about with the work of Simms[7] and Scatchard et al.[8] and was further developed by, e.g., Klotz and Hunston,[9] Gamble et al.,[10,11] Hunston,[12] Buffle et al.[4,13,14] and Nederlof et al.[15-17] Also in the adsorption literature considerable attention was paid to the heterogeneity analysis, see, e.g., the reviews of House[18] and Jaroniec.[19-21] Important for the heterogeneity analysis is the assumed "local binding function", that is the binding function for one type of ligand or surface group, and the difficulties experimental errors can present for the analysis (see, e.g., Turner et al.[22] Fish et al.[23]). For a successful application of any method it is essential to use a sophisticated mathematical technique in order to avoid (1) the occurrence of spurious peaks.[16,17] and (2) unnecessary loss of information.[24]

In this chapter we will first briefly discuss the state of the art for mineral and organic colloidal particles, followed by a general discussion on ion binding to homogeneous and to heterogeneous colloids. It will be shown that the effect of the concentration of the indifferent salt on the proton adsorption behavior can be used to assess an "appropriate" electrostatic model without *a priori* knowledge about the chemical heterogeneity. Having established the electrostatic model it is possible to obtain a proton adsorption curve that is corrected for electrostatic effects. This curve can be used to obtain the intrinsic proton affinity distribution. Various aspects of the heterogeneity analysis will be discussed.

2. MINERAL AND ORGANIC COLLOIDS

Ion binding to clays and metal (hydr)oxides has been studied intensively. For phyllosilicates, interaction of ions with the plate side is to a large extent controlled by purely electrostatic forces. Since the plate side exhibits a permanent negative charge the resulting positive cation adsorption occurs in the diffuse layer, via an ion exchange process. In this exchange process the valency rather than specific ion properties are important. The edge faces of these clay minerals exhibit a pH dependent variable charge. These edges may bind a series of ions through specific (i.e., non-coulombic) interactions between an adsorbing species and a reactive surface group. Depending on the conditions and the type of surface groups present both cations and anions may specifically bind to these edges; especially multivalent ions may adsorb strongly. The ion exchange behavior on the clay plate side is well understood.[25,26] Sound models for specific adsorption on the edge face of clay minerals are scarce.[27-29] Qualitatively speaking it is generally assumed that the clay edges will behave similar to metal (hydr)oxides.

The metal (hydr)oxides are variable charge surfaces with reactive surface groups that can also bind various cations and anions through the formation of surface complexes. Metal hydroxides have been studied extensively. An advantage of crystalline metal (hydr)oxide particles is that in general their shape, size, and specific surface area are well defined and relatively easy to measure by, e.g., electron microscopy and gas adsorption. Moreover, crystallographic information can be used to assess the different types of groups and their site density, once the crystal planes have been identified. In classical modeling studies of the charge development on oxide particles this information is however only partly used because the particles are treated as pseudo-homogeneous. The most common approach is to combine an electrostatic model with a homogeneous site binding model (either "two-pK" or "one-pK"[30]) and to determine the model constants by fitting experimental data to the model. A variety of this type of models, which all have the ability to fit the proton adsorption isotherms very well, has appeared in literature.[31-41] For the more complicated models such as the frequently used two-pK triple layer model, see, e.g., the review of James and Parks,[42] it is difficult to find a unique set of protonation constants.[43] This, and the fact that rather different models can be fitted to the data,[37] shows that by curve fitting only little insight is gained in the adsorption behavior.

Recently Hiemstra et al.[44-46] have presented the MUSIC (multi-site complexation) model which explicitly takes into account the surface heterogeneity and is based on a series of *a priori* estimated proton binding constants of the singly, doubly, and triply coordinated surface oxygen groups. The MUSIC model allows for the prediction of both the point of zero charge (pzc) of a specific metal (hydr)oxide and the charge development without the introduction of adjustable parameters. The extension of the MUSIC model to incorporate specific adsorption is in principle straightforward. However, at the moment, the intrinsic binding constants for specifically adsorbing ions (other than protons) have to be treated as adjustable parameters.

An important result of the MUSIC model is that it predicts that the ΔpK of the two individual protonation constants of each surface oxygen group is around 14 log K units. This implies that in an environmentally relevant pH range (3 to 10) at most only one of the two protonation steps of a certain surface oxygen group is of practical relevance with respect to the variable charge behavior. The pK values of an individual group may also be such that this group does not change its protonation status in the pH range of interest. This is, for instance, the case for the doubly coordinated OH groups at the surface of aluminum (hydr)oxides. On the other hand the singly coordinated oxygen groups on both aluminum (hydr)oxides and iron (hydr)oxides will either occur as singly coordinated SOH or SOH_2 groups, their ratio depending on the pH and the intrinsic protonation constant of the SOH group. It is to be expected that the surface groups of not very well structured amorphous metal (hydr)oxide particles are predominantly singly- and doubly coordinated oxygen groups. If, for the sake of simplicity, one wants to make the approximation of a pseudo-homogeneous surface, a logical consequence of the MUSIC model is that one should favor the "one-pK model"[30,40,47] to describe the basic charging behavior of aluminum and iron (hydr)oxides. Further advantages of the one-pK model over the two-pK model are that (1) the logarithm of the (average) intrinsic proton affinity constant is directly related to the pzc of the metal (hydr)oxide and (2) combination of the one-pK concept with a distribution of proton affinity constants is straightforward.[30,39,47]

Most natural organic macromolecules, such as humic and fulvic acids, can also be treated as variable charge colloidal particles. The groups that are primarily responsible for proton binding are carboxylic and phenolic groups. The binding of multivalent cations to these colloids is often assumed to occur via sites with an acid character. In contrast to crystalline mineral particles, humic and fulvic acids are in general very difficult to characterize. This is partly because the variable charge character of the macromolecules complicates the interpretation of physical chemical measurements like sedimentation, viscosimetry, diffusion, etc., and partly because of the polydisperse nature and the chemical heterogeneity of the materials. Lack of good data on the nature and the dimensions of natural organic colloids has hampered the development of ion binding models. Most insight is obtained by using macromolecular polyelectrolyte models (see, e.g., References 48 through 51). Also Donnan equilibrium models[52-54] have been used.

3. ION BINDING TO HOMOGENEOUS COLLOIDS
3.1 Introduction

We will treat ion binding as ion interaction with reactive groups. Although the treatment is general, examples are chosen which resemble the reactive groups of natural organics like fulvic and humic acids rather than the reactive groups of metal (hydr)oxides. We will assume that the reactive groups interact with protons from the solution, resulting in a pH dependent primary particle charge and electrostatic potential. The complexation of ions other than protons with the reactive groups is influenced by the degree of protonation of reactive groups and hence by the electrostatic potential. The exact nature of the reactive sites and the type of complexes that may be formed is often not well known, the surface groups are therefore simply indicated with the symbol S. First the interaction of the protons with the reactive groups is considered, thereafter specific adsorption of other ions is treated.

3.2 Proton Binding

Although completely homogeneous analogs of natural particles like humic or fulvic acids do not exist, it is possible to simulate the behavior of homogeneous particles with respect to ion binding. The proton binding to a homogeneous organic model colloid, such as, for example, a polystyrene surface with one type of carboxylic group, may be expressed as

$$S^- + H^+ \rightleftarrows SH \tag{1}$$

where S^- and SH represent surface groups. According to Healy and White,[55] who give a statistical thermodynamic foundation of surface ionization equilibria, the equilibrium condition for Equation 1 is

$$\frac{[SH]}{[S^-]a_H} = K_H^{int} \exp(-F\Psi_s/RT) \tag{2}$$

where a_H is the bulk activity of the protons, [SH] and [S^-] are site densities, K_H^{int} is the intrinsic proton affinity constant or the surface acid association constant, and $\exp(-F\Psi_s/RT)$ is expressing the coulombic interaction the proton experiences due to the electric field. F is the Faraday constant, R the gas constant, T the absolute temperature and Ψ_s the potential at each charged site. In the derivation of Equation 2 lateral interactions other than coulombic have been neglected. For sake of convenience Healy and White suggest to use Equation 2 in the form

$$K_H^{int} = \frac{[SH]}{[S^-][H^+]\gamma_H} \tag{3}$$

where [H^+] is the proton concentration and γ_H is a composite activity coefficient.

$$\gamma_H = \gamma_H^b \exp(-F\Psi_s/RT) \tag{4}$$

The activity coefficient γ_H^b is the ordinary activity coefficient for protons in bulk solution. The product [H^+]γ_H can be interpreted as the proton activity in the direct vicinity of the charged sites on the surface. Equation 3 also follows from the application of the mass action law, provided the proton activity is taken in the vicinity of the charged surface. Defining the degree of protonation as

$$\theta_H = \frac{[SH]}{[S^-] + [SH]} \tag{5}$$

followed by a combination of Equations 5 and 3 leads to

$$\theta_H = \frac{K_H^{int}[H^+]\,\gamma_H}{1 + K_H^{int}[H^+]\,\gamma_H} \tag{6}$$

In a more simple notation this can be written as

$$\theta_H = \frac{K_H^{int}[H_s^+]}{1 + K_H^{int}[H_s^+]} \tag{7}$$

where $[H_s^+] = [H^+]\gamma_H$ is the proton activity near the binding sites.

An alternative option, which is frequently followed in literature, is to define the apparent affinity coefficient, K_H^{app} as

$$K_H^{app} = K_H^{app}\gamma_H^b \exp(-F\Psi_s/RT) \tag{8}$$

which leads to

$$\theta_H = \frac{K_H^{app}[H^+]}{1 + K_H^{app}[H^+]} \tag{9}$$

The advantage of Equation 9 is that $[H^+]$ and θ_H are experimentally accessible. Its disadvantage is that K_H^{app} is not a constant; due to the electrostatic interactions K_H^{app} depends on the degree of protonation or pH and the salt concentration.

Note that Equation 6 can also be derived for protonation of a monocarboxylic acid ligand in solution, if for γ_H the activity coefficient for protons in bulk solution is used. In that situation γ_H approaches one for very dilute solutions, whereas for proton binding to a reactive surface group the value of γ_H is (much) larger than one due to the electrostatic attraction. This behavior is illustrated in Figure 1 where θ is plotted as a function of pH for a monoprotic acid in solution (with $\gamma_H = 1$) and a surface acid with the same intrinsic proton affinity constant for two salt levels.

The γ_H in the case of the surface acid is calculated with the Gouy Chapman theory (see, e.g., Hiemenz[56]) which relates the surface potential Ψ_s to the surface charge which, in turn, is related to θ_H. The electrostatic effect is, at a given salt level, maximal for a fully dissociated surface and gradually decreases with increasing proton binding. The decrease in surface charge leads to a decrease in γ_H and K_H^{app}. For a homogeneous surface acid the gradual

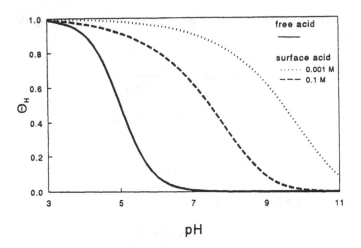

Figure 1. Proton binding curves for a monoprotic acid (solid line) and an equivalent
surface acid (dotted and dashed lines) in aqueous solution. The curves were
calculated using $K_H^{int} = 5$. For the surface acid a site density of 4 sites/nm²
was used. The two curves for the surface acid refer to two different salt levels:
—·— = 10^{-1} mol.dm⁻³ and ... = 10^{-3} mol.dm⁻³ 1-1 electrolyte.

decrease of K_H^{app} with increasing surface protonation can be interpreted as an
apparent (i.e., non-chemical) heterogeneity of the surface groups. The difference between the curves shown in Figure 1 indicates that interpretation of
proton binding to one type of surface acid according to Equation 9 will lead
to a relatively wide distribution of K_H^{app} (see also References 57 and 58).

3.3 Ion Complexation

Ion complexation to homogeneous particles will be illustrated by considering bivalent metal ion adsorption. In general, the surface groups on the
model colloid can not only bind protons but also metal ions, for instance, by
formation of a surface species SM^+ according to

$$S^- + M^{2+} \rightleftarrows SM^+ \tag{10}$$

Following Healy and White[55] the equilibrium condition for Equation 10 equals

$$K_M^{int} = \frac{[SM^+]}{[S^-][M^{2+}]\,\gamma_M} \tag{11}$$

where K_M^{int} is the intrinsic binding constant and

$$\gamma_M = \gamma_M^b \exp(-2F\Psi_S/RT) \tag{12}$$

in which γ_M^b is the bulk solution activity coefficient for species M, and the
number 2 is due to the double charge of M^{2+} (2 is replaced by z in the general
case of M^z).

The relative coverage of the sites by metal ions is defined as:

$$\theta_M = \frac{[SM^+]}{[S^-] + [SH] + [SM^+]} \tag{13}$$

Combination of Equations 3, 11, 13, and the mass balance of surface sites gives:

$$\theta_M = \frac{K_M^{int}[M^{2+}]\, \gamma_M}{1 + K_M^{int}[M^{2+}]\, \gamma_M + K_H^{int}[H^+]\, \gamma_H} \tag{14}$$

Equation 14 may also be expressed in terms of activities near the binding sites:

$$\theta_M = \frac{K_M^{int}[M_s^{2+}]}{1 + K_M^{int}[M_s^{2+}] + K_H^{int}[H_s^+]} \tag{15}$$

where $[M_s^{2+}] = [M^{2+}]\gamma_M$ and $[H_s^+] = [H^+]\gamma_H$.

Note that due to the site competition θ_H is no longer given by Equation 7 but by

$$\theta_H = \frac{K_H^{int}[H_s^+]}{1 + K_M^{int}[M_s^{2+}] + K_H^{int}[H_s^+]} \tag{16}$$

Equation 10 represents the most simple form of complexation. In practice, however, different types of surface species may be formed, for example, a bidentate complex:

$$2S^- + M^{2+} \rightleftarrows S_2M \tag{17}$$

or a metal hydroxy complex:

$$S^- + M^{2+} + H_2O \rightleftarrows SMOH + H^+ \tag{18}$$

This seriously complicates the situation. Unless the types of species present are unequivocally determined by spectroscopic methods, assumptions have to be made with respect to the nature of the metal species being formed. Once these assumptions are made, metal complexation can be described with equations like Equation 14, provided the model parameters are known.

Similarly to proton adsorption the total affinity of the surface groups for metal ions decreases with increasing θ_M, due to electrostatic effects. But the total affinity is also strongly dependent on the pH at which the metal ion adsorption is measured, because the pH strongly affects the surface potential and thus the value of γ_M. If metal ion adsorption is measured at constant pH, constant salt level, and at low values of θ_M, so that $K_M^{int}\gamma_M[M^{2+}] \ll K_H^{int}\gamma_H[H^+]$, then $[H^+]$ and γ_H may be approximated constant and Equation 14 will result in a linear isotherm:

$$\theta_M = K^*[M^{2+}] \tag{19}$$

with

$$K^* = \frac{K_M^{int}\gamma_M}{1 + K_H^{int}\gamma_H[H^+]} \tag{20}$$

This is even the case when different surface species exist, but in that case the expression for K^* is more complicated.

3.4 Ion Exchange

Another approach is to treat metal ion binding as a metal ion/proton exchange reaction as suggested by Kinniburgh et al.[59] for the description of metal ion adsorption on hydrous ferric oxide. The same approach can in principle be applied to metal ion binding to organic variable charge colloids. According to Kinniburgh et al.

$$S(H^+)_p + M^{2+} \rightleftarrows SM^{2+} + p\,H^+ \tag{21}$$

By defining θ_M as

$$\theta_M = \frac{[SM^{2+}]}{[SM^{2+}] + [S(H^+)_p]} \tag{22}$$

and K_{HM} as the metal ion/proton exchange constant, it follows that

$$\theta_M = \frac{K_{HM}[M^{2+}]/[H^+]^p}{1 + K_{HM}[M^{2+}]/[H^+]^p} \tag{23}$$

The parameter p may vary for different metal ions. By using this approach the effect of electrostatic interactions is reduced. For variable charge surfaces the exchange ratio p is, however, in general not equal to 2 and the surface charge must then change as a result of metal ion adsorption. Consequently, for $p \neq 2$, electrostatic interactions still depend on θ_M and K_{HM} is not a constant. Moreover, the effect of a change of the surface potential due to a change in pH is also incorporated in the value of p, meaning that the experimentally measured metal ion/proton exchange ratio is not necessarily equal to the value of p that is used to model the metal ion adsorption (Equation 21).

3.5 Specific Adsorption on Independent Sites

So far, metal ion adsorption was considered to occur on the same sites as protons, leading to metal ion/proton competition for the sites. It is, however, also possible to consider separate sets of sites for protons and specifically adsorbing ions. This situation will be referred to as independent site adsorption. In the literature regarding metal ion binding to natural ligands like humics this approach is quite common (see, e.g., Buffle[4]). With metal ion adsorption on oxides this treatment is rather an exception.[34,38,50,61,62] The metal ion adsorption on independent sites can be described as

$$S + M^{2+} \rightleftarrows SM^{2+} \tag{24}$$

By defining θ_M as

$$\theta_M = \frac{[SM^{2+}]}{[S] + [SM^{2+}]} \tag{25}$$

and K_M^{int} as the intrinsic binding constant, it follows that

$$\theta_M = \frac{K_M^{int}[M^{2+}] \, \gamma_M}{1 + K_M^{int}[M^{2+}] \, \gamma_M} \tag{26}$$

Equation 26 is equivalent to Equation 6 for proton adsorption. An advantage of Equation 26 is that it is far more simple than Equation 14.

Although there is in this approach no direct competition between protons and metal ions for the adsorption sites, the metal adsorption is, in general, still influenced by the pH.[62] For natural variable charge colloids the electrostatic potential near the surface groups is a function of pH, leading to a pH dependency of γ_M and thus to a pH dependent value of θ_M.

4. ION BINDING TO HETEROGENEOUS COLLOIDS
4.1 General Aspects of Heterogeneity

The intrinsic chemical heterogeneity of a natural colloid derives from the presence of different types of reactive groups on the particle. If the behavior of a reactive group is influenced by interaction with other groups present on the particle, it matters how the reactive groups are spatially arranged. If they are randomly arranged over the entire surface (random heterogeneity), only one average interaction term is required to describe the lateral interactions. If the groups are present on clearly separated patches (patchwise heterogeneity), as is the case when different crystal planes are present on the particle surface, the interactions are mainly effective per patch. For instance, with metal (hydr)oxide particles, which may exhibit different crystal planes with different chemical composition, a patchwise treatment of heterogeneity is appropriate. This means that the electrostatic interaction in ion adsorption should be counted per crystal plane. A similar situation occurs when chemically different particles are present in a mixture. A patch may be homogeneous or heterogeneous. In both cases the interactions are considered per patch and an average interaction term (i.e., electrostatic potential) is calculated for the whole patch. In practice the distinction between random and patchwise heterogeneity is not sharp. In the case of electrostatic interactions the "reach" of the interactions is dependent on the salt concentration, as a result it is possible that at low ionic strength a surface has to be considered as random heterogeneous, whereas it shows patchwise behavior at high ionic strength.[63] The heterogeneity of a particle or a surface patch can sometimes be expressed by considering distinctly different reactive groups, each group with its specific affinity, leading to a description in terms of discrete heterogeneity. In the case of groups that vary gradually in affinity the heterogeneity may be described in terms of a continuous affinity distribution.

4.2 Proton Binding to Heterogeneous Particles

The most simple approximation for a heterogeneous surface is to treat it as (pseudo-)homogeneous. The affinity constants for proton and other ions are in that case average quantities for the heterogeneous surface without a thermodynamic meaning. The classical treatment of ion adsorption on metal (hydr)oxides (see, e.g., Westall and Hohl,[37] James and Parks[42]) follows this approach at the expense that only little insight is gained in the true behavior.

A fundamentally better approach is to consider the heterogeneity explicitly. In case of a heterogeneous surface with a few discrete site types the relative overall adsorption of ion, x, $\theta_{t,x}$, is the result of the weighted summation of the relative coverage of the various types of groups. For example, for proton binding one finds

$$\theta_{t,H} = \sum_j f_j \theta_{j,H}(K_{j,H}, \gamma_H^*, [H^+]) \tag{27}$$

where f_j is the fraction of site type j with respect to the total number of sites available, $\theta_{j,H}$ the degree of protonation of subgroup j, and γ_H^* the overall activity coefficient for the protons. Often $\theta_{j,H}$ is called the local isotherm. Equation 27 essentially separates the problem in two components, the heterogeneity expressed through f_j, and the binding characteristics to a group of equal affinity sites expressed by $\theta_{j,H}$. In the absence of competing species $\theta_{j,H}$ is given by Equation 6, with the extra complication that γ_H has to be replaced by γ_H^* and that γ_H^* depends on the type of heterogeneity. For patchwise surfaces the electrostatic interaction acts per patch, i.e., $\gamma_H^* = \gamma_{p,H}$, whereas for random heterogeneous surfaces the electrostatic interactions are counted over the entire surface and $\gamma_H^* = \gamma_{t,H}$.[63] As an experimental model for a patchwise heterogeneous system Gibb and Koopal[47] studied mixtures of rutile and hematite treating pure rutile and hematite as pseudo-homogeneous.

An extra complication arises when a patch itself is randomly or regularly heterogeneous. Such a patch can be treated as a random heterogeneous surface by applying Equation 27 with γ_H^* as electrostatic interaction term for that patch. The adsorption on the entire surface is obtained by applying Equation 27 again, but the summation is now over the contributions of the individual patches. In such a case extensive information is required to calculate the ion binding. This approach has been used in the prediction of the proton binding behavior of metal (hydr)oxides by Hiemstra and Van Riemsdijk.[44-46]

The actual coverage of a site type with protons is now not only a function of its own proton affinity constant and the pH, but through the electrostatic interaction it is also strongly dependent on the presence of other types of groups. When these other groups have lower proton affinity constants they strongly contribute to the negative charge on the patch and therefore strongly suppress the dissociation of higher affinity sites.

Fulvic and humic acids are a mixture of hydrophilic colloidal particles or macromolecules that differ in size, shape, and chemical composition. This situation is comparable to patchwise heterogeneity because in this case the mixture may be thought to be split into different groups of heterogeneous colloidal particles ("patches") with equal physical and chemical characteristics per group of particles. For the same solution conditions, the electrostatic interaction term will, in principle, be different for each group of particles. In general, the information required to deal with this complexity is lacking for fulvic and humic acids.

One way out of this problem is that the electrostatic interactions are not accounted for separately and that the electrostatic effects are incorporated in the overall affinity; as discussed before this leads to a wide apparent affinity distribution. Another approach is to assume that the mixture can be replaced by an ensemble of "equivalent" heterogeneous particles with average properties, i.e., an average heterogeneity, an average particle geometry, and average electrostatic properties. Ideally, these equivalent heterogeneous particles show the same behavior as the complicated mixture. If it is further assumed that there are a discrete number of different site types, Equation 27 applies for the equivalent particle.

Alternatively, one may assume equivalent particles and a continuous distribution of (intrinsic) affinity constants. In view of the effects shown in Table 1 such a continuous distribution seems most appropriate for the description of humic or fulvic acids. The proton adsorption can in that case be described by the following integral equation:

$$\theta_{t,H} = \int_{\Delta H} \theta(K_H^{int}, [H_s^+]) \, f(\log K_H^{int}) \, d\log K_H^{int} \tag{28}$$

where ΔH is the range of intrinsic proton affinity constants, $\theta(K_H^{int}, [H_s^+])$ is given by Equation 6 with $\gamma_H = \gamma_{H,t}$ and $f(\log K_H^{int})$ is the intrinsic affinity distribution. The function $\theta(K_H^{int}, [H_s^+])$ is again called the local isotherm.

4.3 Metal Ion Binding to Heterogeneous Particles

For metal ion binding on natural colloids the same type of reasoning can be applied with respect to the heterogeneity as for the protons. The most simple treatment is not to acknowledge the electrostatic effects explicitly and to assume independent sites for metal ion binding. This approach necessarily relies on random heterogeneity and a continuous distribution of apparent affinities. The integral binding equation therefore becomes

$$\theta_{t,M} = \int_{\Delta M} \theta_M(K_M^{app}, [M]\, \gamma_M^b)\, f(\log K_M^{app})\, d\log K_M^{app} \qquad (29)$$

where ΔM is the range of apparent affinity coefficients, $\theta_M(K_H^{app}, [M]\gamma_M^b)$ is the local binding function and $f(\log K_M^{app})$ is the distribution function of apparent affinities.

Although this type of treatment is frequently followed in literature (see, e.g., References 4, 64-66), it has the disadvantage that the apparent affinity distribution cannot be easily "translated" into a distribution of specific surface groups.

A one-step more complicated approach is to account for the electrostatic effects explicitly on the basis of "equivalent particles" and to assume that the sites for metal ions are fully independent of those for protons. In this case the metal ion adsorption equation is similar to that where only proton adsorption is considered, namely,

$$\theta_{t,M} = \int_{\Delta M} \theta_M(K_M^{int}, [M_S^{2+}])\, f(\log K_M^{int})\, d\log K_M^{int} \qquad (30)$$

As local isotherm in Equation 30 one can use Equation 26.

In the case of continuous distributions and competition between protons and metal ions for the same surface sites a multiple integral equation results, both for proton and metal ion adsorption. The overall adsorption equation for metal ions is

$$\theta_{t,M} = \int_{\Delta M} \int_{\Delta H} \theta_M(K_H^{int}, K_M^{int}, [H_S^+], [M_S^{2+}])\, f(\log K_H^{int}, \log K_M^{int}) \times$$

$$d\log K_H^{int}\, d\log K_M^{int} \qquad (31)$$

For the protons a similar equation applies with $\theta_{t,M}$ and θ_M being replaced by $\theta_{t,H}$ and θ_H, respectively. For the local isotherm $\theta_M(K_H^{int}, K_M^{int}, [H_s^+], [M_s^{2+}])$, Equation 16 may be used if it is assumed that the metal binding on a subset of homogeneous sites can be described with Equations 10 and 11.

Of course, it is also possible to integrate over the total affinities $K_H(=K_H^{app})$ and $K_M(=K_M^{app})$. In this case in Equation 31, the superscripts "int" should be replaced by "app", and instead of $[H_s^+]$ and $[M_s^{2+}]$ the bulk solution activities should be used. It will be clear that the description of metal ion surface complexation with competition is far more complex than that of metal ion binding on independent sites.

In principle, when the distribution function and the local isotherm are known, the overall adsorption $\theta_{t,x}$, where x stands for either H or M, can be calculated with the integral adsorption equation. In general there are no analytical expressions for Equations 28 to 31, but for a few specific distribution functions the equations can be integrated analytically.[20,21,67,68] In most cases no *a priori* information is available on the distribution function, so that $\theta_{t,x}$ cannot be predicted. However, Equations 28 to 31 can in principle be used to assess the distribution function and thus to characterize the particles, when $\theta_{t,x}$ is known or measured. In Section 6 it will be discussed how the affinity distribution functions appearing in Equation 28, 29, or 30 can be assessed on the basis of measured isotherm data in combination with an assumed local isotherm equation.

5. ELECTROSTATIC EFFECTS AND PARTICLE CHARACTERIZATION

5.1 General Aspects

Natural variable charge particles obtain their primary charge by proton adsorption/desorption. The proton binding curves, which can be derived from acid/base titration measurements, reflect both the variable electrostatic effects and the intrinsic site heterogeneity. The electrostatic effect can be considered independently of the (intrinsic) distribution function. It is possible to calculate the γ_H (Equation 4) and to replot the overall isotherm as a function of pH_s by applying a double layer model describing the electrostatic effect. For a series of isotherms measured at various salt levels, the different isotherms should merge into one master curve, when an appropriate electrostatic model is used. In this way an independent check can be made on the model chosen to describe the electrostatic interactions.[69-71]

By applying, for instance, this procedure to the curves shown in Figure 1, the two curves for the surface acid merge with the curve for the dissolved monoprotic ligand (same K_H^{int}); this curve is the master curve for the surface acid, which in this case is identical with a Langmuir isotherm (the free acid curve).

Also in case of proton binding to heterogeneous particles the effect of indifferent salt on the proton binding curves is caused by electrostatic interactions. Calculation of $[H_s^+]$ with a double layer model is in that case also possible. The individual curves can be replotted as a function of pH_s, and in the case of a mixture of identical heterogeneous particles a master curve should result. From the master curve an estimation of the intrinsic proton affinity distribution can be obtained.

Application of this concept to proton binding curves of humic and fulvic acids has been done by De Wit et al.[69-71] and leads to good results as can be seen from Figure 2. Two steps are necessary to obtain such results. First of

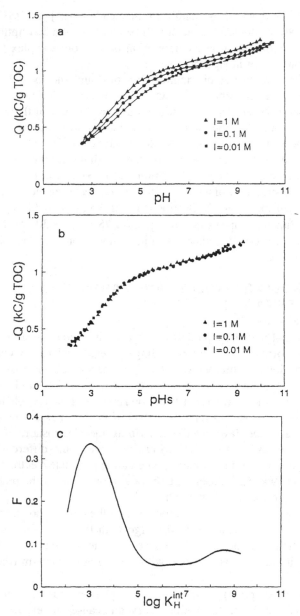

Figure 2. The master curve procedure. (a) Experimental proton charge, Q, in kC/g as
a function of pH measured at three different ionic strength values. Data are
taken from Dempsey and O'Melia[115] (their Figure 3). (b) The Q(pH$_s$) master
curve for the data of Figure 2a. The pH$_s$ ($= -\log[H_s^+]$) is calculated with
$[H_s^+] = \gamma_H[H^+]$ where γ_H is given by Equation 4. The surface potential Ψ_s in
Equation 4 is derived from the proton charge by using a spherical double layer
model with r = 0.7 nm and M = 900 g.mol^{-1} (ρ = 1 g.cm^{-3}). (c) The intrinsic
proton affinity distribution (non normalized) obtained by the application of the
LOGA method to the master curve of Figure 2b. The large peak at log K$_H^{int}$ $\approx$3
is most probably due to carboxylic groups, the small peak at logK$_H^{int}$ $\approx$9 to
phenolic groups (K$_H^{int}$ in dm^3.mol^{-1}).

all one has to invoke the earlier mentioned equivalent heterogeneous particle concept that allows for the calculation of an average electrostatic interaction that is the same for all particles at a certain salt level and $\theta_{t,H}$. Second, one has to assume a certain double layer model. Several options are briefly discussed in the next section.

5.2 Electrostatic Modeling and Particle Characteristics

If it is assumed that the Poisson-Boltzman (PB) equation can describe the electrostatic effects (see, e.g., Fixman[72]), the shape and the size of the particles will affect the electrostatic effect. For large particles the curvature of the surface plays a minor role and a "flat plate" or Gouy-Chapman double layer model (see, e.g., Hiemenz[56]) may be used. For small particles an idealized geometry may be used and algorithms for the solutions of the PB equation are available to calculate the surface potential as a function of charge density and salt level for both spheres[73,74] and cylinders.[75] For low potentials the Debye-Hückel approximation may be used and analytical solutions are available (see, e.g., Tanford[48]).

The electrostatic properties of natural macromolecular organic colloids can also be described, treating the colloid as a gel phase with an approximately constant electrostatic potential in the gel layer. This type of modeling is referred to as "Donnan" models.[25,53,54]

For fulvic and humic acids several models have been used to describe the potential as a function of surface charge. Tipping et al.[76,77] applied an empirical model and in a later study[78] the Debye-Hückel model for spherical particles. De Wit et al.[69-71] have used various solutions of the Poisson-Boltzman equation. The salt dependency of the charging curves of fulvic acids is in general less than that which corresponds with a flat plate type double layer model. This is in accordance with the assumption that the fulvic acid particles are rather small spheres[78] or cylinders.[79] The Donnan model has been applied by Marinsky and Ephraim[52] and De Wit et al.[71]

As stated above, to apply the master curve concept, it is necessary to make the assumption that the mixture of particles can be represented by an ensemble of equivalent particles with an average radius, an average heterogeneity, and an average surface potential, the latter depending on the solution conditions. Apart from this approximation no *a priori* assumption has to be made with respect to the chemical heterogeneity. Numerical simulations using particle size distributions can be used to assess the effect of assuming an average particle dimension on the interpretation of the data.[78]

In all cases the measured data (i.e., charge per gram carbon) should be expressed in charge per unit surface area or per unit volume. Based upon the composition of the humic or fulvic acid, the charge can first be converted into charge per gram humic or fulvic acid which is roughly twice the charge per gram carbon.

The specific surface area ($m^2.g^{-1}$ humic or fulvic acid) can be used to obtain the conversion to charge per unit surface area. Direct measurements of specific surface area of humic and fulvic acid are, however, very difficult. Methods which are based on dried material (EM, BET) probably result in too low values. If no reliable estimate of the surface area is available it may be obtained by the master curve method itself.

In the case of a spherical or cylindrical geometry the specific surface area of a colloid is directly related to the particle radius through the density. If a reasonable value for the particle density is assumed the average particle radius is the only adjustable parameter in the master curve procedure. The radius thus obtained can be compared with values obtained from other methods.[80] In case of a spherical geometry the radius is also related to the specific surface area by means of the molecular weight. If a cylindrical geometry is assumed the average length of the cylinders can be derived from the average molecular weight. In case of a Donnan model assuming a gel phase, the gel volume used should result in a reasonable value for the particle density in order to consider the model as physically realistic. In Table 2 the relationships between the various parameters are given.

TABLE 2. Particle or Gel Properties

Sphere

$$A_s = \frac{4\pi r^2}{(4/3)\ \pi r^3 \rho} = \frac{3}{\rho r}$$

$$A_s = \frac{N_{AV} 4\pi r^2}{M}$$

Cylinder

$$A_s = \frac{2\pi r l}{\pi r^2 l \rho} = \frac{2}{\rho r}$$

$$l = \frac{M}{N_{AV} \pi r^2 \rho}$$

Gel phase

$$\rho = d/\Phi$$

Note: A, specific surface area, r radius, l length, M molecular mass, ρ density of the particles, Φ volume fraction of the gel phase, and d is the particle mass concentration.

If it is possible to obtain a master curve with one or more of the simple electrostatic models discussed above, it can be concluded that a relatively simple model is appropriate to describe the electrostatic effects observed. If none of the models is appropriate, the conclusion must be that one or more of the assumptions are incorrect. A probable reason might be that the particle or gel dimensions depend on the electrostatic potential.

Preliminary analysis of literature data on proton binding to fulvic acids as a function of salt level shows that several data sets reported in literature result in good master curves if it is assumed that the particles are rigid spheres or cylinders.[71] The radii obtained for cylindrical particles are about 1 nm, whereas for spheres radii of approximately 0.4 nm result. The master curve for a specific fulvic acid hardly depends on the chosen geometry. A typical result of the master curve approach is shown in Figure 2. This means that the intrinsic heterogeneity that can be derived from the master curve is hardly sensitive to the geometries studied. For most fulvic acids studied it is not possible to obtain an equally good master curve with the Donnan model as with the cylindrical or spherical PB models. It has been suggested[52] that it is possible to obtain experimentally a protonation curve without the influence of electrostatic interactions by measuring the protonation at high salt level. This curve should thus reflect the master curve. However, calculations show that at 1 M ionic strength the master curve of fulvic acids can still deviate from the experimental curve so that this treatment should be considered with some reservation.[70]

Once a master curve has been obtained, heterogeneity analysis may be performed. The master curve for a homogeneous colloid is identical to the Langmuir isotherm, see Equation 7. When the master curve deviates from the Langmuir isotherm the surface is heterogeneous, and, for instance, the LOGA method[16] can be used to obtain the distribution function. In the following paragraph the problems and possibilities of the heterogeneity analysis are discussed.

6. HETEROGENEITY ANALYSIS
6.1 Introduction
In general all natural reactive colloids exhibit a certain degree of heterogeneity that will affect the interaction of ionic species with those colloids. Direct assessment of the heterogeneity, for instance, by means of crystallographic analysis is seldomly possible. Even if this is possible one needs a model or a theory that translates this chemical heterogeneity into its effect on the binding of species of interest. Therefore, in practice, the heterogeneity is often estimated from the observed binding characteristics of one or more species. The species that is used can be seen as a probe to assess the heterogeneity.

Two main factors influence the intrinsic ion binding characteristics to surface groups on heterogeneous colloids: (1) the local isotherm and (2) the affinity distribution function, see Section 4. To assess the distribution function from the overall binding curves, one or more assumptions are always required. One possibility is to assume both the local isotherm and the type of distribution. In that case the parameters describing the distribution function have to be optimized ("curve fitting"). A well-known example of this approach is the Scatchard technique.[8] As local isotherm the Langmuir model is assumed, i.e., Equation 6 or 26 with all activity coefficients set equal to one, and a distribution function made up of one to three discrete site types. With a graphical technique one can determine up to three individual affinity parameters. Recently a more general technique has been proposed[81] that claims to be able to determine a discrete distribution function composed of a large number of site types using a mathematical procedure instead of a graphical one. In this case a large number of parameters is required, for each site type an affinity constant and a frequency is needed. With the latter type of method one should be aware of the possibility that spurious peaks may appear in the distribution function,[18,82,83] see also Section 6.4.

Another approach which assumes both the local isotherm and the type of distribution function is the application of analytical heterogeneous isotherm equations. This option is discussed in Sections 6.2 and 6.3. In general one does not know the type of distribution function and one should analyze the overall binding curves without making an assumption about the distribution function. In this case only the type of local isotherm is assumed. To obtain the distribution function the integral equation has to be inverted.

6.2 "Monocomponent" Normalized Freundlich Equations

Often it is convenient to use analytical expressions to describe the adsorption; therefore some rather common equations will be discussed below. By using these equations an implicit assumption is made with respect to the heterogeneity.

In order to achieve an analytical solution for θ_t as expressed by Equation 28, 29, or 30 the mathematical expression for the local isotherm should be of the Langmuir type, i.e.:

$$\theta(X) = \frac{KX}{1 + KX} \tag{32}$$

where K may be an intrinsic constant, or a total affinity constant and X may be a single or composite variable as long as it is independent of K. Equations 7, 9, 23, and 26 all fit to Equation 32. Substitution of Equation 32 together with one of a few specific affinity distribution functions in Equation 28, 29, or 30 may result in an analytical solution of the integral equation.[19,21,67,68,84] For a nearly Gaussian distribution function Equation 33 results

$$\theta_t(X) = \frac{(\check{K}X)^m}{1 + (\check{K}X)^m} \tag{33}$$

Equation 33 is called the Langmuir-Freundlich equation. Alternatively $\theta_t(X)$ can be found as

$$\theta_t(X) = \left(\frac{\check{K}X}{1 + \check{K}X}\right)^m \tag{34}$$

Equation 34 is called the generalized Freundlich equation and corresponds with a normalized exponential type distribution function with a high affinity tail. Finally $\theta_t(X)$ may take the form:

$$\theta_t(X) = \frac{\check{K}X}{\{1 + (\check{K}X)^m\}^{1/m}} \tag{35}$$

Equation 35 results from a quasi Gaussian distribution with a low affinity tail and is called the Toth equation.[84]

The $\check{K}$ in Equations 33 through 35 is the weighted average affinity constant. In case of Equation 33 the weighted average affinity corresponds with the peak of the affinity distribution. In general $\check{K}$ determines the position of the affinity distribution at the log K axis. The parameter m should obey the condition,

$$0 \leq m \leq 1 \tag{36}$$

and determines the width of the distribution function. For m = 1 the homogeneous Langmuir equation results for all cases.

In the limit for $\check{K}X \ll 1$ both Equations 33 and 34 reduce to

$$\theta_t(X) = (\check{K}X)^m \tag{37}$$

If the amount adsorbed per unit of sorbent is expressed as Q and the adsorption maximum as Q_{max} Equation 37 may be written as:

$$Q(X) = Q_{max}(\check{K})^m(X)^m \tag{38a}$$

or

$$Q(X) = K_{Fr}(X)^m \tag{38b}$$

where $K_{Fr} = Q_{max}(\check{K})^m$. Equation 38b is the well-known Freundlich equation. It follows directly from this analysis that the Freundlich K, K_{Fr}, is proportional with the adsorption maximum. The composite character of K_{Fr} should be kept in mind if one determines the Freundlich parameters for a certain sample and wants to apply the result to another sample. A generally valid set of Freundlich parameters can thus only be obtained if one can separate K_{Fr} in its two components (Q_{max} and $(\tilde{K})^m$) by determining a quantity that is equal to or proportional with the adsorption maximum. For instance, if the organic matter content dominates the measured adsorption behavior various samples should be compared on the basis of the organic matter content of the samples. Kinniburgh et al.[59] have used Equations 33 through 35 to describe adsorption of calcium and zinc as a function of pH and metal ion concentration on hydrous ferric oxide and have obtained good fits of their results with all three equations. Equation 23 was used as local isotherm, i.e., $X = (M^{2+})/(H^+)^P$ and $\tilde{K} = \tilde{K}_{HM}$. At first sight it seems surprising that the same data set can be described with three equations that are based on quite different affinity distributions. The explanation is that their data cover only a small part of the total isotherm, and consequently only a small part of the distribution is reflected in the data. That part of the distribution that is of relevance with respect to the data is probably quite similar for the three fitted distributions.

It has been argued[23,85] that models based on the presence of a series of discrete sets of sites should be preferred to Equations 33 through 37 because (1) Equations 33 through 37 do not match with general calculation schemes for speciation like MINEQL and Geochem and (2) supposedly these models cannot be extended to include competitive adsorption. Both arguments are incorrect. It is possible to incorporate these types of adsorption models into a general mathematical scheme for the calculation of speciation. The multicomponent competitive analogs of these heterogeneous models can easily be derived if it is assumed that the shape of the affinity distribution is the same for all adsorbing species (same value of m), and that only the position of the distribution on the affinity axis is dependent on the species. A major practical advantage of the analytical equations is that the heterogeneity aspect is introduced at the expense of only one extra parameter (m) compared to a homogeneous model.

6.3 "Multicomponent" Normalized Freundlich Equations

For a homogeneous surface the competitive equivalent of the Langmuir equation can be written as

$$\theta_i(X_1,\ldots,X_n) = \frac{K_i^* X_i}{1 + \sum_{j=1}^{j=n} K_j^* X_j} \tag{39}$$

where both X_i and K_i^* may be composite parameters. The value of n depends on the number of species that compete for the same surface site. An example of application of Equation 39 for metal ion adsorption on metal (hydr)oxides is given by van Riemsdijk et al.[30] The composite parameter K_i^* may be written as,

$$K_i^* = K_{ref}K_i^{**} \tag{40}$$

where K_{ref} is a single affinity constant and K_i^{**} is a (composite) parameter characteristic for species i. Substitution of Equation 40 in Equation 39 results in

$$\theta_i(X_1,\ldots,X_n) = \left(\frac{K_i^{**}X_i}{X^*}\right)\left(\frac{K_{ref}X^*}{1 + K_{ref}X^*}\right) \tag{41}$$

with

$$X^* = \sum_{j=1}^{j=n} K_j^{**}X_j \tag{42}$$

When it is assumed for a heterogeneous surface that only log K_{ref} is characterized by a distribution function, the multiple integral Equation 31 reduces to a single integral with Equation 41 as local isotherm. The integral adsorption equation can then be written as

$$\theta_t(X_1,\ldots,X_n) = \frac{K_i^{**}X_i}{X^*} \int_\Delta \frac{K_{ref}X^*}{1 + K_{ref}X^*} f(\log K_{ref}) \, d\log K_{ref} \tag{43}$$

The integral in Equation 43 has the same form as that for the monocomponent case, thus both solutions are mathematically equivalent. The result for the three types of distribution functions corresponding to Equations 33 through 35 are, respectively:

$$\theta_{t,i}(X_1,\ldots,X_n) = \frac{\tilde{K}_i^*X_i}{\left\{1 + \left(\sum_{j=1}^{j=n} \tilde{K}_j^*X_j\right)^m\right\}\left(\sum_{j=1}^{j=n} \tilde{K}_j^*X_j\right)^{1-m}} \tag{44}$$

$$\theta_{t,i}(X_1,\ldots,X_n) = \frac{\tilde{K}_i^*X_i}{\left(1 + \sum_{j=1}^{j=n} \tilde{K}_j^*X_j\right)^m\left(\sum_{j=1}^{j=n} \tilde{K}_j^*X_j\right)^{1-m}} \tag{45}$$

$$\theta_{t,i}(X_1,\ldots,X_n) = \frac{\tilde{K}_i^*X_i}{\left\{1 + \left(\sum_{j=1}^{i=n} \tilde{K}_j^*X_j\right)^m\right\}^{1/m}} \tag{46}$$

with

$$\tilde{K}_i^* = \tilde{K}_{ref}K_i^{**} \tag{47}$$

The equations may be used in combination with a chosen double layer model in which case $\tilde{K}_i^*$ is a weighted average intrinsic (composite) affinity constant, and X_i represents a (combination of) activitie(s) near the adsorption sites. An alternative is to use activities in solution for X_i, in that case $\tilde{K}_i^*$ is a weighted average (composite) constant and the effect of electrostatics is incorporated in the distribution function. Also the ion exchange model as used by Kinniburgh et al.[59] can be extended to the competitive situation.

It is also possible to assume that the distribution functions for the different species are all of the same type, e.g., Gaussian, but with different positions and different widths. Also in this case the multiple integral can be simplified, but there is no analytical solution to the integral equation. An example of a multicomponent adsorption analysis based on Gaussian affinity distributions is given by Müller et al.[86] The advantage of this treatment is that the individual affinity distributions for the different components may be characterized by a different width, although they all must be Gaussian.

6.4 Numerical Inversion of the Integral Binding Equation

When the correct local isotherm equation is used and when no assumptions are made with respect to the functionality of the energy distribution function, several numerical methods for the inversion of the integral equation are available. These numerical methods for inversion of Equation 28, 29, or 30 stem from the gas adsorption literature and are especially useful if binding data are available over the entire domain of the binding curve. Reviews of these methods have been presented by Jaroniec et al.[20,21] In the early methods (e.g., References 87 through 89) the instability problems encountered with inversion of the integral equation are not treated explicitly, but the so-called regularization methods[82,90-92] and the singular value decomposition method known as CAESAR[93,94] were especially designed to be able to handle both the instability and the experimental error problem. When an appropriate data set is available the regularization methods and CAESAR give good results. Results obtained with CAESAR and the semi-analytical LOGA method (see Section 6.5) for a synthetic data set show that the resolution achieved with CAESAR is better than that with the semi-analytical methods.[24]

So far none of these sophisticated methods has been applied to ion binding data on natural organic particles. This is probably due to the fact that for ion binding it is difficult to obtain a wide range of coverages. If only a "window" of data is available instabilities may easily occur at the outer ends of the distribution function, leading to complications at the interpretation.

6.5 Analytical Inversion of the Integral Equation

A whole series of methods are known which, after some approximations, result in an analytical inversion of Equation 30. Although the mathematical derivations of the various methods are different, the final expressions for the distribution function are quite similar and can often be derived with the local isotherm approximation (LIA) method.[16,17] The essence of the LIA method is that the local isotherm function, i.e., the kernel of the integral equation as given by Equation 9, is replaced by an approximation, which allows an analytical solution for f(logK). The better the LIA function approximates the local binding function, the closer resemblance to the true calculated distribution function. The advantage of these methods is that, just as with the methods discussed in Section 6.4, no *a priori* assumptions have to be made with respect to the distribution function.

Below the various approaches will be discussed in relation to the LIA method. The starting point for the analysis is the integral equation as given by Equations 28, 29, or 30. For sake of simplicity we will omit the subscripts "s", "M", and "H" and the superscripts "int" and "app" in K_H^{app}, K_H^{int}, K_M^{app}, K_M^{int}, H_s, H_s and use the integral equation in the following form:

$$\theta_t([x]) = \int_\Delta \theta(K, [x]) \, f(\log K) \, d\log(K) \tag{48}$$

where [x] denotes the activity and K the affinity.

6.5.1 LIA Methods

A classical and very illustrative LIA method to solve Equation 48 for f(logK) is the condensation approximation (CA).[95,96] In the CA method the local isotherm function is replaced by a step function. This is equivalent with assuming that the heterogeneous binding sites are occupied sequentially, i.e., the sites with the highest affinity first, then those with the second highest affinity, etc. In reality, at a given concentration, all site types have some coverage, the highest affinity sites a relatively high coverage, the lowest affinity sites a low coverage. Nevertheless the CA is a useful simplification which plays a central role in the analytical heterogeneity analysis. The position of the step function replacing the local binding function is dependent on the local binding function and determined by a best fit criterion. In the case of a Langmuir type local isotherm (Equation 7, 9, or 26) the best fit is obtained when the step function intersects the Langmuir isotherm at $\theta = 0.5$. This is illustrated in Figure 3a where the Langmuir isotherm is shown together with the CA local isotherm.

According to the Langmuir local isotherm at $\theta = 0.5$, $K[x] = 1$, so that in the CA the affinity, K_{CA} can be replaced by 1/[x] and the approximation of the local binding function becomes:

$$\theta_{CA} = 0 \quad \text{for} \quad [x] < 1/K_{CA} \tag{49a}$$

$$\theta_{CA} = 1 \quad \text{for} \quad [x] \geq 1/K_{CA} \tag{49b}$$

where θ_{CA} is the approximation of the local binding function. By substitution of Equation 49 into the integral equation one obtains

$$\theta_t(\log[x^*]) = \int_{\log K^*}^{\infty} 1 \cdot f(\log K) \, d\log K \tag{50}$$

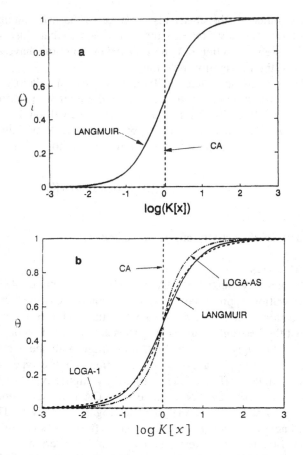

Figure 3. Comparison of several local isotherm approximations with the true local binding function (the Langmuir isotherm). (a) Approximation used in the CA method. (b) Approximations used in the LOGA-1 and LOGA-AS method.

where $x^* = 1/K^*$ is any value of $[x]$. Inversion of the integral is simply done by taking the first derivative of θ_t with respect to $\log[x]$:

$$f_{CA}(\log K_{CA}) = \frac{d\theta_t[x]}{d\log[x]} \tag{51a}$$

$$\log K_{CA} = -\log[x] \tag{51b}$$

Although the CA is a very crude approximation of the Langmuir isotherm the resulting distribution approaches the true distribution well when the true distribution is smooth and wide. For a narrow true distribution the CA, results in a flattened and too wide distribution. An example of the results which can be obtained for highly accurate data is shown in Figure 4a. The true distribution is bi-Gaussian. Figure 4a clearly demonstrates that the CA method has a low resolution power, it can not really discriminate the two peaks. The CA method can thus be used to obtain a first impression on the heterogeneity of the system only, unless the distribution is very wide. An advantage is that only a first derivative of the binding function is required and this makes the method relatively stable towards experimental errors.

Other LIA methods which use a better approximation of the local isotherm than the CA are the "asymptotically correct approximation" (ACA) of the local isotherm suggested by Cerofolini[97,98] and the LOGA method of Nederlof et al.[16] In the ACA method the Langmuir isotherm is substituted by an isotherm with its slope equal to the initial slope of the Langmuir equation. The ACA method leads to an expression for the distribution function which is a combination of the first and second derivative of the binding function. Although the ACA method gives a better resolution than the CA method, it has the disadvantage that the peaks in the distribution function are displaced. In an attempt to improve this behavior Nederlof et al.[16] have suggested a modification of ACA which does not displace the distribution.

A rather close approximation of the local (Langmuir) isotherm is obtained with the so called LOGA isotherm[16] which, in its general form, can be written as

$$\theta_{LOGA} = \alpha(K[x])^{\beta} \qquad \text{for} \quad [x] < 1/K_{LOGA} \qquad (52a)$$

$$\theta_{LOGA} = 1 - \alpha(K[x])^{-\beta} \qquad \text{for} \quad [x] \geq 1/K_{LOGA} \qquad (52b)$$

where α and β can be optimized to give a close fit to the Langmuir equation. For $\alpha = 0.5$ a continuous approximation of the Langmuir isotherm is obtained and we will restrict ourselves to this case. The LOGA isotherm with $\alpha = 0.5$ and β as an optimization parameter leads to the following expression for the distribution function

$$f_{LOGA}(logK_{LOGA}) = \frac{d\theta_t[x]}{dlog[x]} - \frac{0.189}{\beta^2}\frac{d^3\theta_t[x]}{dlog[x]^3} \qquad (53a)$$

$$logK_{LOGA} = -log[x] \qquad (53b)$$

The value of β is determined by a best fit criterion. The closest fit with the Langmuir local isotherm is obtained for $\beta = 0.7$. The resulting local isotherm approximation is indicated in Figure 3b by LOGA-1. Similarly as $f_{CA}(log\ K)$ also $f_{LOGA}(log\ K)$ is an approximation of the distribution function. For relatively smooth and wide distribution functions the LOGA method leads to very

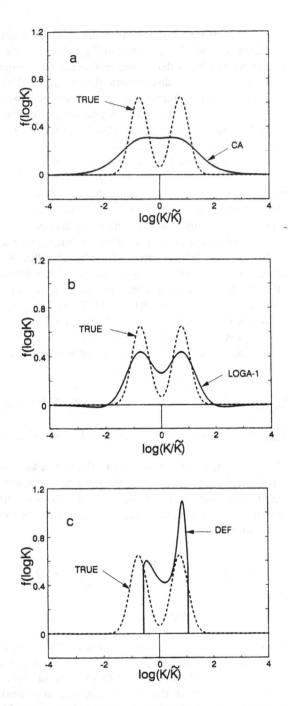

Figure 4. Comparison of the calculated distribution function with the true distribution function for the case of a bi-Gaussian distribution with a difference between the two peak positions of 1.5 logK units and each peak with a width of one logK unit. (a) The CA method. (b) The LOGA-1 method. (c) The DEF-method.

good results. For distribution functions consisting of several narrow peaks which are separated by less than 1.5 pK units the peaks cannot be distinguished.

Figure 4b gives an impression of the best possible results which can be obtained for the bi-Gaussian distribution. Such good results can only be obtained when the experimental error is negligible, since a third derivative of the binding function is required. In practice this is almost never the case and a constrained smoothing of the binding data within the error limits is therefore necessary.[17,24,99-102] Figure 2c can serve as an example of the results which can be achieved in practice. This figure shows the intrinsic proton distribution function of a fulvic acid, calculated with the LOGA method using the master curve data of Figure 2b (see figure caption of Figure 2 for more details). To suppress the effect of experimental errors spline smoothing was used in combination with cross validation.[102] The critical point in the smoothing technique is the determination of the degree of smoothing that is required. The degree of smoothing should be linked to the degree of uncertainty in the data and the latter can be established by cross validation in combination with some physical constraints. The larger the experimental error, the more smoothing is required and less detail can be recovered from the distribution function. Hence, in general two factors limit the "resolution" that can be obtained with the LIA methods: (1) the error inherent to the approximation of the local isotherm and (2) the presence of experimental error in the data.

In relation to the expression for the LOGA distribution function the work of Rudzinski and Jagiello[103,104] should also be mentioned. Hsu et al.,[103] already in 1978, derived an expression for the distribution function, which is equivalent to Equation 53 with $\beta = 0.79$. Their work is based on a series expansion of the distribution function. The results obtained with the RJ method are very close to those obtained with the LOGA-1 method.

6.5.2 Affinity Spectrum

Another series of methods which result in similar expressions as the LIA methods are the affinity spectrum (AS) methods. The AS methods are well known through the work of Hunston,[12] Thakur et al.,[105] and Shuman et al.[64] The basis of the AS methods goes back to the work on the viscoelastic behavior of materials by Nolle,[107] Schwarzl and Staverman[108,109], and Ninomiya and Ferry.[110] Schwarzl and Staverman showed that the equivalent of a distribution function could be obtained by a combination of first and higher order derivatives of the overall elasticity function. Ninomiya and Ferry[110] suggested a specific calculation method to avoid the calculation of these derivatives.

Hunston[12] made the work of Ninomiya and Ferry available for the study of ion binding and introduced the term "affinity spectrum". The first order affinity spectrum (AS_1) is equivalent to the CA method; the difference is that the first derivative of the binding function is replaced by a numerical approximation as suggested by Ninomiya and Ferry.[110] The second-order affinity spectrum (AS_2) has become well known in the following form:[12,64]

$$f_{AS2}(\log K_{AS2} = -\log[x]) = \frac{\theta_t([x]a) - \theta_t([x]/a)}{2\log a} -$$

$$\frac{a}{(a-1)^2} \frac{(\theta_t([x] a^2) - \theta_t([x]/a^2)) - 2(\theta_t([x]a) - \theta_t([x]/a))}{2\log a} \quad (54)$$

where a is the distance between two data points. It is assumed with Equation 54 that the data are equidistantially spaced. Since the expression of the second order affinity spectrum is a combination of the numerical approximations of the first and the third derivative of the binding function, it can be interpreted as a LOGA distribution function (see Equation 53). The corresponding local isotherm is shown in Figure 3, indicated as LOGA-AS. The LOGA-AS curve represents Equation 52 with $\alpha = 0.5$ and $\beta = 1$. The LOGA-AS approximation is a much better approximation than the CA function and only slightly less good than the LOGA-1 isotherm. Therefore, with the AS_2 method it is in principle possible to recover considerably more detail in the affinity distribution than with the CA method, but slightly less than with the LOGA-1 method.

Similarly as with the LOGA method, it is difficult to determine a reliable estimate of a third derivative. The parameter log a in the numerical approximations of the derivatives for the calculation of the distribution function as given by Equation 54 acts as a primitive smoothing parameter and should be in the range 0.2 to 0.4.[110] Despite this smoothing effect the AS_2 method often leads to spurious peaks or artefacts, even when an ordinary extra pre-smoothing of the data has been applied.[23,52,105] These problems have brought some discredit to the affinity spectrum method. As expressed in the discussion of the LOGA method, the error problems can be solved by applying sophisticated smoothing techniques to the experimental data.

6.5.3 DEF Method

Another well-known semi-analytical method to approximate the distribution function is the differential equilibrium function (DEF) method. This method goes back to the work of Simms;[7] it has been developed by Gamble and co-workers[10,11,111,112] and was extended by Buffle and co-workers.[13,14,113,114,116] The method starts by defining an affinity function $\bar{K}$, which is called the weighted average equilibrium function.[10,13] $\bar{K}$ is a function of the degree of binding. For binding of x this function may be written as

$$\bar{K} = \frac{\theta_t}{[x](1 - \theta_t)} \quad (55)$$

In case of a homogeneous system of non-interacting ligands $\overline{K}$ equals the intrinsic affinity constant. For a heterogeneous system a plot of log $\overline{K}$ vs θ_t (or log[x]) gives for each value of θ_t an "equivalent" binding strength between the species and a hypothetical site. According to Gamble[10] $\overline{K}$ gives a poor indication of the underlying heterogeneity, Gamble therefore introduced the differential equilibrium function, K_{DEF}, which was intended to be more closely related to the true individual affinity constants of the system. Moreover K_{DEF} can be determined without knowledge of the total number of sites of the system, which is required for application of Equation 55. The function K_{DEF} is defined by Gamble as

$$K_{DEF} = \frac{d\{\overline{K}(1 - \theta_t)\}}{d(1 - \theta_t)} \qquad (56)$$

The combination of Equations 55 and 56 yields:

$$K_{DEF} = -\frac{d(\theta_t/[x])}{d\theta_t} \qquad (57)$$

Instead of θ_t the experimentally measured concentration of bound ligand can also be used in Equation 57. Hence, K_{DEF} can be derived directly from the experiment.

For the expression of the distribution function of $\log K_{DEF}$ Buffle and co-workers[4,114] simply propose to use the first derivative of the binding function with respect to log K_{DEF}:

$$f_{DEF}(\log K_{DEF}) = -\frac{d\theta_t}{d\log K_{DEF}} \qquad (58)$$

Hence, $f_{DEF}(\log K_{DEF})$ can be found by plotting θ_t vs. $\log K_{DEF}$ and taking the derivative. In Figure 4c the DEF result obtained for the bi-Gaussian distribution function is shown. Although the two peaks are recovered, the peak positions of the weak sites are somewhat shifted to higher affinities. Moreover, the proportion of high affinity sites are amplified compared to reality. Although all methods discussed in Section 6.5 only give an approximation of the true distribution function, the DEF distribution is somewhat more distorted than the LOGA or AS_2 distribution are, because the latter methods do not displace the peak positions in a spectrum, nor do they give different weights to the peaks. An advantage of the DEF method is that the DEF distribution "amplifies" and sharpens the peak corresponding to minor amounts of high affinity sites, which may be of particular importance in practice. With the CA method, which strongly smoothes the distribution function, such sites cannot be found. In this respect combined use of the DEF method with the CA, the AS, or the LOGA method can be helpful. Further illustrations and discussion of DEF distributions can be found in the original literature (e.g.,

Reference 14, 17, 106, and 114). Some distributions obtained for copper binding by several heterogeneous complexants have been calculated by Buffle et al.[14]. Note that although $f_{DEF}(\log K_{DEF})$ is the first derivative of the binding function with respect to $\log K_{DEF}$, it should be realized the K_{DEF} itself is also a derivative, see Equation 57. Therefore, similarly as for the other methods, sophisticated mathematical smoothing techniques are a prerequisite before one can apply the DEF procedure to experimental data, otherwise spurious peaks occur.

7. CONCLUDING REMARKS

The progress made with respect to the estimation of intrinsic proton affinity distributions for humic and fulvic acids is an important step forward in a better characterization of the ion binding to these materials. Future work related to proton binding should be concerned with handling experimental data rather than with new methods for the determination of affinity distributions.

Interpretation of metal ion binding curves in terms of an intrinsic affinity distribution is an extremely complex problem. The various aspects that cause this complexity have been illustrated in this paper. Rigorous solutions to the problem demand very large data sets. For the moment *ad hoc* approximations have to be made for the analysis of metal ion binding data in order to obtain apparent affinity distribution functions. To obtain these functions the methods discussed in Section 6 can be used. Apparent affinity distribution functions, determined under various solution conditions, will be of help for the understanding of the metal ion binding behavior. However, at the moment it is not yet possible to give simple rules for the way to handle metal binding data.

ACKNOWLEDGMENTS

This work was partially funded by the Netherlands Integrated Soil Research Programme under Contract Number PCBB 8948 and partially by the EC Environmental Research Programme on Soil Quality under Contract Number EV4V-0100-NL(GDF). Thanks are due to Mr. M. M. Nederlof for his help and critical comments and to Ms. A. M. de Weerd for typing the manuscript.

REFERENCES

1. Gast, R. G. "Surface and Colloid Chemistry", in *Minerals in Soil Environments*, J. B. Dixon and S. B. Weed, Eds. (Madison, WI, Soil Sci. Soc of Am., 1977) pp. 27–73.

2. Schwertmann, U. "Occurence of Formation of Iron Oxides in Various Pedo-Environments", in *Iron in Soils and Clay Minerals*, J. W. Stucki, B. A. Goodman, and U. Schwertmann, Eds. (Dordrecht: Reidel Publishing, NATO ASI Series Vol. 217, 1988) pp. 267–308.

3. Murphy, P. J., A. M. Posner, and J. P. Quirk. "Characterization of Partially Neutralized Ferric Nitrate solutions," *J. Colloid Interface Sci.* 56:270–283 (1976).

4. Buffle, J. *Complexation Reactions in Aquatic Systems, An Analytical Approach.* Chichester: Ellis Horwood, 1988).

5. Aiken, G. R. and A. H. Gilham. "Determination of Molecular Weights of Humic Substances by Colligative Property Measurements," in *Humic Substances II. In Search of Structure,* M. H. B. Hayes, P. MacCarthy, R. L. Malcolm, and R. S. Swift, Eds. (New York: Wiley & Sons, 1989), chap. 18.

6. Tipping, E. and D. Cooke. "The effects of adsorbed humic substances on the surface charge of goethite (α-FeOOH) in freshwaters," *Geochim. Cosmochim. Acta* 46:75–80 (1982).

7. Simms, H. S. "Dissociation of Polyvalent Substances. I. Relation of Constants to Titration Data," *J. Am. Chem. Soc.* 48:1239–1250 (1926).

8. Scatchard G., I. H. Scheinberg, and S. H. Armstrong, Jr. "Physical Chemistry of Protein Solutions. IV. The Combination of Human Serum Albumin with Chloride Ion," *J. Am. Chem. Soc.* 72:535–540 (1950).

9. Klotz, I. M. and D. L. Hunston. "Properties of Graphical Representations of Multiple Classes of Binding Sites," *Biochemistry* 10:3065–3069 (1971).

10. Gamble, D. S. "Titration Curves of Fulvic Acid: The Analytical Chemistry of a Weak Acid Polyelectrolyte," *Can. J. Chem.* 48:2662–2669 (1970).

11. Gamble, D. S., A. W. Underdown, and C. H. Langford. "Copper (II) Titration of Fulvic Acid Ligand Sites with Theoretical, Potentiometric, and Spectrophotometric Analysis," *Anal. Chem.* 52:1901–1908 (1980).

12. Hunston, D. L. "Two Techniques for Evaluating Small Molecule-Macromolecule Binding in Complex Systems," *Anal. Biochem.* 63:99–109 (1975).

13. Buffle, J. "Natural Organic Matter and Metal-Organic Matter Interactions in Aquatic Systems," in *Circulation of Metals in the Environment,* H. Sigel, Ed. (New York: Marcel Dekker, 1984), pp. 165–221.

14. Buffle, J., R. S. Altmann, M. Filella, and A. Tessier. "Complexation by Natural Heterogeneous Compounds: Site Occupation Distribution Functions, a Normalized Description of Metal Complexation," *Geochim. Cosmochim. Acta* 54:1535–1553 (1990).

15. Nederlof, M. M., W. H. van Riemsdijk, and L. K. Koopal. "Methods to Determine Affinity Distributions for Metal Ion Binding in Heterogeneous Systems," in *Heavy metals in the Hydrological Cycle,* M. Astruc and J. N. Lester, Eds. (London: Selper Ltd, 1988), pp. 361–368.

16. Nederlof, M. M., W. H. Van Riemsdijk, and L. K. Koopal. "Determination of Adsorption Affinity Distributions: A General Framework for Methods Related to Local Isotherm Approximations," *J. Colloid Interface Sci.* 135:410–426 (1990).

17. Nederlof, M. M., W. H. van Riemsdijk, and L. K. Koopal. "Analysis of the Binding Heterogeneity of Natural Ligands using Adsorption Data," J.-P. Vernet, Ed. (Amsterdam: Elsevier Science Publishers, 1991) pp. 365–396.

18. House, W. A. "Adsorption on Heterogeneous Surfaces," in *Colloid Science Specialist Periodical Report 4,* D. H. Everett, Ed. (London: Royal Soc. Chem., 1983) pp. 1–58.

19. Jaroniec, M. "Physical Adsorption on Heterogeneous Solids," *Adv. Colloid Interface Sci.* 18:149–225 (1983).

20. Jaroniec, M. and P. Bräuer. "Recent Progress in Determination of Energetic Heterogeneity of Solids from Adsorption Data," *Surf. Sci. Rep.* 6:65–117 (1986).

21. Jaroniec, M. and R. Madey, "Physical Adsorption on Heterogeneous Solids," (New York: Elsevier, 1988).
22. Turner, D. R., M. S. Varney, M. Whitfield, R. F. C. Mantoura, and J. P. Riley, "Electrochemical Studies of Copper and Lead Complexation by Fulvic Acid. I. Potentiometric Measurements and a Critical Comparison of Metal Binding Models," *Geochim. Cosmochim. Acta* 50:289–297 (1986).
23. Fish, W., D. A. Dzombak, and F. M. M. Morel. "Metal-Humate Interactions. II. Application and Comparison of Models" *Environ. Sci. Tech.* 20:676–683 (1986).
24. Koopal, L. K., M. M. Nederlof, and W. H. van Riemsdijk. "Determination of the Adsorption Energy Distribution Function with the LOGA Method," *Prog. Colloid Polymer Sci.* 82:19–27 (1990).
25. Bolt, G. H. "Theories of Cation Adsorption by Soil Constituents: Distribution Equilibrium in Electrostatic Fields," in: *Soil Chemistry, B. Physico-Chemical Models*, G. H. Bolt, Ed. (Amsterdam: Elsevier, 1979) p. 47–75.
26. Sposito, G. *The Thermodynamics of Soil Solutions.* (Oxford: Clarendon Press, 1981) p. 223.
27. Secor, R. B. and C. J. Radke. "Spillover of the Diffuse Double Layer on Montmorillonite Particles," *J. Colloid Interface Sci.* 103:237–244 (1985).
28. James, A. E. and D. J. A. Williams. "Numerical Solution of the Poisson-Boltzmann Equation," *J. Colloid Interface Sci.* 107:44–59 (1985).
29. Nederlof, M. M., P. Venema, W. H. Van Riemsdijk, and L. K. Koopal. "Modelling Variable Charge Behaviour of Clay Minerals" in *Proc. Euroclay '91.* (Dresden) M. Störr, K. H. Henning, and P. Adolphi, Eds. (Greifswald, Germany: Ernst-Moritz-Arndt Universität, 1991) p. 795–800.
30. Van Riemsdijk, W. H., J. C. M. De Wit, L. K. Koopal, and G. H. Bolt. "Metal Ion Adsorption on Heterogeneous Surfaces: Adsorption Models," *J. Colloid Interface Sci.* 116:511–522 (1987a).
31. Atkinson, R. J., A. M. Posner, and J. P. Quirk. "Adsorption of Potential Determining Ions at the Ferric Oxide-Aqueous Electrolyte Interface" *J. Phys. Chem.* 71:550–558 (1967).
32. Stumm, W., P. C. Huang, and S. R. Jenkins. "Specific Chemical Interaction Affecting the Stability of Dispersed Systems." *Croat. Chem. Acta* 42:223–245 (1990).
33. Schindler, P. W. and H. Gamsjäger. "Acid-Base Reactions of the TiO_2 (Anatase)-Water Interface and the Point of Zero Charge of TiO_2 Suspensions" *Kolloid Z. Z. Polym.* 250:759–763 (1972).
34. Bowden, J. W., M. D. A. Bolland, A. M. Posner, and J. P. Quirk. "Generalized Model for Anion and Cation Adsorption at Oxide Surfaces," *Nature Phys. Sci.* 245:81–83 (1973).
35. Yates, D. E., S. Levine and T. W. Healy. "Site-Binding Model of the Electrical Double Layer at the Oxide/Water Interface," *J. Chem. Soc. Faraday Trans. I.* 70:1807–1808 (1974).
36. Davis. J. A., R. O. James, and J. O. Leckie. "Surface Ionization and Complexation at the Oxide/Water Interface. I. Computation of Electrical Double Layer Properties in Simple Electrolytes," *J. Colloid Interface Sci.* 63:480–499 (1978).
37. Westall, J. and H. Hohl. "A Comparison of Electrostatic Models for the Oxide/Solution Interface," *Adv. Colloid Interface Sci.* 12:265–294 (1980).

38. Barrow, N. J., J. W. Bowden, A. M. Posner, and J. P. Quirk. "An Objective Method for Fitting Models of Ion Adsorption on Variable Charge Surfaces" *Aust. J. Soil Res.* 18:37–47 (1980).

39. van Riemsdijk, W. H., G. H. Bolt, L. K. Koopal, and J. Blaakmeer. "Electrolyte Adsorption on Heterogeneous Surfaces: Adsorption Models," *J. Colloid Interface Sci.* 109:219–228 (1986).

40. Hiemstra, T., W. H. van Riemsdijk, and M. G. M. Bruggenwert. "Proton Adsorption Mechanism at the Gibbsite and Aluminium Oxide Solid/Solution Interface," *Neth. J. Agric. Sci.* 35:281–294 (1987).

41. Dzombak, D. A. and F. M. M. Morel. *Surface Complexation Modeling. Hydrous Ferric Oxide.* New York: John Wiley & Sons, 1990), p. 393.

42. James, R. O. and G. A. Parks. "Characterization of Aqueous Colloids by their Electrical Double Layer and Intrinsic Surface Chemical Properties," in *Surface and Colloid Science,* Vol. 12, E. Matijevic, Ed., 119–216 (1982).

43. Koopal, L. K., W. H. van Riemsdijk, and M. G. Roffey. "Surface Ionization and Complexation Models: A Comparison of Methods for Determining Model Parameters," *J. Colloid Interface Sci.* 118:117–136 (1987).

44. Hiemstra, T., W. H. van Riemsdijk, and G. H. Bolt. "Multi-Site Proton Adsorption Modeling at the Solid/Solution interface of (hydr)oxides: A new approach. I. Model Description and Evaluation of intrinsic reaction constants" *J. Colloid Interface Sci.* 133:91–104 (1989).

45. Hiemstra, T., W. H. van Riemsdijk, and J. C. M. De Wit. "Multi-Site Proton Adsorption Modeling at the Solid/Solution Interface of (Hydr)oxides: A New Approach. II. Application to Various Important (hydr)oxides" *J. Colloid Interface Sci.* 133:105–117 (1989).

46. Hiemstra, T. and W. H. van Riemsdijk, "Physical Chemical Interpretation of Primary Charging Behavior of Metal(hydr)oxides" *Colloids Surf.* 59:7–25 (1991).

47. Gibb, A. W. and L. K. Koopal. "Electrochemistry of a Model for Patchwise Heterogeneous Surfaces; the Rutile-Hematite System," *J. Colloid Interface Sci.* 134:122–138 (1990).

48. Tanford, C. *Physical Chemistry of Macromolecules.* (New York: John Wiley & Sons, 1961), p. 170).

49. Manning, G. S. "Limiting Laws and Counter Ion Condensation in Polyelectrolyte Solutions," *J. Chem. Phys.* 51:924–938 (1969).

50. Oosawa, F. *Polyelectrolytes.* (New York: Marcel Dekker, 1971), p. 159.

51. Katchalski, A. "Polyelectrolytes," *Pure Appl. Chem.* 26:327–373 (1971).

52. Marinsky, J. A. and J. Ephraim. "A Unified Physicochemical Description of the Protonation and Metal Ion Complexation Equilibria of Natural Organic Acids (Humic and Fulvic Acids). I. Analysis of the Influence of Polyelectrolyte Properties on Protonation Equilibria and Ionic Media: Fundamental Concepts," *Environ. Sci. Technol.* 20:349–353 (1986).

53. Marinsky, J. A. "A Two-Phase Model for the Interpretation of Proton and Metal Ion Interaction with Charged Polyelectrolyte Gels and their Linear Analogs," in *Aquatic Surface Chemistry,* W. Stumm, Ed. (John Wiley & Sons, New York, 1987), pp. 49–81.

54. Westall, J. C. "Adsorption Mechanisms in Aquatic Surface Chemistry," in *Aquatic Surface Chemistry,* W. Stumm, E. (New York: John Wiley & Sons, 1987), pp. 3–32.

55. Healy, T. W. and L. R. White, "Ionizable Surface Group Models of Aqueous Interfaces," *Adv. Colloid Interface Sci.* 9:303–345 (1978).

56. Hiemenz, P. C., *"Principles of Colloid and Surface Chemistry,"* 2nd Ed. (New York: Marcel Dekker, 1986)

57. van Riemsdijk, W. H., L. K. Koopal, and J. C. M. De Wit. "Heterogeneity and Electrolyte Adsorption: Intrinsic and Electrostatic Effects" *Neth. J. Agric. Sci.* 35:241–257 (1987).

58. van Riemsdijk, W. H., G. H. Bolt, and L. K. Koopal. "The Electrified Interface of the Soil Solid Phase. B. Effect of Surface Heterogeneity," in *Interactions at the Soil Colloid — Soil Solution Interface,* G. H. Bolt, M. F. de Boodt, M. H. B. Hayes, and M. B. McBride, Eds. (Dordrecht: Kluwer Academic Publishers, 1991), pp. 81–114.

59. Kinniburgh, D. G., J. A. Baker, and M. Whitfield. "A Comparison of Some Simple Adsorption isotherms for describing divalent cation adsorption by ferrihydrite," *J. Colloid Interface Sci.* 95:370–385 (1983).

60. Bowden, J. W., S. Nagarajah, N. J. Barrow, A. M. Posner, and J. P. Quirk. "Describing the Adsorption of Phosphate, Citrate, and Selenite on a Variable Charge Mineral Surface," *Aust. J. Soil Res.* 18:49–60 (1980).

61. Barrow, N. J., "Reactions with Variable-Charge Soils," (Dordrecht: Kluwer Academic Publishers, 1987), ch. 3.

62. Fokkink, L. G. J., A. De Keizer, and J. Lyklema. "Specific Ion Adsorption on Oxides: Surface Charge Adjustment and Proton Stoichiometry," *J. Colloid Interface Sci.* 118:454–462 (1987).

63. Koopal, L. K. and W. H. van Riemsdijk. "Electrosorption on Random and Patchwise Heterogeneous Surfaces: Electrical Double-Layer Effects," *J. Colloid Interface Sci.* 128:188–200 (1989).

64. Shuman, M. S., B. J. Collins, P. J. Fitzgerald, and D. L. Olson. "Distribution of Stability Constants and Dissociation Rates among Binding Sites on Estuarine Copper-Organic Complexes: Rotated Disk Electrode Studies and Affinity Spectrum Analysis of Ion-Selective Electrode and Photometric Data," in *Aquatic and Terrestrial Humic Materials,* R. F. Christman and E. T. Gjessing, Eds. (Ann Arbor, MI: Ann Arbor Science Publishers, 1983) pp. 349–370.

65. Perdue, E. M. and C. R. Lytle. "A Critical Examination of Metal-Ligand Complexation Models: Application to Defined Multiligand Mixtures," in *Aquatic and Terrestrial Humic Materials,* R. F. Christman and E. T. Gjessing, Eds. (Ann Arbor, MI: Ann Arbor Science Publishers, 1983), pp. 259–313.

66. Sposito, G. "Sorption of Trace Metals by Humic Materials in Soils and Natural Waters," *CRC Crit. Rev. Environ. Control* 16:193–229 (1986).

67. Sips, R. "On the Structure of a Catalyst Surface," *J. Chem. Phys.* 16:490–495 (1948).

68. Sips, R. "On the Structure of a Catalyst Surface." II, *J. Chem. Phys.* 18:1024–1026 (1950).

69. De Wit, J. C. M., W. H. van Riemsdijk, and L. K. Koopal, "Proton and Metal Ion Binding on Humic Substances", in *Metals Speciation, Separation and Recovery II,* J. W. Patterson and R. Passino, Eds. (Chelsea, MI: Lewis Publishers, 1990), pp. 329–357.

70. De Wit, J. C. M., W. H. Van Riemsdijk, M. M. Nederlof, D. G. Kinniburgh, and L. K. Koopal. "Analysis of Ion Binding on Humic Substances and the Determination of Intrinsic Affinity Distributions," *Anal. Chim. Acta* 232:189–207 (1990).

71. De Wit, J. C. M., M. M. Nederlof, W. H. van Riemsdijk, and L. K. Koopal, "Determination of Proton and Metal Ion Affinity Distributions for Humic Substances," *Water, Air Soil Pollut.* 57–58:339–349 (1991).

72. Fixman, M., "The Poisson-Boltzmann Equation and its Application to Polyelectrolytes," *J. Chem. Phys.* 70:4995–5005 (1979).

73. Loeb, A. L., J. T. G. Overbeek, and P. H. Wiersma. *The Electrical Double Layer around a Spherical Colloid Particle.* (Cambridge: MIT Press, 1961).

74. Stigter, D. "Functional Representation of Properties of the Electrical Double Layer around a Spherical Colloid Particle," *J. Electroanal. Chem.* 37:61–64 (1972).

75. Van Der Drift, W. P. J. T., A. De Keizer, and J. Th. G. Overbeek. "Electrophoretic Mobility of a Cylinder with High Surface Charge Density," *J. Colloid Interface Sci.* 71:67–78 (1979).

76. Tipping, E., C. A. Backes, and M. A. Hurley. "The Complexation of Protons and Aluminium and Calcium by Aquatic Humic Substances: A Model Incorporating Binding-Site Heterogeneity and Macro-Ionic Effects," *Water Res.* 22:597–611 (1988).

77. Tipping, E. and M. A. Hurley. "A Model of Solid-Solution Interactions in Acid Organic Soils, Based on the Complexation Properties of Humic Substances," *J. Soil Sci.* 39:505–519 (1988).

78. Tipping, E., M. M. Reddy, and M. A. Hurley. "HUMEQ, a Model Describing Electrostatic and Functional Group Heterogeneity Effects on Acid Dissociation of Humic Acids" *Abstr. 5th Int. Meeting IHSS Society,* Nagoya, Japan, (1990) p. 8.

79. Chen, Y. and M. Schnitzer. "Viscosity Measurements on Soil Humic Substances," *Soil Sci. Soc. Am. J.* 40:866–872 (1976).

80. Cameron, R. S., B. K. Thornton, R. S. Swiff, and A. M. Posner. "Molecular Weight and Shape of Humic Acid from Sedimentation and Diffusion Measurements on Fractionated Extracts" *J. Soil Sci.* 23:394–408 (1972).

81. Brassard, P., J. R. Kramer, and P. M. Collins. "Binding Site Analysis Using Linear Programming" *Environ. Sci. Technol.* 24:195–201.

82. Merz, P. H. "Determination of Adsorption Energy Distribution by Regularization and a Characterization of Certain Adsorption Isotherms," *J. Comput. Phys.* 38:64–85 (1980).

83. Papenhuijzen, J. and L. K. Koopal. "Adsorption from Solution onto Heterogeneous Surfaces. Evaluation of Two Numerical Methods to Determine the Adsorption Free Energy Distribution," in *Adsorption from Solution,* R. H. Ottewill, C. H. Rochester, and A. L. Smith, Eds. (London: Academic Press, 1983) p. 211–225.

84. Toth, J., W. Rudzinski, A. Waksmundzki, M. Jaroniec, and S. Sokolowski. "Adsorption of Gases on Heterogeneous Solid Surfaces: The Energy Distribution Function Corresponding to a New Equation for Monolayer Adsorption," *Acta Chim. Acad. Sci. Hung.* 82:11–21 (1974).

85. Dzomback, D. A., W. Fish, and F. M. M. Morel. "Metal-Humate Interactions. I. Discrete Ligand and Continuous Distribution Models," *Environ. Sci. Technol.* 20:676–683 (1986).

86. Müller, G., C. J. Radke, and J. M. Prausnitz. "Adsorption of Weak Organic Electrolytes form Dilute Aqueous Solution onto Activated Carbon. I. Single-Solute Systems," *J. Colloid Interface Sci.* 103:484–492 (1985).

87. Adamson, A. W. *Physical Chemistry of Surfaces.* (New York: Wiley-Interscience, 1976) pp. 607–615.

88. House, W. A. and M. J. Jaycock. "A Numerical Algorithm for the Determination of the Adsorptive Energy Distribution Function from Isotherm Data," *Colloid Polymer Sci.* 256:52–61 (1978).

89. Sacher, R. S. and I. D. Morrison. "An Improved CAEDMON Program for the Adsorption Isotherm of Heterogeneous Surfaces," *J. Colloid Interface Sci.* 70:153–166 (1979).

90. House, W. A. "Adsorption on a Random Configuration of Adsorptive Heterogeneities," *J. Colloid Interface Sci.* 67:166–180 (1978).

91. House, W. A. "Surface Heterogeneity Effects in Adsorption from Solution," *Chem. Phys. Lett.* 60:169–174 (1978).

92. Brown, L. F. and B. J. Travis. "Optimal Smoothing of Site Energy Distributions from Adsorption Isotherms," in *Fundamentals of Adsorption*, A. L. Myers and G. Belfort, Eds. (New York: Engineering Foundation, 1984) pp. 125–134.

93. Koopal, L. K. and C. H. Vos. "Calculation of the Adsorption Energy Distribution from the Adsorption Isotherm by Singular Value Decomposition," *Colloids Surf.* 14:87–95 (1985).

94. Vos, C. H. and L. K. Koopal. "Surface Heterogeneity Analysis by Gas Adsorption: Improved Calculation of the Adsorption Energy Distribution Using a New Algorithm Named CAESAR," *J. Colloid Interface Sci.* 105:183–196 (1985).

95. Roginsky, S. S. *Adsorption Catalysis on Heterogeneous Surfaces.* (Moscow: Academy of Sciences U.S.S.R., 1948).

96. Harris, L. B. "Adsorption on a Patchwise Heterogeneous Surface; Mathematical Analysis of the Step-Function Approximation to the Local Isotherm," *Surface Sci.* 10:129–145 (1968).

97. Cerofolini, G. F. "Adsorption and Surface Heterogeneity," *Surface Sci.* 24:391–403 (1971).

98. Cerofolini, G. F. "Localized Adsorption on Heterogeneous Surfaces," *Thin Solid Films* 23:129–152 (1974).

99. Reinsch, C. H. "Smoothing by Spline Functions," *Numer. Math.* 10:177–183 (1967).

100. Reinsch, C. H. "Smoothing by spline functions II", *Numer. Math.* 16:451–454 (1971).

101. Craven, P. and G. Wahba. "Smoothing Noisy Data with Spline Functions. Estimating the Correct Degree of Smoothing by the Method of Generalized Cross-Validation," *Numer. Math.* 31:377–403 (1979).

102. Woltring, H. J. "A Fortran Package for Generalized Cross-Validatory Spline Smoothing and Differentiation," *Adv. Eng. Software* 8:104–107 (1986).

103. Hsu, C. C., B. W. Wojciechowski, W. Rudzinski, and J. Narkiewicz. "An Improved Method for Evaluating Surface Heterogeneity for Various Models of Local Adsorption," *J. Colloid Interface Sci.* 67:292–303 (1978).

104. Rudzinski, W., J. Jagiello, and Y. Grillet. "Physical Adsorption of Gases on Heterogeneous Solid Surfaces: Evaluation of the Adsorption Energy Distribution from Adsorption Isotherms and Heats of Adsorption," *J. Colloid Interface Sci.* 87:478–491 (1982).

105. Thakur, A. K., P. J. Munson, D. L. Hunston, and D. Rodbard. "Characterization of Ligand-Binding Systems by Continuous Affinity Distributions of Arbitrary Shape," *Anal. Biochem.* 103:240–254 (1980).

106. Nederlof, M. M., W. H. van Riemsdijk, and L. K. Koopal. "Comparison of semi-analytical methods to analyze complexation with heterogeneous ligands," *Environ. Sci. Technol.*, accepted. (1992).

107. Nolle, A. W. "Dynamic Mechanical Properties of Rubberlike Materials," *J. Polym. Sci.* 5:1–54 (1950).

108. Schwarzl, F. and A. J. Staverman. "Higher Approximations of Relaxation Spectra," *Physica* 18:791–798 (1952).

109. Schwarzl, F. and A. J. Staverman. "Higher Approximation Methods for the Relaxation Spectrum from Static and Dynamic Measurements of Visco-Elastic Materials," *Appl. Sci. Res.* 4:127–141 (1953).

110. Ninomiya, K. and J. D. Ferry. "Some Approximate Equations Useful in the Phenomenological Treatment of Linear Viscoelastic Data," *J. Colloid Interface Sci.* 14:36–48 (1959).

111. Burch, R. D., C. H. Langford, and D. S. Gamble. "Methods for the Comparison of Fulvic Acid Samples: The Effects of Origin and Concentration on Acidic Properties," *Can. J. Chem.* 56:1196–1201 (1978).

112. Gamble, D. S. and C. H. Langford. "Complexing Equilibria in Mixed Ligand Systems: Tests of Theory with Computer Simulations," *Environ. Sci. Technol.* 22:1325:1336 (1988).

113. Buffle, J. and R. S. Altmann. "Interpretation of Metal Complexation by Heterogeneous Complexants," in *Aquatic Surface Chemistry: Chemical Processes at the Particle-Water Interface,* W. Stumm, Ed. (New York: John Wiley & Sons, 1987) pp. 351–383.

114. Altmann, R. S. and J. Buffle. "The Use of Differential Equilibrium Functions for Interpretation of Metal Binding in Complex Ligand Systems: Its Relation to Site Occupation and Site Affinity Distributions," *Geochim. Cosmochim. Acta* 52:1505–1519 (1988).

115. Dempsey, B. A. and C. R. O'Melia. "Proton and Calcium Complexation of Four Fulvic Acid Fractions" in *Aquatic and Terrestrial Humic Materials,* R. F. Christman and E. T. Gjessing, Eds. (Ann Arbor, MI: Ann Arbor Science Publishers, 1983) pp. 239–273.

116. Buffle, J., R. S. Altmann, and M. Filella. "Effect of Physico-Chemical Heterogeneity of Natural Complexants. III. Buffering Action and Role of the Background Sites," *Anal. Chim. Acta* 232:225–237 (1990).

CHAPTER 13

Dynamic Aspects of Metal Speciation in Aquatic Colloidal Systems

Herman P. van Leeuwen

Department of Physical and Colloid Chemistry, Wageningen Agricultural University, Wageningen, The Netherlands

TABLE OF CONTENTS

1. Introduction 498

2. Mass Transfer Controlled Adsorption on Dispersed
   Particles 500
   2.1 Diffusion 500
   2.2 Convective Diffusion 502

3. Rates of Chemical Reactions Involved 507

4. Experimental Methods and Results 510
   4.1 Batch Analysis (Abbreviation: B) 510
   4.2 Ligand Exchange (Abbreviation: L) 510

4.3   Crystal Flow Cell with *in situ* Electrochemical
       Detection (Abbreviation: FC)............................511
4.4   Column Method (Abbreviation: C) 511
4.5   Pressure Jump Method (Abbreviation: ΔP) 511
4.6   Conductance Stopped Flow (Abbreviation: ΔK).........512

5.   Evaluation and Conclusion......................................512
5.1   Practical Relaxation Times 512
5.2   Complicating Factors for Environmental Samples 515
       5.2.1   Nonlinear Adsorption 515
       5.2.2   Surface Heterogeneity............................515
       5.2.3   Polydispersity....................................516
       5.2.4   Particle Shape....................................516
5.3   Particle Characterization................................517
5.4   Mass Transport Effects in Studies of Adsorption
       Kinetics 517
5.5   Dynamic Speciation 517

Appendix: List of Symbols and Abbreviations 518

References..519

1. INTRODUCTION

In the practice of the environmental analytical chemistry of (heavy) metal ions, particles play an important role. Most natural samples, whether they are taken from atmosphere, from water, or from soil bodies, contain a variety of particles that may affect the analytical procedure in one way or another. The effects are basically derived from physical and chemical interaction between the particles and the metal ions in the sample. Knowledge of the behavior of metal ions in environmental samples has come to the level where detailed information on the speciation, i.e., the distribution over different species, is essential. Therefore the environmental analytical chemistry of metal ions is now largely concerned with a more or less detailed physical-chemical speciation. This includes — among other things — the distinction between particle-bound metal ions and other forms of metal ions.

Speciation data are implicitly referred to the time-scale of the analytical method, used to obtain them. Hence they are of a non-equilibrium nature if the time-scale of the analytical technique is not much larger than the relaxation times of the equilibration processes involved. Often, analytical information is needed for a specific purpose, that relates to a particular time-scale. For example, the bioavailability of metal ions in a flowing natural system may be limited by the characteristic hydrodynamic time constant of the system. Thus, e.g., the question whether particle-bound heavy metal ions in some river are of any importance to the metal uptake by fish in that river, certainly is a problem with — in principle — dynamic aspects.

Adsorption kinetics of metal ions in disperse systems are usually considered on a more or less empirical level. Sharp distinction between mass transport and chemical conversions is not generally made. Frequent use is made of multi-step kinetic models, the different steps being associated with metal binding sites with different accessibilities. A known alternative is to formulate the kinetics in terms of bimolecular chemical reaction rate equations (see, e.g., Reference 1).

From a physicochemical viewpoint, metal ion adsorption/desorption processes are considered to be composed of a number of basic steps. These are mostly

- Transport to or from the adsorbent interface
- Partial dehydration/hydration reactions
- Surface association/dissociation reactions

Scheme 1 collects these major steps in a scheme where — for simplicity — the metal ion is briefly denoted by M and charges are omitted:

$$
\begin{array}{ll}
M(H_2O)_p\text{,bulk} & \text{(hydrated metal ion in bulk of solution)} \\
\uparrow \downarrow \quad \text{mass transfer} & \\
M(H_2O)_p & \text{(hydrated metal ion at interface)} \\
\uparrow \downarrow \quad \text{(de)hydration} & \qquad\qquad (1) \\
M(H_2O)_{p-q} & \text{(partially dehydrated metal ion at interface)} \\
\uparrow \downarrow \quad \text{surface association/dissociation} & \\
M(H_2O)_{p-q}S & \text{(metal ion in adsorbed state)}
\end{array}
$$

Other steps may be operative and not all of the steps in Scheme 1 may be relevant for every metal ion adsorption process. Much of the scatter in literature on metal ion sorption kinetics seems to be connected with the uncertainties as to which step (or steps) in the scheme is (are) rate-determining. Often, it is not at all clear which is the determining step and yet data are discussed in terms of one step, or, more safely, in terms of an unspecified overall rate constant. We should further point out that Scheme 1 is quite idealistic in that it ignores such frequently occurring phenomena as pore diffusion, penetration of the adsorbate into the (solid) particle, site heterogeneity, etc. Although it will not be possible to disentangle the mass of literature data on overall ion binding rates, we shall try to pinpoint some of the relevant characteristics.

More generally, the intention of this report* is to review the dynamic aspects of ion-particle interaction by:

(i)   Collecting theoretical timing characteristics of adsorption/desorption processes in dispersions
(ii) Critically surveying the existing experimental data on relevant metal ion adsorption/desorption rates
(iii) Outlining the impact of dynamic factors on metal ion analysis procedures, as applied to aquatic samples with adsorbing colloidal constituents

2. MASS TRANSFER CONTROLLED ADSORPTION ON DISPERSED PARTICLES
2.1 Diffusion

The pure diffusion of metal ions M from a homogeneous solution towards a spherical stagnant particle may be described on the basis of the conservation laws (Fick's laws) for M:

$$J_M = -D_M \nabla c_M = -D_M \left(\frac{\partial c_M}{\partial r} \right) \tag{2}$$

$$\frac{\partial c_M}{\partial t} = D_M \left[\frac{\partial^2 c_M}{\partial r^2} + \frac{2}{r} \frac{\partial c_M}{\partial r} \right] \tag{3}$$

where J_M is the flux of metal ions per unit spherical surface area (in $mol.m^{-2}.s^{-1}$). Note that according to its usual definition, a flux towards the interface has a negative sign.

For the limiting transport situation, where c_M at the interface is zero, with negligible depletion of the bulk ($c_M \rightarrow c_M^*$ for $r \rightarrow \infty$), the solution of Equations 2 and 3 is

$$J_M = -D_M c_M^* [(\pi D_M t)^{-1/2} + a^{-1}] \tag{4}$$

The right hand side of this expression shows the two limits: linear diffusion for relatively large particles and small times, and convergent spherical diffusion for the other extreme. The change of the metal ion surface excess concentration Γ_M is directly related to the flux J_M

$$\frac{\partial \Gamma_M}{\partial t} = -J_M \tag{5}$$

Analytical expressions of Γ_M as f(t) are available only for a few relatively simple cases. The problem is that the approach of any new equilibrium is accompanied by a corresponding adjustment of c_M. Thus the time course of

---

* A list of symbols and abbreviations is given at the end of the chapter.

J_M is generally more complicated than Equation 4, which only holds for a constant, specific value of c_M (i.e., zero; with $c_M = $ constant $\neq 0$, Equation 4 holds with c_M^* replaced by $(c_M^* - c_M)$). For the case of essentially linear diffusion, where the second term between brackets in Equation 3 would be negligible, the solution of Equations 2, 3, and 5 yields

$$\Gamma_M(t) = \Gamma_M(0) - \left(\frac{D_M}{\pi}\right)^{1/2} \int_0^t \frac{c_M(0,t) - c_M^*}{(t - u)^{1/2}} \, du \tag{6}$$

where $\Gamma_M(0)$ is the surface excess concentration before the perturbation of the adsorption equilibrium and u is a dummy integration parameter. The variable $c_M(0,t)$ is the metal ion concentration just outside the adsorbed layer which is supposed to be in continuous equilibrium with $\Gamma_M(t)$, implying that the adsorption/desorption reactions are fast. Under these conditions the relation between $c_M(0,t)$ and $\Gamma_M(t)$ is the equilibrium adsorption isotherm. To proceed from Equation 6, the isotherm has to be specified. The simplest case is a linear one, known as the Henry isotherm:

$$\Gamma_M(t) = K_H c_M(0,t) \tag{7}$$

where K_H is the Henry adsorption coefficient (unit m^{+1}).

Solution of Equation 6 for the Henry regime (Equation 7) is conveniently done with the aid of Laplace transformation. The result is

$$\Gamma_M(t) = \Gamma_M^* - [\Gamma_M^* - \Gamma_M(0)] \exp(\xi_M^2 t) \, \text{erfc}(\xi_M t^{1/2}) \tag{8}$$

with

$$\xi_M = \frac{D_M^{1/2}}{K_H} \tag{9}$$

Because of Equation 7, an expression entirely equivalent to Equation 8 holds for $c_M(0,t)$. Figure 1 sketches the development of Γ_M with time, as prescribed by Equation 8, and compares it with the exponential function. The experfc functionality shows a strong initial change and a very gradual approach of the (new) equilibrium value Γ_M^* for $t \to \infty$. It is useful to note that for large argument the experfc approaches a $t^{-1/2}$ functionality:

$$\Gamma_M(t) = \Gamma_M^* - [\Gamma_M^* - \Gamma_M(0)] \frac{1}{\pi^{1/2} \chi_M t^{1/2}} \qquad (\chi_M t^{1/2} >> 1) \tag{10}$$

The argument of the error function, ξ_M, can be seen as a measure of the time constant τ of the adsorption process:

$$\tau = \frac{1}{\chi_M^2} = \frac{K_H^2}{D_M} \tag{11}$$

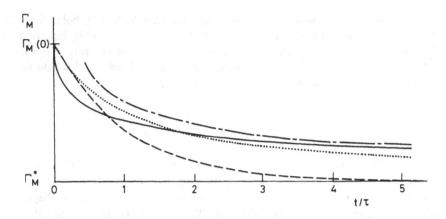

Figure 1. The course of Γ_M as a function of time, according to
(————) exp (t/τ) erfc $(t/\tau)^{1/2}$, Equation 8
(— · — · — ·) $\pi^{-1/2}(t/\tau)^{-1/2}$, Equation 10
(· · · · · · · · ·) $1/(1+t/\tau)$, Equation 25 (complementary form)
(- - - - - -) exp $(-t/\tau)$, Equation 30
Note that $\Gamma_M^* - \Gamma_M(0)$ is positive for an adsorption process and negative for a desorption.

and it should be emphasized that the operation of τ is very different from that of the argument in an exponential function (see Figure 1). In analyzing experimental adsorption data, that are always confined to a certain time window, it is tempting to fit the data to the sum of two or three exponential functions with different arguments. Such fits may seem to be very successful. Yet in cases of diffusion-controlled adsorption (which seldomly are of a nature as simple as above!) the merit of the fit is purely mathematical and no further meaning should be assigned to the corresponding time constants (or the derived "first order sorption rate constants").

2.2 Convective Diffusion

In cases where convection is important, the basis of the metal ion transport is provided by liquid flow and diffusion simultaneously. Suitable starting equations are given by the conservation laws, which for an incompressible liquid read:

$$J_M = -D_M \nabla c_M + v c_M \tag{12}$$

$$\frac{\partial c_M}{\partial t} = D_M \nabla^2 c_M - v \nabla c_M \tag{13}$$

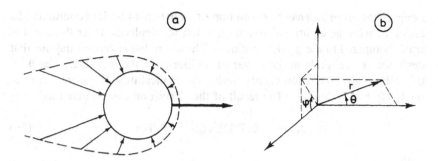

Figure 2. Diffusion to a moving particle (or a particle in a flowing liquid). The family of arrows indicates the depletion of the solution around the particle. The center of the particle is the origin of the coordinate system.

where v is the velocity of the liquid with respect to the particle. Obviously, the cases of a particle moving through a stagnant liquid and of a spatially fixed particle in a flowing liquid are mathematically identical. In the stationary state $\partial c_M/\partial t = 0$ and Equation 13, rewritten in terms of spherical coordinates, reduces to

$$D_M \left[\frac{\partial^2 c_M}{\partial r^2} + \frac{2}{r} \frac{\partial c_M}{\partial r} \right] - v_r \frac{\partial c_M}{\partial r} - \frac{v_\theta}{r} \frac{\partial c_M}{\partial \theta} = 0 \tag{14}$$

where r and θ are the two relevant spherical variables. Figure 2 clarifies that for the geometry of the considered case, the problem is independent of the elevational angle ϕ.

Under limiting transport conditions and for a sufficiently diluted dispersion,* the boundary conditions are

$$r = a \quad : \quad c_M = 0 \tag{15a}$$

$$r \to \infty \quad : \quad c_M = c_M^* \tag{15b}$$

---

* Average distance between particles large compared to thickness of depletion layer.

Levich[2] has given an analytical solution of Equation 14 under conditions 15a and 15b, with the additional assumption that the depletion layer thickness is small compared to the particle radius a. The formulation further implies that depletion is negligible at the center of incidence of the liquid, i.e., at $\theta = 0$. Such an approximation clearly limits the applicability of the treatment to relatively high velocities. The result of the derivation can be given as[2]

$$J_M = -0.79 D_M^{2/3} c_M^* a^{-2/3} v^{1/3} F(\theta) \tag{16}$$

with

$$F(\theta) = \frac{\sin\theta}{\left(\theta - \dfrac{\sin 2\theta}{2}\right)^{1/3}} \tag{17}$$

Inspection of $F(\theta)$ confirms that the flux J_M is maximum for $\theta = 0$. It decreases with increasing θ, while approaching zero for $\theta = \pi$ because the depletion layer thickness is no longer small compared to a. However, the point is not practically important since the fluxes for $\theta = \pi$ are relatively small and hardly count in the total transport rate. It may also be noted that, as a consequence of the approximations in the derivation, the $v \to 0$ limit of Equation 16 does not approach the purely diffusional flux equation.

The total transport rate (in $mol.s^{-1}$) to one particle is found by integration of Equation 16 over the total surface area of the particle:

$$\int J_M dA = 2\pi a^2 \int_0^\pi J_M \sin\theta \, d\theta \tag{18}$$

which, after substitution of Equations 16 and 17, finally gives for the average flux $\bar{J}_M$ ($\int J_M dA / 4\pi a^2$)

$$\bar{J}_M = -0.64 D_M^{2/3} c_M^* a^{-2/3} v^{1/3} \tag{19}$$

It should be emphasized that the resulting Equation 19 holds for *solid particles*. In case of liquid particles, e.g., with emulsions, the convective diffusion process is different due to interfacial momentum transfer which gives rise to a different velocity profile. Consequently, convective diffusion to/from a liquid particle is more effective than for a solid particle. Starting again from Equation 14, the following equivalent of Equation 19 for a moving liquid particle can be obtained[2]

$$\bar{J}_M = -0.62 D_M^{1/2} c_M^* a^{-1/2} v_0^{1/2} \tag{20}$$

with

$$v_0 = \frac{\eta_m v}{2(\eta_m + \eta_p)} \tag{21}$$

where η_m and η_p are the viscosities of the medium and the particle, respectively.

As explained in the previous section, the usual diffusion situation in adsorption processes is complicated by the fact that $c_M(0,t)$ is not a constant. For the case of steady-state convective diffusion it is reasonable to assume that (i) changes of $c_M(0,t)$ are slow as compared to the characteristic time for reaching steady state of the mass transport,* and (ii) spreading of adsorbed metal ions over the particle surface is fast enough to warrant a θ-independent $\Gamma_M(t)$. Then we may postulate the applicability of Equations 19 and 20 to non-limiting transport conditions, provided that c_M^* is replaced by $c_M^* - c_M(0,t)$. Using this, for example, for the solid particle case, we rewrite J_M into

$$J_M = m_s[c_M(0,t) - c_M^*] \tag{22}$$

where m_s is shorthand for the steady-state mass transfer coefficient for a spherical solid particle

$$m_s = 0.64 D_M^{2/3} a^{-2/3} v^{1/3} \tag{23}$$

Let us continue again for the simplest possible adsorption isotherm, i.e., the Henry case (Equation 7), and use the simple steady-state relation

$$\Gamma_M(t) - \Gamma_M(0) = -J_M t \tag{24}$$

---

* If this assumption is not valid, the approach in Section 2.1 should be adopted.

By substitution of Equation 7 into Equation 22 and combination with Equation 24, $\Gamma_M(t)$ (or, equivalently, $c_M(0,t)$) is easily found. The result is conveniently written as

$$\frac{\Gamma_M(t) - \Gamma_M(0)}{\Gamma_M^* - \Gamma_M(0)} = \frac{\dfrac{t}{\tau}}{1 + \dfrac{t}{\tau}} \tag{25}$$

with

$$\tau = \frac{K_H}{m_s} \tag{26}$$

The dependence of Γ_M on t, as given by Equation 25 is included in Figure 1 for comparison. For arguments less than unity, $(1 + t/\tau)^{-1}$ resembles an exponential function but for larger arguments it changes only very gradually. Again we see a long-tailed function, or in other words, a *very gradual* approach of the equilibrium value of Γ_M. It may be noted that the relaxation time τ increases with particle size and decreases with velocity.

Natural convection results from density fluctuations and may be of considerable importance to mass transport in aquatic particle dispersions. It generally starts to overrule pure diffusion control on time-scales beyond the order of 10^2 s. On a strictly empirical basis, steady-state transport by natural convection may be described with the simple Nernst diffusion layer or film layer expressions. One should realize, however, that the idea of a fixed stagnant liquid layer adjacent to the interface seriously oversimplifies the mass transport problem because it spatially separates diffusion and convection processes. In reality, liquid motion persists down to distances much smaller than the depletion layer thickness. Besides, this thickness depends on the diffusion coefficient of the transported species.

Having underlined that only limited meaning can be derived from diffusion layer theory, we briefly indicate its elaboration for the case of a particle dispersion. The basic equation is a special form of Equations 2 or 12:

$$J_M = \frac{D_M[c_M(0,t) - c_M^*]}{\delta} \tag{27}$$

where δ is an unspecified, time-independent parameter which is usually pictured as the thickness of the depletion layer. By comparing Equation 27 to Equation 22 it is clear that there is a strong mathematical analogy with the forced convection case. The derivation represented by Equation 22 to Equation

26 can be applied again and the result can be read from Equation 25 if the time constant τ is taken as

$$\tau = \frac{K_H \delta}{D_M} \qquad (28)$$

In contrast with the previous expressions for τ, Equation 28 contains an empirical parameter, i.e., δ. According to experience, δ has a value on the order of 10^{-4} m under the usual conditions.[2]

We conclude this section with a remark on relatively concentrated dispersions, where the adsorption process gives rise to appreciable depletion of the bulk of solution. In such a case there is a trivial relation between the extent of adsorption and the decrease of c_M^*

$$\Gamma_M(t) - \Gamma_M(0) = \int_0^t \bar{J}_M dt = [c_M^* \text{ (initial)} - c_M^* \text{ (final)}] \frac{V}{A} \qquad (29)$$

where V/A is the (solution)volume/(particle)area ratio of the dispersion. If the adsorption is at equilibrium (no sorption kinetics involved) and the ratio $c_M(0,t)/c_M^*$ is constant, then $d\Gamma_M/dt$ is proportional to c_M^*. Under conditions of a linear isotherm (Equation 7), Γ_M and c_M^* then change with time exponentially

$$\frac{\Gamma_M(t) - \Gamma_M(0)}{\Gamma_M^* - \Gamma_M(0)} = 1 - \exp(-t/\tau) \qquad (30)$$

The time constant τ is again K_H/m_s (compare Equation 26), here with the mass transport coefficient m_s given by the (constant) ratio between the adsorption rate $d\Gamma_M/dt$ and the bulk concentration c_M^*.

3. RATES OF CHEMICAL REACTIONS INVOLVED

The mechanism of metal complex formation reactions in aqueous solutions has appeared to be a fairly simple matter and will not need so much attention as the mass transport aspects treated in the previous section. It has been shown by Eigen et al.[3] that for most metal complexation reactions the stepwise substitution of water molecules from the hydrated metal ion proceeds according to an SN_1 mechanism. The rate-determining step is the removal of the first water molecule(s) from the coordination sphere of the metal ion. Substitution reactions for outer hydration shells are very fast, with characteristic time-scales of less than 10^{-8} s, and rather independent of the nature of the metal ion.

For many different metal ions, the rate constants of complexation reactions are essentially *independent of the nature of the substituting ligand*. This

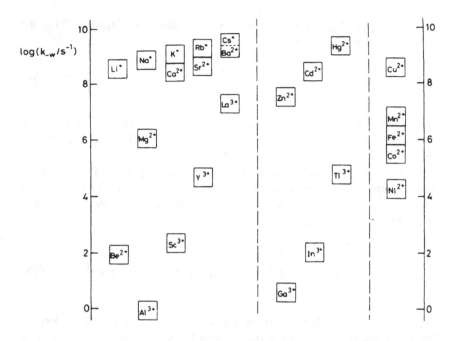

Figure 3. Rate constants for H_2O substitution in the inner coordination sphere of metal ions. (From Eigen, M. and Tamm, K. *Pure Appl. Chem.* 6:97 (1963). With permission.)

remains true, *even for chelating ligands* such as EDTA which replace water molecules from the inner coordination sphere. Ample experimental evidence for these important observations is available.[3,4]

For most divalent metal ions, the rate constants of the H_2O substitution are within the range of 10^3 to 10^9 s^{-1}. They depend on the charge, the radius, and the electron configuration of the bare metal ion. Very globally, there is a linear relation between the log of the dehydration rate constant k_{-w} and the reciprocal radius, which results form the increase in binding strength with decreasing ion radius. For example, the log k_{-w} (in s^{-1}) values are 8.5, 7.5, and 4.5 for Cd^{2+}, Zn^{2+}, and Mg^{2+}, respectively. For a highly charged and relatively small metal ion, k_{-w} may become quite small; e.g., for Al^{3+} it is as small as 1 s^{-1}. A selection of values of rate constants for water substitution from the inner coordination shell is given in Figure 3.

The above knowledge can be extrapolated to the present case of interaction between metal ions and dispersed particles. In as far as specific sites on the particles are responsible for the metal ion binding, the description of the kinetics becomes fully analogous to that of a homogeneous metal ion complexation reaction. Strong support for this is the fact that even chelating ligands with a high number of coordinating sites (up to six in, e.g., EDTA) obey the rule that the complexation kinetics are determined by the dehydration of the central metal ion. Hence, in the steps from $M(H_2O)_p$ to $M(H_2O)_{p-q}S$ as distinguished in Scheme 1, the (partial) dehydration step can be expected to be rate-determining. Consequently, the values from Figure 3 can be used for sorption reactions as well. Some direct evidence from kinetic experiments on dispersions support this.[5,6] In cases where the metal ion adsorption does not involve specific binding sites, the dehydration is generally less severe — often even not involving the primary shell — and the rate constants should be higher than 10^8 s^{-1}. For the present discussion these cases can be considered as infinitely fast.

For sufficiently diluted systems where activity corrections may be neglected, the relaxation time τ of the adsorption/desorption process is related to the rate constants for adsorption and desorption (k_{ads} and k_{des}, respectively) according to the simple relationship[7]

$$\frac{1}{\tau} = k_{des} + k_{ads} \left(c_M^* + \Gamma_S^* \frac{A}{V} \right) \tag{31}$$

where c_M^* is the bulk equilibrium concentration of the hydrated metal ion $M(H_2O)_p$ and Γ_S^* is the equilibrium surface concentration of metal binding sites S (compare Scheme 1). Multiplied by the ratio between the surface area of the particles and the volume of the dispersion, Γ_M^* gives the equivalent volume concentration of free sites. Equation 31 holds for the linear regime of the rate equation; this means that it is limited to small perturbations from equilibrium and that the sign of the perturbation is irrelevant. The shape of the relaxation curve is exponential, with τ as the argument. The adsorption and desorption reactions contribute additively to $1/\tau$; the faster of the two dominates, which is a characteristic feature of chemical relaxation. Experimental evidence for the applicability of Equation 31 to ion/particle interaction in dispersions is available.[8,9] Experiments usually cover conditions where either k_{des} or k_{ads} dominates. For example for small concentrations of the reactants $1/\tau$ approaches k_{des} whereas for high concentrations it comes to the limit $k_{ads}(c_M^* + \Gamma_S^* A/V)$. If only one of the two limits can be practically reached, the equilibrium constant can be used to find the other rate constant.

In the case of a large excess of surface sites over metal ions, Equation 31 reduces to

$$\frac{1}{\tau} = k_{des} + k'_{ads} \tag{32}$$

with

$$k'_{ads} = k_{ads} \Gamma_S^* \frac{A}{V} \tag{33}$$

The apparent volume equilibrium constant $K = k_{ads}/k_{des} = c_{MS}^*/c_M^* c_S^*$ is related to the interfacial sorption equilibrium constant. In the particular regime of an essentially constant c_S^*, the ratio c_{MS}^*/c_S^* is constant and K can be identified as K_H/Γ_S^*, with the Henry constant K_H being defined as Γ_M^*/c_M^* (see Equation 7).

4. EXPERIMENTAL METHODS AND RESULTS
4.1 Batch Analysis[10,11] (Abbreviation: B)

A technique often used for studying the overall ion adsorption kinetics in dispersions of particles is the batch method. It is based on measuring the depletion of the bulk solution and therefore requires fairly high particle concentrations. At certain times after mixing the adsorbent and the adsorbate, samples are withdrawn from the reaction vessel. These samples can be analyzed by any suitable analytical technique. Often the samples are filtered in order to prevent (further) changes in concentrations of the metal ions in solution. Analytical methods used include radioactive tracer determination,[12,13] titration, spectrophotometry, and flameless atomic absorption.[11] The steady-state mass transport conditions in batch analysis experiments are not generally well defined. Hence the value of m_s, as appearing in τ in Equation 30, is uncertain. It is often difficult to separate mass transport control and surface kinetic control on the basis of batch analysis. Comments on the poorness of particle characterization in kinetic sorption experiments (to be given in Section 5) apply especially heavily to batch analysis experiments.

4.2 Ligand Exchange[14-19] (Abbreviation: L)

Metal ion desorption is provoked by addition of a certain competing ligand, usually one that yields a spectrophotometrically active complex. Examples of such complexes are iron (III) — sulfosalicylic acid, aluminum (III) — calcein blue, copper (II) — calcein, and nickel (II) — pyridylazoresorcinol. The experimental procedure is simple. After mixing the dispersion with a solution of the competing ligand, the desorption of the metal ions from the particles is monitored by continuously measuring the spectrophotometric response of the complex formed. The dead period of the technique is determined by the time needed for mixing the solutions in the cuvette. Normally this is of the order of 1 s.

The intrinsic problem of the method is that one introduces another chemical reaction in order to measure the rate of the desorption reaction. It is not at all certain that the undisturbed desorption of metal ions — i.e., the transition from $M(H_2O)_{p-q}$ to $M(H_2O)_p$ in Equation 1 — fully takes place and, if so, that it is rate-determining. According to Hering and Morel,[19] one has to count with both the "disjunctive" pathway (preceding dissociation of the metal from the site on the particle) and the "adjunctive" pathway (direct attack of competing ligand on the initial metal-site surface complex). Pertinent experimental evidence for well-defined model dispersions is badly needed.

4.3 Crystal Flow Cell with *in situ* Electrochemical Detection[20] (Abbreviation: FC)

The crystal flow cell offers a method to determine adsorption and desorption of certain metal species. Solution flows through a channel over a generator electrode, subsequently over a crystal plane of the adsorbent and finally over a collector electrode. The generator electrode consists of the metal which, by oxidation, supplies the metal ions acting as adsorbate. Initially both electrodes are held at a potential negative with respect to the oxidation potential of the adsorbate metal. Then, at a given time, the potential of the upstream electrode is stepped to a value at which the generation of metal ions occurs. The downstream electrode is kept at a potential such that the metal ions are reduced again, so it acts as a detector. If the crystal is absent, there is no adsorbing surface and the recorded signal is only dependent on the concentration of the generated species and the mass transport conditions. If the crystal is present, adsorption of species on the surface reduces the concentration and this is reflected by the decreased detector signal. The response of the current at the downstream electrode in the relevant time range allows the deduction of the rate constants of adsorption and desorption. For Cu^{2+}, this has been done numerically after assuming a Langmuirian adsorption isotherm.[20] The method works under well-defined geometrical and hydrodynamic conditions, but is limited to adsorbate/adsorbent systems that can be realized in an electrochemical flow cell.

4.4 Column Method[11] (Abbreviation C)

By using a column filled with a material which binds metal ions, one can determine the retention time (or volume) needed for the metal to pass the column. Several materials can be used. The particle concentration is very high in this method.

4.5 Pressure Jump Method[5,9,21-25] (Abbreviation: ΔP)

In the pressure jump technique the pressure of a sample is changed from 100 atm to 1 atm within 80 to 100 ms. The method utilizes a burst diaphragm which ruptures spontaneously at a pressure of about 100 atm. The change in electric conductivity is measured. Adsorption and desorption rate constants can be calculated from the relaxation curve. Several types of blank measurements are essential to insure that the registered changes in electrical conductivity are due to adsorption/desorption processes of the metal at the surface of the particles. This includes measuring suspensions without metal ions, solutions without particles, supernatant solutions, etc. The conductance pressure jump method[6] is nearly identical to the pressure jump method. An important difference is that the dead time is much higher, i.e., approximately 1 ms. This is due to the experimental setup.

The electric field pulse method is analogous to the pressure jump method, but has a shorter dead time. For the pressure jump method this time is approximately 80 to 100 ms, the time for the pressure to get down to 1 atm. For the electric field pulse method this time is shorter than 0.1 ms, the duration of the high voltage pulse generally being of the order of tens of milliseconds with rise and decay times much less than 0.1 ms.

4.6 Conductance Stopped Flow[6] (Abbreviation: ΔK)

Two solutions, one containing the adsorbate and the other the adsorbent, are forced to flow into a mixing chamber. The changing conductivity of the mixed solution is measured and in that way the adsorption or desorption of metal to the surface of particles is registered. Using expressions for the reaction rate, the adsorption/desorption rate constants can be determined.

Table 1 collects the experimentally determined values of the desorption rate constants for a number of metal ion/particle systems. Characteristics of the dispersions involved are given wherever these may be taken or derived from the original work. In general, the quantitative information given on physical and chemical properties of the particles is poor, if not completely absent. Consequently the filling of the columns for the PZC (as defined by Sposito[27]) and the geometrical characteristics is grossly incomplete.

5. EVALUATION AND CONCLUSION
5.1 Practical Relaxation Times

For well-defined mass transport conditions and well-defined particles, theory predicts the time course of Γ_M after some perturbation of the adsorption equilibrium. In case of linear diffusion control and linear adsorption, the time constant τ is given by Equation 11. For example, for $K_H = 10^{-3}$ m (fairly strong adsorption) and with D_M being 10^{-9} m^2 s^{-1}, we have $\tau = 10^3$ s. Considering the dependence of Γ_M on t, this means that the adsorption kinetics lasts very long, on the scale of hours, while apparently starting rather rapidly (see Section 2.1). For rather weak adsorption ($K_H < 10^{-5}$ m), τ would be less than 10^{-1} s. Still appreciable relaxation of Γ_M would reach the time-scale of minutes (see the relaxation curve in Figure 1). These considerations are limited to particles which are sufficiently large to render the spherical term in Equation 4 negligible.

In case of convective diffusion, τ is of course always smaller than in the non-convective case. If we calculate τ from the convective expression 26 for the cases where $K_H = 10^{-3}$ m and 10^{-5} m again, while substituting a = 10^{-7} m and $v = 10^{-2}$ m.s^{-1}, the result is 2×10^{-1} and 2×10^{-3} s, respectively. For larger particles, τ becomes larger; e.g., for a = 10^{-5} m the resulting value would be 4 and 4×10^{-2} s, respectively. With natural convection and a δ of 10^{-4} m, the τ values for $K_H = 10^{-3}$ m and 10^{-5} m would, according to Equation 28, equal 10^2 and 1 s, respectively. The latter of the two values is not realistic since natural convection is not yet important on the time-scale of 1 s. This is compounded by comparing it with the τ for pure diffusion, which we just found to be 10^{-1} s. Table 2 gives an overview of mass transport controlled relaxation times, calculated for a number of cases.

Table 1. Experimental Values of Metal Ion Desorption Rate Constants k_{des} (in s^{-1}) for some Metal/Particle Systems

	PZC	Form	pH	I(mM)	T(°C)	Method	Ads. ion	log(k_{des})	Ref.
γ-Al₂O₃	8.3	$a_s = 100 \pm 15$ m²/g, Uniform <1 mm	4.5-5.4	±6	20	ΔP, L, ΔE	Pb²⁺	(1):4.0 (2):1.11	21
γ-Fe₂O₃ Fe₃O₄ TiO₂ silica/alumina	8.4 7.1 — —	— — — —	2.5-5.5	2	25	ΔP	H⁺	γ-Fe₂O₃ : −0.79 Fe₃O₄ : −0.47 TiO₂ :1.11 silica/alumina :1.66	5
α-FeOOH	—	$a_s = 44$ m²/g, $K_a(int) = 1$ mmol/l	3.2-5.5	±1	25	ΔE	Cl⁻ ClO₄⁻	Cl⁻ :4.3 ClO₄⁻ :4.0	22
γ-Al₂O₃	8.3	$a_s = 100 \pm 15$ m²/g, Uniform <1 mm	Cu/Pb:4-6 Zn/Co:5-7	7.5	25	ΔP	Pb²⁺ Cu²⁺ Zn²⁺ Mn²⁺ Co²⁺	Pb²⁺:6.6 Cu²⁺:5.5 Zn²⁺:5.1 Mn²⁺:6.3 Co²⁺:4.8	23
Iron hydrous oxide	—	<0.45 μm	5.8	100	25	ΔK L	Cu²⁺	(1):−1.77 (2):−2.74	26
Goethite	7.1	$a_s = 75.4$ m²/g acicular crystals, 0.5 mm long 0.1 mm wide	4.4/8.5	10	22-24	B, C	Fe(CN)₆³⁻	(1):−0.7	11
Hematite	7.1	crushed and sieved d = 200 mm	—	—	20,30,40	L	As(III)	(1):−1.81 (2):−1.94 (3):−1.95	14
Hydrous ferric oxide	—	amorphous	7.50 ± 0.05	100	25	B	Cd²⁺	(1):−3.18 (2):−2.93	13
Calcite (CaCO₃)	—	crystal (110)plane	7.5-8	100	25	FC	Cu²⁺	(1):2.3	20
α-Fe₂O₃	—	$a_s = 27$ m²/g, powder, d = 0.5mm	2-5	250	30 to 60	B	Sb(V)	only k_{ads}	12
TiS₂	—	TiS₂(M) < 1mm	9.5(?)	—	25	ΔP	Li⁺ Na⁺ K⁺ Rb⁺ Cs⁺	Li:1.11/2.08 Na:1.11/1.51 K:1.0/1.7 Rb:0.95/1.38 Cs:1.28/0.9	24

Table 1. (continued) Experimental Values of Metal Ion Desorption Rate Constants k_{des} (in s^{-1}) for some Metal/Particle Systems

	PZC	Form	pH	I(mM)	T(°C)	Method	Ads. ion	log(k_{des})	Ref.
γ-ZrP	—	d = 4.7 ± 1.1 mm	—	—	25	ΔP	Li$^+$ Na$^+$ K$^+$ Rb$^+$ Cs$^+$	Li:1.23/1.34 Na:1.38/0.68 K:1.70/1.36 Rb:1.40/1.43 Cs:1.48/1.68	25
DOWEX 50W-X8	—	50-100 mesh	—	—	22-25	B	Zn^{2+} Cd^{2+}	(1):1.46 (2):1.48 (3):−4.43	10
Sephadex C25(Li)	—	pK$_a$ = 5.0, <1 mm	6.5-8.6	—	25	ΔP	Li$^+$	(1):−0.82	9
Fulvic acid	—	—	4.6-6.4	±150	24.5	L	Ni^{2+}	(1):−0.17 (2):−0.82 (3):−1.68 (4):−2.59	15
Humic material	—	84 mg/l DOC	7.5	100	25	L	Cu^{2+}	(1):± −1.0 (2):± −1.8	16
Fulvic acid	—	molar mass 900 (mass av.)	±5	100	19.5	L	Al(III)	(1):−1.66 (2):−2.59 (3):−4	17
Fulvic acid	—	—	7.4-8.4	2-100	25	L	Ni^{2+}	(1):−1.58 (2):−2.03 (3):−2.67	18
Humate NTA	—	—	7.4	100	25	L	Cu^{2+}	(1):−4.39 (2):−2.7	19
PVP imidazole	—	molar mass = 77500 (mass av.)	4-6	100	25	Δk	Ni^{2+}	(1):0.83 (2)0.82 (3)0.36 (4)0.40	6

Table 2. Calculated Ranges of Diffusion Controlled Adsorption/Desorption Time Constants τ. Linear Adsorption Isotherm; $D_M = 10^{-9} m^2 s^{-1}$

Type of Mass Transport	τ given by (equation ...)	a/m	K_H/m	τ/s
Linear diffusion	K_H^2/D_M (11)	Irrelevant	$\geq 10^{-3}$	$\geq 10^3$
			$\leq 10^{-5}$	$\leq 10^{-1}$
Convective diffusion		10^{-7}	$\geq 10^{-3}$	$\geq 2 \times 10^{-1}$
			$\leq 10^{-5}$	$\leq 2 \times 10^{-3}$
lin. flow $v = 10^{-2}$ m s^{-1}	K_H/m_s (26,23)	10^{-5}	$\geq 10^{-3}$	≥ 4
			$\leq 10^{-5}$	$\leq 4 \times 10^{-2}$
natural convection	$K_H \delta/D_M$ (28)	Irrelevant	$\geq 10^{-3}$	$\geq 10^2$
			$\leq 10^{-5}$	≤ 1[a]

[a] This value has no practical meaning because the τ for the non-convective diffsion is smaller.

For practical values of τ in real metal ion/particle systems, the adsorption coefficient K_H is of primary importance. Clearly enough, K_H generally depends strongly on experimental conditions such as the type of metal ion, pH, PZC, and ionic strength. For the example of the Cd(II)/rutile system at a pH near 7 and I $= 0.02$ M, the effective K_H can be estimated to be of the order of 10^{-4} m.[28] Using Equation 11, the diffusional time constant in an unstirred system is found to be about 10 s. For a stirred system with $v = 10^{-2}$ m.s^{-1} Equations 23 and 26 yield τ values varying from 2×10^{-2} s to 4×10^{-1} s for particle radii in the range from 10^{-7} to 10^{-5} m.

5.2 Complicating Factors for Environmental Samples

There are a number of circumstances that make the real adsorption/desorption situation in practice a lot more complicated than the simple mass transport as modeled in Section 2. We briefly indicate here the most important factors.

5.2.1 Nonlinear Adsorption

The relation between $\Gamma_M(t)$ and $c_M(0,t)$ is then more involved. For a Langmuirian type of isotherm, analytical solutions for Γ_M as f(t) have been proposed in terms of power series. The first term in such series corresponds to the right-hand side of Equation 8. Whatever the precise nature of the isotherm, deviations from Henry behavior generally imply that $d\Gamma_M/dc_M(0,t)$ is less than K_H. In terms of relaxation rates, such deviations *always* give rise to *slower* adsorption/desorption kinetics and extension of the time range covered.

5.2.2 Surface Heterogeneity

It is well known that the surfaces of particles in aquatic samples are chemically heterogeneous, i.e., the different sites do not possess identical affinities towards the metal ion in solution. Consequently, it is incorrect to describe the ion/particle interaction in terms of one single equilibrium constant (K_H or

K or whatever). Instead, a certain distribution of affinities comes into play[29] and this gives rise to a mean affinity that depends on the metal-to-site ratio of the dispersion. It is easily understood that the change from one metal ion binding constant to a certain distribution of affinities should be paralleled by a corresponding change to distributions of the adsorption/desorption rate constants. This change definitely leads to an extension of the time range of the Γ_M vs. t relaxation curve. In other words, heterogeneity is another practical factor that generates long-tailed Γ_M vs. t curves.

Experimental data can be converted into representations of the affinity distribution function using certain data treatment procedures. The so-called "kinetic spectrum" method evaluates the distribution function H, in the time domain or in the rate constant (frequency) domain, of the adsorption/desorption rate constants for various binding sties on heterogeneous particles.[16] The evaluation of the distribution function H is usually based on some approximation for the kinetic integral

$$\int F(k)\exp(-tk)\ dk$$

The plot of the function H vs. ln t gives a peak shaped-curve. The area under each peak represents the initial concentration of that particular component and the position of the peak on the ln t-axis gives its rate constant (ln k = ln (2/t), for the second-order approximation usually employed).

5.2.3 Polydispersity

Most particle dispersions, and certainly the natural aquatic ones, show a more or less broad distribution of particle sizes. Whether or not this affects the adsorption/desorption rates depends on the situation. Diffusion-controlled relaxation generally depends on particle radius, but reaction-controlled sorption rates are basically independent of a. For the diffusion-controlled case, we may thus expect a certain distribution of relaxation times and consequently extension of the Γ_M vs. t curve over a broader time range. This again compounds with more elementary arguments for long-tailed relaxation curves (see Section 2).

5.2.4 Particle Shape

There is little doubt about the variety in the morphology of particles in aquatic samples. Mass transport towards or from non-spherical particles is extremely complicated, even for well-defined, uniform particles and steady-state conditions. When the depletion layer becomes very thick compared to particle dimensions, the spherical limit of Equation 4 becomes applicable irrespective of the precise shape. Apart from that limit, variations in particle shape give rise to distributions of mass transport relaxation times. Microscopic roughness is generally not important in this respect (it gives rise to an apparent increase of the rates of chemical reactions in the interface), but porosity does the more so. Diffusion through microscopic channels may extend over broad time ranges, thus contributing to the paramount characteristic of adsorbate transport in dispersions, i.e., its extremely broad time window.

5.3 Particle Characterization

It is rather amazing that most metal ion sorption studies, including those that strive for detailed mechanistic interpretations, lack an appropriate characterization of the particles involved. Information on shapes, size distributions, surface roughness, porosity, aggregation, surface charge, etc. is generally very limited, if not absent. Already at this point one may conclude that there is an urgent need for model studies on well-defined dispersions with simple geometrical features (e.g., homodisperse, spherical, non-porous, crystalline, etc.), as well as a need for (further) developing suitable methods of characterization.

5.4 Mass Transport Effects in Studies of Adsorption Kinetics

Diffusion-controlled relaxation in ion/particle dispersions is generally of a quite complicated nature (see Section 2 and the preceding paragraphs of this section). In connection with that, it is most likely (as well as practically impossible to verify) that a good deal of the values published for ion desorption rate constants refer to ion transport processes. It is most unfortunate that mass transport, an essential element in adsorption processes, is often ignored see, e.g., References 13 and 19, or seriously oversimplified by concentrating on a single aspect only. Such oversimplifications include for example assumption of some arbitrarily defined immobile liquid film, which should account for surface irregularities and pore diffusion,[11] and combination of a single exponential relation for the adsorption kinetics with a $\sqrt{t}$ relationship for the pore diffusion.[14] The importance of due attention for ion transport aspects may be illustrated by the findings in a study[26] on Cu(II) adsorption on hydrous oxide particles. The work seems to yield some rather strange results for the ion adsorption; equilibrium data precisely follow a Langmuir isotherm (corresponding to a chemically homogeneous particle surface with a finite number of sites), whereas the kinetic data seem to indicate that "at least four" (!) types of surface groups are present. It leads the authors to making reservations and suggesting that a complicated form of mass transport might be responsible for their findings.

5.5 Dynamic Speciation

Frequently, in metal speciation studies involving dispersed adsorbent particles, the inertness of particle-bound metal is taken as an a priori established fact. Colloid-bound metal ions are identified with "strongly adsorbed" metal that features only an "extremely slow release" (see, e.g., References 30 and 31). It is difficult to accept such *a priori* qualifications if one recalls the extremely fast actual adsorption/desorption reactions (see Section 3) for the metal ions. More sound starting-points have also been chosen see, e.g.,

Reference 32, although these can get lost at some stage of reasoning: the relatively low mobility of particle-bound metal does *not* generally render its contribution to dynamic analytical signals negligible,[32,34] and, one step further, the responses of free and bound metal ions *cannot* be treated as uncoupled[33] if the exchange process is fast.

Several authors (see, e.g., References 10 and 30) have attempted to set up some dynamic speciation scheme on the basis of the time-scale of the analytical method(s) used. These schemes are, apart from being operationally useful, highly welcomed because they substantiate the dynamic nature of the system at a certain point (or in a certain range) in the scheme. Application of the speciation knowledge, e.g., to bioavailability, generally requires speciation data on a time-scale different from the one employed in the analysis, and hence it is very good that thinking in such dynamic terms becomes customary.

It is useful to point out here that the frequently used term "lability" is method-specific and that it is not necessarily of a trivial nature. We mean by this that a system does not necessarily behave as a labile system if $k'_{ads}t$ and $k_{des}t \gg 1$. For example, in voltammetric speciation methods, a flux of ions towards an electrode/solution interface is operative and the lability condition for such a type of method reads[35]

$$\frac{k_{des}}{(k'_{ads})^{1/2}} \left(\frac{D_P}{D_M}\right)^{-1/2} t^{1/2} \gg 1 \tag{34}$$

For the example of Zn^{2+}/humic particle association, with a volume binding constant Kc_S^* of 10^2 (or an equivalent K_H of 10^{-1} m, for reasonable values of $c_S^* = 10^{-1}$ mol.m^{-3} and $A/V = 10^2$ m^{-1}), one calculates $k'_{ads} = 10^4$ s^{-1} — as taken from Section 3 — and hence $k_{des} \approx 10^2$ s^{-1}. With $D_P/D_M \approx 10^{-2}$ (particle diffusion coefficient $\approx 10^{-11}$ m^2.s^{-1}) and a time-scale of 10^{-1} s (which would include pulse and stripping voltammetries) one finds that condition 34 is just hardly fulfilled. For Cd^{2+} and Cu^{2+} ions the adsorption rate constant is practically the same, but the ion/particle interaction strength is generally larger. For $Kc_S^* = 10^3$ (which reduces k_{des} to 10 s^{-1}) and all other parameters being the same as for Zn^{2+}, lability condition 34 is no longer obeyed. The measured signal will then be lower than for Zn^{2+} due to its kinetic nature, although the more trivial condition $k'_{ads}t$, $k_{des}t \gg 1$ is still fulfilled. With increasing K, k_{des} decreases further and when it goes well below unity, the system becomes effectively inert. It goes without further explanation that other cases of non-lability, e.g., due to different adsorption rate constants (see Figure 3), may be found.

APPENDIX: LIST OF SYMBOLS AND ABBREVIATIONS

a   Radius of spherical particle (m)

a_s Specific surface area (m^2 kg^{-1})

A   Surface area (m^2)

B   Batch analysis

c   Concentration (mol m^{-3})

c*   Bulk concentration (mol m^{-3})

C   Column method

d   Diameter (m)

D   Diffusion coefficient (m^2 s^{-1})

ΔE   Electrical field pulse

FC   Crystal flow cell

I   Ionic strength (mol m^{-3})

J   Flux (mol m^{-2} s^{-1})

$\bar{J}$ Flux, averaged over a certain area (mol m^{-2} s^{-1})

k$_{ads}$ Adsorption rate constant (mol^{-1} m^3 s^{-1})

k$'_{ads}$ Given by Equation 33 (s^{-1})

k$_{des}$ Desorption rate constant (s^{-1})

k$_{-w}$ Dehydration rate constant (s^{-1})

K   Volume equilibrium constant $c^*_{MS}/c^*_M c^*_S$ (mol^{-1} m^3)

K$_a$ Acid dissociation constant (mol m^{-3})

K$_H$ Henry adsorption coefficient (m)

L   Ligand exchange

m$_s$ Steady-state mass transfer coefficient (m s^{-1})

ΔP   Pressure jump

PZC   pH value at point of zero charge

r   Distance from center of sphere (m)

t   Time (s)

u   Dummy integration variable

v Velocity (m s^{-1})

v_o Velocity parameter defined by Equation 21 (m s^{-1})

V   Volume (m^3)

Γ Surface excess concentration (mol m^{-2})

$\Gamma(0)$ Γ before perturbation

Γ^* (Eventual) equilibrium value of Γ

δ thickness of depletion layer (m)

∇ $\mathbf{grad}\left(= \mathbf{i}\,\dfrac{\partial}{\partial x} + \mathbf{j}\dfrac{\partial}{\partial y} + \mathbf{k}\,\dfrac{\partial}{\partial z}\right)$

θ Angle in polar coordinate system, defined in Figure 2

ΔK   Conductance stopped flow

χ Parameter defined by equation (9) (s$^{-1/2}$)

τ Time constant (s)

ϕ Angle in polar coordinate system, defined in Figure 2

REFERENCES

1. Florence, T. M. in *Trace Element Speciation: Analytical Methods and Problems,* G. E. Batley, Ed. (Boca Raton, FL: CRC Press, 1989), p. 77.
2. Levich, V. G. *Physicochemical Hydrodynamics.* (Englewood Cliffs, NJ: Prentice-Hall, 1962).

3. Eigen, M. and K. Tamm. "Schallabsorption in Elektrolytlösungen als Folge Chemischer Relaxation. I. Relaxationstheorie der mehrstufigen Dissoziation," *Z. Elektrochem.* 66:93 (1962); M. Eigen. "Fast Elementary Steps in Chemical Reaction Mechanisms," *Pure Appl. Chem.* 6:97 (1963).

4. Maass, G. "Untersuchungen der Schallabsorption von Kupfer-, Zink- und Cadmiumkomplexen in wäßrigen Lösungen als Folge chemischer Relaxation," *Z. Phys. Chem. N.F.* 60:138 (1968).

5. Astumian, R. D., M. Sasaki, T. Yasunage, and Z. A. Schelly. "Proton Adsorption-Desorption Kinetics on Iron Oxides in Aqueous Suspensions, Using the Pressure-Jump Method," *J. Phys. Chem.* 85:3832 (1981).

6. Okubo, T. and A. Enokida. "Conductance Stopped-Flow Study of the Complexation Reaction of Poly(4-vinylpyridine) with Nickel (II) Ions," *J. Chem. Soc. Faraday Trans.* 79:1639 (1983).

7. Bernasconi, C. F. *Relaxation Kinetics* (New York: Academic Press, 1976).

8. Ikeda, T., M. Sasaki, and T. Yasunaga. "Kinetics of the Hydrolysis of Hydroxyl Groups on Zeolite Surfaces Using the Pressure-Jump Relaxation Method," *J. Phys. Chem.* 86:1678 (1982).

9. Hachiya, K., T. Yasunaga, and K. Takeda. "Application of the Pressure-Jump Method to Adsorption-Desorption Kinetics: Adsorption-Desorption of the Lithium Ion on the Carboxylmethyl Group of Sephadex C-25 (Li)," *Colloids Surf.* 37:205 (1989).

10. Muller, F. L. L. and D. R. Kester. "Kinetic Approach to Trace Metal Complexation in Seawater: Application to Zinc and Cadmium," *Environ. Sci. Technol.* 24:234 (1990).

11. Theis, T. L., R. Iyer, and L. W. Kaul. "Kinetic Studies of Cadmium and Ferricyanide Adsorption on Goethite," *Environ. Sci. Technol.* 22:1013 (1988).

12. Ambe, S. "Adsorption Kinetics of Antimony (V) Ions onto α-Fe_2O_3 Surfaces from an Aqueous Solution," *Langmuir* 3:489 (1987).

13. Dzombak, D. A. and F. M. M. Morel. "Sorption of Cadmium on Hydrous Ferric Oxide at High Sorbate/Sorbent Ratios: Equilibrium, Kinetics and Modeling," *J. Colloid Interface Sci.* 112:588 (1986).

14. Singh, D. B., G. Prasad, D. C. Rupainwar, and V. N. Singh. "As (III) Removal from Aqueous Solution by Adsorption," *Water Air Soil Pollut.* 42:373 (1988).

15. Lavigne, J. A., C. H. Langford, and M. K. S. Mak. "Kinetic Study of Speciation of Nickel (II) Bound to a Fulvic Acid," *Anal. Chem.* 59:2616 (1987).

16. Olson, D. L. and M. S. Shuman. "Kinetic Spectrum Method for Analysis of Simultaneous, First-Order Reactions and Application to Copper(II) Dissociation from Aquatic Macromolecules," *Anal. Chem.* 55:1103 (1983).

17. Mak, M. K. S. and C. H. Langford. "A Kinetic Study of the Interaction of Hydrous Aluminum Oxide Colloids with a Well-Characterized Soil Fulvic Acid," *Can. J. Chem.* 60:2023 (1982).

18. Cabaniss, S. E. "pH and Ionic Strength Effects on Nickel-Fulvic Acid Dissociation Kinetics," *Environ. Sci. Technol.* 24:583 (1990).

19. Hering, J. G. and F. M. M. Morel. "Kinetics of Trace Metal Complexation: Ligand-Exchange Reactions," *Environ. Sci. Technol.* 24:242 (1990).

20. Compton, R. G. and K. L. Pritchard. "Kinetics of the Langmuirian Adsorption of Cu(II) Ions at the Calcite/Water Interface," *J. Chem. Soc. Faraday Trans.* 86:129 (1990).

21. Hachiya, K., M. Ashida, M. Sasaki, H. Kan, T. Inoue, and T. Yasunaga. "Study of the Kinetics of Adsorption-Desorption of Pb^{2+} on a γ-Al_2O_3 Surface by Means of Relaxation Techniques," *J. Phys. Chem.* 14:1866 (1979).

22. Sasaki, M., M. Moriya, T. Yasunaga, and R. D. Astumian. "Kinetic Study of Ion-Pair Formation on the Surface of α-FeOOH in Aqueous Suspensions Using the Electric Field Pulse Technique," *J. Phys. Chem.* 87:1449 (1983).

23. Hachiya, K., M. Sasaki, T. Ikeda, N. Mikami, and T. Yasunaga. "Static and Kinetic Studies of Adsorption-Desorption of Metal Ions on a γ-Al_2O_3 Surface. II. Kinetic Study by Means of a Pressure-Jump Technique," *J. Phys. Chem.* 88:27 (1984).

24. Sasaki, M., H. Negishi, H. Ochuchi, M. Inoue, and T. Yasunaga. "Intercalation Kinetics of Alkali Metal Ions into TiS_2 Using the Pressure-Jump Technique," *J. Phys. Chem.* 89:1970 (1985).

25. Mikami, N., M. Sasaki, N. Kawamura, K. F. Hayes, and T. Yasunaga, "Intercalation Kinetics of Alkali Metal Ions into γ-Zirconium Phosphate Using the Pressure-Jump Technique," *J. Phys. Chem.* 90:2757 (1986).

26. Gutzman, D. W. and C. H. Langford. "Multicomponent Kinetic Analysis of Trace Metal Binding Sites on Iron Hydrous Oxide Colloids," *Water Pollut. Res. J. Can.* 23:379 (1988).

27. Sposito, G. "Characterization of Particle Surface Charge," in *Environmental Particles*, J. Buffle and H. P. van Leeuwen, Eds., IUPAC Environmental Analytical and Physical Chemistry Series, Vol. I. (Chelsea, MI: Lewis Publishers), chap. 7.

28. Fokkink, L. G. J., A. de Keizer, and J. Lyklema. "Temperature Dependence of Cadmium Adsorption on Oxides. I. Experimental Observations and Model Analysis," *J. Colloid Interface Sci.* 135:118 (1990).

29. van Riemsdijk, W. H. and L. K. Koopal. "Ion Binding by Natural Heterogeneous Colloids," in *Environmental Particles*, J. Buffle and H. P. van Leeuwen, Eds., IUPAC Environmental Analytical and Physical Chemistry Series, Vol. I. (Chelsea, MI: Lewis Publishers), chap. 12.

30. Figura, P. and M. McDuffie. "Determination of Labilities of Soluble Trace Metal Species in Aqueous Environmental Samples by Anodic Stripping Voltammetry and Chelex Column and Batch Methods," *Anal. Chem.* 52:1433 (1980).

31. Plavsic, M., H. Bilinski, and M. Branica. "Voltammetric Study of Adsorption of Cu(II) Species on Solid Particles Added to Seawater," *Mar. Chem.* 21:151 (1987).

32. Florence, T. M. "Electrochemical Approaches to Trace Element Speciation in Waters. A Review," *Analyst* 111:489 (1986).

33. de Lurdes Simoes Goncalves, M., L. Sigg, and W. Stumm. "Possibilities of Voltammetric Methods for Speciation in the Presence of Particles," *Port. Electrochim. Acta* 5:15 (1987).

34. Dayalan, E., S. Qutubuddin, and A. Hussam. "Electrochemical Investigations in Microemulsion Media. I. Methylviologen Reduction," *Langmuir* 6:715 (1990).

35. de Jong, H. G. and H. P. van Leeuwen. "Voltammetry of Metal Complex Systems with Different Diffusion Coefficients of the Species Involved. II. Behaviour of the Limiting Current and its Dependence on Association/Dissociation Kinetics and Lability" *J. Electroanal. Chem.* 234:17 (1987).

INDEX

Index

Abietic acid, 20, 42
Accelerator mass spectrometry
    (AMS), use in atmospheric
    carbon studies, 28
Accessory pigments, of oceanic
    biogenic particles, 361
Acid-base equilibria, of atmospheric
    aerosols and gases, 75–106
Acid precipitation, 6
Acids
    atmospheric, formation of, 78–83
    in atmospheric aerosols,
        measurement of, 84–89
Acid water bodies, iron oxidation in,
    330
Acronyms, for surface analytical beam
    techniques, 134
Activity coefficient, for protons in
    bulk solution, 462, 464
Adsorption
    in colloidal pumping, 391–393
    coupled to sedimentation, 406–409
    of metal ions on dispersed particles
        mass transfer control of, 500–507

nonlinear, 515
rate constant for, 509
of reductants, 338
sediment metal binding compared
    to, 428–429
of solutes, onto filters, 187–189,
    194, 195, 204, 213
time constant for, 501, 512, 516
Adsorption/desorption process, of ion-
    particle interactions in
    dispersions, 500, 516
Adsorption kinetics, of metal ions,
    499
Aerosol(s)
    atmospheric, acid-base equilibria of,
        75–106
    characterization of Cr(VI)/Cr(III)
        ratio in, 145–168
    chromium in, speciation of,
        155–161
    definition of, 154
    evaporation during sampling, 84
    Fourier transform infrared (FTIR)
        microspectroscopy of, 120

neutralization during transport, 54
particles, see Aerosol particles
semi-volatile inorganic, modeling
 of, 90
Aerosol mass, 12
 apportionment of, see Source
 apportionment
Aerosol particles, 7, 30, 33
 beam techniques for characterization
 of, 107–134
 carbon in, see Carbonaceous
 particles
Affinities, distribution of, in metal ion
 adsorption, 516
Affinity distribution function, 458,
 471, 477
 effect on colloid-ion binding, 475
Affinity spectrum (AS) methods, for
 analytical inversion of the
 integral equation, 485–486
Agglomerative clustering, 43
Aggregates, see also Aggregation
 chemical fixatives for, 241
 correlative microscope technology
 for, 235–237
 electron microscopy of, data
 analysis and interpretation, 273
 enzyme cytochemistry of, 243
 formation at filter surface, 265
 heterogeneity of, 236
 of iron-rich colloids, 262, 345
 loose-type, from colloidal organic
 fibrils, 244
 mineral, 259
 morphology related to physico-
 chemical factors of, 236
 perturbation effects on, 262
 stabilization of, 238, 239
 whole mount preparations of, 245
Aggregation, see also Aggregates
 of colloids, 233, 237, 270, 275,
 320, 391
 of iron particles, 325, 345
 of particles, during filtration, 190,
 194
Air, ambient, chromium levels in,
 151, 153

Aircraft remote sensing colorimetry,
 of aquatic particles, 362
Air masses, transport of, 56, 57
Air pollution, in Los Angeles, see Los
 Angeles
Alaska, atmospheric carbonaceous
 particles over, 29, 30
Albuquerque (New Mexico), source
 apportionment of aerosols
 over, 39, 41, 56
Algae
 cell walls, 274
 in organic colloids, 253, 254,
 255
 silicon in, 268, 269
 phycocolloids from, 243, 244
 as source of fibrils, 248, 249
 viruses of, 253
Alkylammonium cation exchange,
 303–305
Aluminosilicate particles, in marine
 aerosols, 123, 124
Aluminum, 129, 260
Aluminum plants, fluorine tracer for,
 40
Amazon Boundary Layer Experiment,
 individual particle
 characterization in, 47–48
Ammonia
 in atmospheric aerosols, 93–95
 collection on filter packs, 84–85,
 88
Ammonium salts
 in atmospheric aerosols, 83–84,
 93–98
 growth and evaporation of, 88,
 95–97
 from atmospheric sulfur dioxide
 oxidation, 77
 LAMMS detection of, 124
Anapaite, in aquatic iron particles,
 341
ANATEX study, of long-range
 transport, 58–60
Anion exchange capacity, 303
Anoxic waters, particulate iron
 formation in, 340, 341

Antarctica, natural aerosols in, 126
Anthropogenic sources, of
    atmospheric particles, 8
Apparent activity coefficient(s), 463,
    470
Apparent affinity distribution,
    electrostatic effects on, 469
Apparent net proton surface charge
    density, definition of, 306, 310
Apparent net proton surface excess,
    definition of, 306, 310
Apparent volume equilibrium
    constant, 509
Aquatic colloids, see also Colloid(s)
    behavior and interactions of, 270
    common types of, 246–270
    correlative microscope technology
        for, 235–237
    dynamics of metal speciation in
        systems of, 497–521
    electron microscopy of, 231–289
    fibrils, 246–251
    future environmental device for
        study of, 235
    instability of, 237
    iron(II) in, 342
    nonliving organic materials,
        246–253
    types of, 231
Aquatic ecosystems, colloid
    importance in, 233
Archives, human and natural,
    atmospheric particle
    information from, 9
Arctic atmosphere, aerosol studies of,
    126
"Arctic Haze", as combustion
    aerosol, 6, 8
Arsenic
    as aerosol tracer, 14
    sorption onto iron oxyhydroxide,
        440, 442, 444
Artifact-sensitive materials
    electron microscopy of, 233
    iron particles as, 345–346
    minimum perturbation technology
        for, 237, 238, 239, 270–272

Asymptotically correct approximation
    (ACA), of the local isotherm,
    483
Atmospheric compartment, as source
    apportionment characteristic,
    26
Atmospheric particles, 28
    analytical data on, 10
    anthropogenic vs. natural sources
        of, 8–10
    biological data on, 10
    characterization methods for, 5
    compositional information on,
        11–13
    direct and secondary production of,
        9
    effects of, 7–8
    extension of, 6–7
    formation processes for, 11–13
    global and regional budgets for,
        8–9
    importance of different classes of,
        6–8
    meteorological data on, 10
    morphology related to source of,
        10, 21
    source apportionment of, see Source
        apportionment
    transformation of, 7
    visibility data on, 10
Atmospheric titration of particle
    alkalinity, by sulfur dioxide,
    55
Attachment mechanism, of aquatic
    bacteria, 258
Authigenic iron, 344, 345
Authigenic solid phases, isolation of,
    323
Autocatalysis
    role in Fe(II) oxidation, 328, 339
    role in iron dissolution mechanism,
        339

Back-diffusion, in filtration, 195,
    217–218
Background, in EELS, 118, 119

Backscattered electrons, in EPXMA, 109
Bacteria, 244, 258, 260, 321
  magnetosomes of, 265
  role in formation of manganese colloids, 266
  role in redox cycling of iron, 330, 335
  in seawater, 359, 374
  as source of fibrils, 248, 249, 266
Bacteriophages, sizes of, 253
Balance of surface charge, 296
Barium-133, sorption onto hematite, 412
Bases, atmospheric, formation of, 78–83
Batch analysis method, for studies of ion adsorption kinetics, 510
Beam techniques, for characterization of aerosol particles, 107–134
Benzo(ghi)perylene, organic tracers for, 42
Bidentate complexes, surface species of, 465
Binding energy, of photoelectrons, 122
Binding intensity and capacity, effects on competitive sorption models, 438–439
Bioavailability, role in metal ion adsorption by particles, 518
Bioemissions, persistence of, 17
Biofilms, microbial, fibrils in, 249, 250, 258
Biogenic aerosols, sources and composition of, 127–128
Biogenic origin, of marine sediments, 133
Biogeography, of oceanic particles, 373
Biological cells, rupture on membrane surfaces, 192
Blind quality control measurements, use in source apportionment methodology validation, 55
Boise (Idaho), source apportionment of aerosols over, 56

Boston, analysis of aerosols over, 46–47, 52
Bremsstrahlung, 109, 127
Bubble bursting mechanism, sea salt particle formation by, 123, 125, 126

Cadmium
  in aerosol clusters, 47
  distribution coefficient in aquatic systems, 429, 430, 431, 432
  sorption onto iron oxyhydroxide, 440, 442–444
CAESAR method, 480
Calcite, colloidal, 267
Calcium
  in aquatic iron particles, 331, 340, 341
  in mineral colloids, 260, 262, 267
Calcium sulfate, in marine aerosols, 123
Calibration, of membranes, 183–184
California, aerosol component modeling studies on, 90
CAPTEX study, of long-range transport, 58
Carbon-13, carbon-14 plane, for isotopic standards, 29, 31
Carbon-14
  as aerosol tracer, 28, 31, 39, 65
  in atmospheric particles, 28, 29
  calibration/validation for, 41, 42
Carbonaceous particles, 28
  derivation of, 8–9
  fossil and biogenic, 29
  health effects of, 7
  information content of, 17–19
  in marine aerosols, 31–32, 123
  molecular markers for, 19
  pollutant source resolution of, 25, 41
  urban inventory of, 18
"Carbonaceous Particles in the Atmosphere", quadriennial conferences on, 19

Carbonaceous soot, as aerosol
particles, 8
Carbon dioxide, exchange of, six-
reservoir model of, 27
Carbon isotopes
use in quantifying anthropogenic
and natural emissions, 9
use in source apportionment
methodology, 28–32
Carbon monoxide, source
apportionment of, 11
Carbon number maximum, 19
Carbon preference index, 19
Carcinogen, airborne Cr(VI) as, 146
Cation exchange capacity, 303
Cations, adsorption by clay minerals,
294
Cell fragments, organic colloids from,
252–253, 254
Cellulose membranes, use in dialysis,
180
Cell volume, of oceanic biogenic
particles, 359–360
Cenospheres, from coal fly ash
formation, 13, 22
Centrifugation
aquatic particle size measurement
by, 172
iron sampling by, 344–345
Centroid compositions, in aerosol
clusters, 47
Certified Reference Materials (CRMs),
55, 56
Charcoal particles, from Lake
Michigan, 23
Charge transfer, role in dissolution
reactions, 335, 339
Chemical data
derivation from particle classes, 48
real variable project compared to
principal component project of,
50
Chemical element balance, 23, 33
Chemical fixatives
avoidance of use, 245
for colloidal materials, 238, 241,
242, 244

Chemical heterogeneity, role in ion
adsorption, 458
Chemical mass balance (CMB)
of elements and species borne by
fine particles, 16
source-profile complications for,
34–35
in studies of aerosol composition,
10, 24, 33–37, 47, 49, 51, 54,
61, 63, 64
Chemical patterns, use in chemical
mass balance, 33
Chemical reactions, in hybrid methods
for source apportionment, 54
Chemical transformation data, on
atmospheric particles, 10
Chemisorption reactions, in oceanic
scavenging of trace metals,
384
Chemometrics, 54
Chlorite, 318, 319
Chlorophyll
autofluorescence of, 366
in oceanic biogenic particles, 361,
363, 368–370, 372
Chromium
airborne, 146–147, 151–155
chemistry of, 147–151
in filter material, 156
hexavalent, see Chromium(VI)
oxidizability of, 147–148
in particulates, 154, 155
redox potential of, 157
species, standard enthalpy of
formation, Gibbs free energy,
and entropy of, 149, 150
stable chemical forms of, 148
trivalent, see Chromium(III)
valence-specificity of, 155
in water, pH equilibrium diagram
for, 149
Chromium(III), 146
airborne, 154
chemistry of, 147–148
oxidation of, 147, 150–151
in seawater, 149
Chromium(VI), 146

abundance in water samples, 150
in aerosols, 156, 157, 161
chemistry of, 147–148
detection of, 154
determination of, 157, 158, 160
emissions of, 153–154
human exposure to, 155
species of, in aqueous solution, 148
Chromium(VI)/chromium(III) ratio
in aerosols, 145–168
thermodynamics and kinetics of,
    148–151
transformations of, 148, 149–150
Class modeling, of principal aerosol
    components, 48
Clay minerals
association with microbes and
    fibrils, 268
ion binding by, 294, 457, 460, 503
iron in, 262, 318, 319, 320, 340,
    343, 345
TEM of, 268, 269
Cleaning procedure, for filters,
    187–188
Climate, combustion particle effects
    on, 8
Climatology, single-celled organisms
    and, 359
Clogging, of filters, 188, 190–191,
    202, 217
Cloud liquid water, aqueous phase
    oxidation of sulfur dioxide in,
    80, 81
Cluster analysis, 112, 126, 127, 129
display methods for, 45, 46
EPXMA combined with, 112
use in source apportionment
    analysis, 43–49, 54
Clustering algorithms, 43
Coagulation
in high-energy dissipation
    environments, 409–410
of homodisperse particles, 393–395
in marine aerosols, 124–125
on membrane surface during
    filtration, 195, 197, 213–214,
    217–218

role in colloidal pumping, 391–393,
    404–406
Coagulation/flocculation, of colloids,
    233
Coalescence, in marine aerosols, 125
Coal fly ash, 13, 130
Coal mine dust, analysis of, 129
Coastal Zone Color Scanner, 358, 363
Colloid(s)
aggregation of, 233, 276, 342
aquatic, see Aquatic colloids
artifacts from, 265, 271–272, 274
definition of (size), 176, 219, 233
ion binding by, 455–495
living, see Living colloids
role in oceanic transport of
    radionuclides and trace metals,
    379–423
role in particle concentration effect,
    410–414
sinking of, in water, 320
size fractionation of, using filtration
    and ultrafiltration, 171–230
unstable, 239, 265, 272, 274
whole mount preparations of, 245
Colloidal pumping
application to oceanic trace element
    scavenging, 399–415
mechanism of, 395–399
in radionuclide and trace element
    transfer, 391–393
Colloid pool, role in oceanic
    scavenging, 386–390
Column method, use in studies of ion
    adsorption kinetics, 511
Combustion aerosol, 6, 125
information content of, 13–15
in North Sea, 124
Combustion particles, 7, 8
Competing ligands, effects on metal
    ion desorption, 510
Competitive sorption models
binding capacity effects on, 439
binding intensity effects on,
    438–439
component abundance effects on,
    438

problems with, 437–440
for trace metals, 434–440
Complexation reactions, rate constants
of, 507–508
Complexes, stability measurements
on, 181
Component abundance, as problem
with competitive sorption
models, 438
Compound in solution, defined, 176,
220
Computer
use in electron microscopy, 273
use with EELS analysis, 118
use with LAMMS analysis, 115
use with Rutherford back scattering
analysis, 114
use with SPM analysis, 113
Computer databases, for chemical
mass balance, 34
Concentration factor, for membrane
surface, 199, 216
Concentration gradient, in membranes,
193
Concentration polarization
definition of, 220
effects on filtration, 197–201, 217
minimization of, 213–215
role in surface coagulation on
membranes, 194–197, 201,
212
Concentration techniques, in filtration,
179
Condensation
in aqueous aerosol droplets, 97–98
of solid ammonium salt aerosol,
96–97
Condensation approximation (CA), in
analytical inversion of the
integral equation, 481–483,
486
Conductance pressure jump method,
511
Conductance stopped flow method,
512
Conductivity, as basis for pressure
jump method, 511

Confocal laser microscopy, 236
Conformation, of solute, effects on
membrane retention of, 212
Conservative tracers, use in source
apportionment, 53–54
Continental aerosols, remote,
composition of, 126–127
Continental sources, of marine
aerosols, 123, 124–125
Continuous distribution, of affinity
constants, 469
Convection, as force in transport
through membranes, 193
Convective diffusion
forced, 506–507
in metal ion transport, 502–507,
506, 512
Convective flux, of non-reactive
compounds in solution, 193
Conversion reaction, of sodium
chloride in marine aerosols,
124
Copper
in aerosol clusters, 47
distribution coefficient in aquatic
systems, 430, 431
sorption onto iron oxyhydroxide,
440, 442–444
Coprecipitation methods, for Cr(VI)
determination, 158, 159
Core-loss region, of electron energy
loss spectrum, 118
Coulombic interaction, in ion-colloid
binding, 462
Coulter counting, use in
oceanography, 365–366
Counter-ion condensation, 294,
310–311
Counterstains
for iron-rich colloids, 264
for TEM specimens, 242, 243
Criterion elements, as source tracers,
54
Cross-flow ultrafiltration, effect on
marine colloid pool, 389

Cryotechnology techniques, use in
aquatic colloid studies,
244–245, 270, 273
Crystal flow cell, use in studies of ion
adsorption kinetics, 510
Cut-off limit, in filtration, 178
Cyanobacteria, in oceans, 359
Cyclotron, proton beam production
by, 113
Cytochemical techniques, application
to aquatic colloid studies, 241,
260

Databases, for receptor modeling, 64
Dehydration/hydration reactions, 499
Dehydration rate constant, 508
Dehydroabietic acid, as organic tracer
for polycyclic aromatic
hydrocarbons, 42
Dehydrogenation products, as
molecular markers for
carbonaceous particle sources,
19, 20
Deliquescence humidities, of pure and
mixed salts, 90
Dendrogram, use to show aerosol
sample clustering, 45
Denuders
diffusion-type, 85–89
use air collection of chromium, 156
Depletion layer, thickness of, 506,
516
Depletion of bulk layer, by ion
adsorption, 507, 510
Deposition, of particles, 112
Depth filters, 173
particle entrapment in, 204, 213
pore size change in, 198, 202
properties of, 184, 185
Depth profiling, SIMS use for,
116–117
Desorption, rate constant for, 509,
512
Detection limit, comparison among
various microprobe techniques,
110, 112, 133

Detectors, use in SPM analysis, 113
Detritus, in seawater, 359, 374
Diafiltration, 179, 216
Dialysis
fractionation by, 180–181
iron particle studies using, 323
use for aquatic particle studies, 172,
173, 175–176
Diatoms
absorption spectra of, 364
in seawater, 371, 372
Differential equilibrium function, 487
Differential equilibrium function
(DEF) method, for analytical
inversion of the integral
equation, 486–488
Differential optical absorbance
spectroscopy (DOAS), use in
atmospheric acid-base
equilibria studies, 86, 89
Differential staining, of biological
macromolecules, 235
Diffuse attenuation spectrum, of
oceanic biogenic particles,
362, 365
Diffuse layer, ions in, 293, 296
Diffuse-layer surface charge density,
definition of, 296, 311
Diffusion
of dissolved species in water
columns, 321, 323, 324
as force in membrane transport, 193
in metal ion transport, 500–507,
512
shear-enhanced, 199
Diffusional time constant, 515
Diffusion denuders, 85–86, 87–89
Diffusion layer
formation during filtration, 217–218
thickness of, effects of particle size
on, 200
Diffusion models, evaluation by
ANATEX study, 59
Digital imaging microscopy, 236
Digital imaging system, use in fly-ash
emission studies, 130

Direct production, of global particles, 9

Dispersion modeling, in hybrid methods for source apportionment, 54

Dissociation/association rates, of complexes, role in ultrafiltration, 181

Dissolution reactions, of iron oxyhydroxides, 335–340

Dissolved material, discrimination from particulate material, 176–178

Distribution coefficient, of trace metals in aquatic systems, 429–432

Distribution function, 471, 477, 481, 487, 516

Donnan models, of colloids, 473, 474

Double layer, 294, 457

Double layer model, 472, 473, 479

Dry deposition, of atmospheric particles, 123

Dynamic SIMS, 115

EDS, see Energy-dispersive spectroscopy

EDX, see Energy-dispersive spectroscopy

EELS analysis, 129, 234
  comparison with other analytical techniques, 110, 122
  environmental particle analysis by, 117–119
  instrumentation for, 118
  sensitivity of, 110, 119

Effective variance least squares technique, use in chemical mass balance, 34–35

Eigenvector analysis, 49

Elastic recoil detection (ERD) technique, 114

Electric charges, effects on solute-membrane interactions, 208–212

Electric field pulse method, 511–512

Electrochemical methods, for Cr(VI) determination, 159

Electrokinetic effects, role in solute-membrane interactions, 210

Electrolytes, effects on filtration, 217

Electrometric titration, for determination of net proton surface charge, 305–307

Electron(s), interactions in environmental particles, 108

Electron acceptors, role in oxidation of iron, 321

Electron diffraction, of aquatic colloids, 234, 260, 264, 265, 270, 273

Electron energy loss spectroscopy, see EELS analysis

Electron microprobe analysis, of particulate environmental samples, 109, 111–113

Electron microscopes, for TEM and SEM, 234–235

Electron microscopy
  of aquatic colloids, 231–289
  preparatory techniques for, 237–245
  of iron colloids and particles, 318–319, 331, 346

Electron probe X-ray micro analysis, see EPXMA analysis

Electron spectroscopy for chemical analysis, see XPS analysis

Electrostatic model
  for heterogeneity analysis, 459
  particle characteristics and, 473–475

Electrostatic potential, in ion adsorption sites, 457, 462, 467, 468

Electrostatic properties
  effect on ion adsorption, 467
  effect on ion complexation, 465
  effect on ion exchange, 366
  effect on membrane pore size, 194
  particle characterization and, 471–475

Elemental composition, energy-
dispersive spectroscopy of, 234
El Paso (Texas), analysis of aerosol
particles from, 48
Elverum (Norway)
atmospheric carbonaceous particles
over, 29
multivariate particulate data on,
44–46
Emissions
atmospheric, airborne chromium in,
151, 154, 161
of biogenic aerosols, 127
Emissions inventories, of Los Angeles
aerosol, 9
End-member clusters, in frizzy
clustering technique, 48
Energy-dispersive spectroscopy
of calcium-rich particles, 267
of clay particles, 270, 271
of colloids, 234, 260, 273
of iron-rich colloids, 262, 263, 264,
346
of particulate environmental
samples, 109, 346
sample preparation for, 241, 242
Energy dissipation, effects on
coagulation in water, 409–410
Energy loss, by electrons in EELS
analysis, 117
Energy spectra, from Rutherford back
scattering analysis, 114
Enrichment factor(s), 14, 15
Enrichment factor technique, use to
study coal fly ash formation,
13–14
Environmental reference material,
isotopic composition of,
29–30, 31
Enzyme cytochemistry, application to
colloid aggregates, 243
Epifluorescence microscopy, of living
colloids, 236
EPXMA analysis
of biogenic aerosols, 127
comparison with other analytical
techniques, 110, 133

detection limit of, 110, 112
of estuarine/riverine particles, 132
instrumentation for, 109, 111
of marine aerosols, 123, 125
of particulate environmental
samples, 109, 113, 114
of remote continental aerosols, 126
of urban aerosols, 128
EQUILIB model, for atmospheric
production of aerosol
components, 91, 92
Equilibrium constants, ultrafiltration
use to determine, 181
Erosion
role in iron particle formation, 317
as source of atmospheric particles,
13
Estuarine/riverine particles,
composition of, 131
Euclidean distance, use in cluster
analysis, 43
Eulerian models, in simulation of
tracer transport, 59
Evaporation
in aqueous aerosol droplets, 97–98
of solid ammonium salt aerosol,
96–97
Experimental validation, of source
apportionment methodology,
55–56
Exploratory data analysis, principal
component analysis use in, 49
Extractable organic matter (EOM),
apportionment of, 38–40
Extraction methods, for Cr(VI)
determination, 158, 160

Factor analysis, 24, 31, 47, 61
principal component analysis use in,
49–52
Fe/P/Ca globules, in lakewaters, 262,
265, 267, 273
Fe/P/Ca/Si aggregates, in lakewaters,
262
Fe/P/Ca/Si/Al aggregates, in
lakewaters, 262

Ferrihydrite
from aquatic iron oxidation, 331, 332
in aquatic iron particles, 344
dissolution reaction of, 336
Ferromanganese colloid, 266
Ferrous sulfide, in aquatic iron particles, 341–342
"Ferrous wheel", aquatic iron cycle as, 323
Fiber filters, properties of, 184
Fibrillar colloids, 237, 238
Fibrils
as aquatic colloids, 233, 274
attachment to inorganic surfaces of, 261
biological derivation of, 248–249, 266
clay association with, 268
colloidal organic type, 244, 261
effects on aquatic ecosystems, 246–247, 248, 258
role in plant nutrition, 249
SEM analysis of, 249
TEM analysis of, 246–249, 261
Fick's laws, 218, 500
Field emission scanning electron microscope (FESEM), 234
Field flow fractionation (FFF), use for aquatic particle size measurements, 172, 173
Field studies, use in assuring quality control, 56
Filter(s), see also Membranes
characteristics related to size fractionation by, 181–192
cleaning procedure for, 187–188
clogging of, 188, 190–191, 202, 217
contamination of, 187–189, 213
electron microscopy of structure of, 174
for limnology studies, 265, 266, 270, 274
pore size of, see Pore size
Filter(s), preconditioning of, 188
Filter(s)

solute adsorption onto, 187–189
solute-membrane interactions inside pores of, 207–212
types of, 173
Filter load, role in filtration rate, 190, 191, 194, 206
Filter material, chromium content of, 156
Filter packs, use in atmospheric aerosol acid measurement, 84–85, 87–89
Filter sampling, for airborne chromium, 156
Filtration
application in natural water studies, 176–181, 262
artifacts produced by, 189–192
cell rupture during, 192
comparison with other particle separation methods, 175–176
concentration techniques in, 179
cut-off limit in, 178
diafiltration method, 179, 216
effects of concentration polarization in, 197–201
factors affecting retention during, 178, 192–212
field vs. laboratory, 189–190
filtrate concentration in, 217
flow-rate effects on, 214–215, 217
fractionation procedures for, 179–180
gel formation during, 190–191, 213–214, 217
induced aggregates, 265
ion adsorption onto membranes during, 210
of iron-rich colloids, 317
iron sampling by, 344–345
membrane calibration based on retention curves in, 183–184
minimization of sample perturbation in, 216
optimum conditions for, 212–216
parallel procedure for, 179, 180, 201

particle coagulation during,
189–190
pulsed type, 201
secondary effects of, 175
sequential cascade fractionation in,
179–180, 201, 216
sequential size fractionation by,
178–180
stirring effects on, 200, 207,
215–216, 217
surface coagulation during,
190–191, 201–207
tangential flow type, 201
use for size fractionation of aquatic
particles, colloids, and
macromolecules, 171–230
validity of results in, 216–217
washing technique for, 179
water flux modeling for, 194
window for size fractionation in,
215
Fine particle mass (FPM), as response
variable, 11
Fingerprints
derived from particle class balance,
37, 43
from iron-rich crystal diffraction,
265
Flameless atomic absorption, in batch
analysis experiments, 510
Flat plate model, for electrostatic
effects, 473
Flocculation, 251, 260
Flow cytometer, 367
Flow rate
effects on filtration, 214, 217
of polycarbonate filters, 184–185
Fluorescence, of photosynthetic
pigments, 361
Fluorescence flow cytometry, of
oceanic biogenic particles,
361, 363, 366–372
Fluorine, as aerosol tracer, 40, 48
Flux
of metal ions, in diffusion, 500,
504

of non-reactive compounds in
solution, 193–194, 195, 196,
217
Fly ash, 154
in marine aerosols, 123, 124
morphological and chemical
analyses of, 130
oil- and coal-derived, comparison
of, 131
Forest burning, as aerosol source, 127
Forward angle scattering technique
(FAST), 114
Fossil carbon, 42
Fossil fuels
as source of carbonaceous material,
28
sulfuric acid/sulfate emissions from,
79
Fourier transform infrared (FTIR)
microspectroscopy, of aerosols,
120
Fourier transform infrared (FTIR)
spectroscopy
comparison with other analytical
techniques, 110
use in atmospheric acid-base
equilibria studies, 86, 87, 89,
93
use in environmental particle
analysis, 120
Freeze-etching, 275
of fibrils, 249
of humic substances, 252
use in electron microscopy of
colloids, 244–245, 272
Freshwater, iron particles in, 315–355
Freundlich equations
generalized, 477
monocomponent normalized,
476–478
multicomponent normalized,
478–480
Frictional drag, effects on solute-
membrane interactions, 210
Frizzy clustering technique (FCV), 48
Fuel oil fly ash, isotopic tracers for
aerosol from, 40

Fulvic substances, 190, 233, 360
aggregation behavior of, 245, 251
ion binding by, 461, 469, 471–472,
475, 485, 488
TEM analysis of, 251–252

Gas emissions, from forests, 127
Gases, atmospheric, acid-base
equilibria of, 75–106
Gas-to-particle conversion, in marine
aerosols, 124, 126
Gaussian distribution function, 476,
480
Gel formation, during filtration,
190–191, 213–214, 217
Gel layer model, use in concentration
polarization studies, 197–198
Gel phase, colloid treatment as, 473
Genotoxicity, of atmospheric particles,
7, 11
Geochemistry, single-celled organisms
and, 359
Glass filtration apparatus, adsorption
onto, 188
Glauconite, 319
Global aerosols, fly ash in, 130
Global and regional budgets, of
atmospheric particles, 8–9
Global particle production, estimates
of, 9
Global source apportionment, 11, 26,
27
Global Tropospheric Experiment,
aerosol sampling in, 127
Goethite
in aquatic iron particles, 331, 332,
344
dissolution reaction of, 336
Gouy-Chapman theory, 463
Gran titration, 309
Graphitic carbon, see Soot
Gravimetry, membrane use in, 185
Great Smoky Mountains, white haze
over, 8
Greenhouse gases, 8, 11

Greigite, in aquatic iron particles,
333, 343, 344
Groundwaters, lanthanide radionuclide
transport in, 391

Health, atmospheric particle effects
on, 7, 11
Heavy metal staining, of fibrils, 246
Hematite, 396, 397, 405, 412
in aquatic iron particles, 332, 344
dissolution reaction of, 336, 338
Henry isotherm, 501
Heterogeneity, apparent, of surface
groups, 458, 464
Heterogeneity analysis, of ion-binding
data, 459, 475–488
Heterogeneous colloids, 457
ion binding to, 467–471
Heterogenous particles, metal ion
binding to, 470–471
Hierarchical clustering, 43, 44, 126
High-voltage electron microscope
(HVEM)
resolution achieved by, 235
use in colloid studies, 237, 273
High-volume samplers, for aerosol
chromium, 156
Historical records, of atmospheric
particles, 9–10
Homodisperse particles, coagulation
of, 393–395
Homogeneous colloids, ion binding
to, 461–467
Homogeneous subsets, subdivision of
heterogeneous data into, 43
Houston (Texas), atmospheric
carbonacous particles over, 29
Humic substances
ion binding by, 461, 469, 471–472,
488
iron stabilization by, 317, 330, 331,
335
in seawater, 361
TEM analysis of, 251
Hybrid receptor modeling, 17, 25
use in source apportionment, 53–54

Hybrid techniques, for source apportionment, 52–55
Hydration properties
effect on membrane pore size, 194
effects on solute-membrane interactions, 208–212
Hydrocarbon emissions, as source of carbonaceous aerosol, 8–9
Hydrochloric acid, in atmospheric aerosols, 79, 83
Hydrodynamic chromatography, use for aquatic particle size measurements, 172, 173
Hydrogen peroxide, Fe(II) oxidation by, 329

Image analysis, after freeze-etching, 245
Image processing technology, application to colloid studies, 236, 273
Imaging
in SIMS analysis, 116
in XPS analysis, 122, 123
Imaging microscope microanalyzers, 116
Immunoelectron microscopy, of organic particles, 243
Impedance volume, of oceanic biogenic particles, 361, 363, 365–366
Impinger, use for aerosol chromium sampling, 156
Independent site adsorption, of ions, 467, 470, 471
In-depth profiles, from SIMS analysis, 116
In-depth resolution, comparison among various microprobe techniques, 110
Indifferent salt, effect on proton binding curves, 471
Individual particle clustering, 47
Industrial aerosols, sources and composition of, 129–131

Industrial areas, atmospheric chromium levels over, 153
Industrial Revolution, polycyclic aromatic hydrocarbon in sediments from, 9–10
Inner-sphere complex surface charge density, definition of, 295, 311
Inner-sphere surface complex, definition of, 293, 311
Inorganic components, in water, size classification of, 177
Insect viruses, TEM analysis of, 253
Integral adsorption equation, 479
Integral binding equation, 470, 480
Integral equation
analytical inversion of, 480–488
local isotherm approximation (LIA) methods for, 480, 481–485
Integrated Air Cancer Project, 11, 56
Interfacial sorption equilibrium constant, 509
Inter-phase equilibration, effects on atmospheric aerosols, 99
Intrinsic binding constant, 464, 467
Intrinsic chemical heterogeneity, of particles, 458
Intrinsic proton affinity constant, in ion binding, 462, 471
Intrinsic surface charge density, derivation of, 295, 302–303, 311
Ion binding
batch analysis studies of, 510
by natural heterogeneous colloids, 455–495
specific sites for, 509
Ion complexation, to homogeneous particles, 464–466
Ion exchange, in metal ion binding, 466
Ion exchange capacity, definition of, 31
Ion-exchange methods, for Cr(VI) determination, 158
Ion exchange resins, iron colloid separation on, 318–319
Ions

adsorption onto charged
  membranes, 210
interactions in environmental
  particles, 108
primary charge determining type,
  457
specific adsorption type, 457
Iron, 40, 316, 330
  diagenetic, *in situ* collection of, 346
  oxidation products of, 330–335
Iron cycle, in water columns, 325,
  327
Iron hydroxides, 191, 262, 265
Iron(II)
  in iron-rich colloids, 262
  oxidation of, in freshwaters,
    328–330
  in pore waters, 323
Iron(III), as source of iron
    oxyhydroxides, 316
Iron oxides, 123, 261
Iron oxyhydroxides
  as aquatic colloids, 233, 261
  dissolution reactions of, 335–340
  formation in water bodies, 316,
    320–328, 332
  in lake sediments, 320
  reduction of, 339–340
  trace metal sorption onto, 426,
    432–434, 440–446
  in water, ultrafiltration of, 179
Iron oxyhydroxophosphate, in
    eutrophic lake, filtration of,
    201–204, 207
Iron particles
  allochthonous, 345
  crystalline forms of, 344
  dissolution reactions of, 336–340
  electron microscopy of, 319
  filtration of, 318
  in freshwater, 315–355
    characterization of (table),
      332–334
  grain size of, 318, 344
  reduction/dissolution reactions of,
    335–340, 344
  sampling of, 344–348

settling and resuspension of, 320,
  325
from weathering, 317–320
Iron-rich colloids, 260, 261–266, 317,
  318–319
  electron microscopy of, 264,
    318–319
  size classes of, 264
  TEM analysis of, 264, 265
Iron sulfide(s)
  aquatic formation of, 322,
    323–324, 328, 333, 342–343
  diagenetic transformation of, 344
Isoelectric point, 298
Isotherm equations, analytical, 476
Isotopes
  naturally occurring, use in receptor
    modeling, 27
  reference materials for, 29, 31
  use in source apportionment
    methodology, 27–33, 56
Isotopic disequilibrium, oceanic
    scavenging and, 381–385
Isotopic heterogeneity, 29
Italy, aerosol lead studies in, 33
IUPAC Commission on Environmental
    Analytical Chemistry, 5

KEQUILIB model, for atmospheric
    production of aerosol
    components, 92
Kinetic spectrum method, for
    evaluation of distribution
    function, 516

Lability, of metal-particle systems,
  518
Lagrangian models, in simulation of
    tracer transport, 59, 92
Lake waters
  anoxic, redox boundary in, 325
  fibrils in, 247, 249, 251
  iron oxidation in, 331
  iron particles in, 320
  iron-rich colloids in, 261–261

LAMMS analysis, 110
  of biogenic aerosols, 127
  of carbonaceous aerosols, 19, 20,
    21
  environmental particle analysis by,
    113, 114–116
  of fly-ash particles, 131
  of industrial aerosols, 129, 130
  instrumentation for, 115
  of marine aerosols, 123–124
  of remote continental aerosols,
    126–127
  sensitivity of, 110, 114, 116
  SIMS analysis with, 117
  of suspended particles, 132
  of urban aerosols, 128
Langmuir-Freundlich equation, 477
Langmuir isotherm, 471, 475, 476,
    477, 478, 481, 483
Langmuir-type function, 338
Laser, -induced light scattering, 365
Laser microprobe mass spectrometry,
    see LAMMS analysis
Laser probe mass spectrograph
    (LPMS), use in LAMMS
    analysis, 115
Lateral/depth dispersal, of suspended
    marine particles, 132–133
Lateral resolution, comparison among
    various microprobe techniques,
    110
Lawton-Sylvestre estimation of
    unknown profiles, 52
Layer charge, 305, 311
Lead, 47, 128
  as aerosol tracer, 14, 39, 40, 41
  distribution coefficient in aquatic
    systems, 430, 431
  sorption onto iron oxyhydroxide,
    440, 442–444
Lead isotopes, use in source
    apportionment, 33
Lepidocrocite, in aquatic iron
    particles, 330–331, 344
Ligand exchange, 335, 510
Light, effects on iron particles in
    water column, 328, 329, 338

Light absorption, of oceanic biogenic
    particles, 361, 362, 363, 365
Light-emitting diodes (LEDs), use in
    light-scattering studies, 365
Light scattering, of oceanic biogenic
    particles, 361, 363, 365
Limonite, in aquatic iron particles,
    332
Linear adsorption, time constant for,
    512
Linear diffusion control, time constant
    for, 512
Linearly independent factors, in
    principal component analysis,
    51
Linear (receptor) modeling, 25, 43
  bias in, 54
  validation of, 61–64
Lipids, natural, 19
Lipscombite, in aquatic iron particles,
    341
Liquid films, 517
Liquid particles, transport rate of
    metals to, 504–505
Living colloids
  epifluorescence microscopy of, 236
  heavy metal markers for, 243, 244
  TEM analysis of, 253–254, 258,
    259
Local binding function, in
    heterogeneity analysis, 459
Local isotherm, 469, 471, 479
  asymptotically correct
    approximation (ACA) of, 483
  effect on colloid-ion binding,
    475–476
  equation, 480
Local isotherm approximation (LIA)
    methods, 480, 481–485
Local source apportionment, modeling
    of, 26–27
LOGA-AS approximation, 486
LOGA isotherm, derivation of, 483
LOGA method, use to derive the
    distribution function, 475, 485,
    487

Long range transport (LRT), of aerosols, 24, 26, 29, 40, 52, 56, 58–59, 126

Long-range transport (LRT), of marine aerosols, 125

Los Angeles
aerosol emissions in, 18, 19, 88, 98
carbonaceous haze over, 8, 9, 28, 32

Low-loss region, of electron energy loss spectrum, 118

Low-volume samplers, for aerosol chromium, 156

Ludlamite, in aquatic iron particles, 341

Mackinawite, in aquatic iron particles, 333, 343

Macromolecule(s)
definition of (size), 176, 220
size fractionation of, using filtration and ultrafiltration, 171–230

Maghematite, in aquatic iron particles, 344

Magnetite, in aquatic iron particles, 344

Magnetosomes, electron microscopy of, 265

Manganese, 40, 260, 323

Manganese cycle, in water column, 325

Manganese oxides, as scavengers, 261

Manganese oxyhydroxides
as aquatic colloids, 233, 266, 344, 426
sampling for, 179, 346, 347
SEM, TEM, and EDS analyses of, 266
trace metal sorption onto, 432–434

Mapping
comparison among various microprobe techniques, 110
X-ray, in EPXMA and SEM/EDX analyses, 112

Marine aerosols, analyses of, 123–124

Marine sources, of atmospheric pollution, 31–32

Marine surface chemistry, in studies of oceanic transport of radionuclides and trace metals, 381–382

Mass spectra
of fly-ash particles, 130
from SIMS, 116

Mass transfer, 500–507

Mass transport effects, in ion adsorption kinetics, 517

Master curve concept, in electrostatic modeling, 471, 473, 475

Mathematical validation, of source appoprtionment methodology, 56–64

Mechanistic modeling, use with statistical receptor modeling, 52, 54

Membranes, see also Filters
calibration of, based on retention curves, 183–184
cell rupture on surface of, 192
characterization of, 175
choice of, 212–213
comparative tests on, 203
contamination of, 213
dimensions of, 213
electrically charged, 183
electric and hydration properties of, 208–212
flow rate through, 182, 193
ion adsorption onto, 210
pore size of, see Pore size
solute interactions with, 207–212
solute steric conformation effects on, 212
surface coagulation on, 190–191
use in complex stability measurements, 181
use in gravimetry, 185

Mercury light-emitting diodes, use in light-scattering studies, 365

Metal adsorption equation, 470
analytical inversion of, 480–488

Metal complex formation, chemical-reaction rates of, 507–510
Metal (hydr)oxides, ion binding by, 457, 460, 469
Metal hydroxy complexes, surface species of, 465
Metal ion adsorption, particle characterization in, 517
Metal ion adsorption/desorption processes, 499–500
Metal ion/proton exchange reaction, 466
Metal ions
  binding-curve interpretation, 488
  binding sites for, 499
  binding to heterogenous particles, 470–471
  bioavailability of, 499
  complexation of, 464–466
  particle-bound, 498
  specific adsorption to independent sites, 467
Metals
  adsorption onto filters, 188
  particle bound, low mobility of, 518
  sediment binding of, 428–429
Metal speciation, in aquatic colloidal systems, 497–521
Meteorological data, 10, 56
Methane, as greenhouse gas, 11
Methanesulfonic acid, in marine aerosols, 83, 123–124, 126
Microalgae, in seawater, 359
Microanalysis, SIMS use in, 116
Microautotrophs, in oceans, 359
Microchemistry, use in particle behavior studies, 244
Microniches
  aquatic, reduction in, 321
  of water column, pyrite in, 342–343
Microparticles, characterization of, 47
Micro-PIXE analysis, 110
Micro-Raman spectroscopy, 110
  of carbonaceous aerosols, 19, 20

environmental particle analysis by, 120–121
of fly-ash particles, 131
of remote continental aerosols, 126
Mineral colloids
  electron microscopy of, 259–270
  ion binding to, 460–461
Mineral-organic associations, electron microscopy of, 259–270
Minimum perturbation studies, of aquatic colloids, 233, 237, 241, 245, 260, 261, 265, 268, 270–272, 273, 274, 275
Mixed-salt aerosols, modeling of, 90
Modeling, evaluation of accuracy of, 61
Model uniqueness, in principal component analysis, 51
Molecule-specific markers, in aggregate studies, 270
Morphology
  of aerosol particles, 10, 21, 124, 125
  derivation from particle classes, 48
  of environmental particles, 111, 121
  of fly-ash particles, 130
Mossbauer spectroscopy, of iron, 320, 340–341, 341, 346
Multichannel device, in electron microprobe, 109, 111
Multi-method approach, to colloid study, 237, 245, 261, 272, 273
Multiparametric measurement, in fluorescence flow cytometry, 366
Multiphase equilibria, of atmospheric acids and bases, 77
Multiple linear regression, 10, 24, 31
Multivariate data matrix, 24
Multivariate techniques, in studies of aerosol composition, 10, 22, 30–31, 37–52, 61, 112
MUSIC (multi-site complexation) model, for ion binding by colloids, 460–461

Mutagenicity, of atmospheric
    particles, as response variable,
    11

Nanoplast embedding, 275
    of aquatic colloids, 241, 272
    of humic substances, 252
    of mineral colloids, 261, 264, 266
Nanoplast film technique, use for
    whole mount preparations,
    245, 272
Natural aerosols, as baselines for
    composition studies, 126
Natural diffusion, in metal ion
    transport, 512
Natural particles, trace element
    sorption onto, 425–453
Nernst diffusion layer, 506
Net permanent structural surface
    charge density, 294, 311
Net proton surface charge
    determination of, 305–309
    density, derivation of, 294, 307,
    311
Net structural surface charge,
    determination of, 303
Net total particle surface charge
    density, derivation of, 295,
    300–302, 311
Nickel
    as aerosol tracer, 40
    distribution coefficient in aquatic
    systems, 430, 431
    solubility diagram for, 427
    sorption onto iron oxyhydroxide,
    440, 442–444
Nitrate
    in atmospheric aerosols, 77
    coated sea salt particles, 123
Nitric acid
    in atmospheric aerosols, 78–79, 82,
    86–87, 94
    atmospheric nitrate ion from, 77
    collection on diffusion denuders,
    85–89

collection on filter packs, 84–85,
    87–89
Nitrogen Species Intercomparison
    Study (NSIS), 88
NMR shifts, use in studies of
    carbonaceous aerosols, 19
Nontronite, in aquatic iron particles,
    332
North Sea, particle deposition in, 123,
    124
Norway
    atmospheric carbonacous particles
    over, 29, 44
    polycyclic aromatic hydrocarbons
    aerosols over, 48
Numerical trajectories, role in long
    range transport, 56, 57
Nutrient absorption, of oceanic
    biogenic particles, 359–360
Nutrient availability, effects on
    chlorophyll-containing ocean
    particles, 372

Oceanic biogenic particles
    characterization of, 357–376
    fluorescence flow cytometry of,
    361, 363, 366–372
    impedance volume of, 361, 363,
    365–366
    light absorption of, 361, 362, 363,
    365
    light scattering of, 361, 363, 365
    methods for study of, 361–368
    relative sizes of, 359, 371
    time and space distribution of,
    358–360
Ocean optical geography, in studies of
    biogenic particles, 373
Oceans
    particle dynamics in, 400
    radionuclide and trace metal
    transport in, 379–423
Oil burning, as source of urban
    aerosols, 128
One-pK model, of metal (hydr)oxide
    binding, 461

Optical microscopy, use to
    characterize aerosol particles,
    20, 21
Oregon, urban aerosol study on,
    35–36
Organic acids, as redox active
    substances, 335
Organic coatings
    on aquatic colloids, 237–238, 239,
        274, 275
    of mineral colloids, 260–261
Organic colloids
    as aquatic colloids, 244
    ion binding to, 460–461
    from skeletal materials and cell
        fragments, 252–253
Organic compounds
    adsorption onto filters, 189
    as aquatic colloids, ultrafiltration
        of, 179
    calcite adsorption of, 267
    mineral particle association with,
        260–261, 268
    release from organic membranes,
        187
Organic flocs, 248, 260
Organic materials
    as aquatic colloids, 233, 244
    characterization using cytochemical
        techniques, 242–244
    information content of, 15–19
    iron bound to, 319, 339
    nonliving, as aquatic colloids,
        246–253
    trace element binding to, 432–434
    in water, size classification of, 177
Organic pollutants, 251
Organic polyelectrolytes, filtration of,
    211–212
Organo-metallic materials, in aquatic
    environments, 261
Orthokinetic coagulation, causes of,
    199
Osmotic pressure model, 198
Outer-sphere complex surface charge
    density, 295, 311

Outer-sphere surface complex,
    definition of, 293, 311
Overall adsorption equation, 470
Oxalic acid
    in atmospheric aerosols, 79, 83
    Standard Reference Material for, 57
Oxidants, manganese oxides or nitrate
    as, 325
Oxidation/hydrolysis reactions, of iron
    particles in freshwater,
    328–335
Oxidation-reduction process, of iron in
    water bodies, 320–328
Ozone
    tropospheric, formation of, 7–8
    volatile organic compound role in
        production of, 11

Paraffins, natural, 19
Parallel approach method, use in
    studies of aerosol carbon, 14,
    31
Parallel sets of filtration, 179, 180,
    201
Particle(s)
    characterization of, in studies of
        metal ion adsorption, 517
    definition of (size), 176, 219
    effective size of, 194
    metal ion interactions with, in
        aquatic colloid systems, 498
    in ocean, see Oceanic biogenic
        particles
    radius, 474
    role in oceanic transport of
        radionuclides and trace metals,
        379–423
    shape, effect on metal ion
        adsorption, 516–517
    size fractionation of, 171–230
    sizes, distribution of, 516
    surface charge, characterized,
        291–314
Particle charge, role in ion adsorption,
    458

Particle class balance (PCB) method, of source apportionment, 22, 37, 48

Particle concentration, role in membrane gel formation and clogging, 191

Particle concentration effect, colloid role in, 410–414

Particle density, 474

Particle-induced gamma emission. *see* PIGE analysis

Particle mass flux, control by particle coagulation, 414–415

Particle recognition and characterization (PRC), automated, 112

Particulate material
chromium in, 154, 155
discrimination from dissolved material, 176–178

Partition clustering, 43, 47

Passing air collector, for aerosol chromium, 156

Patchwise heterogeneity, in natural colloids, 467, 468

Perfluorocarbon tracers, 65
ANATEX study of, 58, 60

Perikinetic coagulation, causes of, 198

Permeate, definition of, 176, 220

pH, effects on trace metal distribution in aquatic systems, 429, 430, 431

Phase equilibria, of atmospheric aerosol components, 78, 95–99

Phosphate, in aquatic iron particles, 331, 340, 341

Phosphoferrite, in aquatic iron particles, 341

Phosphorus, in mineral colloids, 260, 262, 267

Photochemical oxidation, of Fe(II) in freshwaters, 328–329

Photoelectron signal, in XPS technique, 122

Photons
interactions in environmental particles, 108
role in Raman scattering, 121
in XPS technique, 122

Phytoplankton
role in redox cycling of iron, 335, 339–340
in seawater, 359, 360, 362, 365, 3732

Picoplankton
as aquatic colloids, 253, 267
in surface waters, 254
TEM analysis of, 258, 259

PIGE analysis, of environmental particles, 113

Pimaric acid, dehydrogenation pathway for, 20

PIXE analysis, 110, 113–114
of biogenic aerosols, 128
of marine aerosols, 125
of particulate environmental samples, 109

Plagioclase, 319

Plant roots, as source of fibrils, 248, 249

Plasmon oscillation, in SIMS, 117

Point(s) of zero charge
definition of, 311
derivation of, 296–300
experimental illustrations of, 297

Point of zero net charge, 311

Point of zero net proton charge, 311

Point of zero salt effect, 311

Poisson-Boltzmann equation, 473

Polarography, of iron(II), 342

Pollutant source resolution, 25

Polycarbonate filters, 173–174, 204
merits of, 184–185, 188, 202, 212–213

Polycyclic aromatic hydrocarbons
atmospheric, 3, 77–78
first-order decay reactions of, 54
historical information on, 9
in marine aerosols, 123
organic tracers for, 40, 41–43
pyrosynthetic formation of, 17
in soot, 20, 21
source apportionment of, 48, 54
urban particle standard for, 55

Polydispersity, effects on metal ion
adsorption, 516
Polyethyleneglycol macromolecules,
212
Polymers, Fourier transform infrared
(FTIR) spectroscopy of, 120
Polysaccharides
as aquatic colloids, 233, 274
in fibrils, 246, 261
ruthenium as marker for, 243
of structural colloids of higher
plants, 244
Polysulfides, as redox active
substances, 335
Polysulfone membranes, use in
dialysis, 180
Pore(s), solute-membrane interactions
inside of, 207–212
Pore density, role in size fractionation
of particles, 182
Pore hydration layer, effects on
solute-membrane interactions,
210
Pore radius, water flow rate related to,
186
Pore size(s)
characteristics, of different
membrane types, 184–185
decrease of, 198
by gel formation, 190–191
distribution of, 202, 213
width of, 183
effective, 194
reproducibility of membrane
production related to, 185–187
retention related to, 183
role in filtration, 178, 181–182
of various filter types, 174–175
Pore water peeper, 180
Pore waters
iron(II) in, 323, 341
iron(II) oxidation in, 328, 340
Pore water solutes, of sediments, *in
situ* sampling of, 180
Porosity
of particles, effect on metal ion
adsorption, 516–517

of pores
reproducibility of, 185–186
role in size fractionation of
particles, 182
Portland (Oregon) Aerosol
Characterization Study, 35–36
Potassium, as aerosol tracer, 14, 39,
40, 41
Potency (revertants/μg-EOM), of
woodburning aerosol versus
motor vehicle aerosol, 39
Power plants, fly-ash emission from,
130, 131
Preparatory techniques, for electron
microscopy of aquatic colloids,
237–245
Pressure, role in membrane water flow
rate, 187, 191
Pressure jump method, use in studies
of ion adsorption kinetics,
511–512
Principal component analysis (PCA)
of marine aerosols, 124
use with cluster analysis, 44, 46
use with factor analysis, 49–52
Prochlorophytes, in seawater, 359
Protein-rich cell parts, in aquatic
colloids, 252–253, 256
Proteins, conformation effects on
membrane filtration of, 212
Proton adsorption, 458–459, 460
in ion adsorption phenomena, 457
Protonation, derivation of degree of,
463
Protonation constants, of carboxylic
groups, 458, 459
Proton beams, for SPM analysis, 113
Proton binding
curves, 471
to heterogeneous particles, 468–469
to homogeous colloids, 462–464
metal binding competition with, 470
Proton binding constants, of surface
oxygen groups, 460
Proton-binding curves, analysis for
heterogeneity, 458

Proton-induced X-ray emission, see
    PIXE analysis
Proton microprobe, iron sulfide
    sampling by, 342
Pulsed filtration, 201
Pure component spectra, 52
Pyrite, 317
    in aquatic iron particles, 333–334,
       342–343
    conversion to sulfuric acid, 318
Pyrosynthetic processes, structural
    consequences of, 15, 17
PZC theorems, 299–300, 303, 306,
    307
    definition of, 311

Quail Roost II Receptor Modeling
    Workshop, 61, 64, 65
Quantification, comparison among
    various microprobe techniques,
    110
Quantitative analysis, by
    microanalytical techniques, 133
Quartz, 318, 319

Radioactive tracer determination, 510
Radiocarbon dating, 28
Radionuclides, particle role in oceanic
    transport of, 379–423
Rain, aqueous phase oxidation of
    sulfur dioxide in, 80
Raleigh (North Carolina), source
    apportionment of aerosol in,
    39, 41, 56
Random heterogeneity, in natural
    colloids, 467, 468
Rapid freezing, of organic particles,
    242, 244–245
Rate law, 328
Rayleigh scattering, 121, 361
Receptor modeling, 23
    definition of, 5
    of Los Angeles aerosol, 9
    relation to extensions of the
       atmospheric compartment, 26

Receptor site, 5
Redox boundary
    dissolved and particulate iron at,
       321–322, 331, 343
    migration in freshwaters, 325–326
Red tide dinoflagellates, 372
Reducing environments
    in freshwater systems, 320–321
    iron dissolution in, 317
Reduction/dissolution reactions, of
    aquatic iron particles, 335–340
Regional source apportionment, 26,
    33, 58
Regularization methods, for inversion
    of integral binding equation,
    480
Relaxation time constant, in diffusion
    of metal ions, 506, 507
Relaxation times, in ion adsorption,
    512–515, 516
Remote continental aerosols,
    composition of, 126–127
Remote pollution, carbonaceous, 29
Remote sensing, of ocean particles
    and processes, 358, 360–361,
    362
Residence time data, for polycyclic
    aromatic hydrocarbons, 54
Response variables, in source
    apportioning, 11
Retene, as tracer, 19, 42
Retentate, definition of, 176, 220
Retention, by filters, physicochemical
    factors influencing, 192–212
Retention coefficients, of nonreactive
    compounds in solution,
    193–194
Retention curves, of membranes,
    183–184
Rivers, iron particles in, 317–318,
    320
Roanoke (Virginia), source
    apportionment of aerosols
    over, 56
Roughness, of particles, effect on
    metal ion adsorption, 516

Rural areas, chromium levels in
    ambient air of, 153
Rural pollution, carbonaceous, 29
Rutherford back scattering (RBS)
    analysis, 113, 114, 128

St. Louis, analysis of aerosol samples
    from, 54
Salt aerosols, modeling of, 90
Salt marshes, pyrite formation in,
    343–344
Sampling, for chromium in aerosols,
    156
Satellite remote sensing colorimetry,
    362
Scandium
  as aerosol tracer, 40
  sorption by natural marine parine
    particles, 385–386, 388
Scanning electron microscopy, see
    SEM analysis
Scanning electron microscopy/energy
    dispersive X-ray analysis, see
    SEM/EDX analysis
Scanning laser mass spectrometer
    (SLMS), use in LAMMS
    analysis, 115
Scanning proton microprobe analysis,
    see SPM analysis
Scanning transmission electron
    microscopes, see STEM
    analysis
Scatchard technique, 476
Scavenging capacity, of denuders, 156
Scavenging (oceanic)
  isotopic disequilibrium and,
    381–385
  laboratory experiments on, 385–386
  role in removal of trace elements,
    381, 401
  submicron particle pool role in,
    386–390
Schofield method, description of, 302,
    312
Sea-air interface, marine aerosols in,
    125

Sea salt particles, in marine aerosols,
    123, 124
Sea surface, enrichment of aerosols at,
    125
Seawater
  Cr(VI) in, 150, 161
  organic phosphorus in, 267
Secchi disc measurements, of aquatic
    particles, 362
Secondary electron imaging, in
    microprobe techniques, 111
Secondary ion mass spectrometry, see
    SIMS analysis
Secondary production, of global
    particles, 9
Sedimentary phases, responsible for
    trace element sorption, 432
Sedimentation
  coupled to adsorption, 406–409
  rate of, in seawater, 415
  role in colloidal pumping, 404–406
  use for aquatic particle size
    measurements, 172
Sediments (aquatic)
  clays in, 268, 270
  fibrils in, 249
  iron and manganese colloids in, 261
  iron in, 318, 320, 323, 339, 344
  iron particle sampling of, 346
  manganese oxyhydroxides in, 266
  metal binding properties of,
    428–429
  mineral colloids in, 260
  particle composition of, 131–133
  pore water solutes of, dialysis of,
    180
  redox boundary in, 325
  suspended, in seawater, 359, 374
Sediment traps, for iron sampling,
    324, 344–345
Sediment water interface, oxygen
    consumption at, 321, 325
Selective extraction, in separation of
    chromium species, 157
Selenium, as aerosol tracer, 14
SEM analysis
  of fibrils, 249

hybrids with TEM, 235
instrumentation for, 234
of manganese oxyhydroxides, 266
specimen preparation for, 234, 235,
    244
use to characterize aerosol particles,
    20–21, 22, 48
whole mount preparations for, 245
SEM/EDX analysis
comparison with other analytical
    techniques, 110
of marine aerosols, 125
of particulate environmental
    samples, 109
of suspended particles, 132
of volcanic aerosols, 128
Semi-volatile compounds, in aerosols,
    77–78
Sequential cascade fractionation, in
    filtration, 179–180, 201, 216
Sequential size fractionation, of
    aquatic components, 178–180
SEQUILIB model, for atmospheric
    production of aerosol
    components, 92
Settling, of iron particles in water
    bodies, 320, 321
Si/Al aggregates, in lakewaters, 262
Siderite, in aquatic iron particles, 322,
    323–324, 340, 342
Sieves, nature of, 173
Sieving
comparison with other particle
    separation methods, 175–176
use for aquatic particle size
    measurements, 172, 173
Signal-to-noise ratio, in Raman
    spectroscopy, 121
Silicates, iron in crystal lattices of,
    319
Silicon
from algal frustules, in aquatic
    colloids, 262
in marine complexing particles, 389
in mineral colloids, 260, 262,
    267–268
SIMS analysis

comparison with other analytical
    techniques, 110
detection limit of, 110, 117
of environmental particles, 116–117
of industrial aerosols, 129
instrumentation for, 116, 117
LAMMS use with, 117
lateral resolution in, 110, 117
Singular value decomposition method,
    for inversion of integral
    binding equation, 480
Sinking/buoyancy, of oceanic ocean
    particles, 359–360
Site densities, in ion-colloid binding,
    462
Six-reservoir model, of carbon dioxide
    exchange, 27
Size fractionation of particles
filter characteristic role in, 181–192
membrane choice for, 212–213
role of chemical composition of
    sample in, 182
role of physicochemical properties
    of sample in, 182
use of filtration and ultrafiltration
    for, 171–230
Size ranges, of particulate and
    dissolved material, 176–177
Skeletal materials, organic colloids
    from, 252–253, 257
Smelters, isotopic tracers for aerosol
    from, 40
Smoothing, of ion binding data,
    484–485, 486
Soils, 251, 317
Soil science, electron microscopy
    applied to, 274–275
Solid particles, transport rate of metals
    to, 504
Solute, definition of (size), 176, 220
Soot, carbonaceous, see Carbonaceous
    soot
Sorption, of trace elements on natural
    particles, 425–453
Sorption constants
field measurement of, 440–445

comparison with laboratory-
derived constants, 445–446
of natural particulate material,
estimation of, 440–446
Sorption losses, of aerosol ammonia,
88
Sorption rates
of oceanic radionuclides, 384
reaction-controlled, 516
Source apportionment, 3–74
analytical schemes for, 5, 112
of atmospheric articles, see
Particulate source
apportionment
of bulk particles, 47
of "extractable organic matter" of
atmospheric particles, 11
methodology for, 23–55
validation of, 55–64
multivariate techniques for, 37–52
of organic species, 15–22
of pollutant aerosols, 25
as receptor modeling, 5
simulation of, 62
using chemical mass balance,
33–37
using cluster analysis, 43–49
using hydrid techniques, 52–55
using principal component and
factor analysis, 49–52
using tracer multiple linear
regression, 38–43
Source modeling, definition of, 5
Source profile matrix, in receptor
modeling, 23
Source profiles, 15
South Coast Air Basin, atmospheric
aerosols in, 90, 92, 93
Spain, cluster analysis of aerosols
from, 47
Speciation, of metals in aquatic
colloidal systems, 497–521
Specific absorption coefficient, of
phytoplankton chlorophyll,
365, 372
Specific adsorbed cation and anion
charge, 298

Specific adsorbed ion charge, 295,
312
Specific surface area, 457, 474
Specific surface excess, 295, 298
Specimen damage, avoidance in SPM
analysis, 113
Spectrophotometry
in batch analysis experiments, 510
of chromium(VI), 159
of iron(II), 342
Spectrum, X-ray generated, in
EPXMA and SEM/EDX
analyses, 112
Spline smoothing, of ion binding data,
484–485
SPM analysis
accuracy of, 113
detection limit of, 110, 113, 133
of environmental particles, 113, 114
of fly-ash particles, 131
quantification by, 110, 113
Sputter depth, fly-ash elements as
function of, 130
Sputter rate, in dynamic SIMS, 116
Stains, molecule-specific, 243
Standard Reference Materials (SRMs),
55, 57
Static SIMS analysis, 116
Steady-state convective diffusion, 505
STEM analysis, 235, 241
STEM-EDS analysis, of diagenetic
colloids, 261
Steric rearrangement, role in
dissolution reactions, 335
Stern layer, 294, 295
Stirring, effects on filtration, 200,
207, 215–216, 217
Stratosphere, aerosols in, 128–129
Streams, iron particles in, 317
Strong acids
in atmospheric aerosols, 77, 78–81
gaseous, in atmospheric aerosols,
81–83
vertical distribution of, 99
Subjective decisions, relating to
source apportionment
methodology, 64

Submicron particle pool, role in
    oceanic scavenging, 386–390
Sulfate(s)
  aerosol, dispersion of, 8
  atmospheric, 3
  -coated sea salt particles, 123
  source apportionment of, 11, 52
Sulfide, in bottom waters, 325, 341
Sulfites, as redox active substances,
    335, 340
Sulfur dioxide
  collection on filter packs, 84
  oxidation of, transformation factor
    for, 54–55
  sulfuric acid formation from, in
    aerosols, 80–81
Sulfuric acid
  in aerosols, 78
  from atmospheric sulfur dioxide
    oxidation, 77, 78
Sulfuric acid/sulfate emissions, 79
  secondary formation of, 80–81
Surface acid curves, 471
Surface analysis
  by Fourier transform infrared
    (FTIR) spectroscopy, 120
  of industrial aerosols, 130
  of marine aerosols, 123
  of urban aerosols, 128
  using LAMMS method, 115
  using SIMS method, 116
  XPS use in, 122
Surface analytical beam techniques,
    for characterization and
    analysis of environmental
    particles, 107–134
  acronyms for, 134
Surface association/dissociation
    reactions, 499
Surface charge, 291–314
  components of, 294–296
  definitions of, 293–300
Surface chemical speciation
  definitions and concepts in,
    293–294
  particle charge relation to, 293

Surface coagulation, during filtration,
    190–191, 198–199, 204–206
  experimental evidence for, 201–207
Surface complexation
  of metal ions, 471
  role in dissolution reactions, 335,
    336, 338
Surface complexation models, for
    ocean scavenging, 381
Surface excess concentration, in
    diffusion of metal ions, 500
Surface functional group, 293, 312
Surface heterogeneity
  effects on metal ion adsorption,
    515–516
  role in dissolution reactions, 335
Surface potential, relation to surface
    charge, 463, 465
Surface reactions, on atmospheric
    particles, 108
Suspension effect, in Gran titration,
    309
Suspensions, particle composition of,
    131–133
Syracuse (New York), particle class
    balance studies on, 37

Tangential flow filtration, 201
Target factor analysis (TFA), 51
Target transformation factor analysis
    (TTFA), 51–52, 61, 63
TEM, see also TEM analysis
  of clay minerals, 268, 270
  differential staining of biological
    macromolecules for, 235
  of fibrils, 246–249, 261
  hybrids with SEM, 235
  instrumentation for, 234–235
  of iron-rich colloids, 264, 265
  of manganese oxyhydroxides, 266
  of mineral colloids, 260
  of nonliving organic materials, 246
  of picoplankton, 258, 259
  preparatory techniques for, 237
  resolution achieved by, 235–237

use in multi-method approaches,
    272
use to check membranes, 204–206,
    217
of viruses, 253
whole mount preparations for, 245
TEM analysis
    of aquatic colloids, 233, 240–241,
        242, 244
    of calcium-rich particles, 267
Thickness reduction, of electron
        microscopy samples, 241
Thorium-234
    colloidal pumping of, 398, 399
    oceanic scavenging of, 382, 384,
        391, 402, 404, 410
Time constant, 507, 512, 516
Time-of-flight (TOF) mass
        spectrometer, 115
Tin-113, sorption of, 396–398, 412
Tin radiotracers, 391
Titanium, as aerosol tracer, 40
Titration, in batch analysis
        experiments, 510
Topographical image, in EPXMA
        analysis, 109, 111
Tortuosity factor, role in membrane
        water flow rate, 186
Total variance least squares (TVAR),
    35
Toth equation, 477
Trace elements, sorption onto natural
        particles, 425–453
Trace metals
    in aquatic systems, distribution
        coefficients of, 429–432
    competitive sorption models for,
        434–440
    particle role in oceanic transport of,
        379–423
    role in Fe(II) oxidation, 328
Tracer multiple linear regression
        [MLR(T)] technique
    selection and correction of source
        tracers for, 39–40, 54
    use in source apportionment
        methodology, 38–43, 61, 63

validation of, 40–43
Tracers, for aerosol sources, 39–40,
        see also individual tracers
Transfer, of atmospheric particles, 123
Transformation, of aerosol particles,
        7, 124, 133
Transformation matrix, 51
Transmission electron microscopy, see
        TEM analysis
Transmissometer, 365
Transport
    of particles, 112, 123, 133
    of suspended marine particles, 132
    through membranes,
        physicochemical process
        controlling, 193
Transport model validation exercise
        (ANATEX), 58–59
Transport rate, of metal ions to
        particles, 504
Tropical vegetation, as source of
        atmospheric pollution, 31
Troposphere, 79, 125
Truth, in evaluation of model
        accuracy, 61, 63
Tucson (Arizona), cluster analysis of
        aerosols over, 46–47
Tunable diode laser absorption
        spectroscopy (TDLAS), 86, 89
Tungsten light-emitting diodes, use in
        light-scattering studies, 365
Turbulence, effect on oceanic biogenic
        particles, 372–373
Two-pK model, of metal (hydr)oxide
        binding, 461

Ultrafiltration
    application in natural water studies,
        176–181
    comparison with other particle
        separation methods, 175–176
    cross-flow type, 389
    equilibrium constant determination
        by, 181
    humic substance interference with,
        251

of iron colloids, 318–319
of small compounds, osmotic
    pressure model for, 198
use for size fractionation, 171–230
Ultrathin sections
for electron microscopy of aquatic
    colloids, 241–242, 253
TEM requirement for, 234
Uniqueness assumption, in selection
    and correction of source
    tracers, 40, 52
Unsuspected components, revelation
    by factor analysis, 52
Unweighted least squares, 48
Uranium-238, oceanic scavenging of,
    382
Urban aerosols, composition of, 128
Urban cluster, 45, 46
Urban particle standards, 55, 58
Urban pollution, carbonaceous, 28, 29

Validation, of tracer multiple linear
    regression [MLR(T)]
    technique, 40–43
Vanadium, as aerosol tracer, 14, 40
Van de Graaff accelerator, 113
Variable space, in cluster analysis, 44
Variance-covariance matrix, 24
VARIMAX procedure, use in factor
    analysis, 51, 52
Viruses
as colloids, 233, 253
quantification in aquatic
    environments, 253
TEM analysis of, 253, 258
Visibility, of atmospheric particles,
    10, 11
Vivianite, in aquatic iron particles,
    322, 323–324, 332, 333, 341
Volatile organic compounds (VOC),
    11
Volcanoes
as sources of aerosols, 125, 126,
    128–129
as sources of marine sediments, 133
Voltammetric speciation methods, 518

Washing technique, in filtration, 179
Water
chromium determination in, 158
chromium(VI) in, 150
flow rate through filters, 186, 190,
    191, 197, 204
    particle size effects on, 199–200
size classification of particulate and
    dissolved components in,
    177–178
use of filtration and ultrafiltration
    for fractionation of particles in,
    171–230
Water column
iron sampling in, 344–346
pools of trace-metal binding ligands
    in, 390
Wavelength dispersive X-ray (WDX),
    of particulate environmental
    samples, 109
Weak acids, in atmospheric aerosols,
    77
Weathering, iron particle formation
    from, 317–320
Weighted average
affinity constant, 477
equilibrium function, 486
intrinsic (composite) affinity
    constant, 479–480
Weisz ring oven, 160–161
Wet deposition, of atmospheric
    particles, 123
Whole mount preparations, for
    electron microscopy of aquatic
    colloids, 245, 253
Wind trajectory analysis, in studies of
    aerosol composition, 10,
    48–49, 54
Woodburning
carbon-14 as tracer for, 28
potassium as tracer for, 14, 19, 40,
    41
Workplace aerosols, sources and
    composition of, 129–131
Wustite, in aquatic iron particles, 332

Xenon light-emitting diodes, use in
    light-scattering studies, 365
XPS analysis
    comparison with other analytical
      techniques, 110
    of environmental particles, 121–123
    of industrial aerosols, 130
    of marine aerosols, 123
    quantitative, 110, 122
    scanning system for, 122
    sensitivity of, 110, 121, 122
    spatial resolution of, 122
X-ray(s), in PIXE analysis, 113, 114
X-ray analysis
    of Antarctic aerosols, 126
    EELS compared to, 119
X-ray diffraction
    of lepidocrocite, 331
    of pyrite, 342–342
    of siderite, 341
X-ray fluorescence spectroscopy, of
    iron, 346

X-ray photo-electron spectroscopy, see
    XPS analysis
X-ray photons, measurement of, 109
X-ray spectrometers, use in EPXMA
    and SEM/EDX instruments,
    109

Yellow organics, in oceans, 361

Zero-loss region, of electron energy
    loss spectrum, 118
Zeta potential, 301
Zinc
    in daily diet samples, 50
    distribution coefficient in aquatic
      systems, 430, 431, 432
    solubility diagram for, 427
    sorption onto iron oxyhydroxide,
      440, 442–444

Printed in the United States
by Baker & Taylor Publisher Services

Printed in the United States
by Baker & Taylor Publisher Services